Lecture Notes in Mathematics 1584

Subseries: Fondazione C.I.M.E.

Advisor: Roberto Conti

M. Brokate N. Kenmochi I. Müller
J. F. Rodriguez C. Verdi

Phase Transitions and Hysteresis

Lectures given at the 3rd Session of the Centro Internazionale Matematico Estivo (C.I.M.E.) held in Montecatini Terme, Italy, July 13-21, 1993

Editor: A. Vistini

Fondazione
C.I.M.E.

Springer-Verlag
Berlin Heidelberg New York
London Paris Tokyo
Hong Kong Barcelona
Budapest

Authors

M. Brokate
Institut für Informatik
und Praktische Mathematik
Universität Kiel
24118 Kiel, Germany

Yong Zhong Huo
Institut für Angewandte Mathematik
und Statistik
Technische Universität München
80335 München, Germany

Noboyuki Kenmochi
Department of Mathematics
Chiba University
Inage-Ku
Chiba, Japan

Ingo Müller
Stefan Seelecke
Institut für Thermodynamik
und Reaktionstechnik
Technische Universität Berlin, FB 6
10623 Berlin, Germany

José F. Rodriguez
CMAF
Universidade de Lisboa
1699 Lisboa, Portugal

Claudio Verdi
Dipartimento di Matematica
Università di Milano
20133 Milano, Italy

Editor

Augusto Visintin
Dipartimento di Matematica
Università degli Studi di Trento
38050 Provo, Italy

Mathematics Subject Classification (1991): 35R35, 35K20, 35K60, 47H15

ISBN 3-540-58386-6 Springer-Verlag Berlin Heidelberg New York

CIP-Data applied for

Printed in Germany

Typesetting: Camera ready by author
SPIN: 10130093 46/3140-543210 - Printed on acid-free paper

INTRODUCTION

This volume contains the notes of the courses held at a C.I.M.E. school in Montecatini in July 1993. This was intended as an introduction to physical, analytical and numerical aspects of two classes of phenomena of high applicative interest: phase transitions and hysteresis effects.

About one century ago the macroscopic description of solid-liquid transitions led to the formulation of the (by now classical) *Stefan problem.* This is a boundary value problem for a parabolic partial differential equation. Here the thermal field and the evolution of the interface between the two phases are coupled and unknown: this is a typical example of *free boundary problem.* This model and its many generalizations have then been applied to a multitude of physical phenomena. Mathematical aspects have also been intensively studied in the last thirty years or so; see the monographs of Rubinstein [5] and Meirmanov [4], e.g., and the proceedings of the conferences on free boundary problems which have regularly been held for almost two decades. Relevant results on generalizations of the Stefan problem are dealt with in Kenmochi's and Rodrigues's contributions.

The status of hysteresis modelling is quite different.

Hysteresis can be defined as a *rate independent memory effect.* This is a property of some constitutive laws, which relate an input variable u and an output variable w. *Memory* means that at any instant t, $w(t)$ is determined by the previous evolution of u, and not just by $u(t)$. *Rate independence* means that the curves described in $\mathbf{R}^2$ by the couple (u, w) *(loops,* typically) are invariant for changes of the input rate, such as changes of the frequence, e.g..

Plasticity, ferromagnetism, ferroelectricity are among the most typical examples of hysteresis phenomena. More recently, also so called pseudo-elastic alloys were discovered, where hysteresis appears also as *shape memory;* see Müller's report. Several models have been devised by physicists and engineers to describe hysteresis; in particular, plasticity has a long tradition of mathematical studies. However, no systematic analysis of hysteresis appeared, until in 1970 a group of Russian mathematicians introduced the concept of *hysteresis operator,* and started a detailed investigation of its properties. Krasnosel'skiĭ and Pokrovskiĭ were the most active pioneers in this field, and their research is presented in the monograph [2]. Hysteresis operators are dealt with in Brokate's notes.

In the early 1980's other mathematicians began to study hysteresis phenomena, especially in connection with applications. A monograph of Mayergoyz [3] and the proceedings volume [7] appeared; at this moment the books [1] and [8] are in preparation.

There is a strict relation between phase transitions and hysteresis. For instance, in single-phase systems *supercooling* and *superheating* effects prior to phase nucleation are rate independent, and accordingly can be labelled as hysteresis phenomena. Here surface tension also plays an important role.

Here is a more mathematical example, which illustrates how hysteresis and free boundary problems can be related. The weak formulation of the classical Stefan problem problem involves the *Heaviside graph.* A hysteresis relation is easily obtained by replacing the critical value 0 by two thresholds a, b (with $a < 0 < b$), for downward and upward jumps, respectively. Results have been obtained for this problem, see [6]. This model applies to ferromagnetism more properly than to solid-liquid phase transitions.

Connections between phase transitions and hysteresis appear also by Verdi's contribution, which addresses the numerical treatment of both phenomena.

The school was and this volume is an attempt to cast a bridge between hysteresis and free boundary problems. The reasons for such an interaction are in the phenomena we deal with; but sometimes the mathematical world is moved by different dynamics.

Augusto Visintin

Bibliography

[1] M. Brokate, J. Sprekels: Hysteresis phenomena in phase transitions (Forthcoming monograph). Springer-Verlag

[2] M.A. Krasnosel'skiĭ, A.V. Pokrovskiĭ: Systems with hysteresis. Springer, Berlin 1989 (Russian ed.· Nauka, Moscow 1983).

[3] I.D. Mayergoyz: Mathematical models of hysteresis. Springer, New York 1991.

[4] A.M. Meirmanov: The Stefan problem. De Gruyter, Berlin 1992. (Russian edition: Nauka, Moscow 1986.)

[5] L.I. Rubinstein: The Stefan problem. Trans. Math. Monographs 27. American Mathematical Society, Providence, RI 1971.

[6] A. Visintin: A phase transition problem with delay. Control and Cybernetics, 11 (1982), 5–18.

[7] A. Visintin (ed.): Models of hysteresis. Proceedings of a meeting held in Trento in 1991. Longman, Harlow 1993.

[8] A. Visintin: Differential models of hysteresis (Forthcoming monograph). Springer-Verlag.

TABLE OF CONTENTS

Hysteresis Operators

Martin Brokate *

Contents

1. Introduction

Hysteresis phenomena appear in many branches of science. They usually arise because the underlying process admits more than one stable equilibrium state for certain (or for all) values of the process parameters. For example, let

$$F(x, v) = 0 \tag{1.1}$$

describe the equilibria x of a process in dependence of the parameter v. If the solution set of (1.1) forms a curve like the one in Figure 1, and if the part connecting the points A and B consists of unstable equilibria while the others are stable, then a variation $v = v(t)$ of the parameter in time leads to a relay-type hysteresis relationship $x = x(v)$ as indicated by the arrows in the figure. Nonconvex potentials and nonlinear dynamical systems give rise to numerous variants of this situation, and the tools of nonlinear PDE analysis, bifurcation and singularity (catastrophe) theory provide a lot of structural results. While hysteresis occurs regularly here, it does so rather as an accessory than as an organizing concept, and consequently its role is not emphasized (see e.g. [47], [39]).

Much more complicated hysteresis effects occur in continuum mechanics. Let us consider longitudinal vibrations of a (one-dimensional) rod. Newton's law coupled with the constitutive stress-strain relation, i.e.

$$\partial_{tt}u = \partial_x\sigma\,, \quad \sigma = \mathcal{W}[\varepsilon]\,, \quad \varepsilon = \partial_x u\,, \tag{1.2}$$

*Institut für Informatik und Praktische Mathematik, Universität Kiel, D – 24098 Kiel, Germany

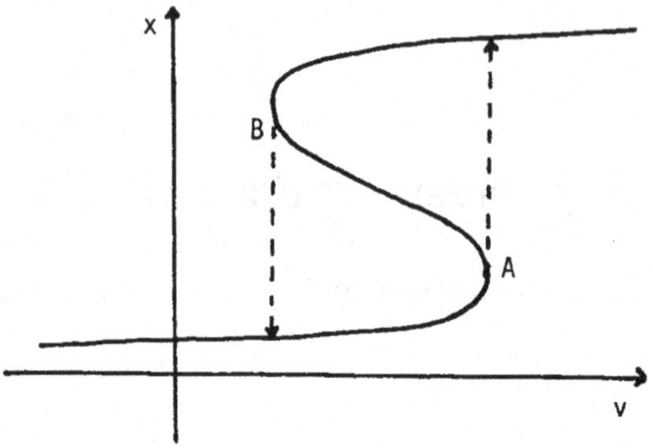

Figure 1: Hysteresis in parameter dependent processes.

together with certain initial and boundary conditions determine the displacement u, the stress σ and the strain ε as functions of time t and space x. Within the elastic range, Hooke's law

$$\mathcal{W}[\varepsilon] = E\varepsilon \tag{1.3}$$

holds, where E denotes the modulus of elasticity. Beyond the elastic limit, many materials exhibit plastic behaviour. Even its simplest description, namely the elastic-perfectly plastic model of Figure 2 with a fixed yield stress $|\sigma| = r$ and pure plastic flow, admits a

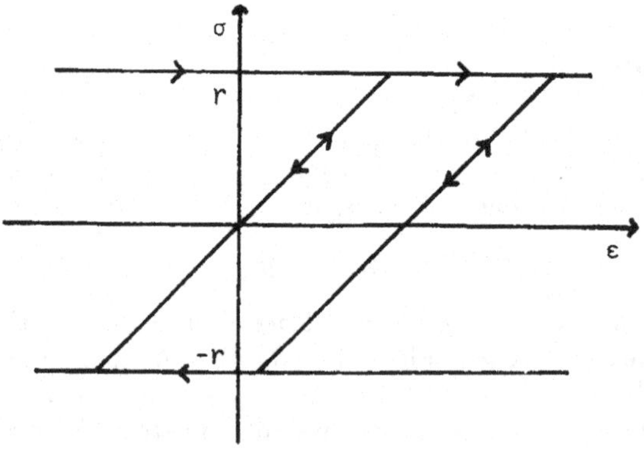

Figure 2: The elastic-perfectly plastic element.

continuum of possible stable states $\{(\sigma,\varepsilon) : \sigma \in [-r,r]\}$ for every value of ε. It will be explained in detail below how Figure 2 gives rise to a *hysteresis operator* $\mathcal{E}_r$ acting on a space of functions; accordingly, the constitutive equation

$$\sigma = \mathcal{E}_r[\varepsilon] \tag{1.4}$$

represents an equation between functions rather than between certain values of stress and strain. In 1928, Prandtl [34] constructed a more elaborate model as the continuous parallel combination of elastic-perfectly plastic elements; in operator notation, it has the form

$$\mathcal{W}[\varepsilon] = \int_0^{\infty} p(r)\mathcal{E}_r[\varepsilon]\, dr\,. \tag{1.5}$$

For certain types of steel at least, Prandtl's model (1.5)[1] yields a reasonable approximation in cyclic plasticity, where one investigates the so-called stabilized elastic-plastic behaviour, namely the behaviour after a large number of load cycles. We will see below how the density function p is related to the stabilized $\sigma-\varepsilon-$curve for initial loading (i.e. starting from $\sigma = \varepsilon = 0$), as well as to the shape of the hysteresis loops.

Prandtl's model allows smaller hysteresis loops embedded into larger ones with an arbitrary depth of nesting; moreover, all outer loops which have not yet been closed may affect the future behaviour. In this manner, the possible stable states of the material not only form a continuum like the interval $[-r, r]$ in the case of the elastic-perfectly plastic element $\mathcal{E}_r$, but require a potentially infinite dimensional memory representation. The same is true for the model developed by Preisach [35] in 1935 in order to describe the hysteresis loops traced out by magnetic field and magnetization in ferromagnetic materials. The operator form of the Preisach model reads

$$\mathcal{W}[v] = \int_0^{+\infty}\int_{-\infty}^{+\infty} \omega(r,s)\, \mathcal{R}_{s-r,\,s+r}[v]\, ds\, dr\,. \tag{1.6}$$

Here, ω denotes some density function like p does in Prandtl's model, and $\mathcal{R}_{x,y}$ denotes the relay with thresholds $x < y$, switching to the value $+1$ when the input attains the value y from below, and to -1 when it attains x from above. In addition, an initial value has to be prescribed for each relay.

All models mentioned so far have one aspect in common: They are *rate independent*. This means that the input-output behaviour does not depend on the speed (or frequency) of the input in the sense that, if the input frequency is doubled while its form is retained, the same is true for the output. This excludes from consideration relaxation effects like viscosity, creep or diffusion, which typically depend on the time scale. We will not enter a discussion of the relative significance of rate independent versus relaxation effects except for the remark that it varies greatly with the application. In the present notes, we will deal exclusively with rate independent models.

Continuum mechanics is not scalar, but takes place in $\mathbf{R}^3$. Although in many special situations a reduction of dimension is possible, the material laws usually are inherently multidimensional. Consequently, one expects that the memory structure has to take into account that the inputs are vectors (or tensors). However, on a level of memory complexity comparable to Prandtl's or Preisach's model, only very few mathematical investigations have been carried out so far – one has to admit, on the other hand, that the experimental basis, which could guide the selection of models, too is much less developed in the true vector case than for situations where one scalar quantity dominates the situation (for example, the tangential stress at the boundary of an interior hole of a two dimensional body). Since it fits well here, we will discuss one particular model which emphasizes memory structure, namely the continuous version of the model due to Mróz [33].

[1] It is also often named after Ishlinskiĭ [16].

The field of partial differential equations presents a particularly challenging area, because of both its difficulty and its importance for the continuum mechanics applications. We will present some results due Hilpert, Krejčí and Visintin; according to the spirit of these notes, we concentrate on the relevant properties of the hysteresis operators.

The approach to hysteresis described in these notes constitutes a mathematical technique whose goal is to analyze systems with hysteresis. A hysteresis operator results from a translation of a hysteresis diagram into a mathematical object, but it does not contribute to an explanation why the hysteresis is there at all. For that reason, hysteresis operators are said to be part of a *phenomenological approach* to hysteresis, and they offer themselves as a natural mathematical tool for a lot of problems in engineering. Nevertheless, there are also connections to the foundation of mechanics, since a hysteresis operator represents a mechanism for the dissipation of energy, if it satisfies an appropriate inequality. We will not explicitly discuss this aspect; it is, however, implicitly present in the analysis of PDE's with hysteresis in the last two sections.

These notes are lecture notes. We will not attempt to review, or even cite, all the relevant literature on the subject. For some time, the basic references have been the monograph of Krasnosel'skii and Pokrovskii [17] and the survey of Visintin [45]; now there is also the survey of Macki, Nistri and Zecca [27]. There will probably soon arrive the monograph of Visintin [46]. Concerning the special topic of optimal control of ODE systems with hysteresis, we also refer to [1]. We also will omit or abridge proofs on several occasions. They are to be found either in the references given or in the forthcoming monograph of Sprekels and the author.

2. Scalar Hysteresis Operators

Given a hysteresis diagram in the $v-w-$plane and an input function $v:[0,T]\to\mathbf{R}$, $T>0$, we want to choose an output function $w:[0,T]\to\mathbf{R}$ such that $(v(t),w(t))$ moves along the curves in the diagram. For such a procedure it is natural to require that the function v is piecewise monotone. Let us denote by $Map[0,T]$ the set of all real-valued functions on $[0,T]$, and by $M_{pm}[0,T]$ and $C_{pm}[0,T]$ the subset of all (respectively, continuous) piecewise monotone functions on $[0,T]$.

Definition 2.1 We say that an operator $\mathcal{W}:C_{pm}[0,T]\to Map[0,T]$ is a *hysteresis operator*, if it is rate independent and has the Volterra property. Rate independence means that

$$\mathcal{W}[v]\circ\varphi=\mathcal{W}[v\circ\varphi] \tag{2.1}$$

holds for all $v\in C_{pm}[0,T]$ and all continuous monotone time transformations $\varphi:[0,T]\to[0,T]$ satisfying $\varphi(0)=0$ and $\varphi(T)=T$. □

The rate independence implies that only the local extremal values of the input function v can have an influence on the memory of the process; consequently, we may replace input functions $v\in C_{pm}[0,T]$ by input strings $(v_0,\dots,v_N)$ with $v_i\in\mathbf{R}$. Let us denote by S the set of all finite strings of real numbers,

$$S=\{(v_0,\dots,v_N):N\in\mathbf{N}_0,\ v_i\in\mathbf{R},\ 0\le i\le N.\},\quad \mathbf{N}_0:=\mathbf{N}\cup\{0\}, \tag{2.2}$$

and by S_H the set of *alternating strings*

$$S_H = \{(v_0, \ldots, v_N) : v_0 \neq v_1 \text{ if } N \geq 1,\ (v_{i+1} - v_i)(v_i - v_{i-1}) < 0,\ 0 < i < N\}. \quad (2.3)$$

For any hysteresis operator $\mathcal{W}$, we define its *final value mapping* $\mathcal{W}_f : S_H \to \mathbf{R}$ by

$$\mathcal{W}_f(v_0, \ldots, v_N) = \mathcal{W}[v](T), \quad (2.4)$$

where $v \in C_{pm}[0,T]$ is any input function having a monotonicity partition $0 = t_0 < \ldots < t_N = T$ such that $v(t_i) = v_i$, $0 \leq i \leq N$. Conversely, any mapping $\mathcal{W}_f : S_H \to \mathbf{R}$ yields a hysteresis operator $\mathcal{W}$ if we set

$$\mathcal{W}[v](t) = \mathcal{W}_f(v(t_0), \ldots, v(t_k)), \quad (2.5)$$

where $0 = t_0 < \ldots < t_k = t$ is a monotonicity partition of $v_{|[0,t]}$ such that the string $(v(t_0), \ldots, v(t_k))$ is alternating. One may check from the definitions that the formulas (2.4) and (2.5) establish a bijective correspondence between the set of all hysteresis operators and the set of all real-valued mappings on S_H. Since we can use (2.4) to define $\mathcal{W}_f$ on all of S, we can interpret any hysteresis operator $\mathcal{W}$ as a mapping $\mathcal{W} : S \to S$ if we set

$$\mathcal{W}(v_0, \ldots, v_N) = (\mathcal{W}_f(v_0), \mathcal{W}_f(v_0, v_1), \ldots, \mathcal{W}_f(v_0, \ldots, v_N)). \quad (2.6)$$

To make a clear formal distinction, we will write $\mathcal{W}[v]$ for functions and $\mathcal{W}(s)$ for strings $s = (v_0, \ldots, v_N)$. We note also that (2.5) makes sense for inputs $v \in M_{pm}[0,T]$. In this manner, we obtain a canonical extension for any hysteresis operator from $C_{pm}[0,T]$ to $M_{pm}[0,T]$.

All scalar hysteresis operators mentioned during the introduction have a common memory structure. Its description involves the hysteresis operator which describes the mechanical play and is called the *play operator*, see Figure 3.

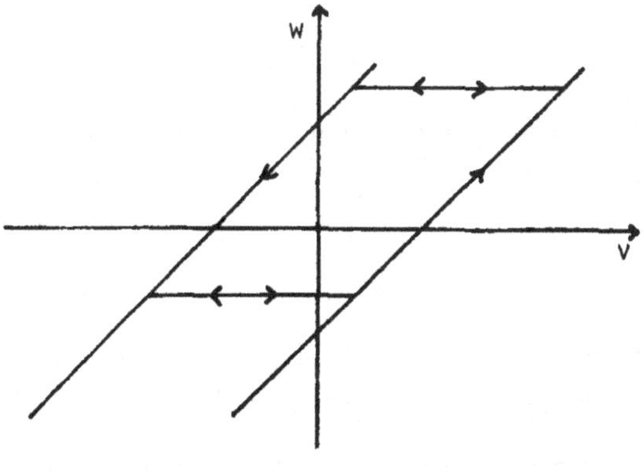

Figure 3: The play operator.

Definition 2.2 Let $r \geq 0$ be given. We define the play operator $\mathcal{F}_r[\,\cdot\,;\, w_{-1}]$ for the initial value[2] $w_{-1} \in \mathbf{R}$,

$$w(t) = \mathcal{F}_r[v\,;\, w_{-1}](t)\,, \tag{2.7}$$

by its corresponding final value mapping $\mathcal{F}_{r,f} : S_H \to \mathbf{R}$ given recursively by

$$\begin{aligned} \mathcal{F}_{r,f}(v_0) &= f_r(v_0, w_{-1})\,, \\ \mathcal{F}_{r,f}(v_0, \ldots, v_N) &= f_r(v_N,\, \mathcal{F}_{r,f}(v_0, \ldots, v_{N-1}))\,, \end{aligned} \tag{2.8}$$

where

$$f_r(v, w) = \max\{v - r\,,\, \min\{v + r, w\}\,\}\,. \tag{2.9}$$

If we do not state the choice of the initial value explicitly, we assume it to be zero. Accordingly, we write $\mathcal{F}_r[v]$ instead of $\mathcal{F}_r[v\,;\, 0]$. □

The final value mapping $\mathcal{R}_{x,y,f}$ of the relay $\mathcal{R}_{x,y}$ with thresholds $x < y$ and initial value $w_{-1}(x, y)$ has the form

$$\mathcal{R}_{x,y,f}(v_0, \ldots, v_N) = \begin{cases} 1\,, & v_N \geq y \\ -1\,, & v_N \leq x \\ \mathcal{R}_{x,y,f}(v_0, \ldots, v_{N-1})\,, & x < v_N < y\,,\ N \geq 1\,, \\ w_{-1}(x, y)\,, & x < v_N < y\,,\ N = 0\,. \end{cases} \tag{2.10}$$

If we do not state the choice of the initial value explicitly, we assume $w_{-1}(x, y) = 1$ if $x + y \leq 0$ and $w_{-1}(x, y) = -1$ otherwise.

The play operator $\mathcal{F}_r$ incorporates the memory of all relays $\mathcal{R}_{x,y}$ with $|x - y| = 2r$.

Lemma 2.3 *For each $r > 0$ and each $s \in \mathbf{R}$ there holds*

$$\mathcal{R}_{s-r,s+r,f}(v_0, \ldots, v_N) = \begin{cases} 1\,, & \mathcal{F}_{r,f}(v_0, \ldots, v_N) > s\,, \\ -1\,, & \mathcal{F}_{r,f}(v_0, \ldots, v_N) < s\,, \end{cases} \tag{2.11}$$

for every $N \geq 0$ and every string $(v_0, \ldots, v_N) \in S$.

Proof. We set $w_N = \mathcal{F}_{r,f}(v_0, \ldots, v_N)$, $\rho_N = \mathcal{R}_{s-r,s+r,f}(v_0, \ldots, v_N)$ and use induction on N. For $N = 0$, the assertion follows from the definitions. We provide the induction step $N - 1 \to N$. Assume that $v_N > v_{N-1}$. By (2.8) and (2.9) we have that

$$w_N = \max\{w_{N-1}, v_N - r\}\,. \tag{2.12}$$

According to the right hand side of (2.11), we distinguish two cases:

- If $s < w_N$, then we have either $w_N = w_{N-1}$ and $\rho_{N-1} = 1$, or $w_N = v_N - r$ and $s + r < v_N$. In both cases, $\rho_N = 1$ follows from (2.10).
- Assume that $s > w_N$. Then we have $s > w_{N-1}$ and hence $\rho_{N-1} = -1$; on the other hand, $s + r > v_N$. Together, this implies that $\rho_N = -1$.

---

[2] w_{-1} represents the internal state *before* $v(0)$ is applied at time $t = 0$.

□

If we insert (2.11) into the defining formula (1.6) of the Preisach operator $\mathcal{W}$, we get

$$\mathcal{W}[v](t) = \int_0^{+\infty} \int_{-\infty}^{\mathcal{F}_r[v](t)} \omega(r,s)\,ds\,dr \;-\; \int_0^{+\infty} \int_{\mathcal{F}_r[v](t)}^{+\infty} \omega(r,s)\,ds\,dr \tag{2.13}$$

for any input $v \in C_{pm}[0,T]$. Therefore, the Preisach operator can be expressed in terms of the play operator as

$$\mathcal{W}[v](t) = \int_0^{\infty} q(r,\, \mathcal{F}_r[v](t))\,dr + q_{00}\,, \tag{2.14}$$

where

$$q(r,s) \;=\; 2\int_0^{s} \omega(r,\sigma)\,d\sigma\,, \tag{2.15}$$

$$q_{00} \;=\; \int_0^{+\infty} \left(\int_{-\infty}^{0} \omega(r,\sigma)\,d\sigma \;-\; \int_0^{+\infty} \omega(r,\sigma)\,d\sigma \right) dr\,. \tag{2.16}$$

Note that $q_{00} = 0$ if $\omega(r,s) = \omega(r,-s)$ for all r and s.

Next on the list is the elastic-plastic element $\mathcal{E}_r$. The following definition formalizes the picture in Figure 2.

Definition 2.4 Let $r \geq 0$ be given. We define the hysteresis operator $\mathcal{E}_r[\,\cdot\,; w_{-1}]$ for the initial value $w_{-1} \in \mathbf{R}$ by its final value mapping $\mathcal{E}_{r,f} : S_H \to \mathbf{R}$ given recursively by

$$\begin{aligned} \mathcal{E}_{r,f}(v_0) &= e_r(v_0 - w_{-1})\,, \\ \mathcal{E}_{r,f}(v_0,\ldots,v_N) &= e_r(v_N - v_{N-1} + \mathcal{E}_{r,f}(v_0,\ldots,v_{N-1}))\,, \end{aligned} \tag{2.17}$$

where

$$e_r(v) = \min\{r, \max\{-r, v\}\}\,. \tag{2.18}$$

Again, we assume $w_{-1} = 0$ if not stated otherwise, and write $\mathcal{E}_r[v]$ instead of $\mathcal{E}_r[v\,;0]$.
□

Lemma 2.5 *We have*

$$\mathcal{F}_r \;+\; \mathcal{E}_r = id\,. \tag{2.19}$$

More precisely, for every $v \in C_{pm}[0,T]$ and every $w_{-1} \in \mathbf{R}$ there holds

$$\mathcal{F}_r[v\,; w_{-1}] \;+\; \mathcal{E}_r[v\,; w_{-1}] = v\,. \tag{2.20}$$

Proof. From the identity

$$v - f_r(v,w) = e_r(v-w)\,, \tag{2.21}$$

which holds for all $v, w \in \mathbf{R}$, one easily computes that

$$\mathcal{F}_{r,f}(v_0,\ldots,v_N) \;+\; \mathcal{E}_{r,f}(v_0,\ldots,v_N) = v_N\,. \tag{2.22}$$

As an immediate consequence, the Prandtl operator $\mathcal{W}$ from (1.5) becomes (note that $\mathcal{F}_0 = id$)

$$\begin{aligned} \mathcal{W}[v](t) &= \int_0^\infty p(r)\,\mathcal{E}_r[v](t)\,dr \\ &= \int_0^\infty p(r)\,dr \cdot \mathcal{F}_0[v](t) - \int_0^\infty p(r)\,\mathcal{F}_r[v](t)\,dr\,. \end{aligned} \tag{2.23}$$

The representations (2.11), (2.14), (2.20) and (2.23) show that thc valucs $(\mathcal{F}_r[v](t))_{r\geq 0}$ play a crucial role in determining the output $\mathcal{W}[v](t)$ for all hysteresis operators considered so far in this section. In fact, these values contain the whole memory information at time t needed to determine the future, i.e. to determine $\mathcal{W}[v]_{[t,T]}$ from $v_{[t,T]}$. In order to see this and to understand the memory evolution, let us consider an arbitrary input string $(v_0, \ldots, v_N) \in S$. It successively generates the memory curves $\psi_k : \mathbf{R}_+ \to \mathbf{R}$,

$$\psi_k(r) = \mathcal{F}_{r,f}(v_0, \ldots, v_k)\,, \quad 0 \leq k \leq N\,, \tag{2.24}$$

through the update

$$\begin{aligned} \psi_k(r) &= f_r(v_k, \psi_{k-1}(r)) \\ &= \max\{v_k - r, \min\{v_k + r, \psi_{k-1}(r)\}\}\,, \quad 0 \leq k \leq N\,. \end{aligned} \tag{2.25}$$

Here, the function $\psi_{-1} : \mathbf{R}_+ \to \mathbf{R}$ represents the initial memory values for the whole family $(\mathcal{F}_r)_{r\geq 0}$. Formula (2.25) shows that, as long as $\|\psi'_{k-1}\|_\infty \leq 1$, the graph of the new memory curve ψ_k consists of a straight line segment with slope sign$(v_{k-1} - v_k)$ originating at the point $(0, v_k)$, and of a portion of the old memory curve ψ_{k-1}, joined at their meeting point. Consequently, if we start with a suitable initial curve ψ_{-1}, the curve ψ_k continuous and consists of at most k straight pieces of finite length with slope alternating between $+1$ and -1, and of a portion of the initial curve extending to infinity to the right. More precisely, we take ψ_{-1} from the set

$$\begin{aligned} \Psi_0 = \{\varphi : \mathbf{R}_+ \to \mathbf{R} \mid\ & |\varphi(r) - \varphi(\bar r)| \leq |r - \bar r|\,, \quad \text{for all } r, \bar r \geq 0\,, \\ & \varphi|_{[\rho,\infty)} = 0\,, \quad \text{for some } \rho \geq 0\}\,, \end{aligned} \tag{2.26}$$

which we call the *set of Preisach memory curves*. It is easy to check that $\psi_k \in \Psi_0$ for all k, if $\psi_{-1} \in \Psi_0$.

In Preisach's original paper [35] we already find a picture of the memory curve and a brief informal description of its evolution. We therefore call *operator of Preisach type* any operator whose memory structure is governed by that memory curve.

Definition 2.6 Let $\mathcal{W} : C_{pm}[0, t_E] \to Map([0, t_E])$ be a hysteresis operator. We say that $\mathcal{W}$ is of *Preisach type* if its final value map $\mathcal{W}_f : S \to \mathbf{R}$ has the form

$$\mathcal{W}_f(v_0, \ldots, v_N) = Q(\psi_f(v_0, \ldots, v_N))\,, \quad (\psi_f(v_0, \ldots, v_N))(r) = \mathcal{F}_{r,f}(v_0, \ldots, v_N)\,, \tag{2.27}$$

for some mapping $Q : \Psi_0 \to \mathbf{R}$, called the output mapping of Q, and some initial condition $\psi_{-1} \in \Psi_0$. Equivalently, we may write (2.27) as

$$\mathcal{W}[v](t) = Q(\psi(t))\,, \quad \psi(t)(r) = \mathcal{F}_r[v\,;\psi_{-1}(r)](t)\,, \quad t \in [0,T]\,, \quad r \geq 0\,. \tag{2.28}$$

If $\psi_{-1} = 0$, we call $\mathcal{W}$ a $\mathcal{P}_0$-operator.

To illustrate Definition 2.6, we specify the mapping Q for the Prandtl model (1.5) and the Preisach model (1.6). For the Prandtl model we have

$$Q(\varphi) = p_0\varphi(0) - \int_0^\infty p(r)\varphi(r)\,dr\,, \quad p_0 = \int_0^\infty p(r)\,dr\,, \tag{2.29}$$

whereas for the Preisach model we get

$$Q(\varphi) = \int_0^\infty q(r,\varphi(r))\,dr + q_{00}\,, \quad q(r,s) = 2\int_0^s \omega(r,\sigma)d\sigma\,, \tag{2.30}$$

where q_{00} is given in (2.16).

Note that the definition of the class of all Prandtl respectively Preisach operators is not quite unique, since one has to specify the class of allowed density functions (or, more generally, measures). No such element of arbitrariness is present in the definition of a $\mathcal{P}_0$-operator.

In the hysteresis diagrams considered so far, all curves are monotone, so the output $\mathcal{W}[v]$ will be monotone on any time interval where the input v is monotone. In that case, $\mathcal{W}$ maps $M_{pm}[0,T]$ into itself.

Definition 2.7 A hysteresis operator $\mathcal{W}$ is called *piecewise (strictly) monotone*, if its *level functions* $l_N : \mathbf{R} \to \mathbf{R}$ defined by

$$l_N(v) = \mathcal{W}_f(v_0, \ldots, v_{N-1}, v) \tag{2.31}$$

are (strictly) increasing functions for any $N \in \mathbf{N}_0$ and any $(v_0, \ldots, v_{N-1}) \in S_H$. □

The operators $\mathcal{F}_r$, $\mathcal{E}_r$ and $\mathcal{R}_{x,y}$ are obviously piecewise monotone. The Prandtl and the Preisach operator are piecewise monotone, if their densities p respectively ω are nonnegative functions - in both cases, however, the nonnegativity is not necessary and can be relaxed, see Proposition 4.8 for the Prandtl operator and Proposition 4.9 for the Preisach operator. A general hysteresis operator of Preisach type $\mathcal{W}$ is piecewise monotone, if its output mapping Q is *order preserving*, i.e. if $\varphi_1 \le \varphi_2$, to be understood pointwise, implies that $Q(\varphi_1) \le Q(\varphi_2)$. Again, this condition is not necessary, as (2.29) immediately demonstrates.

3. Continuity and Regularity

We start with a general remark. In connection with hysteresis operators, estimates of the sup norm and of the total variation are very natural, because both are compatible with the rate independence property in the sense that

$$\|W[v \circ \varphi]\|_\infty = \|W[v] \circ \varphi\|_\infty = \|W[v]\|_\infty \tag{3.1}$$

as well as

$$\mathrm{Var}\,(W[v \circ \varphi]) = \mathrm{Var}\,(W[v] \circ \varphi) = \mathrm{Var}\,(W[v]) \tag{3.2}$$

hold for the time transformations φ considered in the definition of rate independence.

Since the play operator appears here as the basic scalar hysteresis operator, it is natural to study its continuity and regularity properties first.

Proposition 3.1 *Let* $r_1, r_2 \geq 0$. *Then for any input functions* $v_1, v_2 \in M_{pm}[0,T]$, *initial values* $w_{-1,1}, w_{-1,2} \in \mathbf{R}$ *and any* $t \in [0,T]$ *we have*

$$\begin{aligned} | \mathcal{F}_{r_1}[v_1 \,; w_{-1,1}](t) \; - \; \mathcal{F}_{r_2}[v_2 \,; w_{-1,2}](t) \,| \; \leq \\ \leq \; \max \{ \, |r_1 - r_2| \; + \; \sup_{0 \leq s \leq t} |v_1(s) - v_2(s)| \, , \, |w_{-1,1} - w_{-1,2}| \, \} \, . \end{aligned} \tag{3.3}$$

Proof. For any $a, b, c, d \in \mathbf{R}$ we have

$$| \max\{a,b\} \; - \; \max\{c,d\} \,| \leq \max \{ |a-c| \, , \, |b-d| \} \,| \, . \tag{3.4}$$

The estimate (3.4) also holds if we replace "max" with "min" on the left side of (3.4). Applying (3.4) twice, we get

$$|f_{r_1}(x_1, y_1) - f_{r_2}(x_2, y_2)| \leq \max \{ \, |x_1 - x_2| \; + \; |r_1 - r_2| \, , \, |y_1 - y_2| \, \} \, , \tag{3.5}$$

for every $x_1, x_2, y_1, y_2 \in \mathbf{R}$. From (3.5) we obtain the assertion, using induction over a common monotonicity partition for the input functions v_1 and v_2. □

Proposition 3.1 yields the Lipschitz continuity of the play and of the elastic-plastic element in $C[0,T]$.

Theorem 3.2 *For any* $r \geq 0$, *the operators* $\mathcal{F}_r$ *and* $\mathcal{E}_r$ *can be extended uniquely to Lipschitz continuous operators*

$$\mathcal{F}_r \, , \mathcal{E}_r : C[0,T] \times \mathbf{R} \rightarrow C[0,T] \, , \tag{3.6}$$

and the following formulas hold for any $v, v_1, v_2 \in C[0,T]$, *any* $w_{-1}, w_{-1,1}, w_{-1,2} \in \mathbf{R}$, *any* $s \geq 0$, *and any* $0 \leq t \leq T$:

$$\| \mathcal{F}_r[v_1 \,; w_{-1,1}] \; - \; \mathcal{F}_r[v_2 \,; w_{-1,2}] \|_\infty \; \leq \; \max \{ \|v_1 - v_2\|_\infty \, , \, |w_{-1,1} - w_{-1,2}| \, \} \, , \tag{3.7}$$

$$| \mathcal{F}_r[v \,; w_{-1}](t) \; - \; \mathcal{F}_r[v \,; w_{-1}](s) \,| \; \leq \; \sup_{s \leq \tau \leq t} |v(\tau) \; - \; v(s)| \, , \quad \text{if } s \leq t \, , \tag{3.8}$$

$$\| \mathcal{E}_r[v_1] \; - \; \mathcal{E}_r[v_2] \|_\infty \; \leq \; 2 \, \|v_1 - v_2\|_\infty \, . \tag{3.9}$$

Proof. For piecewise monotone inputs, (3.7) follows directly from (3.3), (3.8) follows from (3.7) if we set $v_1 = v$ and $v_2 = v$ on $[0,s]$, $v_2 = v(s)$ on $[s,t]$; (3.9) follows from (3.7) and (2.19). The estimate (3.8) implies in particular that $\mathcal{F}_r[v] \in C[0,T]$ if $v \in C_{pm}[0,T]$. Therefore, $\mathcal{F}_r$ and $\mathcal{E}_r$ are Lipschitz continuous on the dense subset $C_{pm}[0,T] \times \mathbf{R}$ of $C[0,T] \times \mathbf{R}$. Consequently, both operators can be extended to $C[0,T] \times \mathbf{R}$, and formulas (3.7) to (3.9) remain valid for the extension. □

It is easy to see that the constant 2 in (3.9) in general cannot be improved.

We now discuss estimates for the total variation. Since derivatives of input and output functions are involved, we use the space

$$C_{pl}[0,T] = \{ v \in C[0,T] \, : \, v \text{ is piecewise linear} \, \} \tag{3.10}$$

of piecewise linear and continuous functions as the input space.

Lemma 3.3 *Let $r \geq 0$, let $v_1, v_2 \in C_{pl}[0,T]$ be given. Then the following is true:*

(i) *For any $t \in [0,T]$ except for the (finite number of) points where either $\mathcal{F}_r[v_1]'$ or $\mathcal{F}_r[v_2]'$ does not exist, there holds the inequality*

$$\left(\mathcal{F}_r[v_1]'(t) - \mathcal{F}_r[v_2]'(t)\right) \cdot \left(\mathcal{E}_r[v_1](t) - \mathcal{E}_r[v_2](t)\right) \geq 0. \tag{3.11}$$

(ii) *For any $t \in [0,T]$ except for a finite number of points there holds the estimate*

$$|\mathcal{F}_r[v_1]'(t) - \mathcal{F}_r[v_2]'(t)| + |\mathcal{E}_r[v_1] - \mathcal{E}_r[v_2]|'(t) \leq |v_1'(t) - v_2'(t)|. \tag{3.12}$$

Proof. To prove (i), fix any $t \in [0,T]$ where both $\mathcal{F}_r[v_j]'$ exist. Assume that $\mathcal{E}_r[v_1](t) > \mathcal{E}_r[v_2](t)$. We then have on one hand $\mathcal{E}_r[v_1](t) > -r$, hence $\mathcal{F}_r[v_1]'(t) \geq 0$, and on the other hand $\mathcal{E}_r[v_2](t) < r$ and hence $\mathcal{F}_r[v_2]'(t) \leq 0$. To prove (ii), we set (with the momentary convention $\mathrm{sign}(0) = 0$)

$$\rho(t) = \mathrm{sign}\left(\mathcal{E}_r[v_1](t) - \mathcal{E}_r[v_2](t)\right), \tag{3.13}$$

and choose a partition $\Delta = \{t_i\}_{0 \leq i \leq N}$ of $[0, t_E]$ such that, in each open interval (t_i, t_{i+1}), all derivatives occuring in (3.12) exist and the function ρ is either identically or nowhere zero. If ρ is identically zero on such an interval, then (3.12) holds with equality because of (2.20). Otherwise, we conclude from (3.11) that

$$|\mathcal{F}_r[v_1]'(t) - \mathcal{F}_r[v_2]'(t)| = (\mathcal{F}_r[v_1]'(t) - \mathcal{F}_r[v_2]'(t))\,\rho(t). \tag{3.14}$$

We always have

$$|\mathcal{E}_r[v_1] - \mathcal{E}_r[v_2]|'(t) = (\mathcal{E}_r[v_1]'(t) - \mathcal{E}_r[v_2]'(t))\,\rho(t). \tag{3.15}$$

We sum (3.14) and (3.15) and apply (2.20) to obtain the assertion. □

From (3.12) we conclude that the play and the elastic-plastic element are Lipschitz continuous in $W^{1,1}(0,T)$ with respect to its standard norm

$$\|v\|_{BV} = \mathrm{Var}(v) + |v(0)| = \int_0^T |v'(t)|\,dt + |v(0)|. \tag{3.16}$$

Proposition 3.4 *The operators $\mathcal{F}_r$ and $\mathcal{E}_r$ are Lipschitz continuous on $W^{1,1}(0,T)$, and we have the estimate*

$$\|\mathcal{F}_r[v_1] - \mathcal{F}_r[v_2]\|_{BV} \leq \int_0^T |v_1'(t) - v_2'(t)|\,dt + 2|v_1(0) - v_2(0)|, \tag{3.17}$$

$$\|\mathcal{E}_r[v_1] - \mathcal{E}_r[v_2]\|_{BV} \leq 2\|v_1 - v_2\|_{BV}. \tag{3.18}$$

Proof. For arbitrary $v_1, v_2 \in C_{pl}[0,T]$, we obtain from (3.12) that

$$\begin{aligned}\int_0^T |\mathcal{F}_r[v_1]'(t) - \mathcal{F}_r[v_2]'(t)|\,dt + |\mathcal{E}_r[v_1](T) - \mathcal{E}_r[v_2](T)| \\ \leq \int_0^T |v_1'(t) - v_2'(t)|\,dt + |\mathcal{E}_r[v_1](0) - \mathcal{E}_r[v_2](0)| \\ \leq \int_0^T |v_1'(t) - v_2'(t)|\,dt + |v_1(0) - v_2(0)|.\end{aligned} \tag{3.19}$$

Adding $|\mathcal{F}_r[v_1](0) - \mathcal{F}_r[v_2](0)|$ on both sides, the estimate (3.17) follows from (3.7). Since moreover

$$\int_0^T |\mathcal{E}_r[v_1]'(t) - \mathcal{E}_r[v_2]'(t)|\,dt \leq \int_0^T |\mathcal{F}_r[v_1]'(t) - \mathcal{F}_r[v_2]'(t)|\,dt + \int_0^T |v_1'(t) - v_2'(t)|\,dt\,, \tag{3.20}$$

the second estimate (3.18) also follows from (3.19). By density, both estimates extend to $W^{1,1}(0,T)$. □

Due to Theorem 3.2 and Proposition 3.4, formulas involving $\mathcal{F}_r[v]$ or $\mathcal{E}_r[v]$, or their time derivatives, typically carry over to $C[0,T]$ respectively $W^{1,1}(0,T)$.

The play operator also has a *monotone regularization property*. For any function $v : [0,T] \to \mathbf{R}$, let us denote its minimal number of monotonicity intervals by $N_{mon}(v)$; if v is not piecewise monotone, we set $N_{mon}(v) = +\infty$. We also define the (inverse) modulus of continuity of v, if v is continuous, as

$$\eta(\delta\,; v) = \sup_{|t-s|\leq\delta} |v(t) - v(s)|\,, \tag{3.21}$$

$$\eta^{-1}(\epsilon\,; v) = \sup\{\delta : \eta(\delta\,; v) \leq \epsilon\}\,. \tag{3.22}$$

Proposition 3.5 *Let $r > 0$, $v \in C[0,T]$. Then $\mathcal{F}_r[v] \in C_{pm}[0,T]$ and*

$$N_{mon}(\mathcal{F}_r[v]) \leq \frac{T}{\eta^{-1}(2r\,; v)} + 2\,. \tag{3.23}$$

Proof. One exploits the fact that the input has to travel a distance of $2r$ before the output can change its monotonicity direction as well as the uniform continuity of v expressed in the fact that $\eta^{-1}(\epsilon\,; v) > 0$ for $\epsilon > 0$. The details are omitted. □

We now discuss continuity and regularity properties of hysteresis operators of Preisach type. According to (2.28), their memory at time t is described by the memory curve $\psi(t) : \mathbf{R}_+ \to \mathbf{R}$ as

$$\psi(t)(r) = \mathcal{F}_r[v\,; \psi_{-1}(r)](t)\,, \quad r \geq 0\,, \tag{3.24}$$

for some initial condition $\psi_{-1} \in \Psi_0$. The first question is how $\psi(t)$ looks like if the input v is not piecewise monotone. Here, we may use the composition formula for play operators

$$\mathcal{F}_r[\mathcal{F}_s[v\,; y_{-1}]\,; w_{-1}] = \mathcal{F}_{r+s}[v\,; w_{-1}]\,, \tag{3.25}$$

which holds if $r, s \geq 0$ and $|y_{-1} - w_{-1}| \leq r$. There is an elementary, but somewhat tedious, proof of (3.25), which we omit here[3]. As a consequence of (3.25),

$$\psi(t)(s+r) = \mathcal{F}_{r+s}[v\,; \psi_{-1}(s+r)](t) = \mathcal{F}_r[\mathcal{F}_s[v\,; \psi_{-1}(s)]\,; \psi_{-1}(s+r)](t)\,. \tag{3.26}$$

Formula (3.26) says that the shifted curve $\psi^s(t) = \psi(t)(s + \cdot)$ represents the Preisach memory curve associated with the (due to Proposition 3.5) piecewise monotone input $\mathcal{F}_s[v\,; \psi_{-1}(s)]$ and the shifted initial curve $\psi_{-1}^s = \psi_{-1}(s + \cdot)$. In particular, ψ^s has the

[3]For $y_{-1} = w_{-1} = 0$, (3.25) is a special case of Proposition 4.10.

structure described in the paragraph following equation (2.25). Therefore, the portion of $\psi(t)$ not coinciding with ψ_{-1} consists of either finitely or infinitely many straight pieces with slope alternating between $+1$ and -1; if there are infinitely many such pieces, they can only accumulate at the left end $r = 0$. In any case, we have $\psi(t) \in \Psi_0$ if $\psi_{-1} \in \Psi_0$.

The continuity and regularity properties of a Preisach type operator $\mathcal{W}$ are obviously linked to corresponding properties of its output mapping Q. Since the estimates of Theorem 3.2 for the play operator are uniform with respect to r, we work with the quantity

$$\eta(\delta\,; Q) = \sup_{\substack{\varphi,\psi\in\Psi_0 \\ \|\varphi-\psi\|_\infty\le\delta}} |Q(\varphi) - Q(\psi)|\,. \tag{3.27}$$

Proposition 3.6 *Let $\mathcal{W}$ be a hysteresis operator of Preisach type with the output mapping $Q : \Psi_0 \to \mathbf{R}$.*

(i) *If*

$$\lim_{\delta\downarrow 0} \eta(\delta\,; Q) = 0\,, \tag{3.28}$$

then $\mathcal{W}$ is uniformly continuous on $C[0,T]$ and therefore maps bounded subsets of $C[0,T]$ into bounded subsets of $C[0,T]$.

(ii) *If*

$$\eta(\delta\,; Q) \le C\delta\,, \tag{3.29}$$

then $\mathcal{W}$ is Lipschitz continuous on $C[0,T]$ and maps bounded subsets of $W^{1,p}[0,T]$ into bounded subsets of $W^{1,p}[0,T]$, $1 \le p \le \infty$, endowed with their standard norms. Moreover,

$$|\mathcal{W}[v]'(t)| \;\le\; C|v'(t)| \tag{3.30}$$

holds at every $t \in [0,T]$ where both derivatives exist.

Proof. For v, v_1, v_2 with corresponding memory curves $\psi, \psi_1, \psi_2 : [0,T] \to \Psi_0$, Theorem 3.2 yields the estimates

$$|\mathcal{W}[v_1](t) - \mathcal{W}[v_2](t)| \;\le\; \eta(\|\psi_1(t) - \psi_2(t)\|_\infty\,; Q) \;\le\; \eta(\|v_1 - v_2\|_\infty\,; Q)\,, \tag{3.31}$$

$$|\mathcal{W}[v](t) - \mathcal{W}[v](s)| \;\le\; \eta(\|\psi(t) - \psi(s)\|_\infty\,; Q) \;\le\; \eta(\sup_{s\le\tau\le t} |v(\tau) - v(s)|\,; Q) \tag{3.32}$$

for any $t, s \in [0,T]$. The assertions concerning $C[0,T]$ readily follow. If (3.29) holds, (3.32) becomes

$$|\mathcal{W}[v](t) - \mathcal{W}[v](s)| \;\le\; C \sup_{s\le\tau\le t} |v(\tau) - v(s)|\,. \tag{3.33}$$

From (3.33) we see that $\mathcal{W}[v]$ is absolutely continuous, if v is absolutely continuous, and that (3.30) holds almost everywhere. In turn, (3.30) implies the statements concerning $W^{1,p}(0,t_E)$. □

We also refer to [3], [22] and [2], where the reader can find additional material.

Proposition 3.7 *For the Prandtl operator (1.5), condition (3.29) is satisfied if $p \in L^1(\mathbf{R}_+)$. For the Preisach operator, if*

$$C = \frac{1}{2}\int_0^\infty \sup_{s\in\mathbf{R}} |\omega(r,s)|\, dr < \infty\,, \tag{3.34}$$

then (3.29) holds.

Proof. This follows directly from the definition of η. □

4. Hysteresis Memory and Hysteresis Loops

What is remembered and what is forgotten are two complementary aspects of memory. For a general hysteresis operator $\mathcal{W}$, their relation is rather trivial and a direct consequence of the rate independence property. Consider an input string $s = (v_0, \ldots, v_N)$. If $v_i \in [v_{i-1}, v_{i+1}]$ holds, then the value v_i no longer has an impact on the memory (i.e. it is forgotten), when v_{i+1} has been processed. If all such values are deleted from the string s, an alternating string formed by the local extrema of s remains. All those extrema may influence the output through the final value mapping $\mathcal{W}_f$ (see (2.4)) and, if they do, have to be retained within memory in some form.

Hysteresis operators of Preisach type have a more selective memory. For its study, we formulate certain *deletion rules*. We introduce the notation

$$[x, y] = \text{convex hull of } x \text{ and } y\,, \text{ if } x, y \in \mathbf{R}\,, \tag{4.1}$$

since it greatly simplifies the exposition. We call *monotone deletion rule* the string reduction implied by the rate independence, i.e.

$$\begin{aligned} (v_0, v_1) &\mapsto (v_0)\,, && \text{if } v_0 = v_1\,, \\ (v_0, \ldots, v_N) &\mapsto (v_0, \ldots, v_{i-1}, v_{i+1}, \ldots, v_N)\,, && \text{if } v_i \in [v_{i-1}, v_{i+1}]\,. \end{aligned} \tag{4.2}$$

The main distinguishing feature of the Preisach memory is the fact that inner hysteresis loops are forgotten at the moment when they are closed. Already 30 years prior to Preisach, Madelung [28] claimed this property to hold in rate independent ferromagnetism, and Prandtl reported in [34] how a connection to rate independent plasticity was established soon thereafter.

In terms of input strings, an inner hysteresis loop appears as an interval embedded into a larger interval. We say that the pair (v_i, v_{i+1}) is a *Madelung cycle* for the string $s = (v_0, \ldots, v_N)$, if

$$[v_i, v_{i+1}] \subset [v_{i-1}, v_{i+2}]\,, \quad v_i \notin [v_{i-1}, v_{i+1}]\,, \quad v_{i+1} \notin [v_i, v_{i+2}]\,, \tag{4.3}$$

and the corresponding reduction

$$(v_0, \ldots, v_N) \mapsto (v_0, \ldots, v_{i-1}, v_{i+2}, \ldots, v_N) \tag{4.4}$$

we call the *Madelung deletion rule*. We define an ordering $\leq_M$ on the set S of all strings by saying that $s \leq_M s'$, if s' can be reduced to s through a finite sequence of monotone and Madelung deletions. For any such sequence, we may count the Madelung deletions performed during that sequence. To this purpose, we denote by $a(x, y)$ how often the Madelung cycle (x, y) has been deleted, and we call $a : \mathbf{R} \times \mathbf{R} \to \mathbf{N}_0$ the *hysteresis count* of that sequence. A string is called *irreducible w.r.t.* $\leq_M$, if neither the monotone nor the Madelung deletion rule applies to it.

Proposition 4.1 *A string* $s = (v_0, \ldots, v_N)$ *is irreducible w.r.t.* $\leq_M$ *if and only if it is alternating and there exists an index* $J \in \{0, \ldots, N-1\}$ *such that*

$$d_0 < \ldots < d_{J-1}, \quad d_J > \ldots > d_{N-1}, \tag{4.5}$$

holds, where $d_i = |v_{i+1} - v_i|$.

Proof. The "if"-part is obvious. The proof of the converse is based on the observation that (v_i, v_{i+1}) is a Madelung cycle for s if and only if

$$0 < d_i \leq \min\{d_{i-1}, d_{i+1}\}. \tag{4.6}$$

□

Theorem 4.2 *For every* $s = (v_0, \ldots, v_N) \in S$ *there is a unique irreducible string* $s_M \in S$ *with* $s_M \leq_M s$. *Moreover, the symmetric hysteresis count*

$$a_{sym}(x, y) = a(x, y) + a(y, x) \tag{4.7}$$

is the same for all deletion sequences leading from s *to* s_M. *If* $v_i \neq v_j$ *for all* $i \neq j$, *then the same is true for the hysteresis count.*

Proof. Omitted. □

Under the name *rainflow count*, Endo has introduced the hysteresis count within the context of material fatigue analysis (see [32] for a reprint of his original paper). Accordingly, the string s_M in Theorem 4.2 above is also called the *rainflow residual* of s. The algorithm which computes the rainflow count and the rainflow residual for a given string bears the name *rainflow algorithm*. Various versions of it have been developed, see [9], [12], [26], and also [36] [37], [38] for some general considerations concerning counting methods. The rainflow count together with the rainflow residual is an important tool for predicting the life time of mechanical parts as well as for the design of experiments to check those predictions. We illustrate this with a simple example. Given a mechanical part, assume that we know for a given load cycle (x, y) that the part is destroyed after approximately $M(x, y)$ repetitions of the cycle. (Besides an approximate stress-strain law, this is the typical information available from experiments.) We now want to predict the damage produced by an arbitrary load sequence represented by the string $s = (v_0, \ldots, v_N)$. According to the *Palmgren-Miner-Rule* [31] of linear damage accumulation, every cycle (x, y) contributes an amount $1/M(x, y)$ to the total damage. If we use the rainflow algorithm to count the cycles, the value

$$D = \sum_{x,y \in \mathbf{R}} \frac{a(x, y)}{M(x, y)} \tag{4.8}$$

is attributed to the total damage, and destruction of the part is predicted to occur at the moment where D attains the value 1.

We return to the discussion of the memory structure of operators of Preisach type. The monotone and the Madelung deletion rule are not sufficient to characterize it, since a local minimum may also wipe out the memory due to a previous local maximum or vice versa, if the former is large enough. The precise meaning of "large enough", however, depends on the initial condition ψ_{-1}. For simplicity, we restrict ourselves here to the case $\psi_{-1} = 0$; the appropriate *initial deletion rule* is then given by

$$\begin{aligned} (v_0, v_1) &\mapsto (v_1), && \text{if } |v_0| \le |v_1|, \\ (v_0, \dots, v_N) &\mapsto (v_1, \dots, v_N), && \text{if } |v_0| \le \max\{|v_1|, |v_2|\}, \ N \ge 2. \end{aligned} \tag{4.9}$$

We introduce the ordering $\preceq_0$ on S in the same manner as $\preceq_M$, only that we now allow all three deletion rules to be applied during a deletion sequence. A result analogous to Proposition 4.1 and Theorem 4.2 holds.

Proposition 4.3 *For any $s \in S$ there is a unique $s_0 \in S$, irreducible w.r.t. $\preceq_0$, such that $s_0 \preceq_0 s$ holds. The set S_0 of irreducible elements has the form*

$$S_0 = \{(v_0, \dots, v_N) : N \in \mathbf{N}_0, |v_0| > |v_1|, |v_i - v_{i-1}| > |v_{i+1} - v_i| \ \textit{for} \ 0 < i < N\}. \tag{4.10}$$

Proof. Omitted. □

With the three deletion rules stated above, we can precisely characterize hysteresis operators of Preisach type in terms of what they forget.

Definition 4.4 Let $\preceq$ be an ordering on the set S of all strings. We say that a hysteresis operator $\mathcal{W}$ *forgets according to* $\preceq$, if $s' \preceq s$ implies that $\mathcal{W}_f(s) = \mathcal{W}_f(s')$. □

Theorem 4.5 *Let $\mathcal{W}$ be a hysteresis operator. $\mathcal{W}$ is a $\mathcal{P}_0$-operator*[4] *if and only if $\mathcal{W}$ forgets according to the ordering $\preceq_0$ generated by the monotone, Madelung and initial deletion rule.*

Proof. One checks that the play $\mathcal{F}_r$ forgets according to $\preceq_0$; this proves the "only if"-part. For the converse, we note that for $s_0, s_0' \in S_0$ there holds

$$\psi_f(s_0) = \psi_f(s_0') \Leftrightarrow s_0 = s_0'. \tag{4.11}$$

Therefore, we can define an output mapping Q on the set of reachable Preisach memory states by

$$Q(\psi_f(s)) = \mathcal{W}_f(s), \quad s \in S, \tag{4.12}$$

hence $\mathcal{W}$ is a $\mathcal{P}_0$-operator. □

Corollary 4.6 *To every mapping $\mathcal{W}_f : S_0 \to \mathbf{R}$ there corresponds a unique $\mathcal{P}_0$-operator, whose final value mapping equals $\mathcal{W}_f$ on S_0.* □

Remark. The fact that the Preisach operator forgets according to the ordering $\preceq_0$ is called the *wiping out property* of the Preisach operator in [30].

We now turn our attention to the shape of the hysteresis loops. They typically arise when, starting from an input history $(v_0, \dots, v_{N-1})$, we change the input value from v_{N-1}

[4] See Definition 2.6.

to v_N and then back to v_{N-1}. In the input-output plane, we move along the graphs of the level functions

$$\begin{aligned} l_N(x) &= \mathcal{W}_f(v_0, \ldots, v_{N-1}, x), &(4.13)\\ l_{N+1}(x) &= \mathcal{W}_f(v_0, \ldots, v_{N-1}, v_N, x). &(4.14)\end{aligned}$$

For a general hysteresis operator, those functions can be completely arbitrary. If however $\mathcal{W}$ forgets according to the Madelung deletion rule, and if $v_N \in [v_{N-2}, v_{N-1}]$, we return at least to the starting point $(v_{N-1}, \mathcal{W}_f(v_0, \ldots, v_{N-1}))$, since

$$l_{N+1}(v_{N-1}) = \mathcal{W}_f(v_0, \ldots, v_{N-1}, v_N, v_{N-1}) = \mathcal{W}_f(v_0, \ldots, v_{N-1}) = l_{N-1}(v_{N-1}). \quad (4.15)$$

For the Preisach operator (1.6) with zero initial state we get, if $(v_0, \ldots, v_N) \in S_0$,

$$l_N(v) = w_{N-1} + 2\sigma \int_{\Delta(v_{N-1}, v)} \omega(r, s)\, ds\, dr, \quad v \in [v_{N-1}, v_N], \quad (4.16)$$

where $\sigma = \text{sign}(v_N - v_{N-1})$, $w_{N-1} = \mathcal{W}_f(v_0, \ldots, v_{N-1})$, and

$$\begin{aligned} \Delta(v_*, v^*) &= \{(r, s) : 0 \le r \le \frac{v^* - v_*}{2}, v_* + r \le s \le v^* - r\}, \quad \text{if } v_* \le v^*\\ &= \Delta(v^*, v_*) \quad \text{otherwise}, &(4.17)\end{aligned}$$

denotes the triangle with the corners $(0, v_*)$, $(0, v^*)$ and $(|v^* - v_*|/2, (v^* + v_*)/2)$. In particular, the height $w_N - w_{N-1}$ of the hysteresis loop is given by the function h, introduced in 1955 by Everett [13],

$$w_N - w_{N-1} = h(v_{N-1}, v_N) := 2\sigma \int_{\Delta(v_{N-1}, v_N)} \omega(r, s)\, ds\, dr. \quad (4.18)$$

This formula shows that, no matter what the previous history has been, all hysteresis loops for a fixed input cycle (v_{N-1}, v_N) have identical shape and differ only by a vertical translation. Accordingly, this feature is often[5] called the *congruency property* of the Preisach model. These considerations motivate the following definition.

Definition 4.7 Let $\mathcal{W}$ be a $\mathcal{P}_0$–operator. We say that $\mathcal{W}$ has the *height function* $h : \mathbf{R}^2 \to \mathbf{R}$, if there holds

$$\mathcal{W}_f(v_0, \ldots, v_N) = \mathcal{W}_f(v_0, \ldots, v_{N-1}) + h(v_{N-1}, v_N) \quad (4.19)$$

for every $(v_0, \ldots, v_N) \in \bar{S}_0$, where

$$\bar{S}_0 = \{(v_0, \ldots, v_N) \in S : |v_1| \le |v_0| \text{ and } v_{i+1} \in [v_{i-1}, v_i] \text{ for } 1 \le i < N\}. \quad (4.20)$$

□

[5]See e.g. [30].

Obviously, we have

$$\mathcal{W}_f(v_0,\ldots,v_N) = \mathcal{W}_f(v_0) + \sum_{k=1}^{N} h(v_{k-1},v_k)\,. \tag{4.21}$$

for every $(v_0,\ldots,v_N) \in \tilde{S}_0$ and

$$h(v_1,v_2) = -h(v_2,v_1) \qquad \text{for any } v_1,v_2 \in \mathbf{R}\,. \tag{4.22}$$

and in particular $h(v,v) = 0$ for every $v \in \mathbf{R}$.

The operators $\mathcal{F}_r$, $\mathcal{E}_r$ and the Prandtl operator (1.5) have a height function of the form

$$h(v_1,v_2) = g(v_2 - v_1)\,, \tag{4.23}$$

namely

$$g_{\mathcal{F}_r}(v) = \max\{v-2r,0\}\,, \qquad g_{\mathcal{E}_r}(v) = \min\{v,2r\}\,, \tag{4.24}$$

and for the Prandtl operator

$$g(v) = \int_0^\infty p(r) g_{\mathcal{E}_r}(v)\,dr\,. \tag{4.25}$$

The formulas (4.23) - (4.25) are valid for $v \geq 0$ and, due to (4.22), are extended by $g(-v) = -g(v)$ to negative values of v. In all three instances, the initial loading curve l_0 satisfies

$$l_0(v) = \frac{1}{2} g(2v)\,, \quad v \in \mathbf{R}\,, \tag{4.26}$$

so we can describe these operators completely (i.e., for any $s = (v_0,\ldots,v_N) \in S_0$) with the formula

$$\mathcal{W}_f(v_0,\ldots,v_N) = \frac{1}{2} g(2v_0) \; + \; \sum_{k=1}^{N} g(v_k - v_{k-1})\,. \tag{4.27}$$

Formula (4.27) expresses the fact that we obtain the boundary curves of the hysteresis loops if we magnify the initial loading curve with a factor of 2 and attach the former origin to the corner of the hysteresis loop. This is known in engineering as *Masing's law* [29]. In cyclic plasticity, a typical function g, or rather its inverse g^{-1}, is given by the *Ramberg-Osgood equation*

$$\varepsilon = \frac{\sigma}{E} + \left(\frac{\sigma}{K'}\right)^{\frac{1}{n'}}, \tag{4.28}$$

with the material parameters K' and n' in addition to the elasticity modulus E. Note that there is no distinguished yield point and no purely elastic zone in (4.28). A typical value for n' is 0.2.

We can easily determine the relation between the density function p of the Prandtl operator and the function g representing its height function. For $v \geq 0$, partial integration in (4.25) yields

$$g(v) = \int_0^\infty \frac{d}{dr} g_{\mathcal{E}_r}(v) \int_r^\infty p(\rho)\,d\rho\,dr = 2\int_0^{\frac{v}{2}} \int_r^\infty p(\rho)\,d\rho\,dr\,. \tag{4.29}$$

so

$$g'(v) = \int_{\frac{v}{2}}^\infty p(\rho)\,d\rho\,, \quad g''(v) = -\frac{1}{2} p\left(\frac{v}{2}\right). \tag{4.30}$$

Proposition 4.8 *Let $\mathcal{W}$ be the Prandtl operator (1.5), assume that $p \in L^1(\mathbf{R}_+)$. Then the following assertions hold:*
(i) $\mathcal{W}$ is piecewise monotone if and only if

$$\int_r^\infty p(\rho)\,d\rho \geq 0 \qquad \text{for all } r > 0\,. \tag{4.31}$$

(ii) If $p > 0$ everywhere in the interval $(0, R)$, then all hysteresis loops with height not exceeding $2R$ are strictly convex.

Proof. Obvious from (4.30). □

The height function also serves to determine the correct condition for the piecewise monotonicity of the Preisach operator.

Proposition 4.9 *Let $\mathcal{W}$ be the Preisach operator (2.14), (2.15), assume that the density ω is locally bounded and measurable. If there holds*

$$\int_0^r \omega(\rho, s+\rho)\,d\rho \geq 0\,, \quad \int_0^r \omega(\rho, s-\rho)\,d\rho \geq 0\,, \tag{4.32}$$

for all $r \geq 0$, $s \in \mathbf{R}$, then $\mathcal{W}$ is piecewise monotone.

Proof. The inequalities (4.32) imply that $\partial_2 h(v_1, v_2) \geq 0$. □

Mayergoyz has observed that the existence of a height function characterizes the Preisach operator within the class of $\mathcal{P}_0$−operators (in his terminology, a hysteresis operator is Preisach if and only if it has the wiping out and the congruency property, see [30]); there he also states the essential reason why this is true. Besides specifying the class of hysteresis operators under consideration, a formal statement has to relate the regularity of the height function h to the regularity of the Preisach density ω respectively the corresponding measure. An exposition within a framework similar to the one presented here can be found in [4].

One may pose the question whether the composition or the inverse of hysteresis operators is again a hysteresis operator. The first general attempt in this direction has been made by Krejčí in [23]. While the definitions and results in [23] are tailored to invertible operators, the approach described here yields results for the composition of noninvertible operators, too. To roughly summarize those, composition and inverse of Prandtl operators again produces Prandtl operators[6]; the same is true for $\mathcal{P}_0$-operators, but not for Preisach operators. We formally state the result for Prandtl operators. This is most easily done in the representation (4.27),

$$\mathcal{W}_f(v_0, \ldots, v_N) = \frac{1}{2} g(2v_0) + \sum_{k=1}^N g(v_k - v_{k-1})\,, \quad (v_0, \ldots, v_N) \in \bar{S}_0\,, \tag{4.33}$$

repeated for convenience. Let us remind the reader that a function $g : \mathbf{R} \to \mathbf{R}$ is called *odd*, if $g(-x) = -g(x)$ holds for all $x \in \mathbf{R}$.

[6] For the inverse, this is due to [18].

Proposition 4.10

(i) *Let* $g_1, g_2 : \mathbf{R} \to \mathbf{R}$ *be odd functions, assume that* g_1 *is nondecreasing. Let* $\mathcal{W}_1$ *and* $\mathcal{W}_2$ *be the* $\mathcal{P}_0$*-operators defined by (4.33). Then* $\mathcal{W}_2 \circ \mathcal{W}_1$ *is a* $\mathcal{P}_0$*-operator of the form (4.33) with* $g = g_2 \circ g_1$.

(ii) *Let* $g : \mathbf{R} \to \mathbf{R}$ *be continuous, odd and bijective*[7]. *Let* $\mathcal{W}$ *be the* $\mathcal{P}_0$*-operator defined by (4.33). Then* $\mathcal{W}$ *has an inverse, and* $\mathcal{W}^{-1}$ *is again a* $\mathcal{P}_0$*-operator of form (4.33) with* g *replaced by* g^{-1} .

Proof. It is easy to see why assertion (i) should be true. If we compose

$$(w_0, \ldots, w_N) = \mathcal{W}_1(v_0, \ldots, v_N), \quad (z_0, \ldots, z_N) = \mathcal{W}_2(w_0, \ldots, w_N), \tag{4.34}$$

according to (4.33), we get that

$$z_k - z_{k-1} = g_2(w_k - w_{k-1}) = g_2(g_1(v_k - v_{k-1})). \tag{4.35}$$

One has to check, however, that $(w_0, \ldots, w_N) \in \bar{S}_0$ if $(v_0, \ldots, v_N) \in \bar{S}_0$, and that the $\mathcal{P}_0$-operator defined by $g_2 \circ g_1$ actually equals $\mathcal{W}_2 \circ \mathcal{W}_1$. In fact, by virtue of the assumptions on g_1, $v_{k+1} \in [v_{k-1}, v_k]$ implies that

$$w_{k+1} \in [\mathcal{W}_{1,f}(v_0, \ldots, v_k), \mathcal{W}_{1,f}(v_0, \ldots, v_{k-1}, v_k, v_{k-1}] = [w_{k-1}, w_k]. \tag{4.36}$$

A similar argument shows that $|w_1| \le |w_0|$ if $|v_1| \le |v_0|$. Moreover, since $\mathcal{W}_1$ is piecewise monotone, we get that $\mathcal{W}_2 \circ \mathcal{W}_1$ is a hysteresis operator with

$$(\mathcal{W}_2 \circ \mathcal{W}_1)_f(v_0, \ldots, v_N) = \mathcal{W}_{2,f}(\mathcal{W}_1(v_0, \ldots, v_N)). \tag{4.37}$$

The proof of (ii) is similar, but less involved. □

With the aid of (4.29) and (4.30), we may compute the density function of the composition in terms of the density functions of the factors. Note that regularity plays almost no role[8] in Proposition 4.10. Again, we successfully separate the question of memory structure from the question of regularity.

One may use inverse hysteresis operators to define implicit hysteresis operators. In this manner, one obtains a formal description of the so-called moving model proposed by [11], see [5] and [23]. Those references, as well as [3], also discuss the regularity properties of the inverse of a Preisach operator.

5. Vector Hysteresis Operators

As we have seen so far, in the scalar case there are several possibilities to construct complex memory models from simple hysteresis operators; the Preisach operator, for instance, can be represented by a linear superposition of relays as in (1.5), or by a nonlinear superposition of plays as in (2.14). These different viewpoints suggest different ways to

[7] If we consider hysteresis operators with restricted domains, it is sufficient that g is injective.
[8] Except for the continuity assumption in (ii).

define a vector hysteresis operator. The easiest way is to put a family of scalar hysteresis operators into a vector context. In [30], a vector Preisach operator is defined as

$$\mathcal{W}[v](t) = \int \mathcal{W}_e[v \cdot e](t)\, e\, de\,. \tag{5.1}$$

Here, $v : [0,T] \to \mathbf{R}^n$ is a vector input function, $e \in \mathbf{R}^n$ is a unit vector, $\mathcal{W}_e$ is a scalar Preisach operator, and the integration ranges over the unit sphere. The next possibility is to define a *vector relay*, i.e. a switching element reacting to vector inputs, and then use linear superposition as in (1.6), see [10]. While there are several contributions to vector relay hysteresis, in particular recent work of Seidman[9], there has not been much progress in the study of a Preisach type superposition, presumably because it is not easy to identify a tractable memory structure, like the memory curve $\psi(t)$ in the scalar case.

In plasticity, the relay with hysteresis does not appear naturally. The usual approach (in theoretical mechanics, at least; in applied mechanics it can be different) is to describe the elastic-plastic material law by means of a variational inequality. In fact, we could have done the same thing already, since the solution $w : [0,T] \to \mathbf{R}$ of the scalar evolution variational inequality

$$(\dot{v}(t) - \dot{w}(t))(w(t) - x) \geq 0 \text{ a.e. in } [0,T] \quad \forall x \in [-r,r]\,, \tag{5.2}$$

$$w(t) \in [-r,r] \text{ a.e. in } [0,T]\,, \quad w(0) = w_0\,, \tag{5.3}$$

for a given $v : [0,T] \to \mathbf{R}$ is identical to the output $w = \mathcal{E}_r[v; w_{-1}]$ of the elastic-plastic element, if the initial values are related by $w_0 = e_r(v(0) - w_{-1})$. In fact, the multidimensional analogue of $\mathcal{E}_r$ constitutes the fundamental model of plasticity: The interval $[-r,r]$ is replaced by a convex body (for example a ball) whose boundary forms the so-called *yield surface* in the space of deviatoric stress tensors, and the stress-strain evolution is described by the vector version of the variational inequality (5.2), (5.3).

While there is an enormous body of literature concerning the analysis and numerical solution of variational inequalities connected to plasticity, the mathematical community seems to have paid almost no attention to complex memory structures in that context. Exceptions are the papers [44] of Visintin and [24] of Krejčí, where the authors study the *vector Prandtl* model, i.e. the linear superposition of vector versions of $\mathcal{E}_r$ respectively $\mathcal{F}_r$, interpreted as an infinite system of variational inequalities. This is a very natural approach, since one can utilize the highly developed machinery of variational inequalities.

From the standpoint of memory, there is a different way to vectorize the scalar Prandtl model. Chu has found out in [7] and [8] that the continuous version of the multi-yield model due to Mróz [33] has a comparatively simple memory structure. More precisely, one can find for this model a function $\psi(t) : \mathbf{R}_+ \to \mathbf{R}^n$, which connects finitely many corners in $\mathbf{R}^n$ by straight lines and completely incorporates the memory at time t, thus yielding a rather striking analogy to the scalar case. However, the graph of $\psi(t)$ no longer separates the halfspace $\mathbf{R}_+ \times \mathbf{R}^n$ into different regions if $n > 1$, so the connection to a relay model is lost.

Since it fits well into our approach of emphasizing memory properties, we will use the remainder of this section to discuss that model. In order to understand its mechanism,

[9] See [40], [41].

we return for a moment to the particular scalar rate independent stress-strain law shown in Figure 4. The left diagram represents the Bauschinger effect, i.e. during plastic deformation along the upper arc, the elastic range $[-r, r]$ is shifted to larger values of stress, i.e. it becomes $[-r + \Delta r, r + \Delta r]$ for some $\Delta r > 0$. This phenomenon is usually called *kinematic hardening*.

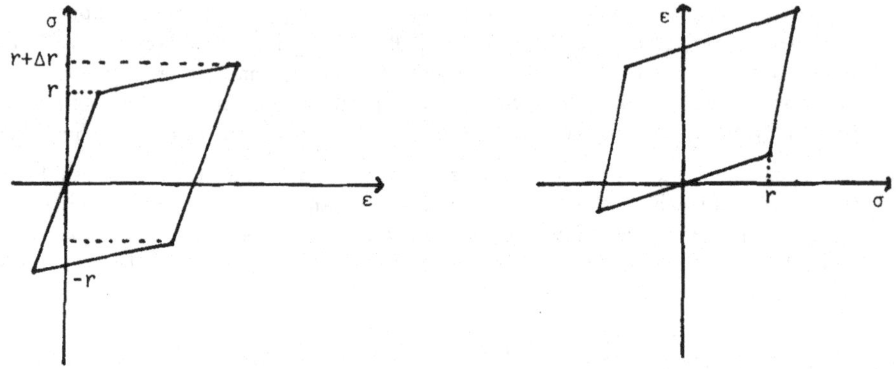

Figure 4: Scalar kinematic hardening.

We translate the inverse diagram on the right to a hysteresis operator as

$$\varepsilon = \mathcal{W}[\sigma] = \varepsilon^e + \varepsilon^p = \frac{1}{E}\sigma + \kappa \mathcal{F}_r[\sigma]. \tag{5.4}$$

For a general initial loading curve, as for example that described by the Ramberg-Osgood equation (4.28), we replace (5.4) with an inverse Prandtl operator, namely

$$\varepsilon(t) = \frac{1}{E}\sigma(t) + \int_0^\infty \kappa(r)\mathcal{F}_r[\sigma](t)\, dr. \tag{5.5}$$

We may think of (5.5) as incorporating a one parameter family of "elastic range intervals" (the intervals of the plays) contained within each other and moving together. The continuous version of the Mróz model carries over this feature to the vector case, i.e. it postulates a one parameter family of balls $B_r(t)$ with midpoints $\psi(t, r)$, parametrized by their radius r, which are all contained within each other and whose boundaries

$$S_r(t) = \{u \in U : |u - \psi(t, r)| = r\} \tag{5.6}$$

define the continuous family of yield surfaces. Here, V denotes the linear space of deviatoric stress tensors and $|\cdot|$ the Euclidean norm in V. An input function $\sigma^d : [0, T] \to V$ determines the movement of the balls, and therefore of the yield surfaces, as follows. We fix the zero initial condition $\psi_{-1}(r) = 0$. We require that the stress input must not lie outside any yield surface, so

$$|\sigma^d(t) - \psi(t, r)| \leq r, \quad t \in [0, T], \quad r \geq 0. \tag{5.7}$$

In particular, (5.7) implies that

$$\psi(t,0) = \sigma^d(t)\,, \quad t \in [0,T]\,. \tag{5.8}$$

Next, the yield surface can move only while the stress deviator lies on the yield surface. This means that

$$\partial_t \psi(t,r) = 0\,, \quad \text{if } r > |\sigma^d(t) - \psi(t,r)|\,, \quad t \in [0,T]\,. \tag{5.9}$$

The condition that all balls lie within each other can be conveniently formalized as

$$|\psi(t,r_1) - \psi(t,r_2)| \le r_2 - r_1\,, \quad 0 \le r_1 \le r_2\,, \quad t \in [0,T]\,. \tag{5.10}$$

Consequently, different yield surfaces can meet at a single point only, and they have a common tangent and outer normal in this case.

We claim that the conditions (5.7) - (5.10) uniquely determine the evolution of the midpoint curve, which represents the memory as in the scalar case. We introduce the space Ψ of admissible memory states,

$$\begin{aligned} \Psi \;=\; & \{\varphi \,|\, \varphi : [0,\infty) \to V\,,\, |\varphi(r) - \varphi(s)| \;\le\; |r-s| \text{ for any } r,s \ge 0, \\ & \text{and there exists } R > 0 \text{ with } \varphi(r) = 0 \text{ for any } r \ge R\}\,. \end{aligned} \tag{5.11}$$

We want to define the discrete update rule corresponding to (2.25) in the scalar case. To this end, we set

$$\alpha(v,\varphi) = \min\{r \ge 0 : |\varphi(r) - v| = r\}\,. \tag{5.12}$$

One easily checks that $\alpha : V \times \Psi \to \mathbf{R}$ is well defined, and that

$$r < |\varphi(r) - v| \quad \text{if and only if} \quad 0 \le r < \alpha(v,\varphi)\,. \tag{5.13}$$

This means that the spheres with radius less than $\alpha(v,\varphi)$ have to move, while the others do not, since for them the straight line segment connecting $\varphi(0)$ and v is contained in their interior. Moreover, due to (5.7) - (5.10) the new position $\varphi_*(r)$ of the spheres with radius $r < r_* = \alpha(v,\varphi)$ has to satisfy

$$|\varphi_*(r) - v| \le r\,, \quad |\varphi_*(r) - \varphi(r_*)| \le r_* - r\,. \tag{5.14}$$

Because $|\varphi(r_*) - v| = r_*$, their centers have to arrange themselves along the straight line connecting $\varphi(r_*)$ and v. These considerations show that the discrete evolution according to (5.7) - (5.10) is uniquely determined, and that we may describe it in the following way.

Definition 5.1 We define an operator $G : U \times \Psi \to \Psi$ by

$$G(v,\varphi)(r) = \begin{cases} \varphi(r)\,, & \text{if } r \ge \alpha(v,\varphi) \\ v + \dfrac{r}{\alpha(v,\varphi)} \Big(\varphi(\alpha(v,\varphi)) - v \Big)\,, & \text{otherwise} \end{cases} \tag{5.15}$$

for any $r \ge 0$ and any $v \in V$, $\varphi \in \Psi$, where $\alpha(v,\varphi)$ is defined in (5.12). □

The operator G describes a single memory update step. If now $(\sigma_0^d, \dots, \sigma_N^d)$ is an input string in V, we define the corresponding sequence of memory states as

$$\psi_k = G(\sigma^d, \psi_{k-1}), \quad \psi_{-1} = 0. \tag{5.16}$$

If $\sigma^d : [0,T] \to V$ is piecewise linear with respect to some partition (t_k) of $[0,T]$, we set

$$\psi(t,r) = G(\sigma^d(t), \psi(t_k, \cdot))(r), \quad t \in (t_k, t_{k+1}], \quad r \geq 0. \tag{5.17}$$

In this manner, we obtain an operator $\mathcal{F}$ mapping input functions to memory state functions,

$$\psi = \mathcal{F}[\sigma^d], \tag{5.18}$$

which is defined on the set of piecewise linear functions $\sigma^d : [0,T] \to V$. We call $\mathcal{F}$ the *Mróz hardening rule.*

It is obvious from the definition that the memory states $\psi(t,\cdot)$ generated by (5.17) are piecewise linear curves in V with finitely many corners. Their length is

$$L(\psi(t,\cdot)) = \sup_{0 \leq \tau \leq t} |\sigma^d(\tau)|, \tag{5.19}$$

and they satisfy

$$|\partial_r \psi(t,r)| = 1, \qquad \text{if } r < L(\psi(t,\cdot)), \tag{5.20}$$

except in corners, of course. For $r > L(\psi(t,\cdot))$, we must have $\psi(t,r) = 0$ because of the initial condition. As a result, the memory state $\psi(t,\cdot)$ is actually determined by the position of its corners.

We present two regularity results concerning the hardening rule $\mathcal{F}$. For the proofs we refer to [6].

Theorem 5.2 *The operator $\mathcal{F}$ defined in (5.18) can be extended to an operator*

$$\mathcal{F} : C(0,T;V) \to C(0,T;\Psi), \tag{5.21}$$

and we have for $\psi_1 = \mathcal{F}[\sigma_1^d]$, $\psi_2 = \mathcal{F}[\sigma_2^d]$, where $\sigma_1^d, \sigma_2^d \in C(0,T;V)$, the estimate

$$\max_{\substack{0 \leq t \leq T \\ r \geq 0}} |\psi_1(t,r) - \psi_2(t,r)| \leq \sqrt{2R \max_{0 \leq t \leq T} |\sigma_1^d(t) - \sigma_2^d(t)|}, \tag{5.22}$$

where

$$R = \max\{\|\sigma_1^d\|_\infty, \|\sigma_2^d\|_\infty\}. \tag{5.23}$$

□

Theorem 5.3 *For any piecewise linear $\sigma^d \in W^{1,1}(0,T;V)$, the function*

$$t \mapsto \psi(t,r) = \mathcal{F}[\sigma^d](t,r) \tag{5.24}$$

is an element of $W^{1,1}(0,T;U)$ and satisfies

$$\int_0^T |\partial_t \psi(t,r)|\, dt \leq 3 \int_0^T |(\sigma^d)'(t)|\, dt. \tag{5.25}$$

□

To complete the description of the continuous version of the Mróz model, we have to specify the output mapping, also called *flow rule* in plasticity, since it determines the actual plastic flow ε^p respectively $\dot{\varepsilon}^p$. In analogy to the scalar case (5.5), we set

$$\varepsilon^p(t) = \int_0^\infty \kappa(r)\psi(t,r)\,dr\,, \quad \psi = \mathcal{F}[\sigma^d]\,, \tag{5.26}$$

where $\kappa \geq 0$ still represents the density of the scalar model. It turns out that this definition leads to a model which is thermodynamically correct in the sense that no cyclic process generates mechanical energy. The proof mainly rests on an energy inequality for the Mróz hardening rule, namely

$$\begin{aligned}\frac{1}{2}\left(|\psi(t,r)|^2 - |\psi(s,r)|^2\right) \quad &- \quad \langle\psi(t,r),\sigma^d(t)\rangle \; + \; \langle\psi(s,r),\sigma^d(s)\rangle \\ &+ \int_s^t \langle\psi(\tau,r),\dot{\sigma}^d(\tau)\rangle\,d\tau \quad \leq \; 0\,,\end{aligned} \tag{5.27}$$

which holds for every $0 \leq s < t \leq T$ and every $r \geq 0$. Again, we refer to [6] for a proof.

6. Hysteresis in the Heat Equation

Parabolic equations with a hysteresis operator in the principal part have been studied extensively by Visintin from different standpoints. We restrict ourselves to an exposition of his original approach to the existence proof in [42] and [43], and to Hilpert's uniqueness result in [15].

We consider the initial-boundary value problem given by

$$\partial_t y(x,t) + \partial_t w(x,t) - \Delta y(x,t) = f(x,t)\,, \qquad (x,t) \in \Omega \times (0,T)\,, \tag{6.1}$$

$$y(x,t) = 0\,, \qquad (x,t) \in \partial\Omega \times (0,T)\,, \tag{6.2}$$

$$y(x,0) = y_0(x)\,, \qquad x \in \Omega\,, \tag{6.3}$$

$$w(x,\cdot) = \mathcal{W}[y(x,\cdot)]\,, \quad x \in \Omega\,. \tag{6.4}$$

Assumption 6.1
(i) $\Omega \subset \mathbf{R}^n$ *is open and bounded,* $\partial\Omega$ *is smooth.*
(ii) $y_0 \in H_0^1(\Omega)$, $f \in L^2(\Omega \times (0,T))$.
(iii) $\mathcal{W}$ *is a piecewise monotone hysteresis operator.*
(iv) $\mathcal{W} : C[0,T] \to C[0,T]$ *is continuous.*
(v) The level curves $l_N(v) = \mathcal{W}(v_0,\ldots,v_{N-1},v)$ *of* $\mathcal{W}$ *are measurable and satisfy*

$$|l_N(v)| \leq c_0 + c_1|v|\,, \tag{6.5}$$

where the constants c_0 *and* c_1 *do not depend on* N *and* $(v_0,\ldots,v_{N-1})$. □

Theorem 6.2 *Let Assumption 6.1 hold. Then there exists a weak solution*

$$y \in L^\infty(0,T;H_0^1(\Omega)) \cap H^1(0,T;L^2(\Omega))\,, \quad w \in H^1(0,T;H^{-1}(\Omega)) \cap L^2(\Omega;C[0,T])\,, \tag{6.6}$$

of (6.1) - (6.4) in the sense that it satisfies (6.3) and (6.4) for a.e. $x \in \Omega$ *as well as*

$$\frac{d}{dt}\int_\Omega (y+w)(x,t)\varphi(x)\,dx + \int_\Omega \nabla y(x,t)\nabla\varphi(x,t)\,dx = \int_\Omega f(x,t)\varphi(x)\,dx \quad \forall \varphi \in H_0^1(\Omega) \tag{6.7}$$

for a.e. $t \in [0,T]$.

Remark. There is a compact embedding

$$L^\infty(0,T;H_0^1(\Omega)) \cap H^1(0,T;L^2(\Omega)) \subset L^2(\Omega;C[0,T]). \tag{6.8}$$

Therefore, (6.4) makes sense for the weak solution.

Idea of the existence proof: We test (6.1) with $\partial_t y$ (only formally!) and get (arguments x,t suppressed)

$$\int_\Omega (\partial_t y)^2\,dx + \int_\Omega \partial_t y \cdot \partial_t w\,dx + \int_\Omega \nabla y \cdot \nabla(\partial_t y)\,dx = \int_\Omega \partial_t y \cdot f\,dx. \tag{6.9}$$

The piecewise monotonicity of $\mathcal{W}$ should imply that $\partial_t y \cdot \partial_t w \geq 0$. We integrate (6.9) in time and apply Hölder's inequality to the right hand side to obtain the a priori estimate

$$\int_0^T \int_\Omega (\partial_t y)^2(x,t)\,dx\,dt + \sup_t \int_\Omega |\nabla y(x,t)|^2\,dx \leq C(f,y_0). \tag{6.10}$$

A semidiscretization of (6.7) in time followed by a passage to the limit with the aid of the embedding (6.8) turns these estimates as well as the use of the piecewise monotonicity of $\mathcal{W}$ into a genuine proof.

Sketch of the Proof of Theorem 6.2: Let $\Delta t > 0$ and $N \in \mathbf{N}$ with $N\Delta t = T$ be given. We consider the semidiscrete problem

$$\begin{aligned}\frac{1}{\Delta t}\int_\Omega (y^n(x) - y^{n-1}(x))\varphi(x)\,dx + \frac{1}{\Delta t}\int_\Omega (w^n(x) - w^{n-1}(x))\varphi(x)\,dx \\ + \int_\Omega \nabla y^n(x)\nabla\varphi(x)\,dx = \int_\Omega f^n(x)\varphi(x)\,dx \qquad \forall \varphi \in H_0^1(\Omega),\end{aligned} \tag{6.11}$$

$$w^n(x) = \mathcal{W}(y^0(x),\ldots,y^n(x)), \quad x \in \Omega. \tag{6.12}$$

Here, $y^0,\ldots,y^{n-1},w^{n-1}$ are already known from previous steps, and

$$f^n(x) = \frac{1}{\Delta t}\int_{(n-1)\Delta t}^{n\Delta t} f(x,t)\,dt. \tag{6.13}$$

Formula (6.11) represents a semilinear elliptic variational equation, which we may write as

$$a(y^n,\varphi) + \int_\Omega b^n(x,y^n(x))\varphi(x)\,dx = 0 \quad \forall \varphi \in H_0^1(\Omega), \tag{6.14}$$

where a stands for the scalar product of $H_0^1(\Omega)$, and $b^n : \Omega \times \mathbf{R} \to \mathbf{R}$,

$$b^n(x,y) = \frac{1}{\Delta t}(y + \mathcal{W}(y^0(x),\ldots,y^{n-1}(x),y)) - \frac{1}{\Delta t}(y^{n-1}(x) + w^{n-1}(x)) - f^n(x), \tag{6.15}$$

is measurable in x as well as continuous and strictly increasing in y; again we use the piecewise monotonicity of $\mathcal{W}$. Therefore, (6.14) has a unique solution $y^n \in H_0^1(\Omega)$ (see e.g. [48], Proposition 28.9, p.634). We define w^n by (6.12) and obtain $w^n \in L^2(\Omega)$ because of Assumption 6.1(v). We now test (6.11) with $\varphi = y^n - y^{n-1}$ and perform the discrete analogue of the computation leading from (6.9) to the a priori estimate (6.10) in order to obtain the semidiscrete version of (6.10), namely

$$\sum_{n=1}^{N} \Delta t \int_\Omega \left(\frac{y^n(x) - y^{n-1}(x)}{\Delta t}\right)^2 dx + \sup_{n\le N} \int_\Omega |\nabla y^n(x)|^2\, dx \le C(f, y_0)\,. \tag{6.16}$$

Let y_N and w_N denote the linear interpolates in time of the functions y^n and w^n, while $\hat{y}_N$ stands for the piecewise constant extension, i.e. $\hat{y}_N(x,t) = y^n(x)$, if $(n-1)\Delta t < t \le n\Delta t$. The semidiscrete a priori estimate (6.10) implies that

$$\begin{aligned} &y_N \text{ is uniformly bounded in } L^\infty(0,T;H_0^1(\Omega)) \cap H^1(0,T;L^2(\Omega))\,,\\ &w_N \text{ is uniformly bounded in } H^1(0,T;H^{-1}(\Omega))\,,\\ &\hat{y}_N \text{ is uniformly bounded in } L^\infty(0,T;H_0^1(\Omega))\,.\end{aligned} \tag{6.17}$$

In the final step one proves that the weak star limits yield a solution. We only note that Assumption 6.1(iv) together with the compact embedding (6.8) guarantees that the hysteresis equation (6.4) is satisfied in the limit. □

Idea of the uniqueness proof: Let (y_1, w_1) and (y_2, w_2) be solutions of the variational problem for a.e. $t \in [0,T]$,

$$\int_\Omega (\partial_t y_i + \partial_t w_i)(x,t)\varphi(x)\, dx + \int_\Omega \nabla y_i(x,t)\cdot\nabla\varphi(x)\, dx = \int_\Omega f_i(x,t)\varphi(x)\, dx \quad \forall\varphi \in H_0^1(\Omega)\,, \tag{6.18}$$

$$w_i(x,\cdot\,) = \mathcal{W}[y_i(x,\cdot\,]\,, \quad \text{a.e. in } \Omega\,. \tag{6.19}$$

The difference $(y,w) = (y_2 - y_1, w_2 - w_1)$ satisfies (arguments (x,t) omitted)

$$\int_\Omega (\partial_t y + \partial_t w)\varphi\, dx + \int_\Omega \nabla y\cdot\nabla\varphi\, dx = \int_\Omega (f_2 - f_1)\varphi\, dx \quad \forall\varphi \in H_0^1(\Omega)\,. \tag{6.20}$$

(Of course, we do not have $w = \mathcal{W}[y]$.) We fix t and test (6.20) with $\varphi(x) = H(y(x,t))$, where H denotes the Heaviside function defined by $H(v) = 1$, if $v > 0$, and $H(v) = 0$ otherwise. We get

$$\int_\Omega \partial_t y\cdot H(y)\, dx + \int_\Omega \partial_t w\cdot H(y)\, dx + \int_\Omega \nabla y\cdot\nabla(H(y))\, dx = \int_\Omega (f_2 - f_1)H(y)\, dx\,. \tag{6.21}$$

Since $\partial_t y \cdot H(y) = \partial_t y_+$, where as usual $y_+ = \max\{y, 0\}$, and because (again purely formal)

$$\nabla y\cdot\nabla(H(y)) = |\nabla y|^2\cdot H'(y) \ge 0\,, \tag{6.22}$$

integration of (6.21) in time yields

$$\int_\Omega y_+(x,t)\, dx + \int_\Omega\int_0^t \partial_t w\cdot H(y)\, d\tau\, dx \le \int_0^t\int_\Omega (f_2 - f_1)H(y)\, dx\, d\tau + \int_\Omega y_+(x,0)\, dx\,. \tag{6.23}$$

It will turn out that there holds, if e.g. $\mathcal{W}$ is the play operator,

$$\partial_t w_+ \le \partial_t w\cdot H(y)\,. \tag{6.24}$$

Using (6.24), we estimate the second integral on the left side of (6.23) from below and obtain L^1-stability and, if the data for the two solutions are equal, uniqueness.

Proposition 6.3 *Let $\mathcal{W}$ be the hysteresis operator given by*

$$\mathcal{W}[v](t) = q(\mathcal{F}_r[v](t)), \tag{6.25}$$

where $q \in W^{1,\infty}_{loc}$ is nondecreasing. Then for all $v_1, v_2 \in W^{1,1}(0,T)$ there holds

$$\frac{d}{dt} w_+(t) \leq w'(t) \cdot H(v(t)), \quad \text{a.e. in } [0,T], \tag{6.26}$$

for the differences $v = v_2 - v_1$ and $w = \mathcal{W}[v_2] - \mathcal{W}[v_1]$.

Proof. For a fixed $t \in [0,T]$, inequality (6.26) is equivalent to

$$\begin{aligned} 0 &\leq w'(t) \cdot H(v(t)), \quad \text{if } w_2(t) < w_1(t), \\ w'(t) &\leq w'(t) \cdot H(v(t)), \quad \text{if } w_1(t) < w_2(t). \end{aligned} \tag{6.27}$$

In turn, (6.27) is equivalent to

$$\begin{aligned} w_2'(t) &\geq w_1'(t), \quad \text{if } w_2(t) < w_1(t), \; v_2(t) > v_1(t), \\ w_2'(t) &\leq w_1'(t), \quad \text{if } w_1(t) < w_2(t), \; v_2(t) \leq v_1(t). \end{aligned} \tag{6.28}$$

For (6.28) to hold it is sufficient to prove that

$$w_2(t) < w_1(t), \; v_2(t) \geq v_1(t) \quad \Rightarrow \quad w_2'(t) \geq 0, \; w_1'(t) \leq 0. \tag{6.29}$$

One checks from the definition or from Figure 3 that the implication (6.29) holds for the play operator $\mathcal{W} = \mathcal{F}_r$. Since $q' \geq 0$, (6.29) remains valid a.e. if we replace $w_i(t)$ by $q(w_i(t))$. □

Theorem 6.4 *Consider (6.1) - (6.4) with the data $(y_{0,1}, f_1)$ and $(y_{0,2}, f_2)$. Assume that (6.1) holds and that, in addition, $\mathcal{W}$ maps $W^{1,1}(0,T)$ into itself and satisfies (6.26) as well as*

$$|\mathcal{W}[v]'(t)| \leq C|v'(t)| \tag{6.30}$$

for all $v \in W^{1,1}(0,T)$ and some constant C independent of v[10]. *Then any pair of weak solutions (y_i, w_i) in the sense of Theorem 6.2 satisfies, for a.e. $t \in [0,T]$,*

$$\int_\Omega |y_2 - y_1|(x,t)\,dx + \int_\Omega |w_2 - w_1|(x,t)\,dx \leq$$
$$\leq \int_\Omega |y_{0,2} - y_{0,1}|(x)\,dx + \int_\Omega |w_2 - w_1|(x,0)\,dx + \int_0^t \int_\Omega |f_2 - f_1|(x,\tau)\,dx\,d\tau \tag{6.31}$$

Proof. Let $H_\epsilon : \mathbf{R} \to \mathbf{R}$ be the regularized Heaviside function

$$H_\epsilon(v) = \begin{cases} 1, & v \geq \epsilon, \\ \frac{1}{\epsilon} v, & 0 \leq v \leq \epsilon, \\ 0, & v \leq 0. \end{cases} \tag{6.32}$$

[10]Compare Proposition 3.6 and 3.7.

As before we set $y = y_2 - y_1$, $w = w_2 - w_1$, $w_i(x,\cdot) = \mathcal{W}[v_i(x,\cdot)]$. We have

$$\begin{aligned} &H_\epsilon \circ y \in L^\infty(\Omega \times (0,T)) \cap L^\infty(0,T; H_0^1(\Omega)), \\ &\partial_t y, \ \partial_t w \in L^1(\Omega \times (0,T)), \\ &y(x,\cdot) \in C[0,T]; \ \partial_t y(x,\cdot), \ \partial_t w(x,\cdot) \in L^1(0,T), \quad \text{a.e. in } \Omega. \end{aligned} \tag{6.33}$$

We may therefore test the time integral of the two variational equations, which arise from (6.7) for the data $(y_{0,1}, f_1)$ and $(y_{0,2}, f_2)$, with $\varphi = H_\epsilon \circ y$ and obtain (arguments (x,τ) omitted)

$$\begin{aligned} &\int_0^t \int_\Omega \partial_t y \cdot H_\epsilon(y)\,dx\,d\tau + \int_0^t \int_\Omega \partial_t w \cdot H_\epsilon(y)\,dx\,d\tau + \int_0^t \int_\Omega \nabla y \cdot \nabla(H_\epsilon \circ y)\,dx\,d\tau = \\ &\qquad = \int_0^t \int_\Omega (f_2 - f_1) \cdot H_\epsilon(y)\,dx\,d\tau. \end{aligned} \tag{6.34}$$

The chain rule in $H_0^1(\Omega)$ yields $\nabla y \cdot \nabla(H_\epsilon \circ y) = H_\epsilon'(y) \cdot |\nabla y|^2 \geq 0$. We therefore conclude from (6.34) that

$$\int_0^t \int_\Omega (\partial_t y + \partial_t w)(x,\tau) H_\epsilon(y(x,\tau))\,dx\,d\tau \leq \int_0^t \int_\Omega (f_2 - f_1)(x,\tau) H_\epsilon(y(x,\tau))\,dx\,d\tau. \tag{6.35}$$

We may now pass to the limit $\epsilon \downarrow 0$ and get

$$\begin{aligned} &\int_\Omega y_+(x,t)\,dx + \int_\Omega \int_0^t \partial_t w(x,\tau) H(y(x,\tau))\,d\tau\,dx \\ &\qquad \leq \int_\Omega y_+(x,0)\,dx + \int_0^t \int_\Omega (f_2 - f_1)(x,\tau) H(y(x,\tau))\,dx\,d\tau. \end{aligned} \tag{6.36}$$

We use (6.26) to estimate the second integral from below and get

$$\begin{aligned} &\int_\Omega y_+(x,t)\,dx + \int_\Omega w_+(x,t)\,dx \\ &\qquad \leq \int_\Omega y_+(x,0)\,dx + \int_\Omega w_+(x,0)\,dx + \int_0^t \int_\Omega (f_2 - f_1)(x,\tau) H(y(x,\tau))\,dx\,d\tau. \end{aligned} \tag{6.37}$$

We add (6.37) to the corresponding inequality which results from an interchange of the indices 1 and 2. Since $0 \leq H(y) + H(-y) \leq 1$, the assertion is proved. □

Corollary 6.5 *Under the assumptions of Theorem 6.4, the weak solution in the sense of Theorem 6.2 is unique.* □

In general, the Preisach operator does not satisfy the inequality (6.26). However, a weaker form of (6.26), which will be sufficient for the uniqueness proof, still holds. Let us consider the Preisach operator in the form (2.14) with $q_{00} = 0$, i.e.

$$\mathcal{W}[v](t) = \int_0^\infty q(r, \mathcal{F}_r[v](t))\,dr, \tag{6.38}$$

and assume that its density $\omega = \frac{1}{2}\partial_s q(r,s)$ is nonnegative. We apply Proposition 6.3 to the hysteresis operators

$$\mathcal{W}_r[v](t) = q(r, \mathcal{F}_r[v](t)), \quad r \geq 0, \tag{6.39}$$

and obtain for the differences $w = w_2 - w_1$ and $v = v_2 - v_1$ that

$$\begin{aligned} w'(t)H(v(t)) &= \int_0^\infty \partial_t(q(r,\mathcal{F}_r[v_2](t)) - q(r,\mathcal{F}_r[v_1](t))) \cdot H(v(t))\,dr \\ &\geq \int_0^\infty \partial_t\left([q(r,\mathcal{F}_r[v_2](t)) - q(r,\mathcal{F}_r[v_1](t))]_+\right) dr\,. \end{aligned} \tag{6.40}$$

Therefore, we can bound the time integral of the left side of (6.40) from below as

$$\begin{aligned} \int_0^t w'(\tau)H(v(\tau))\,d\tau &\geq \int_0^\infty \int_0^t \partial_t\left([q(r,\mathcal{F}_r[v_2](\tau)) - q(r,\mathcal{F}_r[v_1](\tau))]_+\right) d\tau\,dr \\ &\geq -\int_0^\infty [q(r,\mathcal{F}_r[v_2](0)) - q(r,\mathcal{F}_r[v_1](0))]_+\,dr\,. \end{aligned} \tag{6.41}$$

From these computations, we obtain a variant of Theorem 6.4 for the Preisach operator.

Theorem 6.6 *Let $\mathcal{W}$ be the Preisach operator defined as in (2.14) and (2.15), assume that its density function ω is nonnegative and measurable, and that there holds*

$$\int_0^\infty \sup_{s\in\mathbf{R}} |\omega(r,s)|\,dr < \infty\,. \tag{6.42}$$

Then any pair of weak solutions (y_i, w_i) in the sense of Theorem 6.2 satisfies, for a.e. $t \in [0,T]$,

$$\begin{aligned} \int_\Omega |y_2 - y_1|(x,t)\,dx &\leq \int_0^t \int_\Omega |f_2 - f_1|(x,\tau)\,dx\,d\tau + \int_\Omega |y_{0,2} - y_{0,1}|(x)\,dx + \\ &\quad + \int_0^\infty |q(r,\mathcal{F}_r[y_2(x,\cdot)](0)) - q(r,\mathcal{F}_r[y_1(x,\cdot)](0))|\,dr\,. \end{aligned} \tag{6.43}$$

Proof. The proof is the same as that of Theorem 6.4, except that we use (6.41) to estimate the second integral in (6.36) from below. Proposition 3.7 shows that $\mathcal{W}$ has the properties required in Theorem 6.4. □

Corollary 6.7 *Under the assumptions of Theorem 6.6, the weak solution in the sense of Theorem 6.2 is unique.* □

7. Hysteresis in the Wave Equation

Hyperbolic equations with a hysteresis operator in the principal part have been studied extensively by Krejčí. He was mainly interested in the dissipative properties of such systems and in the existence of periodic solutions. We restrict ourselves here to the discussion of the initial boundary value problem in one space dimension. Our exposition is based upon [19] and [25]. Let us return to the system (1.2) governing the elastic-plastic evolution in a one-dimensional rod

$$\partial_{tt}u = \partial_x\sigma\,, \quad \sigma = \mathcal{W}[\varepsilon]\,, \quad \varepsilon = \partial_x u\,. \tag{7.1}$$

One can put (7.1) into the form of a wave equation with hysteresis as

$$\partial_{tt}u - \partial_x(\mathcal{W}[\partial_x u]) = 0\,, \tag{7.2}$$

or alternatively as

$$\partial_t(\mathcal{W}^{-1}[\partial_t z]) - \partial_{xx} z = 0\,, \tag{7.3}$$

for the variable z representing the time integral of the stress,

$$z(x,t) = \int_0^t \sigma(x,\tau)\,d\tau + z^0(x)\,, \quad z^0(x) = -\int_x^1 \partial_t u(\xi,0)\,d\xi\,. \tag{7.4}$$

We consider the first order system

$$\partial_t u(x,t) = \partial_x z(x,t)\,, \tag{7.5}$$

$$\partial_t z(x,t) = \sigma(x,t)\,, \quad \varepsilon(x,t) = \partial_x u(x,t)\,, \tag{7.6}$$

$$\sigma(x,t) = \mathcal{W}[\varepsilon(x,\cdot)](t)\,, \tag{7.7}$$

for $(x,t) \in \Omega := (0,1) \times (0,T)$, together with the initial conditions

$$u(x,0) = u^0(x)\,, \quad z(x,0) = z^0(x)\,, \quad x \in (0,1)\,, \tag{7.8}$$

and the boundary conditions

$$u(0,t) = 0\,, \quad z(1,t) = 0\,, \quad t \in (0,T)\,. \tag{7.9}$$

The condition (7.9) means that the rod is fixed at the left end and free (with no boundary forces) at the right end.

Idea of the existence proof: The main idea is to relate the strict convexity of the hysteresis loops of $\mathcal{W}$ (which will be assumed) to an inequality for the input v and the output $w = \mathcal{W}[v]$ involving the second derivative, namely

$$\frac{1}{2}w'(t)v'(t) - \frac{1}{2}w'(s)v'(s) \le \int_s^t w'(\tau)v''(\tau)\,d\tau - \frac{\gamma}{2}\int_s^t |v'(\tau)|^3\,d\tau\,. \tag{7.10}$$

In a purely formal manner, we form the time derivative in (7.10) and translate to (7.7) to obtain

$$\partial_t(\frac{1}{2}\partial_t\sigma \cdot \partial_t\varepsilon) \le \partial_t\sigma \cdot \partial_{tt}\varepsilon - \frac{\gamma}{2}|\partial_t\varepsilon|^3 \tag{7.11}$$

Newton's law (7.1) yields

$$\partial_t(\frac{1}{2}(\partial_{tt}u)^2 = \partial_x\sigma \cdot \partial_{tx}\sigma\,. \tag{7.12}$$

We sum (7.11) and (7.12) and integrate in space. Since $\partial_{tt}\varepsilon = \partial_{xx}\sigma$, we may perform partial integration and obtain the inequality

$$\partial_t \left[\int \frac{1}{2}(\partial_{tt}u)^2\,dx + \int \frac{1}{2}\partial_t\sigma \cdot \partial_t\varepsilon\,dx\right] + \frac{\gamma}{2}\int |\partial_t\varepsilon|^3\,dx \le 0\,. \tag{7.13}$$

Integrating in time, we arrive at the a priori estimate one wants to prove. The actual proof works with a semidiscretization (this time in space) and utilizes the compact embedding

$$U = \{u \in L^1(\Omega) : \partial_t u \in L^3(\Omega)\,,\ \partial_x u \in L^2(\Omega)\} \quad \subset \quad C(\bar{\Omega}) \tag{7.14}$$

to enable passage to the limit in the hysteresis equation (7.7). Note that we may interchange the role of ∂_t and ∂_x in (7.14). □

Let us derive the inequality (7.10) in a precise manner. We set

$$P(t) = \frac{1}{2}w'(t)v'(t), \tag{7.15}$$

and consider first the smooth superposition of smooth functions, i.e.

$$w(t) = l(v(t)), \quad l \in C^2(\mathbf{R}), \quad v \in W^{2,1}(0,T). \tag{7.16}$$

We compute the derivative and eliminate w'' to obtain

$$P'(t) = w'(t)v''(t) + \frac{1}{2}l''(v(t))v'(t)^3. \tag{7.17}$$

Let us traverse a strictly convex hysteresis loop in the mathematically negative sense (as in the case of $\mathcal{E}_r$). If l represents the upper boundary, where $v' \geq 0$, there should hold $l'' \leq \gamma < 0$; for the lower boundary we would have $v' \leq 0$ and $l'' \geq \gamma > 0$. In both cases, (7.17) yields

$$P'(t) \leq w'(t)v''(t) - \frac{\gamma}{2}|v'(t)|^3. \tag{7.18}$$

Now assume that $x = v(t)$ is a point of discontinuity of l'. We have the implications

$$l'(x+) \leq l'(x-), \quad v'(t) \geq 0 \quad \Rightarrow \quad P(t+) \leq P(t-), \tag{7.19}$$

$$l'(x+) \geq l'(x-), \quad v'(t) \leq 0 \quad \Rightarrow \quad P(t+) \leq P(t-). \tag{7.20}$$

Consequently, if $v' \geq 0$ (or $v' \leq 0$) in some interval $[t_-, t_+]$ and if $l \in C(\mathbf{R})$ is piecewise C^2 and has the properties stated above, then we have

$$P(t_-) - P(t_+) \leq \int_{t_-}^{t_+} w'(t)v''(t)\,dt - \frac{\gamma}{2}\int_{t_-}^{t_+} |v'(t)|^3\,dt. \tag{7.21}$$

Proposition 7.1 *Let $\mathcal{W}$ be a hysteresis operator of Preisach type. We assume that $\mathcal{W}$ is piecewise strictly monotone and satisfies (3.29), and that the memory level functions $l_N : \mathbf{R} \to \mathbf{R}$,*

$$l_N(x) = \mathcal{W}_f(v_0, \ldots, v_{N-1}, x), \tag{7.22}$$

have the following properties:
(i) $l_N \in C(\mathbf{R}) \cap C^2_{pm}(\mathbf{R})$,
(ii) $l'_N(x+) \leq l'_N(x-)$ if $x > v_{N-1}$, $l'_N(x+) \geq l'_N(x-)$ if $x < v_{N-1}$.
(iii) For every M there exists a γ such that

$$\begin{aligned} l''_N &\leq -\gamma < 0 \quad \textit{for} \quad v_{N-1} < x \leq M, \\ l''_N &\geq \gamma > 0 \quad \textit{for} \quad v_{N-1} > x \geq -M. \end{aligned} \tag{7.23}$$

Then for every $v \in W^{2,1}(0,T)$ the function

$$P(t) = \frac{1}{2}w'(t)v'(t), \quad w = \mathcal{W}[v], \tag{7.24}$$

is defined a.e. in $(0,T)$ and satisfies, in all points s,t where it is defined,

$$P(t) - P(s) \leq \int_s^t w'(\tau)v''(\tau)\,d\tau - \frac{\gamma}{2}\int_s^t |v'(\tau)|^3\,d\tau, \tag{7.25}$$

where γ is chosen according to (7.23) for $M = \|v\|_\infty$.

Proof. Set
$$D_w = \{t \in (0,T) : w'(t) \text{ exists }\}. \tag{7.26}$$
By Proposition 3.6, we have
$$\{w' = 0\} \subset \{v' = 0\} \subset D_w . \tag{7.27}$$
Since v' is continuous on $[0,T]$, we can write the set $\{v' \neq 0\}$ as a countable disjoint union of open intervals (a_j, b_j); at their endpoints except possibly 0 and T we have $v'(a_j) = v'(b_j) = 0$, hence also $P(a_j) = P(b_j) = 0$. Because of the structure of memory of Preisach type operators, for every $t \in (a_j, b_j)$ the memory state $\psi(t)$ has only finitely many corners, so we have
$$w(\tau) = \mathcal{W}_f(v_0, \ldots, v_{N-1}, v(\tau)) = l_N(v(\tau)), \quad \tau \in [t, b_j), \tag{7.28}$$
for some memory level function. We are now in the situation discussed in (7.18) - (7.21) above, so the assertion (7.25) holds for any $a_j < s < t \le b_j$. Since $w'(a_j) = 0$, any sequence $s_k \downarrow a_j$ with $s_k \in D_w$ yields $P(s_k) \to P(a_j) = 0$, so (7.25) also holds for $a_j = s < t \le b_j$. Now let $s, t \in D_w$ be arbitrary. We define
$$t^* = \sup\{\tau : \tau \le t,\, v'(\tau) = 0\}, t_* = \inf\{\tau : \tau \ge s,\, v'(\tau) = 0\}. \tag{7.29}$$
Then
$$\begin{aligned} P(t) - P(s) &= P(t) - P(t^*) \quad + \sum_{(a_j,b_j)\subset(t_*,t^*)} (P(b_j) - P(a_j)) + P(t_*) - P(s) \\ &\le \int_s^t w'(\tau)v''(\tau)\,d\tau - \frac{\gamma}{2}\int_s^t |v'(\tau)|^3\,d\tau, \end{aligned} \tag{7.30}$$
since the integrals on the right hand side are 0 on $\{v' = 0\}$. □

Theorem 7.2 *Let the assumptions of Proposition 7.1 hold, assume moreover that $\mathcal{W}$ has an inverse $\mathcal{W}^{-1}$ which also satisfies (3.29), and that the initial data have the regularity $u^0, z^0 \in H^2(0,1)$. Then the initial boundary value problem (7.5) - (7.9) has a weak solution $u, \sigma, \varepsilon : \Omega \to \mathbf{R}$ with $\partial_{tt}u, \partial_{tx}u, \partial_{tt}z, \partial_{tx}z \in L^\infty(0,T; L^2(0,1))$ and $\partial_{tx}u, \partial_{tt}z \in L^3(\Omega)$ and $\sigma, \varepsilon \in C(\bar{\Omega})$.*

Proof. For $n \in \mathbf{N}$, we define a semidiscretization in space with stepsize Δx, $n\Delta x = 1$. Let us denote the forward difference quotient of f_j by
$$\Delta f_j = n(f_{j+1} - f_j), \tag{7.31}$$
and in the same way for the other variables. We consider the ODE system
$$\begin{aligned} u_j'(t) &= \Delta z_j(t), & (7.32)\\ z_j'(t) &= \sigma_j(t), & (7.33)\\ \sigma_j(t) &= \mathcal{W}[\varepsilon_j](t), \quad \varepsilon_j(t) = \Delta u_{j-1}(t), & (7.34) \end{aligned}$$
for $1 \le j \le n-1$, together with the initial conditions
$$u_j(0) = u^0\left(\frac{j}{n}\right), \quad z_j(0) = z^0\left(\frac{j}{n}\right). \tag{7.35}$$

According to the boundary conditions, we set

$$u_0 = 0\,, \quad z_n = 0\,, \quad u_n = u_{n-1}\,, \quad z_0 = z_1\,. \tag{7.36}$$

Because of the regularity of $\mathcal{W}$ given by Proposition 3.6, the initial value problem (7.32) - (7.35) has a unique solution

$$\sigma_j \in W^{1,\infty}(0,T)\,, \quad z_j \in W^{2,\infty}(0,T)\,, \quad u_j, \varepsilon_j \in W^{3,\infty}(0,T)\,. \tag{7.37}$$

We define the discrete energy

$$E^n(t) = \frac{1}{n}\sum_{j=1}^{n-1}\left[\frac{1}{2}u_j''(t)^2 + \frac{1}{2}\varepsilon_j'(t)\sigma_j'(t)\right]\,. \tag{7.38}$$

We compute with the aid of (7.32) - (7.36) that

$$\frac{d}{dt}\left(\frac{1}{n}\sum_{j=1}^{n-1}\frac{1}{2}u_j''(t)^2\right) = \frac{1}{n}\sum_{j=1}^{n-1}\varepsilon_j''(t)\sigma_j'(t)\,. \tag{7.39}$$

Using (7.39), we conclude that for a.e. $s,t \in (0,T)$, $s < t$, there holds

$$E^n(t) - E^n(s) = \frac{1}{n}\sum_{j=1}^{n-1}\left(-\int_s^t \varepsilon_j''(\tau)\sigma_j'(\tau)\,d\tau + \left[\frac{1}{2}\varepsilon_j'(\tau)\sigma_j'(\tau)\right]_{\tau=s}^{\tau=t}\right)\,. \tag{7.40}$$

From Proposition 7.1 we get that $E^n(t) \le E^n(s)$. In order to estimate E^n near 0, we use first (3.30) and then the ODE system (7.32) - (7.34) to derive

$$E^n(t) \le \frac{1}{2n}\sum_{j=1}^{n-1} u_j''(t)^2 + c\varepsilon_j'(t)^2\,, \qquad \text{a.e. in } (0,T)\,, \tag{7.41}$$

$$u_j''(t) = n(\mathcal{W}[\Delta u_j](t) - \mathcal{W}[\Delta u_{j-1}](t))\,, \tag{7.42}$$

$$\varepsilon_j'(t) = n(\Delta z_j(t) - \Delta z_{j-1}(t))\,. \tag{7.43}$$

Choose a sequence $t_k \downarrow 0$ such that (7.41) holds in t_k. We get

$$\begin{aligned}
&\limsup_{k\to\infty} E^n(t_k) \le \\
&\le \frac{1}{2n}\sum_{j=1}^{n-1}[n(\mathcal{W}_f(\Delta u_j(0)) - \mathcal{W}_f(\Delta u_{j-1}(0)))]^2 + [n(\Delta z_j(0) - \Delta z_{j-1}(0))]^2 \\
&\le C(u^0, z^0)\,,
\end{aligned} \tag{7.44}$$

where the bound does not depend on n. This proves that

$$E^n(t) \le C \tag{7.45}$$

at every point where it is defined, so moreover (note that $\sigma_j'\varepsilon_j' \ge 0$, and (3.30) holds for $\mathcal{W}$ and $\mathcal{W}^{-1}$)

$$\frac{1}{n}\sum_{j=1}^{n-1} u_j''(t)^2 \le C\,, \quad \frac{1}{n}\sum_{j=1}^{n-1}\sigma_j'(t)^2 \le C\,, \quad \frac{1}{n}\sum_{j=1}^{n-1}\varepsilon_j'(t)^2 \le C\,. \tag{7.46}$$

Using the ODE system once more, we get

$$|\sigma_j(t)| = |z_j'(t)| = |\frac{1}{n}\sum u_i''(t)| \leq C\,, \tag{7.47}$$

so Proposition 3.6 implies that $\|\varepsilon_j\|_\infty$ is uniformly bounded. Therefore, we can fix a γ according to assumption (iii) of Proposition 7.1, and improve (7.45) to

$$E^n(t) + \frac{\gamma}{2n}\sum_{j=1}^{n-1}\int_0^t |\varepsilon_j'(\tau)|^3\,d\tau \leq C\,. \tag{7.48}$$

This completes the discrete a priori estimates. We now proceed in the standard way and define the linear and piecewise constant interpolates u^n, z^n respectively $\tilde{u}^n, \tilde{z}^n$ as

$$u^n(x,t) = u_j(t) + (x - \frac{j}{n})\Delta u_j[t]\,, \quad \frac{j}{n} \leq x < \frac{j+1}{n}\,, \tag{7.49}$$

the same for z^n, and

$$\tilde{u}^n(x,t) = u_j(t)\,, \quad \tilde{z}^n(x,t) = z_{j+1}(t)\,, \quad \frac{j}{n} \leq x < \frac{j+1}{n}\,. \tag{7.50}$$

The interpolates are solutions of the system

$$\partial_t \tilde{u}^n(x,t) = \partial_x z^n(x,t)\,, \tag{7.51}$$

$$\partial_t \tilde{z}^n(x,t) = \mathcal{W}[\partial_x u^n(x,\cdot)](t)\,. \tag{7.52}$$

The discrete a priori estimates translate into

$$\int_0^1 |\partial_{tt} u^n(x,t)|^2\,dx \leq C\,, \quad \int_0^1 |\partial_{tx} u^n(x,t)|^2\,dx \leq C\,, \tag{7.53}$$

$$\int_0^1 |\partial_{tt} z^n(x,t)|^2\,dx \leq C\,, \quad \int_0^1 |\partial_{tx} z^n(x,t)|^2\,dx \leq C\,, \tag{7.54}$$

$$\int_0^T\int_0^1 |\partial_{tx} u^n(x,t)|^3\,dx\,dt \leq C\,, \quad \int_0^T\int_0^1 |\partial_{tt} z^n(x,t)|^3\,dx\,dt \leq C\,. \tag{7.55}$$

One uses weak star convergence, the embedding (7.14) and the continuity of $\mathcal{W}^{-1}$ to pass to the limit. □

Remark. For the Prandtl operator, Krejčí has proved uniqueness in [25]. The proof uses a monotonicity argument and is based on the inequality (3.11), which can be extended to the Prandtl operator due to the linearity of the output mapping Q. For the Preisach operator, a similar argument "almost" works, but there is a gap in regularity (see [25]).

References

[1] Brokate, M.: *Optimale Steuerung von gewöhnlichen Differentialgleichungen mit Nichtlinearitäten vom Hysteresis-Typ*, Lang 1987. In German; English translation in: Automation and Remote Control **52** (1991), 1639 - 1681, and **53** (1992), 1 - 33.

[2] Brokate, M.: *Some BV properties of the Preisach hysteresis operator*, Applicable Analysis **32** (1989), 229 - 252.

[3] Brokate, M., Visintin, A.: *Properties of the Preisach model for hysteresis*, J. Reine Angew. Math. **402** (1989), 1 - 40.

[4] Brokate, M.: *On a characterization of the Preisach model for hysteresis*, Rend. Sem. Mat. Padova **83** (1990), 153 - 163.

[5] Brokate, M.: *On the moving Preisach model*, Math. Methods Appl. Sci. **15** (1992), 145 - 157.

[6] Brokate, M., Dreßler, K., Krejčí, P.: *On the Mróz model*, submitted.

[7] Chu, C.C.: *A three-dimensional model of anisotropic hardening in metals and its application to the analysis of sheet metal formability*, J. Mech. Phys. Solids **32** (1984), 197 - 212.

[8] Chu, C.C.: *The analysis of multiaxial cyclic problems with an anisotropic hardening model*, Int. J. Solids Structures **23** (1987), 567 - 579.

[9] Clormann, U.H., Seeger, T.: *RAINFLOW-HCM: Ein Zählverfahren für Betriebsfestigkeitsnachweise auf werkstoffmechanischer Grundlage*, Stahlbau **55** (1986), 65 - 71. (In German.)

[10] Damlamian, A., Visintin, A.: *Une généralization vectorielle du modèle de Preisach pour l'hystérésis*, C. R. Acad. Sci. Paris, Série I, **297** (1983), 437 - 440.

[11] Della Torre, E.: *Effect of interaction on the magnetization of single domain particles*, IEEE Trans. Audio Electroacoustics **AU 14** (1966), 86 - 93.

[12] Dreßler, K., Krüger, W.: *The optimal stochastic reconstruction of loading histories from a rainflow matrix*, submitted.

[13] Everett, D.H.: *A general approach to hysteresis. Part 1* (with Whitton, W.I.); *Part 2: Development of the domain theory* (with Smith, F.W.); *Part 3: A formal treatment of the independent domain model of hysteresis*; *Part 4: An alternative formulation of the domain model*, Trans. Faraday Soc. **48** (1952), 749 - 757; **50** (1954), 187 -197; **50** (1954), 1077 - 1096; **51** (1955), 1551 - 1557.

[14] Fuchs, H.O., Stephens, R.I.: *Metal fatigue in engineering*, Wiley 1980.

[15] Hilpert, M.: *On uniqueness for evolution problems with hysteresis*, in: Mathematical models for phase change problems (ed. J.F. Rodrigues), Birkhäuser 1989, 377 -388.

[16] Ishlinskii, A.Yu.: *Some applications of statistical methods to describing deformations of bodies*, Izv. AN SSSR, Techn.Ser., no.9(1944), 580 – 590. (In Russian.)

[17] Krasnosel'skii, M.A., Pokrovskii, A.V.: *Systems with hysteresis*, Springer 1989. Russian edition: Nauka 1983.

[18] Krejčí, P.: *Hysteresis and periodic solutions to semilinear and quasilinear wave equations*, Math. Z. **193** (1986), 247 - 264.

[19] Krejčí, P.: *Existence and large time behaviour of solutions to equations with hysteresis*, Preprint no. 21, Institute of Math., Czechoslovakian Academy of Science, 1986.

[20] Krejčí, P.: *A monotonicity method for solving hyperbolic problems with hysteresis*, Apl. Mat **33** (1988), 197 – 203.

[21] Krejčí, P.: *On Maxwell equations with the Preisach hysteresis operator: The one-dimensional time-periodic case*, Apl. Mat. **34** (1989), 364 – 374.

[22] Krejčí, P., Lovicar, V.: *Continuity of hysteresis operators in Sobolev spaces*, Apl. Mat. **35** , 60 – 66.

[23] Krejčí, P.: *Hysteresis memory preserving operators*, Applications of Math. **36** (1991), 305 - 326.

[24] Krejčí, P.: *Vector hysteresis models*, European J. Appl. Math. **2** (1991), 281 – 292.

[25] Krejčí, P.: *Global behaviour of solutions to the wave equations with hysteresis*, submitted.

[26] Krüger, W., Scheutzow, M., Beste, A., Petersen, J.: *Markov- und Rainflowrekonstruktion stochastischer Beanspruchungszeitfunktionen*, VDI-Report, series 18, no. 22, 1985. (In German.)

[27] Macki, J.W., Nistri, P., Zecca, P.: *Mathematical models for hysteresis*, SIAM Review **35** (1993), 94 - 123.

[28] Madelung, E.: *Über Magnetisierung durch schnellverlaufende Ströme und die Wirkungsweise des Rutherford-Marconischen Magnetdetektors*, Ann. Phys. **17** (1905), 861 – 890. (In German.)

[29] Masing, G.: *Eigenspannungen und Verfestigung bei Messing*, in: Proc. 2nd Int. Congress of Appl. Mech., 1926, 332 – 335. (In German.)

[30] Mayergoyz, I.D.: *Mathematical models of hysteresis*, Springer 1991.

[31] Miner, M.A.: *Cumulative damage in fatigue*, J. Appl. Mech. **12** (1945), A 159 – A 164.

[32] Murakami, Y. (ed.): *The rainflow method in fatigue*, Butterworth & Heinemann, Oxford 1992.

[33] Mróz, Z.: *On the description of anisotropic workhardening*, J. Mech. Phys. Solids **15** (1967), 163 - 175.

[34] Prandtl, L.: *Ein Gedankenmodell zur kinetischen Theorie der festen Körper*, ZAMM **8** (1928), 85 - 106. (In German.)

[35] Preisach, F.: *Über die magnetische Nachwirkung*, Z. Physik **94** (1935), 277 - 302. (In German.)

[36] Rychlik, I.: *A new definition of the rainflow cycle counting method*, Int. J. Fatigue **9** (1987), 119 - 121.

[37] Rychlik, I.: *Rainflow cycles in Gaussian loads*, Fatigue Fract. Engng. Mater. Struct. **15** (1992), 57 - 72.

[38] Rychlik, I.: *Note on cycle counts in irregular loads*, Fatigue Fract. Engng. Mater. Struct. (to appear)

[39] Saunders, P.T.: *An introduction to catastrophe theory*, Cambridge University Press 1980.

[40] Seidman, T.I.: *Switching systems and periodicity*, in: Nonlinear semigroups, partial differential equations and attractors, LN Math. **1394** (1989), 199 - 210.

[41] Seidman, T.I.: *Switching systems I*, Control Cybernet. **19** (1990), 63 - 92.

[42] Visintin, A.: *A model for hysteresis of distributed systems*, Ann. Mat. Pura Appl. **131** (1982), 203 - 231.

[43] Visintin, A.: *On the Preisach model for hysteresis*, Nonlinear Anal. **9** (1984), 977 - 996.

[44] Visintin, A.: *Rheological models and hysteresis effects*, Rend. Sem. Matem. Univ. Padova **77** (1987), 213 -243.

[45] Visintin, A.: *Mathematical models of hysteresis*, in: Topics in nonsmooth mechanics (eds. J.J. Moreau, P.D. Panagiotopoulos, G. Strang), Birkhäuser 1988, 295 - 326.

[46] Visintin, A.: *Differential models of hysteresis*, to appear.

[47] Zeeman, E.C.: *Catastrophe theory*, Scientific American **234** (1976), 65 - 83. In expanded form in: Catastrophe theory, Addison-Wesley 1977.

[48] Zeidler, E.: *Nonlinear functional analysis and its applications*, vol. II B, Springer 1990.

[49] Ziegler, H.: *An introduction to thermomechanics*, 2nd edition, North Holland 1983.

Systems of Nonlinear PDEs Arising from Dynamical Phase Transitions

Nobuyuki Kenmochi
Department of Mathematics, Faculty of Education
Chiba University, Inage-ku, Chiba, 263 Japan

Introduction

In this paper we propose some models for diffusive phase transition dynamics in non-isothermal case and discuss them from the point of time-dependent subdifferential operators in Hilbert spaces.

The first part of this paper is concerned with a model for solid-liquid phase change, which is a coupled system of nonlinear second order parabolic PDEs as follows:

$$(\rho(u)+w)_t - \Delta u = f(t,x) \qquad \text{in } Q := (0,T)\times\Omega,$$

$$\nu w_t - \kappa\Delta w + \beta(w) + g(w) - u \ni 0 \qquad \text{in } Q,$$

$$\frac{\partial u}{\partial n} + \alpha u = h_o(t,x), \quad \frac{\partial w}{\partial n} = 0 \qquad \text{on } \Sigma := (0,T)\times\Gamma,$$

$$u(0,\cdot) = u_o, \quad w(0,\cdot) = w_o \qquad \text{in } \Omega.$$

Here Ω is a bounded domain in $\mathbf{R}^N$ ($N = 2$ or 3) with smooth boundary Γ, ρ is a monotone increasing and bi-Lipschitz continuous function on $\mathbf{R}$, β is a maximal monotone graph in $\mathbf{R}\times\mathbf{R}$ with bounded domain, g is a smooth function on $\mathbf{R}$, κ, ν, α are positive constants and f, h_o, u_o, w_o are prescribed data. This system is called a phase field model with constraint, denoted by (PFC); in this context, $\theta = \rho(u(t,x))$ represents the temperature and $w = w(t,x)$ the order parameter, which indicates the physical situation of the material occupying Ω at time t and position x.

In particular, when $\rho(u) = u$ and $\beta(w) \equiv 0$, the system is the standard phase field model which has been studied in [10,19,20,48], and this case is also able to be handled in the framework of nonlinear abstract evolution equations (cf. [3]). One of most important characteristics of our model (PFC) is the nonlinear term $\beta(w)$ (double obstacles) which allows the coexistence of pure phases in the dynamical phase transition process. In this point we may consider (PFC) as a new model for diffusive phase transitions, and recently this model has been studied independently by several authors (cf. [5,14,22,25,26,27,34,35,43]).

The objective of the first part is to reformulate (PFC) as a nonlinear evolution equation of the form

$$U'(t) + \partial\varphi^t(U(t)) + p(U(t)) \ni \ell(t), \quad 0 < t < T, \tag{1}$$

in an adequate Hilbert space H; here $\partial\varphi^t$ is the subdifferential of a time-dependent convex function φ^t on H and p is a Lipschitz continuous operator in H.

The second part is concerned with a model for diffusive phase separations, for instance the component separation in a binary system of alloys, in non-isothermal case. Our model is described as a coupled system of a nonlinear second order parabolic PDE and a nonlinear fourth order parabolic PDE with constraint, denoted by (PSC), as follows:

$$(\rho(u) + w)_t - \Delta u = f(t,x) \qquad \text{in } Q,$$

$$\nu w_t - \Delta\{-\kappa\Delta w + \beta(w) + g(w) - u\} \ni 0 \qquad \text{in } Q,$$

$$\frac{\partial u}{\partial n} + \alpha u = h_o(t,x), \quad \frac{\partial w}{\partial n} = 0 \qquad \text{on } \Sigma,$$

$$\frac{\partial}{\partial n}\{-\kappa\Delta w + \beta(w) + g(w) - u\} \ni 0 \qquad \text{on } \Sigma,$$

$$u(0,\cdot) = u_o, \quad w(0,\cdot) = w_o \qquad \text{in } \Omega.$$

Here $\rho, \beta, \nu, \kappa, f, h_o, u_o, w_o$ are as in the case of (PFC).

The parabolic PDE, which is obtained by putting $u =$ const. and $\beta(w) \equiv 0$ in the second equation of (PSC), is called the Cahn-Hilliard equation; we refer to [18,44,46,47] for works treating it and to [6,17,31] for recent works treating the Cahn-Hilliard equation with constraint. We have noticed several papers on mathematical analysis of duffusive phase separations in non-isothermal case (cf. [1,2,42]), but only a few papers on that of (PSC) (cf. [5,28,29,30]). Just as in the case (PFC), we shall reformulate it as an evolution equation of the form

$$(AU)'(t) + \partial\varphi^t(U(t)) + p(U(t)) \ni \ell(t), \quad 0 < t < T, \tag{2}$$

in an adequate Hilbert space H; here $\partial\varphi^t$ and p are as in the case of (1) and A is a linear, monotone, positive and continuous operator in H. The basic idea for the reformulation as above is found in [13,37,45].

For the physical interpretation of thermodynamical approach to phase transitions and of non-smooth constraint $\beta(w)$ we refer to [15,40].

The third part is devoted to the solvability of the evolution equation (2), which includes (1) as a specisal case, in an abstract Hilbert space; we shall give an existence-uniqueness theorem for (2) and a theorem on the asymptotic stability of the solution, and furthermore as their direct applications discuss existence and uniqueness of global solutions of (PFC) and (PSC) with asymptotic stability as $t \to +\infty$.

Notations. In general, for a real Banach space Y we denote by $|\cdot|_Y$ the norm and by Y^* the dual space with the dual norm of $|\cdot|_Y$.

Let H be a Hilbert space with inner product $(\cdot,\cdot)_H$. For a proper l.s.c. and convex function φ on H, the effective domain is

$$D(\varphi) := \{z \in H; \varphi(z) < +\infty\}$$

and the subdifferential of φ at $z \in H$ is

$$\partial\phi(z) := \{y \in H; (y, v - z)_H \leq \varphi(v) - \varphi(z) \text{ for all } v \in H\};$$

the domain of the subdifferential $\partial\varphi$ is the set

$$D(\partial\varphi) := \{z \in H; \partial\varphi(z) \neq \emptyset\};$$

the range of the subdifferential $\partial\varphi$ is the set

$$R(\partial\varphi) := \cup_{z\in H}\partial\varphi(z).$$

We refer to [7,36] for the basic concepts and fundamental facts in the convex analysis and nonlinear monotone operator theory, for instance, maximal monotone operators, subdifferentials of convex functions, duality mappings and Yosida-approximations etc.

I. Phase Field Models with Constraints

1. Stefan problem (solid-liquid phase change)

Let us consider a material which may be in either solid or liquid state, for instance, ice and water for H_2O–based system. Suppose that the material occupies a bounded domain Ω, with smooth boudary Γ, in $\mathbf{R}^3$ (or $\mathbf{R}^2$), and at each time $t \geq 0$ the domain Ω is devided into three parts of the liquid, solid regions and their interface. These regions are variable in time t, since the phenomenon of freezing or melting takes place on the interface and the interface is variable in time. We use the following notations:

$S(t)$: the interface at t, $S := \cup_{t>0}\{t\} \times S(t)$,
$\Omega_\ell(t)$: the liquid region at t, $Q_\ell := \cup_{t>0}\{t\} \times \Omega_\ell(t)$,
$\Omega_s(t)$: the solid region at t, $Q_s := \cup_{t>0}\{t\} \times \Omega_s(t)$.

In order to get a mathematical description as a system of PDEs for such a physical process we postulate that

(1) the process has the melting temperature (= the freezing temperature) θ_* which is a constant in $(t,x) \in [0,T] \times \Omega$; this is the only one temperature at which phase transition takes place, i.e. the material changes its state from solid into liquid (or from liquid into

solid), and $Q_\ell = \{\theta > \theta_*\}, Q_s = \{\theta < \theta\}$, where $\theta = \theta(t,x)$ denotes the temperature distribution,

(2) the amount ℓ_o (= const. > 0) of heat is required for the phase transition per unit volume and time; ℓ_o is called the "latent heat",

(3) the interface S is a smooth hypersurface in the (t,x)–space, and $S = \{\theta = \theta_*\}$.

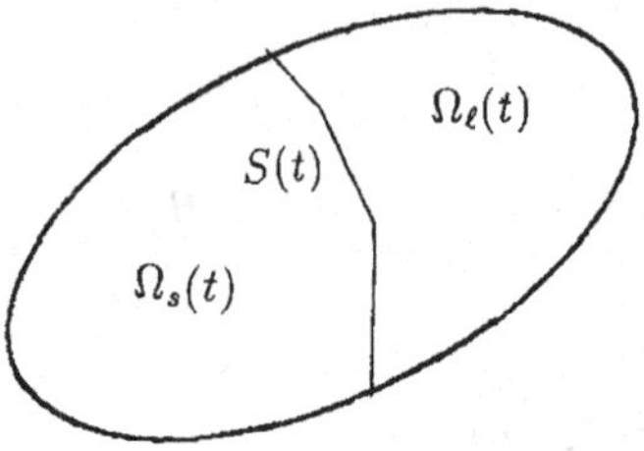

$$\Omega = \Omega_\ell(t) \cup S(t) \cup \Omega_s(t)$$

Figure 1

In the liquid region Q_ℓ and solid region Q_s the heat equations are satisfied:

$$\theta_t - c_\ell \Delta\theta = f(t,x) \qquad \text{in } Q_\ell, \tag{1.1}$$

$$\theta_t - c_s \Delta\theta = f(t,x) \qquad \text{in } Q_s, \tag{1.2}$$

where c_ℓ and c_s are positive constants, which are respectively the specific heats in the liquid region Q_ℓ and solid region Q_s, and f is a prescribed heat source. Equations (1.1) and (1.2) are written in the form

$$\rho(u)_t - \Delta u = f(t,x) \qquad \text{in } Q - S, \tag{1.3}$$

where $Q := (0,T) \times \Omega$,

$$u := \begin{cases} c_\ell(\theta - \theta_*) & \text{in } Q_\ell, \\ 0 & \text{on } S, \\ c_s(\theta - \theta_*) & \text{in } Q_s, \end{cases}$$

and

$$\rho(u) := \left.\begin{cases} \dfrac{1}{c_\ell} u & \text{for } u > 0 \\ 0 & \text{for } u = 0 \\ \dfrac{1}{c_s} u & \text{for } u < 0 \end{cases}\right\} (= \theta - \theta_*).$$

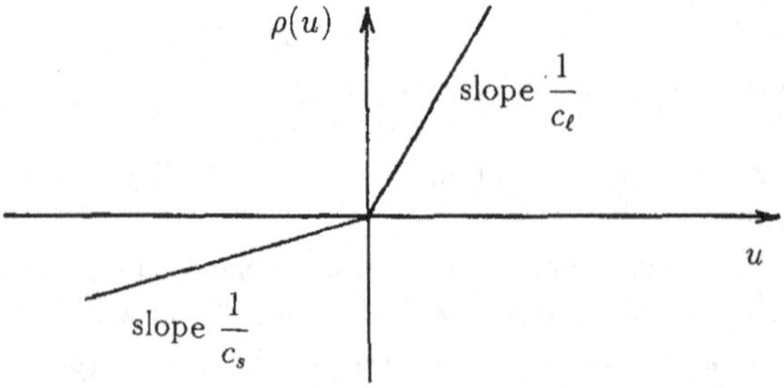

Figure 2

On the interface, where the phase transition takes place, we have

$$u = 0 \text{ (i.e. } \theta = \theta_\star) \qquad \text{on } S \tag{1.4}$$

and the following energy balance equation

$$\ell_o \vec{n} \cdot \vec{t} = \sum_{i=1}^{3} [\frac{\partial u}{\partial x_i}]_S \vec{n} \cdot \vec{x_i} \qquad \text{on } S, \tag{1.5}$$

where $\vec{n} := \vec{n}(t,x)$ is the unit vector normal to S, oriented from Q_s toward Q_ℓ, at $(t,x) \in S$; $\vec{t}$ and $\vec{x_i}$ are the unit vectors on the t–axis and x_i–axis, respectively; $[\cdot]_S$ denotes the jump of $[\cdot]$ across S from Q_s toward Q_ℓ.

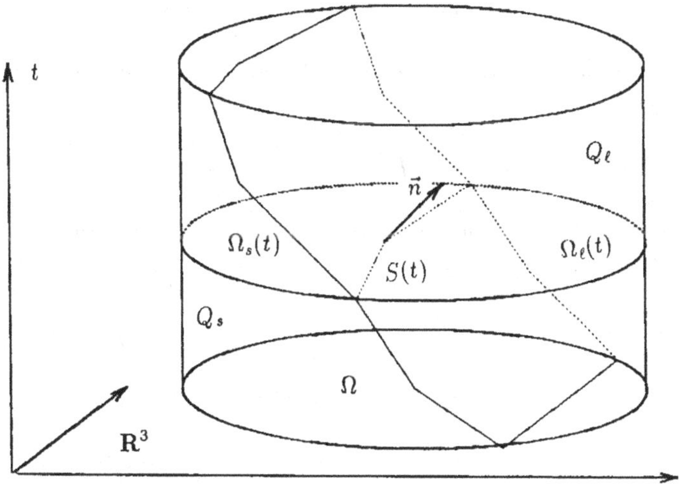

Figure 3

A couple of (1.4) and (1.5) is called the free boundary condition, and (1.5) can be interpreted as follows. The left hand side $\ell_o \vec{n} \cdot \vec{t}$ of (1.5) is the amount of heat which is required in order for the interface to move per unit volume and time, and hence it should be balanced with the difference of heat flux on S (= the right hand side of (1.5)). Prescribing the initial and boundary conditions, we arrive at the following system:

$$(H) \qquad \rho(u)_t - \Delta u = f \qquad \text{in } Q - S,$$

$$(F) \qquad u = 0, \quad \ell_o \vec{n} \cdot \vec{t} = \sum_{i=1}^{3} [\frac{\partial u}{\partial x_i}]_S \vec{n} \cdot \vec{x_i} \qquad \text{on } S,$$

$$(BC) \qquad \frac{\partial u}{\partial n} + \alpha u = h_o \qquad \text{on } \Sigma,$$

$$(IC) \qquad u(0,\cdot) = u_o, \quad S_{|t=0} = S_o \qquad \text{in } \Omega,$$

where $\Sigma := (0,T) \times \Gamma$, α is a given positive constant, h_o a given function on Σ, $n = n(x)$ is the unit outward normal vector to Ω at $x \in \Gamma$, u_o a given function on Ω and S_o a given hypersurface in Ω. This system {(H),(F),(BC),(IC)} is the classical formulation for the Stefan problem.

Now, following the approach due to [23,39], we introduce the phase function

$$w(t,x) := \begin{cases} \frac{\ell_o}{2} & \text{for } u(t,x) > 0, \\ 0 & \text{for } u(t,x) = 0, \\ -\frac{\ell_o}{2} & \text{for } u(t,x) < 0. \end{cases} \tag{1.6}$$

Then, since $u > 0$ on Q_ℓ and $u < 0$ on Q_s, (H) is written in the form

$$(\rho(u) + w)_t - \Delta u = f \qquad \text{in } Q - S. \tag{1.7}$$

Multiplying the both sides of (1.7) by any test function $\eta \in D(Q)$ and integrating the resultant over Q_ℓ and Q_s, we have

$$\int_{Q_\ell} \{(\rho(u) + w)_t - \Delta u\}\eta dxdt + \int_{Q_s} \{(\rho(u) + w)_t - \Delta u\}\eta dxdt = \int_Q f\eta dxdt,$$

since the volume of S is zero. Using the Green's formula, we observe that

$$\begin{aligned} \int_{Q_\ell} \{\rho(u) + w)_t - \Delta u\}\eta dxdt &= -\int_{Q_\ell} (\rho(u) + w)\eta_t dxdt + \int_{Q_\ell} \nabla u \cdot \nabla \eta dxdt \\ &+ \int_S (\frac{\ell_o}{2}\eta\vec{n} \cdot \vec{t})dS - \int_S \sum_{i=1}^{3} \frac{\partial u}{\partial x_i}\eta\vec{n} \cdot \vec{x}_i dS \end{aligned}$$

and

$$\begin{aligned} \int_{Q_s} \{(\rho(u) + w)_t - \Delta u\}\eta dxdt &= -\int_{Q_s} (\rho(u) + w)\eta_t dxdt + \int_{Q_s} \nabla u \cdot \nabla \eta dxdt \\ &+ \int_S (\frac{\ell_o}{2}\eta\vec{n} \cdot \vec{t})dS + \int_S \sum_{i=1}^{3} \frac{\partial u}{\partial x_i}\eta\vec{n} \cdot \vec{x}_i dS. \end{aligned}$$

Adding thses equalities and taking account of free boundary condition (F), we obtain that

$$-\int_Q (\rho(u) + w)\eta_t dxdt + \int_Q \nabla u \cdot \nabla \eta dxdt = \int_Q f\eta dxdt,$$

that is,

$$(\rho(u) + w)_t - \Delta u = f \qquad \text{in } Q \text{ (distribution sense)}.$$

Therefore the following system is obtained as a weak formulation of the Stefan problem, denoted by (SP):

$$(\rho(u) + w)_t - \Delta u = f \qquad \text{in } Q, \tag{1.8}$$

$$\beta(w) \ni u \qquad \text{in } Q, \tag{1.9}$$

$$\frac{\partial u}{\partial n} + \alpha u = h_o \qquad \text{on } \Sigma, \tag{1.10}$$

$$(\rho(u)+w)(0)=\rho(u_o)+\frac{\ell_o}{2}\chi_{\{u_o>0\}}-\frac{\ell_o}{2}\chi_{\{u_o<0\}}, \tag{1.11}$$

where β is a maximal monotone graph in $\mathbf{R}\times\mathbf{R}$ defined by

$$\beta(w):=\begin{cases}[0,+\infty) & \text{if } w=\dfrac{\ell_o}{2},\\ \{0\} & \text{if } |w|<\dfrac{\ell_o}{2},\\ (-\infty,0] & \text{if } w=-\dfrac{\ell_o}{2},\\ \emptyset & \text{otherwise.}\end{cases} \tag{1.12}$$

Here $\chi_{\{u_o>0\}}$ (resp. $\chi_{\{u_o<0\}}$) is the characteristic function of the set $\{u_o>0\}$ (resp. $\{u_o<0\}$), and for initial data u_o and S_o the compatibility condition (1.13) below is assumed:

$$\begin{cases}S_o \text{ devides } \Omega \text{ into two parts } \Omega_{\ell,o} \text{ and } \Omega_{s,o} \text{ such that}\\ \Omega_{\ell,o}=\{u_o>0\}, \quad \Omega_{s,o}=\{u_o<0\}, \quad S_o=\{u_o=0\}.\end{cases} \tag{1.13}$$

It is easy to check that (1.6) implics (1.9) and the initial condition (IC) implies (1.11) under (1.13).

We should notice that the above weak formulation (SP) does not include explicitly the free boundary condition (F). For some works on (SP), see [12,21,23,39].

2. Phase field model with constraint

Real phenomena are however more complicated than the mathematical descriptions mentioned in section 1. For instance, in the real processes of solid-liquid phase transitions one observes very often supercooling and superheating phenomena, namely the temperature of liquid is sometimes below $\theta_\star$ or that of solid is sometimes above $\theta_\star$. These are neglected in the formulation of the Stefan problem.

In this section, taking account of supercooling or superheating effects, let us consider a mathematical description of the interface between solid and liquid. For this purpose the basic ideas are the following (a) and (b).

(a) The interface is no longer a hypersurface, but a region $\Omega_i(t)$, $t\geq 0$, of a finite thickness in which the phase function $w(t,x)$ continuously changes from $w=-\frac{\ell_o}{2}$ to $w=\frac{\ell_o}{2}$.

The liquid region $\Omega_\ell(t)$, solid region $\Omega_s(t)$ and interface $\Omega_i(t)$ are respectively defined by means of the phase function w as follows:

$$\Omega_\ell(t)=\{x\in\Omega; w(t,x)=\frac{\ell_o}{2}\},$$

$$\Omega_s(t)=\{x\in\Omega; w(t,x)=-\frac{\ell_o}{2}\},$$

$$\Omega_i(t) = \{x \in \Omega; |w(t,x)| < \frac{\ell_o}{2}\}.$$

Along the cross-section L (see Figure 4), the behaviour of w is as shown by Figure 5.

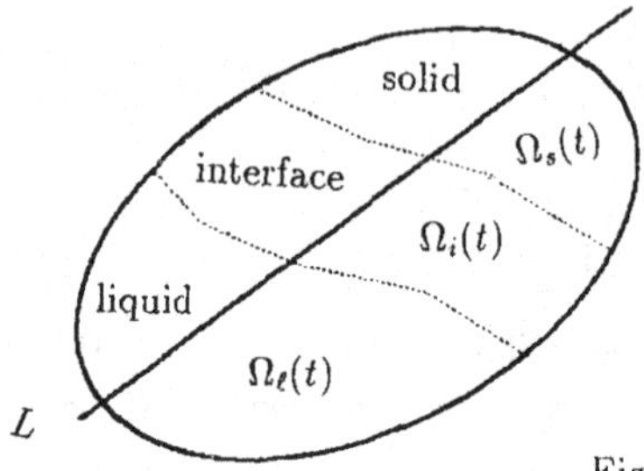

Figure 4

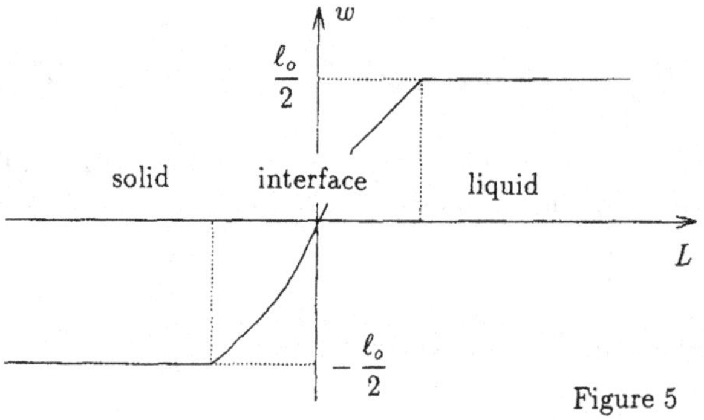

Figure 5

(b) We suppose that the dynamics of the interface is interpreted by a free energy functional of the form

$$F_\Omega(u,w) := \int_\Omega \{\frac{\kappa}{2}|\nabla w|^2 + \hat{\beta}(w) + \hat{g}(w) - uw\}dx \tag{2.1}$$

for any $u \in L^2(\Omega)$ and $w \in H^1(\Omega)$,

where κ is a positive constant, $\hat{\beta}$ is the indicator function associated to the interval $[-\frac{\ell_o}{2}, \frac{\ell_o}{2}]$, i.e.

$$\hat{\beta}(w) = \begin{cases} 0 & \text{for } |w| \le \frac{\ell_o}{2}, \\ +\infty & \text{otherwise,} \end{cases}$$

and $\hat{g}$ is any non-negative smooth function on $\mathbf{R}$ such that the graph has double wells at $w = \pm\frac{\ell_o}{2}$ (see Figure 6). Note that $F_\Omega(u,w) = +\infty$ if $\hat{\beta}(w) \notin L^1(\Omega)$, or equivalently w does not satisfy "$|w| \le \frac{\ell_o}{2}$ a.e on Ω".

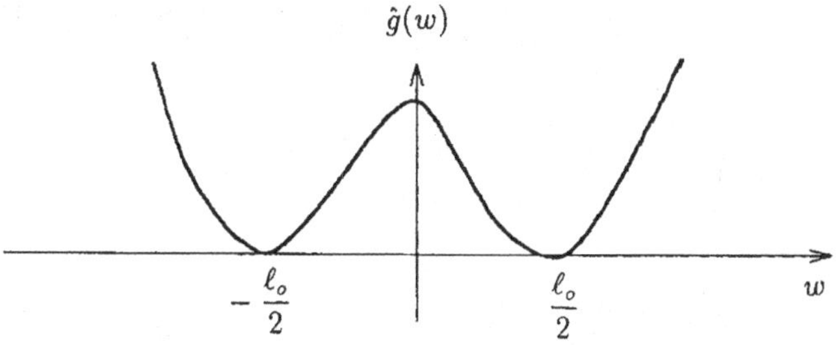

Figure 6

Prior to the evolution case of our process we consider the equilibrium case.

The equilibrium case

For a fixed $u = u_\infty(x)$ (hence the temperature $\theta = \theta_\infty := \rho(u_\infty) + \theta_\star$ is fixed, too) the function w, which corresponds to an equilibrium, might be a local minimizer $w_\infty(x)$ of the free energy functional $F_\Omega(u_\infty, \cdot)$ on $H^1(\Omega)$ and its characterization is given by the following theorem.

Theorem 2.1. (Equilibrium) *Let $u_\infty \in L^2(\Omega)$ and w_∞ be any local minimizer of $F_\Omega(u_\infty, \cdot)$ on $H^1(\Omega)$, i.e. there exists an open ball $U_r(w_\infty) := \{w \in H^1(\Omega); |w - w_\infty|_{H^1(\Omega)} < r\}$ for a number $r > 0$ such that*

$$+\infty > F_\infty(u_\infty, w_\infty) = \min_{w \in U_r(w_\infty)} F_\Omega(u_\infty, w). \tag{2.2}$$

Then $w_\infty \in H^2(\Omega)$ and it is a unique solution of

$$\begin{cases} -\kappa \Delta w_\infty + \xi_\infty + g(w_\infty) - u_\infty = 0 \quad a.e. \ in \ \Omega, \\ \xi_\infty \in L^2(\Omega), \quad \xi_\infty \in \beta(w_\infty) \quad a.e. \ in \ \Omega, \\ \dfrac{\partial w_\infty}{\partial n} = 0 \quad a.e. \ on \ \Gamma, \end{cases} \tag{2.3}$$

where g is the derivative of $\hat{g}$, i.e. $g = \hat{g}'$, and β is the subdifferential of $\hat{\beta}$ in $\mathbf{R}$; note that β is the maximal monotone graph in $\mathbf{R} \times \mathbf{R}$ given by (1.12).

The proof of this theorem will be given in section 3.

Remark 2.1. As will be seen from the proof of Theorem 2.1 given in section 3, the assertion of the theorem is still valid for any maximal monotone graph β in $\mathbf{R} \times \mathbf{R}$ with bounded domain in $\mathbf{R}$.

Next, consider the non-equilibrium case.

The non-equilibrium case

In the non-equilibrium case, the phase function w no longer satisfies that

$$-\kappa \Delta w + \xi + g(w) - u = 0, \quad \xi \in \beta(w), \quad \text{a.e. in } Q,$$

but $-\kappa\Delta w+\xi+g(w)-u$ is proportional to the time derivative w_t with a relaxation time ν (positive constant), i.e.

$$\begin{cases} \nu w_t - \kappa\Delta w + \xi + g(w) - u = 0 & \text{in } Q, \\ \xi \in \beta(w) \quad \text{in } Q, \\ \dfrac{\partial w}{\partial n} = 0 \quad \text{on } \Sigma. \end{cases}$$

Moreover u and w satisfy the same heat equation (H) as in section 1. Therefore we have the following coupled system, denoted by (PFC), in the non-equilibrium case:

$$(\rho(u)+w)_t - \Delta u = f \qquad \text{in } Q, \tag{2.4}$$

$$\nu w_t - \kappa\Delta w + \xi + g(w) - u = 0 \qquad \text{in } Q, \tag{2.5}$$

$$\xi \in \beta(w) \qquad \text{in } Q, \tag{2.6}$$

$$\frac{\partial u}{\partial n} + \alpha u = h_o, \quad \frac{\partial w}{\partial n} = 0 \qquad \text{on } \Sigma, \tag{2.7}$$

$$u(0,\cdot) = u_o, \quad w(0,\cdot) = w_o \qquad \text{in } \Omega. \tag{2.8}$$

This is called the phase field model with constraint. In the case when $\beta \equiv 0$, this system was proposed by Fix [20] and Caginalp [10], and subsequenetly studied by Zheng [48] and Elliott-Zheng [19].

Remark 2.2. When $\nu = 0, \kappa = 0, g \equiv 0$ and the interface $\Omega_i(t)$ is a smooth hypersurface, (2.5) and (2.6) are formally written in the form "$u \in \beta(w)$ in Q", namely, (PFC) is nothing but the Stefan problem (SP).

Example 2.1. We consider the case when $\Omega := (-2,2)\times(-2,2)\times(-2,2)$, $\nu = \kappa = \frac{\ell_o}{2} = \alpha = 1$, $\hat{g}(w) := \frac{\pi^2}{16}(w^2-1)^2$ and

$$u_\infty(x) := \begin{cases} \dfrac{\pi^2}{4} & \text{for } 2 > x_1 > 1, \\ \dfrac{\pi^2}{4}\sin^3(\dfrac{\pi}{2}x_1) & \text{for } |x_1| \le 1, \\ -\dfrac{\pi^2}{4} & \text{for } -1 > x_1 > -2, \end{cases}$$

where $x = (x_1, x_2, x_3) \in \Omega$. Then, as is easily seen, the function

$$w_\infty(x) := \begin{cases} 1 & \text{for } 2 > x_1 > 1, \\ \sin(\dfrac{\pi}{2}x_1) & \text{for } |x_1| \le 1, \\ -1 & \text{for } -1 > x_1 > -2, \end{cases} \tag{2.9}$$

satisfies that $w_\infty \in H^2(\Omega)$ and

$$-\Delta w_\infty + \xi_\infty + g(w_\infty) - u_\infty = 0 \quad \text{a.e. in } \Omega,$$

$$\frac{\partial w_\infty}{\partial n} = 0 \quad \text{a.e. on } \Gamma,$$

where

$$\xi_\infty(x) := \left\{ \begin{array}{ll} \frac{\pi^2}{4} & \text{for } 2 > x_1 > 1 \\ 0 & \text{for } |x_1| \leq 1 \\ -\frac{\pi^2}{4} & \text{for } -1 > x_1 > -2 \end{array} \right\} \in \beta(w_\infty(x)) \text{ for a.e. } x \in \Omega.$$

By the way, u_∞ satisfies that $u_\infty \in H^2(\Omega)$ and

$$-\Delta u_\infty = f_\infty \qquad \text{a.e. in } \Omega,$$

$$\frac{\partial u_\infty}{\partial n} + u_\infty = h_{o,\infty} \qquad \text{a.e. on } \Gamma,$$

where

$$f_\infty(x) := \begin{cases} 0 & \text{for } 2 > x_1 > 1, \\ \frac{9\pi^4}{16}\sin^3(\frac{\pi}{2}x_1) - \frac{3\pi^4}{8}\sin(\frac{\pi}{2}x_1) & \text{for } |x_1| \leq 1, \\ 0 & \text{for } -1 > x_1 > -2, \end{cases}$$

and

$$h_{o,\infty}(x) := \begin{cases} \frac{\pi^2}{4} & \text{on } \Gamma \cap \{x_1 = 2\}, \\ -\frac{\pi^2}{4} & \text{on } \Gamma \cap \{x_1 = -2\}, \\ \frac{\pi^2}{4}\sin^3(\frac{\pi}{2}x_1) & \text{otherwise.} \end{cases}$$

In this equilibrium case, by (2.9) the liquid region $\Omega_\ell(\infty)$, solid region $\Omega_s(\infty)$ and interface $\Omega_i(\infty)$ are respectively given by

$$\Omega_\ell(\infty) = \{x \in \Omega; 1 \leq x_1 < 2\},$$

$$\Omega_s(\infty) = \{x \in \Omega; -2 < x_1 \leq -1\},$$

and

$$\Omega_i(\infty) = \{x \in \Omega; |x_1| < 1\}.$$

Note that they have positive measures, and moreover that w_∞, u_∞ are in $W^{2,\infty}(\Omega)$, but not in $C^2(\Omega)$.

It is not difficult to give a non-equilibrium example in which for every time $t > 0$, the three unknown regions $\Omega_\ell(t)$, $\Omega_s(t)$ and $\Omega_i(t)$ have strictly positive measures.

In the formulations mentioned so far, the functions ρ, β and g were respectively specified. But, in the sequel, we consider (PFC):= $\{(2.4)-(2.8)\}$ as a mathematical problem for ρ, β, g in more general classes. In the rest of this paper we suppose the following conditions (ρ),(β) and (g):

(ρ) $\rho : \mathbf{R} \to \mathbf{R}$ is an increasing bi-Lipschitz continuous function; we denote by $C(\rho)$ a common Lipschitz constant of ρ and ρ^{-1}.

(β) β is a maximal monotone graph in $\mathbf{R} \times \mathbf{R}$ such that for some numbers σ_*, σ^* with $-\infty < \sigma_* < \sigma^* < +\infty$

$$\overline{D(\beta)} = [\sigma_*, \sigma^*];$$

note in this case that $R(\beta) = \mathbf{R}$, so that there is a non-negative proper l.s.c. convex function $\hat{\beta}$ on $\mathbf{R}$ whose subdifferential $\partial\hat{\beta}$ coincides with β in $\mathbf{R}$.

(g) $g : \mathbf{R} \to \mathbf{R}$ is a Lipschitz continuous function with compact support $supp(g)$ in $\mathbf{R}$; in this case note that there is a non-negative primitive $\hat{g}$ of g.

Remark 2.3. As is easily understood from (2.6) of the formulation (PFC), the function w of any solution $\{u, w\}$ has the constraint "$\sigma_* \le w \le \sigma^*$ a.e. on Q" under (β), so that w does not depend upon the behaviour of g on the outside of $[\sigma_*, \sigma^*]$. Therefore condition (g) is not essential and it is enough to suppose that g is locally Lipschitz continuous on $\mathbf{R}$.

Moreover, we suppose that

$$\kappa > 0, \ \nu > 0 \text{ and } \alpha > 0 \text{ are constants}$$

and

(f) $f \in L^2_{loc}(\mathbf{R}_+; L^2(\Omega))$,

(h_o) $h_o \in W^{1,2}_{loc}(\mathbf{R}_+; L^2(\Gamma))$.

We use also some standard notations:

$$(v, z) := \int_\Omega vz\,dx, \qquad v, z \in L^2(\Omega),$$

$$a(v, z) := \int_\Omega \nabla v \cdot \nabla z\,dx, \qquad v, z \in H^1(\Omega),$$

$$(v, z)_\Gamma := \int_\Gamma vz\,d\Gamma, \qquad v, z \in L^2(\Gamma).$$

3. A lemma

We prepare a lemma on the minimization problem for the functional

$$J(z) := \begin{cases} \frac{\kappa}{2}|\nabla z|^2_{L^2(\Omega)} + \int_\Omega \hat{\gamma}(x, z)dx + \int_\Omega \hat{\beta}(z)dx - (\ell, z) \\ \qquad\qquad \text{if } z \in H^1(\Omega) \text{ and } \hat{\beta}(z) \in L^1(\Omega), \\ +\infty \qquad \text{otherwise} \end{cases} \tag{3.1}$$

on $L^2(\Omega)$. Here we suppose that

(1) $\ell \in L^2(\Omega)$,

(2) $\hat{\beta}$ is a proper l.s.c. convex function on $\mathbf{R}$ as given in condition (β),

(3) $\hat{\gamma} : \Omega \times \mathbf{R} \to \mathbf{R}$ is a function such that $\hat{\gamma}(x,r)$ is of C^1-class and convex in $r \in \mathbf{R}$ for a.e. $x \in \Omega$, measurable in $x \in \Omega$ for each $r \in \mathbf{R}$ and

$$0 \le \hat{\gamma}(x,r) \le C_o|r|^2 + C_o' \quad \text{for all } r \in \mathbf{R} \text{ and a.e. } x \in \Omega,$$

and such that $\gamma(x,r) := \frac{\partial}{\partial r}\hat{\gamma}(x,r)$ is strictly increasing in $r \in \mathbf{R}$ for a.e. $x \in \Omega$ and

$$|\gamma(x,r)| \le C_1|r| + C_1' \quad \text{for all } r \in \mathbf{R} \text{ and a.e. } x \in \Omega, \tag{3.2}$$

where C_o, C_o', C_1, C_1' are positive constants.

In this case the functional J which is defined by (3.1) is proper, l.s.c. and coercive on $L^2(\Omega)$, and strictly convex on its effective domain. Therefore there exists one and only one minimizer of J on $L^2(\Omega)$.

Lemma 3.1. *The following three statements (a), (b), (c) are equivalent to each other:*

(a) *v_o is the minimizer of J on $L^2(\Omega)$, i.e.*

$$J(v_o) = \min_{z \in L^2(\Omega)} J(z).$$

(b) *$v_o \in H^1(\Omega)$, $\hat{\beta}(v_o) \in L^1(\Omega)$ and*

$$\kappa a(v_o, v_o - z) + (\gamma(\cdot, v_o), v_o - z) + \int_\Omega \hat{\beta}(v_o)dx$$
$$\le (\ell, v_o - z) + \int_\Omega \hat{\beta}(z)dx \qquad \text{for all } z \in H^1(\Omega).$$

(c) *$v_o \in H^2(\Omega)$ and*

$$\begin{cases} -\kappa\Delta v_o + \gamma(x, v_o) + \xi = \ell & \text{a.e. in } \Omega, \\ \xi \in L^2(\Omega), \quad \xi \in \beta(v_o) & \text{a.e. in } \Omega, \\ \dfrac{\partial v_o}{\partial n} = 0 & \text{a.e. on } \Gamma. \end{cases} \tag{3.3}$$

Proof. The equivalence "(a)↔(b)" is standard as well as the implication "(c)→(b)". Therefore it is enough to show the existence of a solution $v_o \in H^2(\Omega)$ of problem (3.3) in order to accomplish the proof of the lemma. For this purpose, for each $0 < \lambda \le 1$, consider the problem

$$\begin{cases} -\kappa\Delta v_\lambda + \gamma(x, v_\lambda) + \beta_\lambda(v_\lambda) = \ell & \text{a.e. in } \Omega, \\ \dfrac{\partial v_\lambda}{\partial n} = 0 & \text{a.e. on } \Gamma, \end{cases} \tag{3.4}$$

where β_λ is the Yosida-approximation of β, i.e.

$$\beta_\lambda(r) := \frac{r - (I + \lambda\beta)^{-1}r}{\lambda}, \qquad r \in \mathbf{R}.$$

Note that β_λ is of C^1-class and Lipschitz continuous on $\mathbf{R}$, there are positive constants C_2, C_2' independent of λ such that

$$|\beta_\lambda(r)| \geq C_2|r| - C_2' \qquad \text{for all } r \in \mathbf{R}, \tag{3.5}$$

and that $\beta_\lambda(r_o) = 0$ if $\beta(r_o) \ni 0$. By the semilinear theory of elliptic PDEs (cf. [8]) problem (3.4) admits a solution $v_\lambda \in H^2(\Omega)$, and by the strictly monotonicity of $\gamma(x, r)$ in $r \in \mathbf{R}$ the solution is unique.

Now, multiply the equation of (3.4) by $\beta_\lambda(v_\lambda)(= \beta_\lambda(v_\lambda) - \beta_\lambda(r_o))$ and integrate the resultant over Ω to obtain

$$\kappa \int_\Omega \beta_\lambda'(v_\lambda)|\nabla v_\lambda|^2 dx + (\gamma(\cdot, v_\lambda) - \gamma(\cdot, r_o), \beta_\lambda(v_\lambda)) + |\beta_\lambda(v_\lambda)|^2_{L^2(\Omega)} \tag{3.6}$$

$$= (\ell - \gamma(\cdot, r_o), \beta_\lambda(v_\lambda)).$$

Moreover, note from the monotonicity of β_λ and $\gamma(x, \cdot)$ that $\beta_\lambda'(v_\lambda) \geq 0$ and $(\gamma(\cdot, v_\lambda) - \gamma(\cdot . r_o), \beta_\lambda(v_\lambda)) \geq 0$. Then it follows that $\{\beta_\lambda(v_\lambda)\}$ is bounded in $L^2(\Omega)$, so that $\{v_\lambda\}$ and $\{\gamma(\cdot, v_\lambda)\}$ are bounded in $L^2(\Omega)$ by (3.2) and (3.5). This implies that $\{v_\lambda\}$ is bounded in $H^2(\Omega)$. Therefore there is a sequence $\{\lambda_n\}$ with $\lambda_n \to 0$ (as $n \to +\infty$) and a function $v_o \in H^2(\Omega)$ such that

$$v_n := v_{\lambda_n} \to v_o \qquad \text{weakly in } H^2(\Omega)$$

and

$$\beta_{\lambda_n}(v_n) \to \xi \qquad \text{weakly in } L^2(\Omega).$$

By the standard argument of maximal monotone operators we see that $\xi \in \beta(v_o)$ a.e. in Ω. Letting $\lambda = \lambda_n \to 0$ in (3.4), we conclude that v_o is a solution of (3.3). ◇

Remark 3.1. As is easily seen from (3.6) in the proof of Lemma 3.1, the function ξ in (3.3) satisfies bound

$$|\xi|_{L^2(\Omega)} \leq |\ell|_{L^2(\Omega)} + |\gamma(\cdot, r_o)|_{L^2(\Omega)}$$

and hence there is a constant $R(\gamma) \geq 0$, dependent only on the function γ, such that

$$\kappa|v_o|_{H^2(\Omega)} \leq R(\gamma)(|\ell|_{L^2(\Omega)} + 1).$$

Proof of Theorem 2.1: Let w_∞ be a local minimizer of $F_\Omega(u_\infty, \cdot)$ on $H^1(\Omega)$. Then, for any small $0 < \delta < 1$ and $z \in H^1(\Omega)$ with $|z| \leq \frac{\ell_o}{2}$ a.e. in Ω,

$$\frac{1}{\delta}\{F_\Omega(u_\infty, w_\infty + \delta(z - w_\infty)) - F_\Omega(u_\infty, w_\infty)\} \geq 0.$$

Letting $\delta \downarrow 0$ gives

$$\kappa a(w_\infty, w_\infty - z) + (g(w_\infty) - u_\infty, w_\infty - z) \leq 0$$

for all $z \in H^1(\Omega)$ with $|z| \le \frac{\ell_0}{2}$ a.e. in Ω. Now, apply Lemma 3.1, when $\hat{\gamma}(x,r) = \frac{1}{2}r^2$, $\hat{\beta}$ is the indicator function of the interval $[-\frac{\ell_0}{2}, \frac{\ell_0}{2}]$, $v_o = w_\infty$ and $\ell = u_\infty - g(w_\infty) - w_\infty$. Then, by the equivalence "(b)↔(c)", $w_\infty \in H^2(\Omega)$ and it satisfies (2.3). ◇

4. Evolution operators associated with (PFC)

We introduce some function spaces for the abstract setting of (PFC). We denote by $V = H^1(\Omega)$ the Hilbert space with norm

$$|z|_V := \{|\nabla z|^2_{L^2(\Omega)} + \alpha |z|^2_{L^2(\Gamma)}\}^{\frac{1}{2}}, \tag{4.1}$$

by V^* the dual space of V and by $\langle \cdot, \cdot \rangle$ the duality pairing between V^* and V; we have by identifying $L^2(\Omega)$ with its dual

$$V \subset L^2(\Omega) \subset V^*$$

with compact injections. Also we denote by F the duality mapping from V onto V^* which assigns to each $v \in V$ the element $Fv \in V^*$ satisfying

$$\langle Fv, z \rangle = a(v,z) + \alpha(v,z)_\Gamma \qquad \text{for all } z \in V.$$

Clearly, F is an isometric operator from V onto V^*. In particular, if $q := Fv \in L^2(\Omega)$, then $v \in H^2(\Omega)$ and v is the unique solution of

$$-\Delta v = q \qquad \text{a.e. in } \Omega, \qquad \frac{\partial v}{\partial n} + \alpha v = 0 \qquad \text{a.e. on } \Gamma.$$

Next, for the boundary function h_o we consider a function $h : \mathbf{R}_+ \to H^1(\Omega)$ such that for each $t \ge 0$

$$a(h(t), z) + (\alpha h(t) - h_o(t), z)_\Gamma = 0 \qquad \text{for all } z \in H^1(\Omega); \tag{4.2}$$

by assumption (h_o) we see that $h \in W^{1,2}_{loc}(\mathbf{R}_+; H^1(\Omega))$.

For the initial data u_o, w_o we suppose that

$$u_o \in L^2(\Omega), \quad w_o \in L^\infty(\Omega) \text{ with } \sigma_* \le w_o \le \sigma^* \text{ a.e. in } \Omega. \tag{4.3}$$

Definition 4.1. (Weak formulation of (PFC)) Let $0 < T < +\infty$. Then a couple of functions u and w is called a weak solution of (PFC) on $[0,T]$, if the following conditions (w1) - (w3) are satisfied:

(w1) $\rho(u) \in C([0,T]; V^*) \cap W^{1,2}_{loc}((0,T]; V^*) \cap L^2(0,T; L^2(\Omega))$, $u \in L^2_{loc}((0,T]; V)$,
$w \in C([0,T]; L^2(\Omega)) \cap W^{1,2}_{loc}((0,T]; L^2(\Omega)) \cap L^2(0,T; H^1(\Omega))$,
$\hat{\beta}(w) \in L^1(0,T; L^1(\Omega))$.

(w2) $\rho(u)(0) = \rho(u_o)$ and

$$\langle \rho(u)'(t) + w'(t), z \rangle + a(u(t) - h(t), z) + \alpha(u(t) - h(t), z)_\Gamma = (f(t), z)$$

$$\text{for all } z \in V \text{ and a.e. } t \in [0,T],$$

where the prime $'$ denotes the derivative in time.

(w3) $w(0) = w_o$ and there is $\xi \in L^2_{loc}((0,T]; L^2(\Omega))$ such that

$$\xi(t) \in \beta(w(t)) \quad \text{a.e. in } \Omega \text{ for a.e. } t \in [0.T],$$

$$\nu(w'(t),\eta) + \kappa a(w(t),\eta) + (\xi(t) + g(w(t)),\eta) = (u(t),\eta)$$

$$\text{for all } \eta \in H^1(\Omega) \text{ and a.e. } t \in [0,T].$$

We are going to give an expression of (PFC) in the form

$$\begin{cases} U'(t) + \partial\varphi^t(U(t)) + G(U(t)) \ni \tilde{f}(t), & t > 0, \\ U(0) = U_o, \end{cases} \tag{4.4}$$

in a Hilbert space X, where $\varphi^t(\cdot)$ is a time-dependent convex function on X and $G(\cdot)$ is a Lipschitz continuous operator in X.

We take as a Hilbert space

$$X := V^* \times L^2(\Omega)$$

with inner product $(\cdot,\cdot)_X$ given by

$$([e_1, w_1], [e_2, w_2])_X := \langle e_1, F^{-1}e_2\rangle + \nu(w_1, w_2)$$

$$\text{for all } [e_i, w_i] \in X,\ i = 1, 2.$$

It is clear that X is a Hilbert space with this inner product $(\cdot,\cdot)_X$. Next, for each $t \geq 0$, we define a function $\varphi^t(\cdot)$ on X by

$$\varphi^t([e,w]) := \begin{cases} \int_\Omega \widehat{\rho^{-1}}(e-w)dx + \frac{\kappa}{2}|\nabla w|^2_{L^2(\Omega)} + \int_\Omega \hat{\beta}(w)dx - (h(t), e) \\ \qquad\qquad \text{if } [e,w] \in L^2(\Omega) \times H^1(\Omega) \text{ with } \hat{\beta}(w) \in L^1(\Omega), \\ +\infty \qquad\qquad \text{otherwise,} \end{cases}$$

where $\widehat{\rho^{-1}}$ is a non-negative primitive of ρ^{-1}.

Theorem 4.1. *(a) For each $t \geq 0$, φ^t is proper, l.s.c. and convex on X.*

(b) The subdifferential $\partial\varphi^t$ of φ^t in X is characterized as follows: $[e^*, w^*] \in \partial\varphi^t([e,w])$ if and only if

$$e^* = F(\rho^{-1}(e-w) - h(t)) \tag{4.6}$$

and there is $\xi \in L^2(\Omega)$ with $\xi \in \beta(w)$ a.e. in Ω such that

$$\nu(w^*,\eta) = \kappa a(w,\eta) + (\xi - \rho^{-1}(e-w),\eta) \qquad \text{for all } \eta \in H^1(\Omega). \tag{4.7}$$

(c) If $[e_i^\star, w_i^\star] \in \partial\varphi^t([e_i, w_i]), i = 1, 2,$ *then*

$$([e_1^\star, w_1^\star] - [e_2^\star, w_2^\star], [e_1, w_1] - [e_2, w_2])_X$$

$$= (\rho^{-1}(e_1 - w_1) - \rho^{-1}(e_2 - w_2), (e_1 - w_1) - (e_2 - w_2)) + \kappa|\nabla(w_1 - w_2)|^2_{L^2(\Omega)} + (\xi_1 - \xi_2, w_1 - w_2) \tag{4.8}$$

$$\geq \frac{1}{C(\rho)}|(e_1 - w_1) - (e_2 - w_2)|^2_{L^2(\Omega)} + \kappa|\nabla(w_1 - w_2)|^2_{L^2(\Omega)},$$

where $\xi_i, i = 1, 2,$ *are the function* ξ *as in (4.7).*

Proof. Assertion (a) is clear. Assume $[e^\star, w^\star] \in \partial\varphi^t(e, w])$. Then, by definition,

$$([e^\star, w^\star], [e_1, w_1] - [e, w])_X (= \langle e^\star, F^{-1}(e_1 - e)\rangle + \nu(w^\star, w_1 - w))$$

$$\leq \varphi^t([e_1, w_1]) - \varphi^t([e, w]) \qquad \text{for all } [e_1, w_1] \in X. \tag{4.9}$$

Here, take as $[e_1, w_1]$ of (4.9) any element of the form

$$\varepsilon[\tilde{e}, w] + (1 - \varepsilon)[e, w], \quad \tilde{e} \in L^2(\Omega), \quad 0 < \varepsilon < 1,$$

and pass to the limit $\varepsilon \downarrow 0$ after deviding the resultant by ε. Then

$$\langle e^\star, F^{-1}(\tilde{e} - e)\rangle \leq (\rho^{-1}(e - w) - h(t), \tilde{e} - e).$$

Since $\langle e^\star, F^{-1}(\tilde{e} - e)\rangle = \langle \tilde{e} - e, F^{-1}e^\star\rangle = (F^{-1}e^\star, \tilde{e} - e)$, it follows from the arbitrariness of $\tilde{e} \in L^2(\Omega)$ that

$$F^{-1}e^\star = \rho^{-1}(e - w) - h(t), \quad \text{i.e. } e^\star = F(\rho^{-1}(e - w) - h(t)).$$

Thus (4.6) holds. Next, take as $[e_1, w_1]$ of (4.9) any element of the form

$$[e, \tilde{w}], \qquad \tilde{w} \in H^1(\Omega).$$

Then we have

$$\nu(w^\star, \tilde{w} - w) \leq \varphi^t([e, \tilde{w}]) - \varphi^t([e, w]) \qquad \text{for all } \tilde{w} \in H^1(\Omega),$$

which is written in the form

$$J_1(w) = \min_{\tilde{w} \in L^2(\Omega)} J_1(\tilde{w}),$$

where

$$J_1(\tilde{w}) := \begin{cases} \frac{\kappa}{2}|\nabla\tilde{w}|^2_{L^2(\Omega)} + \int_\Omega \hat{\rho^{-1}}(e - \tilde{w})dx + \int_\Omega \hat{\beta}(\tilde{w})dx - \nu(w^\star, \tilde{w}) \\ \qquad\qquad \text{if } \tilde{w} \in H^1(\Omega) \text{ and } \hat{\beta}(\tilde{w}) \in L^1(\Omega), \\ +\infty \qquad\qquad \text{otherwise.} \end{cases}$$

Therefore it follows from Lemma 3.1 that there exists $\xi \in L^2(\Omega)$ with $\xi \in \beta(w)$ a.e. in Ω such that (4.7) holds. Finally, assertion (c) is straightforward. $\diamondsuit$

Remark 4.1. By virtue of "(b)↔(c)" of Lemma 3.1, the assertion (b) of Theorem 4.1 can be stated as follows: $[e^*, w^*] \in \partial\varphi^t([e,w])$ if and only if (4.6) holds, $w \in H^2(\Omega)$ and

$$\begin{cases} \nu w^* = -\kappa\Delta w + \xi - \rho^{-1}(e-w) & \text{a.e. in } \Omega, \\ \xi \in L^2(\Omega), \quad \xi \in \beta(w) & \text{a.e. in } \Omega, \\ \dfrac{\partial w}{\partial n} = 0 \quad \text{a.e. on } \Gamma. \end{cases} \tag{4.10}$$

Remark 4.2. The abstract seeting in the product space $V^* \times L^2(\Omega)$ was earlier found in Visintin [45], where the problem with $\kappa = 0$ was discussed as a model of the Stefan problem with phase relaxation. Subsequently, this idea was extensively used in Damlamian-Kenmochi-Sato [13,14] to obtain Theorem 4.1.

We define an operator $G : X \to X$ by

$$G([e,w]) := [0, \frac{1}{\nu}g(w)] \qquad \text{for all } [e,w] \in X.$$

Clearly G is Lipschitz continuous in X; in fact, with Lipschitz constant $C(g)$ of g, we have

$$\begin{aligned} |G([e_1,w_1]) - G([e_2,w_2])|_X &\le \nu^{-\frac{1}{2}}C(g)|w_1 - w_2|_{L^2(\Omega)} \\ &\le \nu^{-1}C(g)|[e_1,w_1] - [e_2,w_2]|_X, \end{aligned}$$

since $|[e,w]|_X^2 = |e|_{V^*}^2 + \nu|w|_{L^2(\Omega)}^2$.

Now it is easy to reformulate (PFC) as an evolution equation of the form (4.4) in X by using $\partial\varphi^t$ and G.

Corollary 4.1 to Theorem 4.1. *Let $\{u,w\}$ be a weak solution of (PFC) on $[0,T]$. Then $U(t) := [\rho(u(t)) + w(t), w(t)]$, $0 \le t \le T$, satisfies that*

$$U \in C([0,T];X) \cap W^{1,2}_{loc}((0,T];X), \quad \varphi^t(U) \in L^1(0,T), \tag{4.11}$$

and it is a solution of

$$\begin{cases} U'(t) + \partial\varphi^t(U(t)) + G(U(t)) \ni \tilde{f}(t) & \text{in } X \text{ for a.e. } t \in [0,T], \\ U(0) = U_o := [\rho(u_o) + w_o, w_o], \end{cases} \tag{4.12}$$

where $\tilde{f}(t) := [f(t), 0]$ for $t \in [0,T]$. Conversely, if $U(t) := [e(t), w(t)]$ is a solution of (4.12) having the regularity property (4.11), then $\{u,w\}$ with $u = \rho^{-1}(e-w)$ is a weak solution of (PFC) on $[0,T]$.

Proof. The first assertion of the corollary immediately follows from Theorem 4.1 and Definition 4.1. Conversely, let $U(t) := [e(t), w(t)]$ be a solution of (4.12) having property (4.11). Then, by Theorem 4.1 (a),

$$e'(t) - f(t) = F(\rho^{-1}(e(t) - w(t)) - h(t)),$$

and by (4.11), $e' - f \in L^2_{loc}((0,T];V^*)$. Hence $\rho^{-1}(e-w) - h \in L^2_{loc}((0,T];V)$. Since $h \in W^{1,2}(0,T;V)$, this implies $u := \rho^{-1}(e-w) \in L^2_{loc}((0,T];V)$. Any other conditions in (w1) and (w2) of Definition 4.1 are easily checked from (4.11) and (4.12). Next, by Theorem 4.1 (a) again, for a.e. $t \in [0,T]$ there is $\xi(t) \in L^2(\Omega)$ such that $\xi(t) \in \beta(w(t))$ a.e. in Ω and

$$(-\nu w'(t) - g(w(t)), \eta) = \kappa a(w(t), \eta) + (\xi(t) - u(t), \eta) \qquad \text{for all } \eta \in H^1(\Omega).$$

Moreover, by Remark 3.1,

$$|\xi|_{L^2(\Omega)} \leq R(\gamma)(\nu|w'(t)|_{L^2(\Omega)} + |g(w(t))|_{L^2(\Omega)} + |u(t)|_{L^2(\Omega)} + 1) \qquad \text{for a.e. } t \in [0,T],$$

which implies that $\xi \in L^2_{loc}((0,T];L^2(\Omega))$. Thus (w3) holds. ◇

II. Phase Separation Models with Constraints

5. Phase separation in a binary mixture

In this section we propose a model for diffusive phase separations, for instance, in a binary system of alloys with components A, B, in non-isothermal case.

We suppose that the system occupies a bounded domain $\Omega \subset \mathbf{R}^3$ (or $\mathbf{R}^2$); denoting by $|\Omega|$, c_A, c_B the volume of Ω, the total volumes of A and B, respectively, we have

$$|\Omega| = c_A + c_B.$$

To indicate any situation of mixture we define a function

$$\begin{aligned} w_A(t,x) \ &:= \ \text{the local ratio of the component A at } (t,x) \\ &:= \ \lim_{r \downarrow 0} \frac{\text{the total volume of A in } B_r(x) \text{ at } (t,x)}{|B_r(x)|}, \end{aligned}$$

where $x \in \Omega$ and $B_r(x)$ is the ball with center x and radius r. Clearly, $0 \leq w_A(t,x) \leq 1$ on $\mathbf{R}_+ \times \Omega$ and

$$\int_\Omega w_A(t,x)dx = c_A \qquad \text{for all } t \geq 0. \tag{5.1}$$

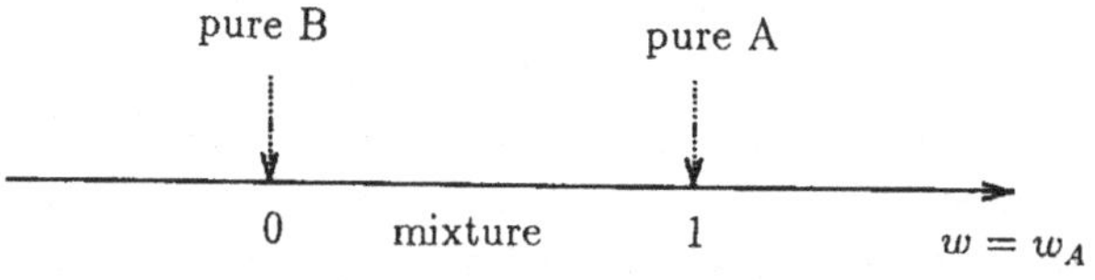

Figure 7

By the parameter w_A, which is called the conserved order parameter because of (5.1), the physical situation at each point $x \in \Omega$ and time t is indicated; for instance, the region $\Omega_A(t) := \{x \in \Omega; w_A(t,x) = 1\}$ is of pure A, the region $\Omega_B(t) := \{x \in \Omega; w_A(t,x) = 0\}$ is of pure B and the region $\Omega_m(t) := \{x \in \Omega; 0 < w_A(t,x) < 1\}$ is of mixture.

The phase separation phenomenon is roughly explained as follows. In many cases, one can obtain the homogeneous mixture of A and B, provided that the temperature is very high. But, after rapid quenting of temperature, the homogeneous mixture is no longer stable, and the same kind of molecules intends to be in an aggregation very slowly in time. The molecules of components A and B exchange the positions each other, and after long time a pattern is formed (see Figures 8, 9).

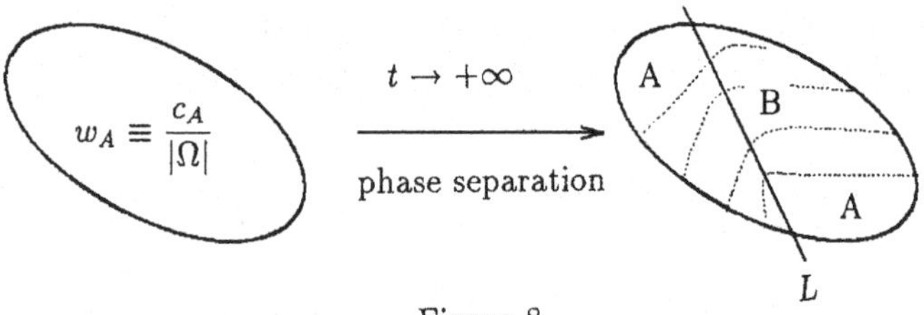

Figure 8

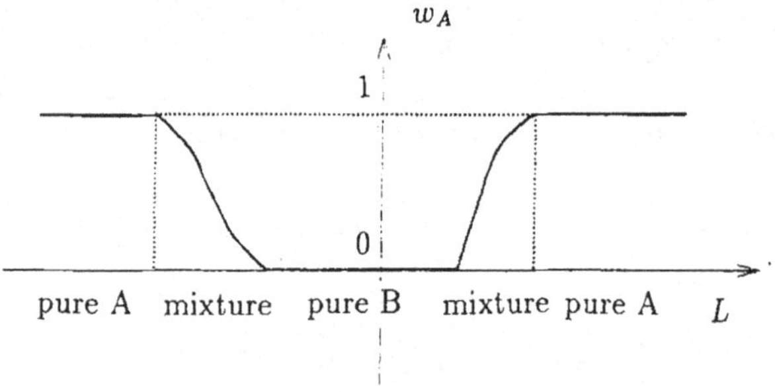

Figure 9

For simplicity we write w for w_A. The dynamics of phase separation is also interpreted, just as in the phase field model, through a free energy functional. Now suppose that the free energy functional $F_\Omega(u,w)$ depends upon the function u, where $\rho(u) = \theta - \theta_*$ for a function $\rho : \mathbf{R} \to \mathbf{R}$ ($\theta = \theta(t,x)$ is the temperature and $\theta_* =$ const. is the critical temperature), and it is of the form

$$F_\Omega(u,w) := \int_\Omega \{\frac{\kappa}{2}|\nabla w|^2 + \hat{\beta}(w) + \hat{g}(w) - uw\}dx, \tag{5.2}$$

where $\hat{\beta}$ is a non-negative proper l.s.c. convex function on $\mathbf{R}$ with $\overline{D(\hat{\beta})} = [0,1]$, $\hat{g}$ is a non-negative smooth function on $\mathbf{R}$ and $\kappa > 0$ is a constant; note that $F_\Omega(u,w) = +\infty$, if $\hat{\beta}(w) \notin L^1(\Omega)$. The form $F_\Omega(u,w)$ is the same as in the case of (PFC), but in the present case it should be defined for any $u \in L^2(\Omega)$ and any w in the set

$$K := \{z \in H^1(\Omega); \int_\Omega z dx = c_A\}$$

because of constraint (5.1). Also we denote by π_o the projection from $L^2(\Omega)$ onto

$$L^2(\Omega)_o := \{z \in L^2(\Omega); \int_\Omega z dx = 0\}.$$

For a fixed $u_\infty := u_\infty(x) \in L^2(\Omega)$, a function $w_\infty \in K$ which gives an equilibrium in our phase separation process is a local minimizer of $F_\Omega(u_\Omega, \cdot)$ on K.

Theorem 5.1. *Let $u_\infty \in L^2(\Omega)$ and $w_\infty \in K$ be a local minimizer of $F_\Omega(u_\infty, \cdot)$ on K, i.e. there is an open ball $U_r(w_\infty) := \{w \in H^1(\Omega); |w - w_\infty|_{H^1(\Omega)} < r\}$ for a number $r > 0$ such that*

$$+\infty > F_\Omega(u_\infty, w_\infty) = \min_{w \in U_r(w_\infty) \cap K} F_\infty(u_\infty, w).$$

Then $w_\infty \in H^2(\Omega)$ and

$$\begin{cases} -\kappa\Delta w_\infty + \pi_o[\xi_\infty + g(w_\infty) - u_\infty] = 0 & a.e. \text{ in } \Omega, \\ \xi_\infty \in L^2(\Omega), \quad \xi_\infty \in \beta(w_\infty) & a.e. \text{ in } \Omega, \\ \dfrac{\partial w_\infty}{\partial n} = 0 \quad a.e. \text{ on } \Gamma, & \\ \int_\Omega w_\infty dx = c_A, & \end{cases} \tag{5.3}$$

where $g = \hat{g}'$ and $\beta = \partial\hat{\beta}$.

This theorem will be proved in the next section.

Now let us consider the non-equilibrium case. Following the thermodynamical approach to phase separations in [1,2,41], we see that the chemical potential difference (in a generalized sense) is given by

$$\frac{\delta}{\delta w} F_\Omega(u, w) := \{-\kappa\Delta w + \pi_o[\xi + g(w) - u]; \xi \in \beta(w) \text{ in } \Omega\} \tag{5.4}$$

and the mass flux (in a generalized sense) is given by

$$\nabla \frac{\delta}{\delta w} F_\Omega(u, w),$$

where $\frac{\partial}{\partial w}$ means a functional derivative with respect to w in a generalized sense. Therefore, its kinetic equation is of the form

$$\nu w_t \in div\{\nabla \frac{\delta}{\delta w} F_\Omega(u, w)\}$$

for a positive constant ν, i.e.

$$\nu w_t - \Delta\{-\kappa\Delta w + \xi + g(w) - u\} = 0, \quad \xi \in \beta(w), \quad \text{in } Q.$$

In the non-isothermal case, the same type of heat equation as in the case of (PFC) is satisfied, i.e.

$$(\rho(u)+w)_t - \Delta u = f \qquad \text{in } Q.$$

Accordingly, as a mathematical description for non-isothermal diffusive phase separations, we propose the following system, which is called "Phase separation model with constraint" and is denoted by (PSC):

$$(\rho(u)+w)_t - \Delta u = f(t,x) \qquad \text{in } Q, \tag{5.5}$$

$$\nu w_t - \Delta\{-\kappa\Delta w + \xi + g(w) - u\} = 0 \qquad \text{in } Q, \tag{5.6}$$

$$\xi \in \beta(w) \qquad \text{in } Q, \tag{5.7}$$

$$\frac{\partial u}{\partial n} + \alpha u = h_o(t,x) \qquad \text{on } \Sigma, \tag{5.8}$$

$$\frac{\partial w}{\partial n} = 0, \quad \frac{\partial}{\partial n}\{-\kappa\Delta w + \xi + g(w) - u\} = 0 \qquad \text{on } \Sigma, \tag{5.9}$$

$$u(0,\cdot) = u_o, \quad w(0,\cdot) = w_o \qquad \text{in } \Omega. \tag{5.10}$$

Remark 5.1. (i) It should be noticed for the order parameter w that the equation corresponding to any equilibrium is a second order PDE, but the evolution equation for the non-equilibrium case is a fourth order PDE.

(ii) The boundary condition

$$\frac{\partial}{\partial n}\{-\kappa\Delta w + \xi + g(w) - u\} = 0 \quad \text{on } \Sigma$$

corresponds to the constraint (5.1); in fact, under this boundary condition we have by integrating (5.6) over Ω

$$\frac{d}{dt}\int_\Omega w(t,x)dx = 0, \quad \text{i.e.} \int_\Omega w(t,x)dx = \int_\Omega w_o(x)dx \ (= c_A) \quad \text{for all } t \geq 0.$$

In the sequel, (PSC) is discussed for ρ, β, g, f, h_o satisfying conditions (ρ), (β), (g), (f), (h_o), respectively, mentioned in section 2.

6. A minimization problem

In this section, consider a minimization problem for the functional J_o defined on $L^2(\Omega)$ by

$$J_o(z) := \begin{cases} \frac{\kappa}{2}|\nabla z|^2_{L^2(\Omega)} + \int_\Omega \hat{\gamma}(x,z)dx + \int_\Omega \hat{\beta}(z)dx - (\ell, z) \\ \qquad\qquad \text{if } z \in H^1(\Omega), \int_\Omega z dx = c \text{ and } \hat{\beta}(z) \in L^1(\Omega), \\ +\infty \qquad \text{otherwise,} \end{cases} \tag{6.1}$$

where $\kappa, \hat{\gamma}, \hat{\beta}$ and ℓ are the same as in section 3, and c is a constant with

$$\sigma_* < \frac{c}{|\Omega|} < \sigma^*. \tag{6.2}$$

We note that (6.2) implies

$$\frac{c}{|\Omega|} \in int.D(\beta),$$

where $\beta = \partial\hat{\beta}$ in $\mathbf{R}$. For simplicity we put

$$K_c := \{z \in L^2(\Omega); \int_\Omega z dx = c\}.$$

The functional J_o is proper, l.s.c. and coercive on $L^2(\Omega)$ and is strictly convex on $H^1(\Omega) \cap K_c \cap \{z \in L^2(\Omega); \hat{\beta}(z) \in L^1(\Omega)\}$ $(\neq \emptyset)$, so that there is one and only one minimizer v_o of J_o.

Proposition 6.1. *Assume (6.2) holds. Then the following three statements (a), (b), (c) are equivalent to each other:*

(a) *v_o is a (unique) minimizer of J_o on $L^2(\Omega)$, i.e,*

$$J_o(v_o) = \min_{z \in L^2(\Omega)} J_o(z).$$

(b) *$v_o \in H^1(\Omega) \cap K_c, \hat{\beta}(v_o) \in L^1(\Omega)$ and*

$$\kappa a(v_o, v_o - z) + (\gamma(\cdot, v_o), v_o - z) + \int_\Omega \hat{\beta}(v_o) dx$$
$$\leq (\ell, v_o - z) + \int_\Omega \hat{\beta}(z) dx \qquad \text{for all } z \in H^1(\Omega) \cap K_c.$$

(c) *$v_o \in H^2(\Omega)$ and*

$$\begin{cases} -\kappa \Delta v_o + \pi_o[\xi + \gamma(\cdot, v_o) - \ell] = 0 & a.e.\ in\ \Omega, \\ \xi \in L^2(\Omega), \quad \xi \in \beta(v_o) & a.e.\ in\ \Omega, \\ \dfrac{\partial v_o}{\partial n} = 0 & a.e.\ on\ \Gamma, \\ \int_\Omega v_o dx = c. \end{cases} \tag{6.3}$$

The equivalence "(a)↔(b)" and implication "(c)→(b)" in Proposition 6.1 are easily obtained. In order to complete its proof it is enough to show the existence of a solution $v_o \in H^2(\Omega)$ of (6.3).

Lemma 6.1. *Let $q \in L^2(\Omega)$. Then there exists one and only one solution $v \in H^2(\Omega)$ of*

$$\begin{cases} -\kappa \Delta v = \pi_o(q) & a.e.\ in\ \Omega, \\ \dfrac{\partial v}{\partial n} = 0 & a.e.\ on\ \Gamma, \\ \int_\Omega v dx = c. \end{cases} \tag{6.4}$$

Moreover, v is given as a unique solution of

$$\begin{cases} v \in H^1(\Omega) \cap K_c, \\ \kappa a(v, v - z) \leq (q, v - z) \qquad \text{for all } z \in H^1(\Omega) \cap K_c, \end{cases} \tag{6.5}$$

or equivalently

$$\begin{cases} v \in H^1(\Omega) \cap K_c, \\ \kappa a(v, \eta) = (q, \eta) \qquad \text{for all } z \in H^1(\Omega) \text{ with } \int_\Omega \eta dx = 0. \end{cases}$$

Proof. We define a bilinear form $b(\cdot,\cdot)$ on $H^1(\Omega) \times H^1(\Omega)$ by

$$b(z_1, z_2) := \kappa a(z_1, z_2) + (\int_\Omega z_1 dx)(\int_\Omega z_2 dx).$$

Clearly it holds for some positive constants C_b, C_b' that

$$b(z, z) \geq C_b |z|^2_{H^1(\Omega)} \qquad \text{for all } z \in H^1(\Omega)$$

and

$$|b(z_1, z_2)| \leq C_b' |z_1|_{H^1(\Omega)} |z_2|_{H^1(\Omega)} \qquad \text{for all } z_1, z_2 \in H^1(\Omega).$$

Therefore, for the element of $H^1(\Omega)^*$ corresponding to the functional $z \to (\pi_o(q), z) + c\int_\Omega z dx$ there exists $v \in H^1(\Omega)$ such that

$$(\pi_o(q), z) + c\int_\Omega z dx = b(v, z) := \kappa a(v, z) + (\int_\Omega v dx)(\int_\Omega z dx) \tag{6.6}$$

$$\text{for all } z \in H^1(\Omega).$$

Now, take $z \equiv 1$ in (6.6). Then, since $\int_\Omega \pi_o(q) dx = 0$, it follows that

$$\int_\Omega v dx = c.$$

Hence (6.6) yields that

$$(\pi_o(q), z) = \kappa a(v, z) \qquad \text{for all } z \in H^1(\Omega),$$

which implies that $v \in H^2(\Omega)$, $-\kappa \Delta v = \pi_o(q)$ a.e. in Ω and $\frac{\partial v}{\partial n} = 0$ a.e. on Γ. Thus v is a required solution of (6.4) and v clearly satisfies (6.5). Since the solution of (6.5) is unique, the solution of (6.4) is characterized by that of (6.5). ◇

Now, for each $0 < \lambda \leq 1$, let us consider the minimization problem of determining $v_\lambda \in H^1(\Omega) \cap K_c$ such that

$$J_{o\lambda}(v_\lambda) = \min_{z \in L^2(\Omega)} J_{o\lambda}(z), \tag{6.7}$$

where

$$J_{o\lambda}(z) := \begin{cases} \frac{\kappa}{2}|\nabla z|^2_{L^2(\Omega)} + \int_\Omega \hat{\gamma}(x,z)dx + \int_\Omega \hat{\beta}_\lambda(z)dx - (\ell, z) \\ \qquad \text{for } z \in H^1(\Omega) \cap K_c, \\ +\infty \qquad \text{otherwise} \end{cases}$$

and $\hat{\beta}_\lambda$ is the Yosida-approximation of $\hat{\beta}$, i.e.

$$\hat{\beta}_\lambda(r) := \inf_{r' \in \mathbf{R}} \{\frac{1}{2\lambda}|r - r'|^2 + \hat{\beta}(r')\}, \qquad r \in \mathbf{R}.$$

Then, since $J_{o\lambda}$ is proper, l.s.c. and convex on $L^2(\Omega)$ and strictly convex on $H^1(\Omega) \cap K_c$, the problem (6.7) has one and only one solution $v_\lambda \in H^1(\Omega) \cap K_c$. Clearly, v_λ satisfies the variational inequality

$$\kappa a(v_\lambda, v_\lambda - z) + (\gamma(\cdot, v_\lambda) + \beta_\lambda(v_\lambda), v_\lambda - z) \le (\ell, v_\lambda - z) \tag{6.8}$$

$$\text{for all } z \in H^1(\Omega) \cap K_c.$$

This implies by Lemma 6.1 that $v_\lambda \in H^2(\Omega)$ and

$$\begin{cases} -\kappa\Delta v_\lambda + \pi_o[\gamma(\cdot, v_\lambda) + \beta_\lambda(v_\lambda) - \ell] = 0 \qquad \text{a.e. in } \Omega, \\ \frac{\partial v_\lambda}{\partial n} = 0 \qquad \text{a.e. on } \Gamma, \\ \int_\Omega v_\lambda dx = c. \end{cases} \tag{6.9}$$

Next we give some uniform estimates for $\{v_\lambda\}$.

Lemma 6.2. *There is a positive constant M_o such that*

$$|v_\lambda|_{H^1(\Omega)} + \kappa|\Delta v_\lambda|_{L^2(\Omega)} + |\pi_o[\beta_\lambda(v_\lambda)]|_{L^2(\Omega)} \le M_o \qquad \textit{for all } 0 < \lambda \le 1. \tag{6.10}$$

Proof. Since $\frac{c}{|\Omega|} \in H^1(\Omega) \cap K_c$ and $0 \le \hat{\beta}_\lambda(\frac{c}{|\Omega|}) \le \hat{\beta}(\frac{c}{|\Omega|}) < +\infty$, we see that

$$\begin{aligned} J_{o\lambda}(\frac{c}{|\Omega|}) &= \int_\Omega \hat{\gamma}(x, \frac{c}{|\Omega|})dx + \int_\Omega \hat{\beta}_\lambda(\frac{c}{|\Omega|})dx - \frac{c}{|\Omega|}\int_\Omega \ell dx \\ &\ge J_{o\lambda}(v_\lambda) \end{aligned}$$

and that $\{J_{o\lambda}(v_\lambda)\}$ is bounded. Hence $\{v_\lambda\}$ is bounded in $H^1(\Omega)$, so that there is a constant $M_1 > 0$ such that

$$|v_\lambda|_{H^1(\Omega)} + |\gamma(\cdot, v_\lambda)|_{L^2(\Omega)} \le M_1 \qquad \text{for all } 0 < \lambda \le 1.$$

Next, multiplying the first equation of (6.9) by $-\Delta v_\lambda$ and using the inequality

$$(\pi_o[\beta_\lambda(v_\lambda)], -\Delta v_\lambda) = (\beta_\lambda(v_\lambda), -\Delta v_\lambda) = \int_\Omega \beta'_\lambda(v_\lambda)|\nabla v_\lambda|^2 dx \ge 0,$$

we have

$$\kappa|\Delta v_\lambda|_{L^2(\Omega)} \le |\gamma(\cdot, v_\lambda)|_{L^2(\Omega)} + |\ell|_{L^2(\Omega)} \le M_1 + |\ell|_{L^2(\Omega)}$$

and

$$\begin{aligned} |\pi_o[\beta_\lambda(v_\lambda)]|_{L^2(\Omega)} &\le \kappa|\Delta v_\lambda|_{L^2(\Omega)} + |\gamma(\cdot, v_\lambda)|_{L^2(\Omega)} + |\ell|_{L^2(\Omega)} \\ &\le 2(M_1 + |\ell|_{L^2(\Omega)}). \end{aligned}$$

Thus (6.10) holds for $M_o = 3M_1 + 2|\ell|_{L^2(\Omega)}$. ◇

Lemma 6.3. *$\{\beta_\lambda(v_\lambda)\}$ is bounded in $L^2(\Omega)$.*

Proof. By Lemma 6.2 it is enough to show that $\{\int_\Omega \beta_\lambda(v_\lambda)dx\}$ is bounded in $\mathbf{R}$. Fix r_o with $\beta(r_o) \ni 0$, so $\beta_\lambda(r_o) = 0$. Next, choose a small number $\delta > 0$ so that

$$\sigma_\star < \frac{c}{|\Omega|} - \delta < \frac{c}{|\Omega|} + \delta < \sigma^\star;$$

this is possible by assumption (6.2). Put

$$m_1 := \max\{\frac{c}{|\Omega|} + \delta, r_o\}, \qquad m_2 := \min\{\frac{c}{|\Omega|} - \delta, r_o\};$$

note that $m_1, m_2 \in D(\beta)$. Let us multiply the first equation of (6.9) by $v_\lambda - \frac{c}{|\Omega|}$ to get

$$\kappa|\nabla v_\lambda|^2_{L^2(\Omega)} + (\beta_\lambda(v_\lambda), v_\lambda - \frac{c}{|\Omega|}) = (\ell - \gamma(\cdot, v_\lambda), v_\lambda - \frac{c}{|\Omega|}).$$

This implies by Lemma 6.2 that

$$(\beta_\lambda(v_\lambda), v_\lambda - \frac{c}{|\Omega|}) \le M_2 \qquad \text{for all } 0 < \lambda \le 1, \tag{6.11}$$

where M_2 is a positive constant. Also we have

$$\begin{aligned} (\beta_\lambda(v_\lambda), v_\lambda - \frac{c}{|\Omega|}) &= (\int_{\{v_\lambda < m_1\}} + \int_{\{v_\lambda > m_2\}} + \int_{\{m_2 \le v_\lambda \le m_1\}})\{\beta_\lambda(v_\lambda)(v_\lambda - \frac{c}{|\Omega|})\}dx \\ &=: I_1(\lambda) + I_2(\lambda) + I_3(\lambda). \end{aligned}$$

Observe here that

$$I_1(\lambda) \ge \delta \int_{\{v_\lambda > m_1\}} |\beta_\lambda(v_\lambda)|dx$$

and

$$I_2(\lambda) \ge \delta \int_{\{v_\lambda < m_2\}} |\beta_\lambda(v_\lambda)|dx,$$

since

$$\beta_\lambda(v_\lambda) \ge 0, \quad v_\lambda - \frac{c}{|\Omega|} \ge \delta \qquad \text{on } \{v_\lambda > m_1\}$$

and

$$\beta_\lambda(v_\lambda) \le 0, \quad v_\lambda - \frac{c}{|\Omega|} \le -\delta \qquad \text{on } \{v_\lambda < m_2\}.$$

Further, by the minimun-norm property of the Yosida-approximation,

$$|\beta_\lambda(r)| \le \min\{|r'|; r' \in \beta(r)\}, \qquad r \in D(\beta).$$

This implies that

$$|\beta_\lambda(v_\lambda)| \le M_3 \qquad \text{a.e. on } \{m_2 \le v_\lambda \le m_1\},$$

where

$$M_3 := \min\{|r'|; r' \in \beta(m_1)\} + \min\{|r'|; r' \in \beta(m_2)\} \ (< +\infty).$$

Hence

$$\begin{aligned} I_3(\lambda) &\ge -M_3(|m_1 - \frac{c}{|\Omega|}| + |m_2 - \frac{c}{|\Omega|}|)|\Omega| =: M_4 \\ &\ge \delta \int_{\{m_2 \le v_\lambda \le m_1\}} |\beta_\lambda(v_\lambda)|dx - \delta M_3|\Omega| + M_4. \end{aligned}$$

By combining the estimates $I_k(\lambda), k = 1,2,3$, with (6.11), we conclude that

$$M_2 + \delta M_3|\Omega| - M_4 \ge \delta \int_\Omega |\beta_\lambda(v_\lambda)|dx.$$

This completes the proof of the lemma. ◇

By Lemmas 6.2 and 6.3, there is a sequence $\{\lambda_n\}$ with $\lambda_n \to 0$ (as $n \to +\infty$) and functions $v_o \in H^2(\Omega), \xi \in L^2(\Omega)$ such that

$$v_n := v_{\lambda_n} \to v_o \qquad \text{weakly in } H^2(\Omega),$$

$$(\text{hence } v_n \to v_o \qquad \text{in } H^1(\Omega))$$

and

$$\beta_{\lambda_n} \to \xi \qquad \text{weakly in } L^2(\Omega).$$

Clearly $\xi \in \beta(v_o)$ a.e. in Ω. Passing to the limit as $\lambda = \lambda_n \to 0$ in (6.9), we see that v_o and ξ satisfy (6.3). Thus we obtain Proposition 6.1.

Remark 6.1. As is easily checked from the proofs of Lemmas 6.2 and 6.3 the functions v_o and ξ in (c) of Proposition 6.1 satisfy that

$$\kappa|v_o|_{H^2(\Omega)} + |\xi|_{L^2(\Omega)} \le R_1(\beta,\gamma,c)(|\ell|_{L^2(\Omega)} + 1),$$

where $R_1(\beta,\gamma,c)$ is a positive constant dependent only on β, γ and the constant c.

Proof of Theorem 5.1: Let w_∞ be a local minimizer of $F_\Omega(u_\infty,\cdot)$ on K. Then w_∞ satisfies that $w_\infty \in H^1(\Omega) \cap K, \hat\beta(w_\infty) \in L^1(\Omega)$ and

$$\kappa a(w_\infty, w_\infty - z) + (g(w_\infty) - u_\infty, w_\infty - z) + \int_\Omega \hat\beta(w_\infty)dx \le \int_\Omega \hat\beta(z)dx$$

$$\text{for all } z \in H^1(\Omega) \cap K.$$

Applying Proposition 6.1 with

$$c = c_A, \quad \hat{\gamma}(x,r) := \frac{1}{2}|r|^2, \quad \ell := u_\infty - g(w_\infty) + w_\infty,$$

we see that $w_\infty \in H^2(\Omega)$ and (5.3) holds. $\diamondsuit$

7. Evolution operators associated with (PSC)

Let V, $F : V \to V^*$ and $\langle \cdot,\cdot \rangle : V^* \times V \to \mathbf{R}$ be as in section 4. Further let

$$V_o := \{z \in H^1(\Omega); \int_\Omega z dx = 0\}$$

be the Hilbert space with norm

$$|z|_{V_o} := |\nabla z|_{L^2(\Omega)};$$

denote by $\langle \cdot,\cdot \rangle_o$ the duality pairing between V_o^* and V_o, and by F_o the duality mapping from V_o onto V_o^*. Note that

$$V_o \subset L^2(\Omega)_o \subset V_o^*$$

with compact injections; we use still the same nototaions $L^2(\Omega)_o$ and π_o for the space $\{z \in L^2(\Omega); \int_\Omega z dx = 0\}$ and the projection from $L^2(\Omega)$ onto $L^2(\Omega)_o$, respectively.

For the data f, h_o, u_o, w_o we suppose that conditions (f), (h_o) hold, and

$$u_o \in L^2(\Omega),\ w_o \in L^\infty(\Omega) \text{ with } \sigma_* \le w_o \le \sigma^* \text{ a.e. in } \Omega, \quad \int_\Omega w_o dx = c, \tag{7.1}$$

where c is a constant with

$$\sigma_* < \frac{c}{|\Omega|} < \sigma^*, \qquad \text{i.e. } \frac{c}{|\Omega|} \in int.D(\beta). \tag{7.2}$$

Associated to the boundary function h_o, the function $h : \mathbf{R}_+ \to H^1(\Omega)$ is defined by (4.2) as well.

Definition 7.1. (Weak formulation for (PSC)). Let $0 < T < +\infty$. Then a couple of functions u and w is called a weak solution of (PSC) on $[0,T]$, if the following conditions (w1) - (w3) are fulfilled:

(w1) $\rho(u) \in C([0,T]; V^*) \cap W^{1,2}_{loc}((0,T]; V^*) \cap L^2(0,T; L^2(\Omega)), u \in L^2_{loc}((0,T]; V)$,
$w - \frac{c}{|\Omega|} \in C([0,T]; L^2(\Omega)_o) \cap L^2(0,T; V_o), w \in L^2_{loc}((0,T]; H^2(\Omega))$,
$w' \in L^2_{loc}((0,T]; V_o^*), \hat{\beta}(w) \in L^1(0,T; L^1(\Omega))$,

(w2) $\rho(u)(0) = \rho(u_o)$ and

$$\langle \rho(u)'(t) + w'(t), z\rangle + a(u(t) - h(t), z) + \alpha(u(t) - h(t), z)_\Gamma = (f(t), z)$$
$$\text{for all } z \in V \text{ and a.e. } t \in [0,T], \tag{7.3}$$

(w3) $w(0) = w_o$, $\dfrac{\partial w(t)}{\partial n} = 0$ a.e. on Γ for a.e. $t \in [0,T]$,

and there is $\xi \in L^2_{loc}((0,T];L^2(\Omega))$ such that

$$\xi(t) \in \beta(w(t)) \quad \text{a.e. in } \Omega \text{ for a.e. } t \in [0,T],$$

$$\nu\langle w'(t),\eta\rangle_o + \kappa(\Delta w(t),\Delta\eta) - (\xi(t) + g(w(t)) - u(t),\Delta\eta) = 0 \tag{7.4}$$

for all $\eta \in H^2(\Omega)\cap L^2(\Omega)_o$ with $\dfrac{\partial \eta}{\partial n} = 0$ a.e. on Γ and for a.e. $t \in [0,T]$.

Remark 7.1. (1) In the above definition, since $w' \in L^2_{loc}((0,T];V_o^*)$ and $\int_\Omega w(t,x)dx = c$ for all $t \in [0,T]$, w' can be considered as a function in $L^2_{loc}((0,T];V^*)$ by $\langle w'(t),z\rangle = \langle w'(t),\pi_o(z)\rangle_o$ for all $z \in V$.

(2) With the inverse F_o^{-1} of F_o the variational identity (7.4) can be written in the form

$$\nu(F_o^{-1}w'(t),z) + \kappa a(w(t),z) + (\xi(t) + g(w(t)) - u(t),z) = 0 \tag{7.5}$$

$$\text{for all } z \in V_o \text{ and a.e. } t \in [0,T].$$

In fact, for any $z \in V_o$ we put $\eta = F_o^{-1}z$, i.e. $F_o\eta = z$. By definition, $\langle z,\tilde\eta\rangle_o(=(z,\tilde\eta)) = a(\eta,\tilde\eta)$ for all $\tilde\eta \in V_o$. Now, it follows from Lemma 6.1 that $\eta \in H^2(\Omega), F_o\eta = -\Delta\eta$ and $\frac{\partial\eta}{\partial n} = 0$ a.e. on Γ, which shows that η is one of test functions for (7.4). Thus we have

$$\nu\langle w'(t),F_o^{-1}z\rangle_o - \kappa(\Delta w(t),z) + (\xi(t) + g(w(t)) - u(t),z) = 0,$$

so that (7.5) is obtained, since $\langle w'(t),F_o^{-1}z\rangle_o = \langle z,F_o^{-1}w'(t)\rangle_o = (F_o^{-1}w'(t),z)$ and $-(\Delta w(t),z) = a(w(t),z)$. Further, by Lemma 6.1 again, (7.5) is written in the form

$$\nu F_o^{-1}w'(t) + \kappa F_o[\pi_o(w(t))] + \pi_o[\xi(t) + g(w(t)) - u(t)] = 0 \tag{7.6}$$

$$\text{in } V_o^* \text{ for a.e. } t \in [0,T];$$

note that expression (7.6) (or (7.5)) includes the boundary condition "$\frac{\partial w(t)}{\partial n} = 0$ a.e. on Γ for a.e. $t \in [0,T]$", since $\kappa F_o[\pi_o(w(t))] = -\nu F_o^{-1}w'(t) - \pi_o[\xi(t) + g(w(t)) - u(t)] \in L^2(\Omega)_o$ for a.e. $t \in [0,T]$.

For an abstract setting of (PSC), we consider a Hilbert space

$$X_o := V^* \times L^2(\Omega)_o$$

with inner product $(\cdot,\cdot)_{X_o}$ given by

$$([e_1,v_1],[e_2,v_2])_{X_o} := \langle e_1,F^{-1}e_2\rangle + \nu(v_1,v_2) \quad \text{for any } [e_i,v_i] \in X_o, i = 1,2;$$

the corresponding norm is given by

$$|[e,v]|_{X_o} := \{|e|^2_{V^*} + \nu|v|^2_{L^2(\Omega)}\}^{\frac{1}{2}} \quad \text{for } [e,v] \in X_o.$$

Now, for each $t \geq 0$ we define a function $\varphi_o^t(\cdot)$ on X_o by

$$\varphi_o^t([e,v]) := \begin{cases} \int_\Omega \hat{\rho^{-1}}(e - v - \frac{c}{|\Omega|})dx + \frac{\kappa}{2}|\nabla v|^2_{L^2(\Omega)} + \int_\Omega \hat{\beta}(v + \frac{c}{|\Omega|})dx - (h(t), e) \\ \qquad \text{if } [e,v] \in L^2(\Omega) \times V_o \text{ and } \hat{\beta}(v + \frac{c}{|\Omega|}) \in L^1(\Omega), \\ +\infty \qquad \text{otherwise.} \end{cases} \tag{7.7}$$

Theorem 7.1. *(a) For each $t \geq 0$, φ_o^t is proper, l.s.c. and convex on X_o.*

(b) The subdifferential $\partial\varphi_o^t$ of φ_o^t in X_o is characterized as follows: $[e^*, v^*] \in \partial\varphi_o^t([e,v])$ if and only if

$$e^* = F(\rho^{-1}(e - v - \frac{c}{|\Omega|}) - h(t)) \tag{7.8}$$

and there is $\xi \in L^2(\Omega)$ with $\xi \in \beta(v + \frac{c}{|\Omega|})$ a.e. in Ω such that

$$\nu v^* = \kappa F_o v + \pi_o[\xi - \rho^{-1}(e - v - \frac{c}{|\Omega|})] \qquad \text{in } V_o^* \tag{7.9}$$

i.e.

$$\nu \langle v^*, \eta \rangle_o = \kappa a(v, \eta) + (\xi - \rho^{-1}(e - v - \frac{c}{|\Omega|}), \eta) \qquad \text{for all } \eta \in V_o.$$

(c) If $[e_i^*, v_i^*] \in \partial\varphi_o^t([e_i, v_i]), i = 1, 2,$ then

$$\begin{aligned} &([e_1^*, v_1^*] - [e_2^*, v_2^*], [e_1, v_1] - [e_2, v_2])_{X_o} \\ = \; &(\rho^{-1}(e_1 - v_1 - \frac{c}{|\Omega|}) - \rho^{-1}(e_2 - v_2 - \frac{c}{|\Omega|}), (e_1 - v_1) - (e_2 - v_2)) \\ &+ \kappa|\nabla(v_1 - v_2)|^2_{L^2(\Omega)} + (\xi_1 - \xi_2, v_1 - v_2) \\ \geq \; &\frac{1}{C(\rho)}|(e_1 - v_1) - (e_2 - v_2)|^2_{L^2(\Omega)} + \kappa|\nabla(v_1 - v_2)|^2_{L^2(\Omega)}, \end{aligned}$$

where $\xi_i, i = 1, 2,$ are the function ξ as in (7.9).

We omit the proof of Theorem 7.1, since it is quite similar to that of Theorem 4.1.

Next, let A_o be an operator in X_o defined by

$$A_o([e,v]) := [e, \kappa F_o^{-1} v], \qquad [e,v] \in X_o,$$

and G_o defined by

$$G_o([e,v]) := [0, \frac{1}{\nu}\pi_o[g(v + \frac{c}{|\Omega|})]], \qquad [e,v] \in X_o.$$

Then it is easy to see that A_o is linear, continuous, selfadjoint in X_o and

$$(A_o([e,v]), [e,v])_{X_o} = |e|^2_{V^*} + \kappa\nu|v|^2_{V_o^*} \qquad \text{for all } [e,v] \in X_o. \tag{7.10}$$

Hence A_o is positive in X_o. Further G_o is clearly Lipschitz continuous in X_o.

Corollary 7.1 to Theorem 7.1. *Let $\{u,w\}$ be a weak solution of (PSC) on $[0,T]$. Then the function $U(t) := [\rho(u(t)) + w(t), w(t) - \frac{c}{|\Omega|}]$ satisfies that*

$$U \in C([0,T]; X_o), \quad U' \in L^2_{loc}((0,T]; V^* \times V_o^*), \quad \varphi_o^t(U) \in L^1(0,T), \tag{7.11}$$

and

$$\begin{cases} \frac{d}{dt}A_oU(t) + \partial\varphi_o^t(U(t)) + G_o(U(t)) \ni \tilde{f}(t) & \text{in } X_o \text{ for a.e. } t \in [0,T], \\ U(0) = U_o := [\rho(u_o) + w_o, w_o - \frac{c}{|\Omega|}], \end{cases} \tag{7.12}$$

where $\tilde{f}(t) := [f(t), 0]$ for a.e. $t \in [0,T]$. Conversely, if $U(t) := [e(t), v(t)]$ satisfies (7.11) and (7.12), then the couple $\{u,w\}$ with $u = \rho^{-1}(e - v - \frac{c}{|\Omega|})$ and $w = v + \frac{c}{|\Omega|}$ is a weak solution of (PSC) on $[0,T]$.

Proof. The first assertion is easily obtained from Definition 7.1 with the help of Theorem 7.1. Conversely, let $U(t) := [e(t), v(t)]$ be a solution of (7.12) satisfying (7.11). Then all the regularity properties of $u := \rho^{-1}(e - v - \frac{c}{|\Omega|})$ and $w := v + \frac{c}{|\Omega|}$, except $u \in L^2_{loc}((0,T]; V)$ and $w \in L^2_{loc}((0,T]; H^2(\Omega))$, in (w1) of Definition 7.1 are immediately follow from (7.11). Further $u \in L^2_{loc}((0,T]; V)$ can be obtained in the same way as that in the proof of Corollary 4.1. Also, applying Theorem 7.1, we see from (7.12) that for a.e. $t \in [0,T]$ there is $\xi(t) \in L^2(\Omega)$ with $\xi(t) \in \beta(w(t))$ a.e. in Ω and

$$(-\nu F_o^{-1} w'(t) - \pi_o[g(w(t))], \eta)_o = \kappa a(w(t), \eta) + (\xi(t) - u(t), \eta) \qquad \text{for all } \eta \in V_o. \tag{7.13}$$

We note that (7.13) is written in the form (7.4) (cf. Remark 7.1) and $\xi \in L^2_{loc}((0,T]; L^2(\Omega))$ and $w \in L^2_{loc}((0,T]; H^2(\Omega))$ (cf. Remark 6.1). Thus $\{u,w\}$ is a weak solution of (PSC) on $[0,T]$ in the sense of Definition 7.1. $\diamondsuit$

III. Solvability of Evolution Equations Associated with (PFC) and (PSC)

8. Abstract evolution equations in Hilbert spaces

Throughout this section, let H be a (real) Hilbert space with inner product $(\cdot,\cdot)_H$ and norm $|\cdot|_H$, and $\psi^t(\cdot)$ be a proper l.s.c. convex function on H for each $t \in \mathbf{R}_+$. Now, let us consider the abstract Cauchy problem

$$CP(f, v_o) \qquad \begin{cases} (Av)'(t) + \partial\psi^t(v(t)) + p(v(t)) \ni f(t) & \text{in } H, \quad t > 0, \\ v(0) = v_o, \end{cases}$$

where A is a linear operator in H, p is a nonlinear operator in H, $f \in L^2_{loc}(\mathbf{R}_+; H)$ and $v_o \in \overline{D(\psi^0)}$. This problem is discussed for a family $\{\psi^t\}$ in $\Psi_H(a; K_o)$, specified below by a function $a \in W^{1,1}_{loc}(\mathbf{R}_+)$ and a constant $K_o > 0$.

We denote by $\Psi_H(a; K_o)$ the class of all families $\{\psi^t\}_{t\geq 0}$ of proper l.s.c. convex functions on H which satisfy the following conditions (Ψ1) - (Ψ3):

(Ψ1) $\psi^t(z) \ge K_o|z|_H^2$ for all $z \in H$ and $t \ge 0$.

(Ψ2) $D(\psi^t) = D(\psi^0)$ for all $t > 0$, and

$$|\psi^t(z) - \psi^s(z)| \le |a(t) - a(s)|(1 + \psi^s(z))$$

for all $s, t \ge 0$ and $z \in D(\psi^0)$.

(Ψ3) For each $r \ge 0$, the set $\cup_{t\ge 0}\{z \in H; \psi^t(z) \le r\}$ is relatively compact in H.

Further we suppose that p is a Lipschitz continuous operator in H, the range $R(p)$ of p is bounded in H and there is a non-negative potential $P : H \to \mathbf{R}$ such that $\nabla P = p$; in this case, if $v \in W^{1,2}(0,T;H)$, then $P(v) \in W^{1,1}(0,T)$ and

$$\frac{d}{dt}P(v(t)) = (\nabla P(v(t)), v'(t))_H \qquad \text{for a.e. } t \in [0,T].$$

Also, we suppose that A is a linear, continuous, positive (i.e. $(Az,z)_H > 0$ if $z \ne 0$) and selfadjoint operator in H; in this case, the fractional power $\frac{1}{2}$ of A, denoted by $A^{\frac{1}{2}}$, is defined as a linear, continuous, positive and selfadjoint operator in H again, and A is the subdifferential of the continuous convex function

$$j_A(z) := \frac{1}{2}|A^{\frac{1}{2}}z|_H^2, \qquad z \in H.$$

For uniqueness of a solution to $CP(f, v_o)$ we require the following condition ($\star$):
($\star$) For each $\varepsilon > 0$ there is a number $C(\varepsilon) > 0$ such that

$$|z_1 - z_2|_H^2 \le \varepsilon(z_1^\star - z_2^\star, z_1 - z_2)_H + C(\varepsilon)|A^{\frac{1}{2}}(z_1 - z_2)|_H^2$$

for all $z_i \in D(\psi^t)$, $z_i^\star \in \psi^t(z_i), i = 1, 2$, and $t \ge 0$.

Definition 8.1. Let $0 < T < +\infty$, $f \in L^2(0,T;H)$ and $v_o \in \overline{D(\psi^0)}$. Then a function $v : [0,T] \to H$ is a solution of $CP(f, v_o)$ on $[0,T]$, if $A^{\frac{1}{2}}v \in C([0,T];H) \cap W_{loc}^{1,2}((0,T];H)$, $\psi^t(v) \in L^1(0,T)$, $(A^{\frac{1}{2}}v)(0) = A^{\frac{1}{2}}v_o$ and

$$f(t) - p(v(t)) - (Av)'(t) \in \partial\psi^t(v(t)) \qquad \text{for a.e. } t \in [0,T].$$

Remark 8.1. In Definition 8.1, note that $(Av)'(t) = A^{\frac{1}{2}}[(A^{\frac{1}{2}}v)'(t)] \in H$ for a.e. $t \in [0,T]$, since $(A^{\frac{1}{2}}v)'(t) \in H$ for a.e. $t \in [0,T]$.

Now the solvability of $CP(f, v_o)$ is mentioned in the following theorem.

Theorem 8.1. *Assume that $\{\psi^t\} \in \Psi_H(a, K_o)$ and p is as above. Let $0 < T < +\infty$, $f \in W^{1,2}(0,T;H)$ and $v_o \in \overline{D(\psi^0)}$. Then $CP(f, v_o)$ admits one and only one solution v on $[0,T]$ such that*

$$t^{\frac{1}{2}}(A^{\frac{1}{2}}v)' \in L^2(0,T;H), \quad t\psi^t(v) \in L^\infty(0,T).$$

In particular, if $v_o \in D(\psi^0)$*, then*

$$A^{\frac{1}{2}}v \in W^{1,2}(0,T;H), \qquad \psi^t(v) \in L^\infty(0,T).$$

Remark 8.2. In the case when $A = I$, for any $f \in L^2(0,T;H)$ we have the same results as in Theorem 8.1.

Approximate Problems for $CP(f, v_o)$

In order to construct a solution of $CP(f, v_o)$, let us consider the approximate problems with parameter $\lambda > 0$:

$$CP_\lambda \qquad \begin{cases} [Av_\lambda + \lambda v_\lambda]'(t) + \partial\psi_\lambda^t(v_\lambda(t)) + p(v_\lambda(t)) = f(t), \quad 0 < t < T, \\ v_\lambda(0) = v_o, \end{cases}$$

where ψ_λ^t is the Yosida-approximation of ψ^t, i.e.

$$\psi_\lambda^t(z) := \inf_{w\in H} \{\frac{1}{2\lambda}|z - w|_H^2 + \psi^t(w)\}, \qquad z \in H.$$

It is well known (cf. [7]) that the resolvent $J_\lambda^t := (I + \lambda\partial\psi^t)^{-1}$ is contractive in H and

$$\psi_\lambda^t(z) = \frac{1}{2\lambda}|z - J_\lambda^t z|_H^2 + \psi^t(J_\lambda^t z), \qquad z \in H.$$

For any $\{\psi^t\} \in \Psi(_H(a; K_o)$ the following statements (A) - (C) hold (cf. [24;Chapters 1, 2], [32,33]).

(A) $\partial\psi_\lambda^t(z)$ is Lipschitz continuous in $z \in H$ for a.e. $t \geq 0$, measurable in $t \geq 0$ for each $z \in H$ and bounded in H on $\{(t,z); 0 \leq t \leq T, |z|_H \leq r\}$ for each finite $T > 0$ and $r > 0$.

(B) There is $0 < \lambda_o \leq 1$ such that

$$\psi_\lambda^t(z) \geq \frac{K_o}{2}|z|_H^2 \qquad \text{for all } z \in H \text{ and } 0 < \lambda \leq \lambda_o;$$

actually we can take $\lambda_o := \min\{1, \frac{1}{2K_o}\}$.

(C) If $v \in W^{1,2}(t_o, t_1; H)$ for $0 \leq t_o < t_1 < +\infty$, then $\psi_\lambda^t(v)$ is absolutely continuous on $[t_o, t_1]$ and

$$\frac{d}{dt}\psi_\lambda^t(v(t)) - (\partial\psi_\lambda^t(v(t)), v'(t))_H \leq |a'(t)|(1 + \psi_\lambda^t(v(t))) \tag{8.1}$$

$$\text{for a.e. } t \in [t_o, t_1].$$

Now write CP_λ in the normal form with $u_\lambda := [A + \lambda I]v_\lambda$ and $B_\lambda := [A + \lambda I]^{-1}$:

$$\begin{cases} u_\lambda'(t) + \partial\psi_\lambda^t(B_\lambda u_\lambda(t)) + p(B_\lambda u_\lambda(t)) = f(t), \quad 0 < t < T, \\ u_\lambda(0) = Av_o + \lambda v_o. \end{cases} \tag{8.2}$$

Since B_λ is Lipschitz continuous in H, it follows from (A) that the function $Y(t,z) := \partial\psi^t_\lambda(B_\lambda z)+p(B_\lambda z)$ is bounded on each bounded subset of $\mathbf{R}_+ \times H$, is Lipschitz continuous in $z \in H$ for a.e. $t \in [0,T]$ and is measurable in $t \in [0,T]$ for each $z \in H$. Therefore, by the standard theory of abstract differential equations in Hilbert spaces, (8.2) has one and only one solution $u_\lambda \in W^{1,2}(0,T;H)$, so does CP_λ; in fact, $v_\lambda = B_\lambda u_\lambda \in W^{1,2}(0,T;H)$ is the unique solution of CP_λ.

Lemma 8.1. *(i) There is a positive constant R_o, depending only on $|v_o|_H$ and $|f|_{L^2(0,T;H)}$, such that*

$$|A^{\frac{1}{2}}v_\lambda|^2_{C([0,T];H)} + \lambda|v_\lambda|^2_{C([0,T];H)} + |\psi^t_\lambda(v_\lambda)|_{L^1(0,T)} \le R_o \quad \textit{for all } 0<\lambda\le\lambda_o \tag{8.3}$$

and hence

$$|v_\lambda|^2_{L^2(0,T;H)} \le \frac{2R_o}{K_o} \quad \textit{for all } 0<\lambda\le\lambda_o. \tag{8.4}$$

(ii) There is a positive constant R_1, depending only on $\psi^0(v_o)$ and $|f|_{W^{1,2}(0,T;H)}$, such that

$$|[A^{\frac{1}{2}}v_\lambda]'|^2_{L^2(0,T;H)} + \lambda|v'_\lambda|^2_{L^2(0,T;H)} + \sup_{0\le t\le T}\psi^t_\lambda(v_\lambda(t)) \le R_1 \tag{8.5}$$

$$\textit{for all } 0<\lambda\le\lambda_o$$

and hence

$$|v_\lambda|_{C([0,T];H)} \le \{\frac{2R_1}{K_o}\}^{\frac{1}{2}} \quad \textit{for all } 0<\lambda\le\lambda_o. \tag{8.6}$$

(iii) There is a positive constant R_2, depending only on $|v_o|_H$ and $|f|_{W^{1,2}(0,T;H)}$, such that

$$|t^{\frac{1}{2}}[A^{\frac{1}{2}}v_\lambda]'|^2_{L^2(0,T;H)} + \lambda|t^{\frac{1}{2}}v'_\lambda|^2_{L^2(0,T;H)} + \sup_{0\le t\le T} t\psi^t_\lambda(v_\lambda(t)) \le R_2 \tag{8.7}$$

$$\textit{for all } 0<\lambda\le\lambda_o$$

and hence

$$|t^{\frac{1}{2}}v_\lambda|_{C([0,T];H)} \le \{\frac{2R_2}{K_o}\}^{\frac{1}{2}} \quad \textit{for all } 0<\lambda\le\lambda_o. \tag{8.8}$$

Proof. Multiply the equation of CP_λ by $v_\lambda - z_o$ with $z_o \in D(\psi^0)$. Then, since $R(p)$ is bounded in H by assumption,

$$\begin{aligned}
&\frac{1}{2}\frac{d}{dt}\{|A^{\frac{1}{2}}(v_\lambda(t)-z_o)|^2_H + \lambda|v_\lambda(t)-z_o|^2_H\} + \psi^t_\lambda(v_\lambda(t))\\
\le\ & \psi^t(z_o) + (f(t)-p(v_\lambda(t)), v_\lambda(t)-z_o)_H\\
\le\ & \frac{K_o}{4}|v_\lambda(t)|^2_H + k_1(|f(t)|^2_H+1) + \psi^t(z_o),
\end{aligned}$$

where k_1 is a positive constant independent of λ, f, v_o. Integrating this over $[0,s]$, $0 < s \le T$, we have by (B)

$$|A^{\frac{1}{2}}(v_\lambda(s)-z_o)|^2_H + \lambda|v_\lambda(t)-z_o|^2_H + \int_0^s \psi^t_\lambda(v_\lambda)dt$$

$$\le 2\int_0^T \{k_1(|f|^2_H+1) + \psi^t(z_o)\}dt + |A^{\frac{1}{2}}(v_o-z_o)|^2_H + \lambda|v_o-z_o|^2_H.$$

Thus (8.3) and (8.4) hold for some constant R_o depending only on $|v_o|_H$ and $|f|_{L^2(0,T;H)}$.

Next, multiply the equation of CP_λ by v'_λ and use (8.1) in (C). Then

$$|A^{\frac{1}{2}}v'_\lambda(t)|_H^2 + \lambda|v'_\lambda(t)|_H^2 + \frac{d}{dt}\{\psi_\lambda^t(v_\lambda(t)) + P(v_\lambda(t)) - (f(t),v_\lambda(t))_H\} \tag{8.9}$$

$$\leq |a'(t)|(1+\psi_\lambda^t(v_\lambda(t))) - (f'(t),v_\lambda(t))_H \qquad \text{for a.e. } t \in [0,T],$$

so that for all $s \in [0,T]$

$$\int_0^s \{|A^{\frac{1}{2}}v'_\lambda|_H^2 + \lambda|v'_\lambda|_H^2\}dt + \psi_\lambda^s(v_\lambda(s)) + P(v_\lambda(s)) - (f(s),v_\lambda(s))_H \tag{8.10}$$

$$\leq \psi^0(v_o) + P(v_o) - (f(0),v_o)_H + \int_0^s |a'|(1+\psi_\lambda^t(v_\lambda))dt + \int_0^s |f'|_H|v_\lambda|_H dt.$$

Applying the Gronwall's lemma to this inequality and using condition ($\Psi1$) and the estimates in (i), we see that (8.5) and (8.6) hold for some R_1 depending only on $\psi^0(v_o)$ and $|f|_{W^{1,2}(0,T;H)}$.

Multiplying (8.9) by t, we have

$$t|A^{\frac{1}{2}}v'_\lambda(t)|_H^2 + \lambda t|v'_\lambda(t)|_H^2 + \frac{d}{dt}\{t\psi_\lambda^t(v_\lambda(t)) + tP(v_\lambda(t)) - t(f(t),v_\lambda(t))_H\}$$

$$\leq t|a'(t)|(1+\psi_\lambda^t(v_\lambda(t))) - t(f'(t),v_\lambda(t))_H + \psi_\lambda^t(v_\lambda(t)) + P(v_\lambda(t)) - (f(t),v_\lambda(t))_H$$

$$\text{for a.e. } t \in [0,T].$$

Integrating of this inequality over $[0,s]$ yields

$$\int_0^s \{t|A^{\frac{1}{2}}v'_\lambda|_H^2 + \lambda t|v'|_H^2\}dt + s\psi^s(v_\lambda(s)) + sP(v_\lambda(s)) - s(f(s),v_\lambda(s))_H$$

$$\leq \int_0^s \{t|a'|\psi_\lambda^t(v_\lambda) + t|a'| - t(f',v_\lambda)_H + \psi_\lambda^t(v_\lambda) + P(v_\lambda) - (f,v_\lambda)_H\}dt$$

$$\leq \int_0^s |a'|\{t\psi_\lambda^t(v_\lambda) + tP(v_\lambda) - t(f,v_\lambda)_H\}dt$$

$$+ \int_0^s \{t|a'|(P(v_\lambda) + |(f,v_\lambda)_H|) + t|a'| - t(f',v_\lambda)_H + \psi_\lambda^t(v_\lambda) + P(v_\lambda) - (f,v_\lambda)_H\}dt$$

for all $s \in [0,T]$. Applying again the Gronwall's lemma to this inequality and using condition ($\Psi1$) and the estimates in (i), we have the estimates (8.7) and (8.8) for some constant R_2 depending only on $|v_o|_H$ and $|f|_{W^{1,2}(0,T;H)}$. $\diamondsuit$

Convergence of v_λ as $\lambda \to 0$ in case $v_o \in D(\psi^0)$

According to the estimates in (ii) of Lemma 8.1, there is a sequence $\{\lambda_n\}$ with $\lambda_n \to 0$ (as $n \to +\infty$) with a function $\Lambda \in W^{1,2}(0,T;H)$ such that

$$\lambda_n v_n \to 0 \qquad \text{in } L^2(0,T;H),$$

$$A^{\frac{1}{2}}v_n \to \Lambda \qquad \text{weakly in } W^{1,2}(0,T;H), \tag{8.11}$$

and

$$\frac{K_o}{2}|v_n|_H^2 \le \psi_{\lambda_n}^t(v_n) \le R_1 \qquad \text{on } [0,T]. \tag{8.12}$$

Since

$$\psi_{\lambda_n}^t(v_n(t)) = \frac{1}{2\lambda_n}|v_n(t) - J_{\lambda_n}^t v_n(t)|_H^2 + \psi^t(J_{\lambda_n}^t v_n(t)),$$

it follows from (8.12) and condition (Ψ3) that

$$|v_n(t) - J_{\lambda_n}^t v_n(t)|_H^2 \le 2\lambda_n R_1, \quad \psi^t(J_{\lambda_n}^t v_n(t)) \le R_1 \qquad \text{for all } t \in [0,T], \tag{8.13}$$

and $\{J_{\lambda_n}^t v_n(t); n \ge 1, t \in [0,T]\}$ is relatively compact in H. Now we show that

$$\{J_{\lambda_n}^t v_n(t)\}_{n=1}^\infty \text{ converges in } H \text{ for each } t \in [0,T].$$

In fact, let z_i, $i = 1,2$, be any cluster points of $J_{\lambda_n}^t v_n(t)$ as $n \to +\infty$, and choose subsequences $\{n_{k(i)}\}$ of $\{n\}$ so that $J_{\lambda_{n_{k(i)}}}^t v_{n_{k(i)}}(t) \to z_i$ in H as $k(i) \to +\infty$. Then, since

$$A^{\frac{1}{2}}(v_{n_{k(i)}}(t)) = A^{\frac{1}{2}}(v_{n_{k(i)}}(t) - J_{\lambda_{n_{k(i)}}}^t v_{n_{k(i)}}(t)) + A^{\frac{1}{2}}(J_{\lambda_{n_{k(i)}}}^t v_{n_{k(i)}}(t)),$$

we derive from (8.11) and (8.13) that

$$\Lambda(t) = A^{\frac{1}{2}}(z_i), \ i = 1,2, \text{ hence } A^{\frac{1}{2}}(z_1 - z_2) = 0.$$

Combining this with the positivity of A, we see that $z_1 = z_2$. Thus $J_{\lambda_n}^t v_n(t)$ converges in H and hence, by (8.13),

$$v(t) := \lim_{n\to+\infty} v_n(t) \quad \text{exists in } H \text{ for each } t \in [0,T]. \tag{8.14}$$

We have clearly

$$v_n \to v \qquad \text{in } L^2(0,T;H) \text{ and weakly* in } L^\infty(0,T;H) \tag{8.15}$$

as well as $\Lambda = A^{\frac{1}{2}}v \in W^{1,2}(0,T;H)$, $(A^{\frac{1}{2}}v)(0) = A^{\frac{1}{2}}v_o$,

$$p(v_n) \to p(v) \qquad \text{in } L^2(0,T;H) \tag{8.16}$$

and

$$(Av_n)' \to (Av)' \qquad \text{weakly in } L^2(0,T;H). \tag{8.17}$$

We observe from the convergences obtained above that

$$\partial\psi_{\lambda_n}^t(v_n) = f - p(v_n) - (Av_n)' - \lambda_n v_n' \to f - p(v) - (Av)' \quad \text{weakly in } L^2(0,T;H). \tag{8.18}$$

This shows that $f(t) - p(v(t)) - (Av)'(t) \in \partial\psi^t(v(t))$ for a.e. $t \in [0,T]$. Thus v is a solution of $CP(f,v_o)$ on $[0,T]$ such that $\psi^t(v) \in L^\infty(0,T)$ and $A^{\frac{1}{2}}v \in W^{1,2}(0,T;H)$.

Uniqueness in the general case of $v_o \in \overline{D(\psi^0)}$

Let v_1, v_2 be any solutions of $CP(f,v_o)$ on $[0,T]$. Then

$$\frac{1}{2}\frac{d}{dt}|A^{\frac{1}{2}}(v_1(t) - v_2(t))|_H^2 + (v_1^*(t) - v_2^*(t), v_1(t) - v_2(t))_H \tag{8.19}$$

$$+(p(v_1(t)) - p(v_2(t)), v_1(t) - v_2(t))_H = 0 \qquad \text{for a.e. } t \in [0,T],$$

where $v_i^*(t) = f(t) - p(v_i(t)) - (Av_i)'(t) \in \partial\psi^t(v_i(t))$, $i = 1,2$. By assumptions on p in $(\star)$, there is a constant $N_1 > 0$ such that

$$\begin{aligned} &|(p(v_1(t)) - p(v_2(t)), v_1(t) - v_2(t))_H| \\ &\leq \frac{1}{2}(v_1^*(t) - v_2^*(t), v_1(t) - v_2(t))_H + N_1|A^{\frac{1}{2}}(v_1(t) - v_2(t))|_H^2 \end{aligned} \tag{8.20}$$

It follows from (8.19) and (8.20) that

$$\frac{1}{2}\frac{d}{dt}|A^{\frac{1}{2}}(v_1(t) - v_2(t))|_H^2 \leq N_1|A^{\frac{1}{2}}(v_1(t) - v_2(t))|_H^2 \quad \text{for a.e. } t \in [0,T], \tag{8.21}$$

which implies $A^{\frac{1}{2}}(v_1 - v_2) = 0$, i.e. $v_1 = v_2$ on $[0,T]$.

Existence in the general case $v_o \in \overline{D(\psi^0)}$

Choose a sequence $\{v_{on}\} \subset D(\psi^0)$ and denote by v_n the solution of $CP(f, v_{on})$ on $[0,T]$. Then, just as (8.21),

$$\frac{1}{2}\frac{d}{dt}|A^{\frac{1}{2}}(v_n(t) - v_m(t))|_H^2 \leq N_1|A^{\frac{1}{2}}(v_n(t) - v_m(t))|_H^2 \quad \text{for a.e. } t \in [0,T],$$

so that

$$|A^{\frac{1}{2}}(v_n(t) - v_m(t))|_H \leq e^{N_1 t}|A^{\frac{1}{2}}(v_{on} - v_{om})|_H \quad \text{for all } t \in [0,T].$$

Accordingly, $A^{\frac{1}{2}}v_n$ converges in $C([0,T];H)$ for some function $\Lambda \in C([0,T];H)$. Also, on account of (8.7) in Lemma 8.1,

$$\psi^t(v_n(t)) \leq \frac{R_2}{t} \qquad \text{for all } t \in (0,T].$$

Therefore, we can prove in a silimar way to that in the convergence proof of v_λ that

$$v(t) := \lim_{n\to+\infty} v_n(t) \quad \text{exists in } H \text{ for any } 0 < t \leq T,$$

so that

$$v_n \to v \qquad \text{in } L^2(t_o,T;H) \text{ for every } 0 < t_o < T \text{ and weakly in } L^2(0,T;H),$$

$$t^{\frac{1}{2}}(A^{\frac{1}{2}}v_n)' \to t^{\frac{1}{2}}(A^{\frac{1}{2}}v)' \qquad \text{weakly in } L^2(0,T;H),$$

$$\frac{R_2}{t} \geq \liminf_{n\to+\infty} \psi^t(v_n(t)) \geq \psi^t(v(t)) \qquad \text{for all } t \in (0,T].$$

Hence $\Lambda = A^{\frac{1}{2}}v \in C([0,T];H)$ with $(A^{\frac{1}{2}}v)(0) = A^{\frac{1}{2}}v_o$, and $f - p(v_n) - (Av_n)' \to f - p(v) - (Av)'$ weakly in $L^2([t_o,T];H)$ for every $0 < t_o < T$. From these facts it follows that $f - p(v) - (Av)' \in \partial\psi^t(v)$ a.e. on $[0,T]$ and thus v is a solution of $CP(f, v_o)$ having the required properties.

We have obtained Theorem 8.1, and end this section with the following remarks.

Remark 8.3. (Energy inequality) Let v be the solution of $CP(f, v_o)$ on $[0,T]$, which is given by Theorem 8.1. Then, for any $0 < s < t \leq T$ it holds that

$$\int_s^t |(A^{\frac{1}{2}}v)'(\tau)|_H^2 d\tau + \psi^t(v(t)) + P(v(t)) - (f(t), v(t))_H$$

$$\leq \psi^s(v(s)) + P(v(s)) - (f(s), v(s))_H \tag{8.22}$$

$$+ \int_s^t |a'(\tau)|(1 + \psi^\tau(v(\tau))d\tau - \int_s^t (f'(\tau), v(\tau))_H d\tau$$

In fact, this energy inequality is proved as follows. First, assume that $v_o \in D(\psi^0)$. Let v_λ be the approximate solution which was considered above. Since $v_\lambda \to v$ in $L^2(0,T;H)$ (cf. (8.15)) and $\partial\psi_\lambda^{(\cdot)}(v_\lambda)$ is bounded in $L^2(0,T;H)$ (cf. (8.18)) and $\psi_\lambda^{(\cdot)}(v_\lambda)$ is uniformly bounded in $(t,\lambda) \in [0,T] \times (0,\lambda_o]$ (cf. (8.12)), it follows from the basic properties of the Yosida-approximation that

$$\liminf_{\lambda \to 0} \psi_\lambda^\tau(v_\lambda(\tau)) \geq \psi^\tau(v(\tau)) \qquad \text{for all } \tau \in [0,T] \tag{8.23}$$

and

$$\lim_{\lambda \to 0} \psi_\lambda^\tau(v_\lambda(\tau)) = \psi^\tau(v(\tau)) \qquad \text{for a.e. } \tau \in [0,T]. \tag{8.24}$$

Pass to the limit as $\lambda \to 0$ in (8.10). Then, by (8.14),(8.23) and (8.24), (8.22) holds true for $0 = s < t \leq T$. In the general case, since $v(s) \in D(\psi^0)$ for any $s \in (0,T]$, taking s as the initial time and repeating the same argument as above, we see that (8.22) holds.

Remark 8.4. In [11,16], an abstract nonlinear equation of the form $Lv + Bv \ni f$, where L is a linear or nonlinear and B is a nonlinear operator in Hilbert or reflexive Banach soaces, was treated, and some slight extensions of their results might be applicable to the existence proof for $CP(f, v_o)$.

9. Asymptotic behaviour as $t \to 0$

In the previous section we showed that $CP(f, v_o)$ has one and only one solution v on $[0,T]$ for any finite $T > 0$, that is, $CP(f, v_o)$ has a unique global (in time) solution v, provided that $f \in W_{loc}^{1,2}(\mathbf{R}_+; H)$ and $v_o \in \overline{D(\psi^0)}$.

Let p be as in the previous section and $\{\psi^t\} \in \Psi_H(a; K_0)$ and further suppose that

$$f' \in L^1(\mathbf{R}_+; H); \; f^\infty := \lim_{t \to +\infty} f(t) \text{ in } H, \tag{9.1}$$

$$a' \in L^1(\mathbf{R}_+), \tag{9.2}$$

$$\psi^t \to \psi^\infty \quad \text{(in the sense of Mosco [38]) as } t \to +\infty, \tag{9.3}$$

where ψ^∞ is a non-negative proper l.s.c. convex function, with $D(\psi^\infty) = D(\psi^0)$, on H. Here, by "$\psi^t \to \psi^\infty$ (in the sense of Mosco) as $t \to +\infty$" we mean that the following two conditions (M1) and (M2) are fulfilled:

(M1) if $t_n \to +\infty$ and $z_n \to z$ weakly in H, then

$$\liminf_{n\to+\infty} \psi^{t_n}(z_n) \geq \psi^\infty(z).$$

(M2) For any $z \in D(\psi^\infty)$ there is a function $w : \mathbf{R}_+ \to H$ such that

$$w(t) \to z \quad \text{in } H, \quad \psi^t(w(t)) \to \psi^\infty(z) \quad \text{as } t \to +\infty.$$

From the definition of the convergence in the sense of Mosco we immediately see that

$$\psi^\infty(z) \geq K_o|z|_H^2 \qquad \text{for all } z \in H,$$

that is, ψ^∞ is coercive on H, and for each $r > 0$ the set $\{z \in H; \psi^\infty(z) \leq r\}$ is compact in H. Therefore, the stationary problem

$$\partial\psi^\infty(v_\infty) + p(v_\infty) \ni f^\infty \qquad \text{in } H \tag{9.4}$$

has at least one solution v_∞ and the set of all solutions is compact in H; a solution of (9.4) is not unique in general, since p is not monotone in H.

Theorem 9.1. *Let p be as in section 8 and $\{\psi^t\} \in \Psi_H(a; K_o)$, and suppose that $v_o \in \overline{D(\psi^0)}$ and (9.1)-(9.3) hold. Then, for the global solution v of $CP(f, v_o)$,*
(i) $(A^{\frac{1}{2}}v)' \in L^2(t_o, +\infty; H)$ for each finite $t_o > 0$.
Moreover, the $\omega-$limit set

$$\omega(v) := \{z \in H; v(t_n) \to z \text{ in } H \text{ for some } t_n \text{ with } t_n \to +\infty\}$$

satisfies that
(ii) $\omega(v)$ is non-empty, connected and compact in H,
(iii) any point $v_\infty \in \omega(v)$ is a solution of (9.4),
(iv) for any $v_\infty \in \omega(v)$,

$$\lim_{t\to+\infty} \{\psi^t(v(t)) + P(v(t)) - (f(t), v(t))_H\} = \psi^\infty(v_\infty) + P(v_\infty) - (f^\infty, v_\infty)_H.$$

Proof. We put

$$Y(t) := \psi^t(v(t)) + P(v(t)) - (f(t), v(t))_H \qquad \text{for } t > 0.$$

Then, by the energy inequality (8.22) in Remark 8.3,

$$\frac{1}{2}\int_s^t |(A^{\frac{1}{2}}v)'|_H^2 d\tau + Y(t) \leq Y(s) + \int_s^t \{|a'|(1 + \psi^\tau(v)) + |f'|_H|v|_H\} d\tau \tag{9.5}$$

for all $0 < s < t$. Here, note from (Ψ1) and (9.1) that

$$K_o|v(\tau)|_H^2 \leq \psi^\tau(v(\tau)) \leq k_2 Y(\tau) + k_3 \qquad \text{for all } \tau > 0,$$

where k_2, k_3 are positive constants. Therefore we derive from (9.5) that there are positive constants N_2, N_3 such that

$$\frac{1}{2}\int_s^t |(A^{\frac{1}{2}}v)'|_H^2 d\tau + Y(t) \le Y(s) + N_2\int_s^t (|a'| + |f'|_H) Y d\tau + N_3\int_s^t (|a'| + |f'|_H) d\tau \quad (9.6)$$

for all $0 < s < t$. Applying the Gronwall's inequality to (9.6), we see that Y is bounded on $[s, +\infty)$ and the function

$$Y(t) + \frac{1}{2}\int_s^t |(A^{\frac{1}{2}}v)'|_H^2 d\tau - N_2\int_s^t (|a'| + |f'|_H) Y d\tau - N_3\int_s^t (|a'| + |f'|_H) d\tau$$

is non-increasing on $[s, +\infty)$, $(A^{\frac{1}{2}}v)' \in L^2(s, +\infty; H)$ for each $s > 0$ and $\lim_{t\to+\infty} Y(t)$ exists, since $|a'| + |f'|_H \in L^1(\mathbf{R}_+)$. Moreover, by ($\Psi$3), $\{v(t); t \ge s\}$ is relatively compact in H for each $s > 0$.

Now, let $\{t_n\}$ be any sequence in $\mathbf{R}_+$ such that $v(t_n) \to v_\infty$ in H for some $v_\infty \in H$; such a sequence $\{t_n\}$ exists by the above facts; namely $v_\infty \in \omega(v) \ne \emptyset$. Put

$$v_n(t) := v(t_n + t), \quad \psi_n^t := \psi^{t_n + t} \qquad \text{for } t \in [0, 1].$$

Since $(A^{\frac{1}{2}}v_n)' \to 0$ in $L^2(0, 1; H)$, we have

$$A^{\frac{1}{2}}v_n \to A^{\frac{1}{2}}v_\infty \qquad \text{in } C([0, 1]; H). \quad (9.7)$$

Then we have

$$v_n \to v_\infty \qquad \text{in } C([0, 1]; H). \quad (9.8)$$

We show it by contradiction. In fact, assume that there are a number $\varepsilon_o > 0$ with a subsequence $\{n_k\}$ of $\{n\}$ and a convergent sequence $\{\tau_k\}$ in $[0, 1]$ such that

$$|v_{n_k}(\tau_k) - v_\infty|_H \ge \varepsilon_o \qquad \text{for all } k. \quad (9.9)$$

By the relative compactness of $\{v_{n_k}(\tau_k)\}$ in H we may assume that $z_o := \lim_{k\to+\infty} v_{n_k}(\tau_k)$ exists in H. In this case, by (9.7) and (9.9), $A^{\frac{1}{2}}z_o = A^{\frac{1}{2}}v_\infty$ and $|z_o - v_\infty|_H \ge \varepsilon_o$. But the positivity of A implies that $z_o = v_\infty$ which yields a contradiction. Thus (9.8) must holds.

Now, we prove that v_∞ is a solution of (9.4). First, choose a sequence $\{s_n\} \subset [0, 1]$ such that

$$\partial\psi_n^{s_n}(v_n(s_n)) \ni f(t_n + s_n) - p(v_n(s_n)) - (A^{\frac{1}{2}}v_n)'(s_n) \to f^\infty - p(v_\infty) \quad \text{in } H.$$

Let z be any element in $D(\psi^\infty)$ and $\{z_n\} \subset H$ such that

$$z_n \to z \quad \text{in } H, \quad \psi_n^{s_n}(z_n) \ (= \psi^{t_n + s_n}(z_n)) \to \psi^\infty(z).$$

Then,

$$(f(t_n + s_n) - p(v_n(s_n)) - (A^{\frac{1}{2}}v_n)'(s_n), z_n - v_n(s_n))_H \le \psi_n^{s_n}(z_n) - \psi_n^{s_n}(v_n(s_n)).$$

Since $d^* := \lim_{n\to+\infty} \psi_n^{s_n}(v_n(s_n))$ exists and $d^* \ge \psi^\infty(v_\infty)$, letting $n \to +\infty$ yields

$$(f^\infty - p(v_\infty), z - v_\infty)_H \le \psi^\infty(z) - d^* \le \psi^\infty(z) - \psi^\infty(v_\infty).$$

This shows that $f^\infty - p(v_\infty) \in \partial\psi^\infty(v_\infty)$. Moreover, by taking v_∞ as z we have $d^* \le \psi^\infty(v_\infty)$, hence $d^* = \lim_{t\to+\infty}\psi^\infty(v_\infty)$. Therefore (iii) and (iv) hold. It is quite standard to prove the connectedness and compactness of $\omega(v)$ in H. Thus Theorem 9.1 has been obtained. $\diamondsuit$

10. Applications to (PFC) and (PSC)

In the final section of this paper we give applications of the abstract results proved in sections 8 and 9 to systems (PFC) and (PSC).

Application to (PFC)

We use the same notations as in sections 3 and 4; for the data $\rho, \beta, g, f, h_o, u_o$ and w_o we assume that $(\rho), (\beta), (g), (f), (h_o)$ and (4.3) hold, and $\kappa > 0, \nu > 0$ and $\alpha > 0$ are constants as well.

In order to apply the abstract results of sections 8 and 9, we take

$$X := V^* \times L^2(\Omega)$$

as the Hilbert space H and the identity mapping in X as A; in this case condition $(\star)$ is trivially satisfied. Further suppose that

$$f' \in L^1(\mathbf{R}_+; L^2(\Omega)); \quad f^\infty := \lim_{t\to+\infty} f(t) \quad \text{in } L^2(\Omega) \tag{10.1}$$

and

$$h_o' \in L^1(\mathbf{R}_+; L^2(\Gamma)); \quad h_o^\infty := \lim_{t\to+\infty} h_o(t) \quad \text{in } L^2(\Gamma). \tag{10.2}$$

By (10.2), $h' \in L^1(\mathbf{R}_+; H^1(\Omega))$ and $h^\infty := \lim_{t\to+\infty} h(t)$ in $H^1(\Omega)$ for the function h which is defined by (4.2). Also, we observe from the definition of φ^t (see section 4) that

$$\varphi^t([e,w]) \ge a_1(|e|^2_{L^2(\Omega)} + |w|^2_{H^1(\Omega)}) - a_2 \qquad \text{for all } [e,w] \in L^2(\Omega) \times H^1(\Omega), \tag{10.3}$$

where a_1, a_2 are appropriate positive constants, and

$$|\varphi^t([e,w]) - \varphi^s([e,w])| = |(h(t) - h(s), e)| \le |a(t) - a(s)|(1 + \varphi^s([e,w]) + a_2) \tag{10.4}$$

$$\text{for all } [e,w] \in L^2(\Omega) \times H^1(\Omega),$$

where

$$a(t) := (1 + \frac{1}{a_1}) \int_0^t |h'(\tau)|_{L^2(\Omega)} d\tau, \qquad t \ge 0; \tag{10.5}$$

note that $a' := (1 + \frac{1}{a_1})|h'|_{L^2(\Omega)} \in L^1(\mathbf{R}_+)$. It is easy to see from (10.3) and (10.4) that the family $\{\psi^t\}$ with $\psi^t(\cdot) := \varphi^t(\cdot) + a_2$ belongs to $\Psi_X(a; K_o)$, where $a(\cdot)$ is the function given by (10.5) and K_o is a positive constant satisfying

$$a_1(|e|_{L^2(\Omega)} + |w|_{H^1(\Omega)}) \ge K_o(|e|_{V^*} + \nu|w|^2_{L^2(\Omega)}) \qquad \text{for all } [e,w] \in L^2(\Omega) \times H^1(\Omega).$$

Next, let us define a proper l.s.c. convex function φ^∞ on X by φ^t with $h(t)$ replaced by h^∞. Then, since $h(t) \to h^\infty$ in $H^1(\Omega)$ as $t \to +\infty$, we see that

$$\psi^t := \varphi^t + a_2 \to \psi^\infty := \varphi^\infty + a_2 \text{ on } X \text{ (in the sense of Mosco) as } t \to +\infty. \tag{10.6}$$

On account of Theorem 4.1 for $\tilde{f}^\infty := [f^\infty, 0] \in X$ the corresponding steady-state problem

$$\partial\psi^\infty([e_\infty, w_\infty]) + G([e_\infty, w_\infty]) \ni \tilde{f}^\infty \qquad \text{in } X \tag{10.7}$$

is equivalent to the following system (10.8) and (10.9) for $u_\infty := \rho^{-1}(e_\infty - w_\infty) \in V$ and $w_\infty \in H^1(\Omega)$:

$$a(u_\infty - h^\infty, \eta) + \alpha(u_\infty - h^\infty, \eta)_\Gamma = (f^\infty, \eta) \quad \text{for all } \eta \in V \tag{10.8}$$

and

$$\begin{cases} \kappa a(w_\infty, \eta) + (\xi_\infty + g(w_\infty) - u_\infty, \eta) = 0 \quad \text{for all } \eta \in H^1(\Omega), \\ \xi_\infty \in L^2(\Omega), \quad \xi_\infty \in \beta(w_\infty) \text{ a.e. in } \Omega. \end{cases} \tag{10.9}$$

It sholud be noted here that problem (10.8) does not include w_∞, so that it has one and only one solution u_∞ in V, while (10.9) includes u_∞ and its solution is not unique in general because of non-monotonicity of g.

Now, applying Theorems 8.1 and 9.1, we have:

Theorem 10.1. *Assume that $(\rho), (\beta), (g), (f), (h_o)$, (4.3), (10.1) and (10.2) hold. Then (PFC) admits one and only one global (in time) weak solution $\{u, w\}$. Moreover, the following statements hold.*

(a) *For every finite $T > 0$,*

$$t^{\frac{1}{2}}\rho(u) \in L^\infty(0, T; L^2(\Omega)), \quad t^{\frac{1}{2}}\rho(u)_t \in L^2(0, T; V^*),$$
$$t^{\frac{1}{2}}w \in L^\infty(0, T; H^1(\Omega)), \quad t^{\frac{1}{2}}w_t \in L^2(0, T; L^2(\Omega)),$$
$$t^{\frac{1}{2}}\xi \in L^2(0, T; L^2(\Omega)), \quad t^{\frac{1}{2}}\hat{\beta}(w) \in L^\infty(0, T; L^1(\Omega)),$$

where ξ is the function as in (w3) of Definition 4.1.

(b) *For every finite $T > 0$,*

$$\rho(u) \in L^\infty(T, +\infty; L^2(\Omega)), \quad w \in L^\infty(T, \infty; H^1(\Omega)), \quad \hat{\beta}(w) \in L^\infty(T, +\infty; L^1(\Omega)),$$
$$\rho(u)_t \in L^2(T, +\infty; V^*), \quad w_t \in L^2(T, +\infty; L^2(\Omega)).$$

(c) *$u(t) \to u_\infty$ in $L^2(\Omega)$ as $t \to +\infty$, where u_∞ is the solution of (10.8).*

(d) *The ω-limit set $\omega(w)$ of w as $t \to +\infty$, i.e.*

$$\omega(w) := \{z \in L^2(\Omega); w(t_n) \to z \text{ in } L^2(\Omega) \text{ for some } t_n \text{ with } t_n \to +\infty\}$$

is non-empty, connected and compact in $L^2(\Omega)$, and furthermore any $w_\infty \in \omega(w)$ satisfies (10.9).

Proof. As a direct consequence of Theorem 8.1, for every finite $T > 0$ the evolution problem $CP(\tilde{f}, U_o)$, with $U_o := [\rho(u_o) + w_o, w_o]$, has one and only one solution $U(t) := [e(t), w(t)]$ on $[0,T]$ and it satisfies that

$$t^{\frac{1}{2}}U' \in L^2(0,T;X), \quad t\psi^t(U) \text{ (hence } t\varphi^t(U)) \in L^\infty(0,T). \tag{10.10}$$

By virtue of Corollary 4.1, the couple of functions $\{u, w\}$ with $u := \rho^{-1}(e - w)$ is a unique weak solution of (PFC). Moreover (10.10) implies that (a) holds.

Next, apply Theorem 9.1 to $CP(\tilde{f}, U_o)$. Then (b) is straightforward, and the ω–limit set $\omega(U)$ of U in X as $t \to +\infty$ is non-empty, connected and compact in X. Let $U_\infty := [e_\infty, w_\infty] \in \omega(U)$ and let $\{t_n\}$ be a sequence in $\mathbf{R}_+$ with $t_n \to +\infty$ such that $e(t_n) \to e_\infty$ in V^* and $w(t_n) \to w_\infty$ in $L^2(\Omega)$. Then it follows from Theorem 9.1 that

$$\lim_{n\to+\infty} \varphi^{t_n}(U(t_n)) = \varphi^\infty(U_\infty), \tag{10.11}$$

and $u_\infty := \rho^{-1}(e_\infty - w_\infty)$ and w_∞ satisfy (10.8) and (10.9). Hence, by (b) and by the lower semicontinuity of the functionals in expression (7.7) of $\varphi^t(\cdot)$, (10.11) implies that

$$\int_\Omega \hat{\rho^{-1}}(e(t_n) - w(t_n))dx \to \int_\Omega \hat{\rho^{-1}}(e_\infty - w_\infty)dx.$$

so that

$$\begin{aligned}
0 &= \lim_{n\to+\infty} \int_\Omega \hat{\rho^{-1}}(e(t_n) - w(t_n))dx - \int_\Omega \hat{\rho^{-1}}(e_\infty - w_\infty)dx \\
&= \lim_{n\to+\infty} \int_0^1 (\rho^{-1}(r\{e(t_n) - w(t_n)\} - e_\infty + w_\infty), e(t_n) - w(t_n) - e_\infty + w_\infty)dr \\
&= \lim_{n\to+\infty} \int_0^1 (\rho^{-1}(r\{e(t_n) - w(t_n)\} - e_\infty + w_\infty) - \rho^{-1}(e_\infty - w_\infty), \\
&\qquad\qquad e(t_n) - w(t_n) - e_\infty + w_\infty)dr \\
&\geq \lim_{n\to+\infty} \frac{1}{C(\rho)} \int_0^1 r|e(t_n) - w(t_n) - e_\infty + w_\infty|^2_{L^2(\Omega)}dr \\
&= \frac{1}{2C(\rho)} \lim_{n\to+\infty} |e(t_n) - w(t_n) - e_\infty + w_\infty|^2_{L^2(\Omega)}.
\end{aligned}$$

Hence, $e(t_n) - w(t_n) \to e_\infty - w_\infty$ in $L^2(\Omega)$, namely $u(t_n) \to u_\infty$ in $L^2(\Omega)$. This shows that $u(t) \to u_\infty$ in $L^2(\Omega)$ as $t \to +\infty$, since u_∞ is a unique solution of (10.8), and thus (c) holds. Noting that $\omega(U) = \{(u_\infty, w_\infty); w_\infty \in \omega(w)\}$, we have (d). $\diamond$

Application to (PSC)

We use the same notations as in sections 7 and 8. In the case of (PSC) we take

$$X_o := V^* \times L^2(\Omega)_o$$

as the Hilbert space H and the operator A_o in X_o as A. Let V, V_o and $F : V \to V^*$, $F_o : V_o \to V_o^*$ be as in section 7 as well. By a compactness theorem (cf. [36;Chapter 1, Lemma 5.1]), for each $\varepsilon > 0$ there is a constant $C(\varepsilon)$ depending only on ε such that

$$|v|^2_{L^2(\Omega)} \leq \varepsilon|v|^2_{V_o} + C(\varepsilon)|v|^2_{V_o^*} \quad \text{for all } v \in V_o.$$

This together with Theorem 7.1 (c) implies that condition ($\star$) is satisfied, since

$$|A_o^{\frac{1}{2}}[e,v]|_{X_o}^2 = |e|_{V^*}^2 + \nu\kappa|v|_{V_o^*}^2 \quad \text{for all } [e,v] \in V^* \times V_o.$$

Moreover, for the data ρ, β, g, f, h_o, u_o and w_o suppose that (ρ), (β), (g), (f), (h_o), (10.1), (10.2) and (7.1) hold; $\kappa > 0$, $\nu > 0$ and $\alpha > 0$ are constants as well. Then, just as in the previous case, we observe that the proper l.s.c. convex function φ_o^t on X_o satisfies the similar properties as (10.3) and (10.4), and hence the family $\{\varphi_o^t + a_2\} \in \Psi_{X_o}(a; K_o)$ for appropriate positive constants a_2, K_o and a function $a(\cdot) \in W_{loc}^{1,1}(\mathbf{R}_+)$ having the regularlity a' in $L^1(\mathbf{R}_+)$. Also,

$$\varphi_o^t + a_2 \to \varphi_o^\infty + a_2 \quad \text{(in the sense of Mosco) as } t \to +\infty, \tag{10.12}$$

where φ_o^∞ is the proper l.s.c. convex function on X_o which is defined by φ_o^t with $h(t)$ replaced by h^∞.

By virtue of Theorem 7.1 and Remark 7.1 we note that for $\tilde{f}^\infty := [f^\infty, 0] \in X_o$ the corresponding steady-state problem

$$\partial\varphi^\infty([e_\infty, w_\infty]) + G_o([e_\infty, w_\infty]) \ni \tilde{f}^\infty \quad \text{in } X_o$$

is equivalent to the following system (10.13) and (10.14) for $u_\infty := \rho^{-1}(e_\infty - w_\infty) \in V$ and $w_\infty \in H^1(\Omega)$:

$$a(u_\infty - h^\infty, \eta) + \alpha(u_\infty - h^\infty, \eta)_\Gamma = (f^\infty, \eta) \quad \text{for all } \eta \in V. \tag{10.13}$$

$$\begin{cases} \kappa a(w_\infty, \eta) + (\xi_\infty + g(w_\infty) - u_\infty, \eta) = 0 \quad \text{for all } \eta \in V_o, \\ w_\infty - \dfrac{c}{|\Omega|} \in V_o, \\ \xi_\infty \in L^2(\Omega), \quad \xi_\infty \in \beta(w_\infty) \text{ a.e. in } \Omega. \end{cases} \tag{10.14}$$

Now, applying Theorems 8.1 and 9.1, we have:

Theorem 10.2. *Assume that (ρ), (β), (g), (f), (h_o), (7.1), (7.2), (10.1) and (10.2) hold. Then (PSC) admits one and only one global (in time) weak solution $\{u, w\}$. Moreover, the following statements hold.*

(a) *For every finite $T > 0$,*

$$t^{\frac{1}{2}}\rho(u) \in L^\infty(0,T;L^2(\Omega)), \quad t^{\frac{1}{2}}\rho(u)_t \in L^2(0,T;V^*),$$

$$t^{\frac{1}{2}}(w - \frac{c}{|\Omega|}) \in L^\infty(0,T;V_o), \quad t^{\frac{1}{2}}w_t \in L^2(0,T;V_o^*),$$

$$t^{\frac{1}{2}}\xi \in L^2(0,T;L^2(\Omega)), \quad t^{\frac{1}{2}}\hat{\beta}(w) \in L^\infty(0,T;L^1(\Omega)),$$

where ξ is the function as in (w3) of Definition 7.1.

(b) *For every finite $T > 0$,*

$$\rho(u) \in L^\infty(T,+\infty;L^2(\Omega)), \quad w - \frac{c}{|\Omega|} \in L^\infty(T,\infty;V_o), \quad \hat{\beta}(w) \in L^\infty(T,+\infty;L^1(\Omega)),$$

$$\rho(u)_t \in L^2(T,+\infty;V^*), \quad w_t \in L^2(T,+\infty;V_o^*).$$

(c) *$u(t) \to u_\infty$ in $L^2(\Omega)$ as $t \to +\infty$, where u_∞ is the solution of (10.13).*

(d) *The ω−limit set $\omega(w)$ of w as $t \to +\infty$, i.e.*

$$\omega(w) := \{z \in L^2(\Omega); w(t_n) \to z \text{ in } L^2(\Omega) \text{ for some } t_n \text{ with } t_n \to +\infty\}$$

is non-empty, connected and compact in $L^2(\Omega)$, and furthermore any $w_\infty \in \omega(w)$ satisfies (10.14).

Proof. Noting (cf. (7.10)) that for $U(t) := [e(t), w(t) - \frac{c}{|\Omega|}]$

$$|A_o^{\frac{1}{2}}U(t)|^2_{X_o} = |e(t)|^2_{V^*} + \kappa\nu|w(t) - \frac{c}{|\Omega|}|^2_{V_o^*},$$

we can prove the theorem in a way similar to that of the proof of Theorem 10.1. ◇

Remark 10.1. We refer to [4,9,25,27,29,44] for related works on the asymptotic stability of global solutions of phase field models and phase separation models as $t \to +\infty$.

References

1. H. W. Alt and I. Pawlow, Dynamics of non-isothermal phase separation, in *Free Boundary Value Problems*, K.H. Hoffmann and J.Sprekels ed., ISNM **95**, Birkhäuser, Basel, 1990, pp. 1-26.

2. H. W. Alt and I. Pawlow, Existence of solutions for non-isothermal phase separation, Adv. Math. Sci. Appl. **1**(1992), 319-409.

3. H. Amann, Nonhomogeneous linear and quasilinear elliptic and parabolic boundary value problems, preprint, 1993.

4. P. W. Bates and S. Zheng, Inertial manifolds and inertial sets for the phase-field equations, J. Dynamics Diff. Eqns 4(1992), 375-398.

5. J. F. Blowey and C. M. Elliott, Curvature dependent phase boundary motion and parabolic double obstacle problems, to appear in the proceedings of the IMA (Minneapolis) workshop "Degenerate Diffusions" ed. Wei-Ming Ni, L. A. Peletier and J. L. Vazquez, Springer Verlag, New York.

6. F. E. Blowey and C. M. Elliott, The Cahn-Hilliard gradient theory for phase separation with non-smooth free energy, Part I: Mathematical analysis, European J. Appl. Math. **2**(1991), 233-280.

7. H. Brézis, *Opérateurs Maximaux Monotones et Semi-groupes de Contractions dans les Espaces de Hilbert*, North-Holland, Amsterdam, 1973.

8. H. Brézis, M. Crandall and A. Pazy, Perturbations of nonlinear maximal monotone sets, Comm. Pure Appl. Math., **23**(1970),123-144.

9. D. Brochet, X. Chen and D. Hilhorst, Finite Dimensional exponential attractor for the phase field model, IMA Preprint series No. 858, Univ. Minnesota, 1991.

10. G. Caginalp, An analysis of a phase field model of a free boundary, Arch. Rat. Mech. Anal., **92**(1986), 205-245.

11. P. Colli and A. Visintin, On a class of doubly nonlinear evolution equations, Comm. Partial Differential Equations **15**(1990), 737-756.

12. A. Damlamian, Some results on the multi-phase Stefan problem, Communs Partial Diff. Eqns. 2(1977), 1017-1044.

13. A. Damlamian, N. Kenmochi and N. Sato, Subdifferential operator approach to a class of nonlinear systems for Stefan problems with phase relaxation, to appear in Nonlinear Anal. TMA.

14. A. Damlamian, N. Kenmochi and N. Sato, Phase field equations with constraints, to appear in the proceedings of the conference on *"Nonlinear Mathematical Problems in Industry"*, Gakkōtosho, Tokyo, 1993.

15. S. R. DeGroot and P. Mazur, *Non-Equilibrium Thermodynamics*, Dover Publ., New York, 1984.

16. E. DiBenedetto and R. E. Showalter, A pseudo-parabolic variational inequality and Stefan problem, Nonlinear Anal.TMA **6**(1982),279-291.

17. C. E. Elliott, The Cahn-Hilliard model for the kinetics of phase separation, in *Mathematical Models for Phase Change Problems*, J. F. Rodrigues ed., ISNM **88**, Birkhäuser, Basel, 1989, pp. 35-73.

18. C. M. Elliott and S. Zheng, On the Cahn-Hilliard equation, Arch. Rat. Mech. Anal. **96**(1986), 339-357.

19. C. Elliott and S. Zheng, Global existence and stability of solutions to the phase field equations, *Free Boundary Problems*, pp. 48-58, Intern. Ser. Numer. Math. Vol.**95**, Birkhäuser, Basel, 1990.

20. G. J. Fix, Phase field models for free boundary problems, *Free Boundary Problems: Theory and Applications*, pp. 580-589, Pitman Research Notes Math. Ser. **79**, Longman, London, 1983.

21. A. Friedman, The Stefan problem in several space variables, Trans. Amer. Math. Sco. 133(1968),51-87.

22. W. Horn, J. Sprekels and S. Zheng, Global existence of smooth solutions to the Penrose-Fife model for Ising ferromagnets, preprint, Univ. GH Essen, 1993.

23. S. L. Kamenomostskaja, On Stefans's problem, Mat. Sb. 53(1961), 489-514.

24. N. Kenmochi, Solvability of nonlinear evolution equations with time-dependent constraints, Bull. Fac. Education, Chiba Univ., 30(1981), 1-87.

25. N. Kenmochi and M. Niezgódka, Systems of variational inequalities arising in nonlinear diffusion with phase change, *Free Boundary Problems in Continuum Mechanics*, pp. 149-157, Intern. Ser. Numer. Math. Vol. **106**, Birkhäuser, Basel, 1992.

26. N. Kenmochi and M. Niezgódka, Evolution systems of nonlinear variational inequalities arising from phase change problems, to appear in Nonlinear Anal. TMA.

27. N. Kenmochi and M. Niezgódka, Systems of nonlinear parabolic equations for phase change problems, Adv. Math. Sci. Appl. 3(1993/94), 89-117.

28. N. Kenmochi and M. Niezgódka, Nonlinear system for non-isothermal diffusive phase separation, to appear in J. Math. Anal. Appl.

29. N. Kenmochi and M. Niezgódka, Large time behaviour of a nonlinear system for phase separation, *Progreee in PDEs: the Metz Surveys 2*, pp. 12-22, Pitman Reseach Notes Math. **296**, Longman, New York, 1993.

30. N. Kenmochi and M. Niezgódka, A perturbation model for non-isothermal diffusive phase separations, preprint, Chiba Univ., 1993.

31. N. Kenmochi, M. Niezgódka and I. Pawlow, Subdifferential operator appraoch to the Cahn-Hilliard equation with constraint, to appear in J. Differential Equations.

32. N. Kenmochi and I. Pawlow, A class of nonlinear elliptic-parabolic equations with time-dependent constraints, Nonlinear Anal. TMA., **10**(1986), 1181-1202.

33. N. Kenmochi and I. Pawlow, Asymptotic behaviour of solutions to parabolic-elliptic variational inequalities, Nonlinear Anal. TMA 13(1989), 1191-1213.

34. Ph. Laurençot, A double obstacle problem, to appear in J. Math. Anal. Appl.

35. Ph. Laurençot, Weak solutions to a Penrose-Fife phase-field model for phase transitions, preprint, 1993.

36. J. L. Lions, *Quelques méthodes de résolution des problémes aux limites non linéaires*, Dunod, Gauthier-Villars, Paris, 1969.

37. S. Luckhaus and A. Visintin, Phase transition in multicomponent systems, Manuscripta Math. 43(1983), 261-288.

38. U. Mosco, Convergence of convex sets and of solutions of variational inequalities, Advances Math. 3(1969), 510-585.

39. O. A. Oleinik, On a method of solving the general Stefan problem, Soviet Math. Dokl. 1(1960), 1350-1354.

40. Y. Oono and S. Puri, Study of the phase separation dynamics by use of cell dynamical systems, I. Modelling., Phys. Rev. A 38(1988), 434-453

41. O. Penrose and P. C. Fife, Thermodynamically consistent models of phase-field type for the kinetics of phase transitions, Physica D, **43**(1990), 44-62.

42. W. Shen and S. Zheng, On the coupled Cahn-Hilliard equations, preprint.

43. J. Sprekels and S. Zheng, Global smooth solutions to a thermodynamically consistent model of phase-field type in higher space dimensions, to appear in J. Math. Anal. Appl.

44. R. Temam, *Infinite Dimensional Dynamical Systems in Mechanics and Physics*,Springer Verlag, Berlin, 1988.

45. A. Visintin, Stefan problems with phase relaxation. IMA J. Appl. Math. 34(1985), 225-245.

46. W. von Wahl, On the Cahn-Hilliard equation $u' + \Delta^2 u - \Delta f(u) = 0$, Delft Progress Report 10(1985), 291-310.

47. S. Zheng, Asymptotic behaviour of the solution to the Cahn-Hilliard equation, Applicable Anal. 23(1986), 165-184.

48. S. Zheng, Global existence for a thermodynamically consistent model of phase field type, Differential Integral Eq., **5**(1992), 241-253.

QUASIPLASTICITY AND PSEUDOELASTICITY IN SHAPE MEMORY ALLOYS

Y. Huo *, I. Müller, S. Seelecke

FB 10 - Thermodynamik - TU Berlin
*TU München

TABLE OF CONTENTS

1. PHENOMENA

1.1. Quasi-Plasticity and Pseudo-Elasticity (Schematic)

The best known property of shape memory alloys - from which their name is derived - is the recovery of the original shape upon heating after a (quasi) plastic deformation. Figure 1.1 explains that behaviour in a schematic picture: A straight wire is bent into a spiral at low temperature and by heating recovers its straight shape which it retains after cooling to the initial temperature. It is *as if* the wire remembered its original shape.

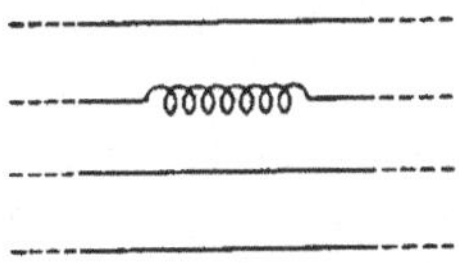

before deformation	at 20° C
after deformation	at 20° C
after heating	at 60° C
after cooling	at 20° C

Figure 1.1. The shape memory effect.

This is a spectacular effect indeed and yet it is only one of the amazing properties of shape memory alloys. We may say that such materials are characterized by a strong dependence of the load deformation diagrams upon temperature. Figure 1.2. illustrates that schematically. It shows the load **P** as a function of deformation **d** for four different temperatures $\mathbf{T_1}$ through $\mathbf{T_4}$.

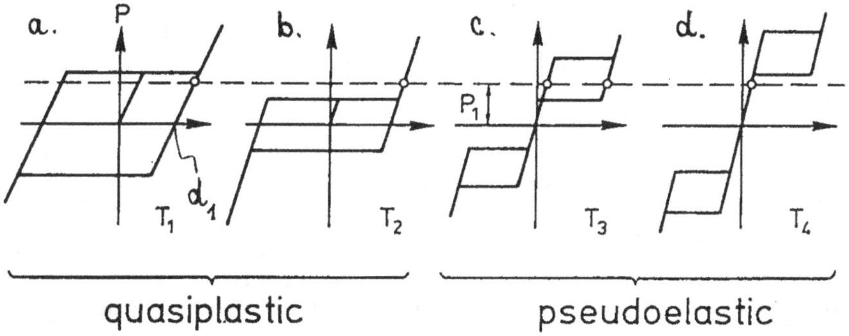

Figure 1.2. Quasi-plasticity (a,b) and pseudo-elasticity (c,d).

In tensile loading **(P > 0)** at low temperatures beginning at **(P,d)=(0,0)** we first have an elastic deformation which is followed by yield and, upon unloading, by a residual deformation, just as in a plastic body. Unlike plasticity there is a limit to the yield, and the possibility to load elastically - after the limit has been reached - far beyond the yield load. For compressive loading **(P < 0)** we observe analogous phenomena, as illustrated in Figure 1.2a., and consequently the state of a specimen runs through a *hysteresis loop*

around the origin in a process that alternates between tensile and compressive loading, provided that the yield load is surpassed. Figure 1.2b. shows that the yield load is decreased at a slightly higher temperature. Because of the similarity of the **(P,d)**-behaviour to that of a plastic body we say that the body is quasi-plastic at low temperatures.
At higher temperatures the **(P,d)** curves are rather different, see Figures 1.2c. and 1.2d. To be sure, there is still an elastic line through the origin, a yield load and a limit to the yield which may be followed by an elastic deformation up to large values of **P**; all this is similar to the quasi-plastic case. *But*, upon unloading, when the load falls below a recovery load, the body recovers what it yielded before. It thus returns to the original elastic line and, upon complete unloading, to the origin. This behaviour is called pseudo-elastic. It is *elastic* in the sense that the state of the body returns to the origin in a loading-unloading cycle; we call it *pseudo-elastic*, because in such a cycle the state traces out a *hysteresis loop.* That loop lies in the first quadrant for tensile loads and in the third quadrant for compressive loads. The hysteresis loops move away from the origin for higher temperatures without significantly changing their size.

The curves of Figure 1.2. imply the shape memory effect. Indeed, suppose that at low temperature we have given the body with the residual deformation $\mathbf{d_1}$ (see Figure 2a). If we increase the temperature to $\mathbf{T_3}$ (say), P=0 and $\mathbf{d=d_1}$ is no longer a possible state of the body. Infact, the only unloaded state at $\mathbf{T_3}$ occurs for **d=0** and therefore the body returns to the origin. Typically a recoverable strain is about 5% and the temperature interval $\mathbf{T_4}$-$\mathbf{T_1}$ may be about 50° C around room temperature.

While in this series of lectures we shall treat both quasi-plasticity and pseudo-elasticity we first proceed by looking more closely at pseudo-elasticity. This phenomenon is easier to observe, because it does not require compression, and it is also more interesting thermodynamically than quasi-plasticity, because it is close - in some sense to be specified - to thermodynamic equilibrium.

1.2. Pseudo-Elasticity and Ideal Pseudo-Elasticity

Fu has conducted extensive experiments with CuZnAl single crystals. The tests were uniaxial tensile tests with specimens as shown in Figure 1.3. The total length is 5 cm, the width of the narrow part is 0,8 cm and the thickness is 0,15 cm. Striations appear on the surface of the specimen during the test - up to 40 per mm - as indicated by the micrograph; they will be interpreted later.

The pseudo-elastic hysteresis loops measured by Fu are shown on the left hand side of Figure 1.4. On the right hand side we have idealized the observed behaviour. The ideal curve is characterized by two steep parallel elastic lines and two horizontal lines for yield and recovery. We call such behaviour *ideally pseudo-elastic*; it permits the formulation of a simple analytic mathematical model as we shall see.

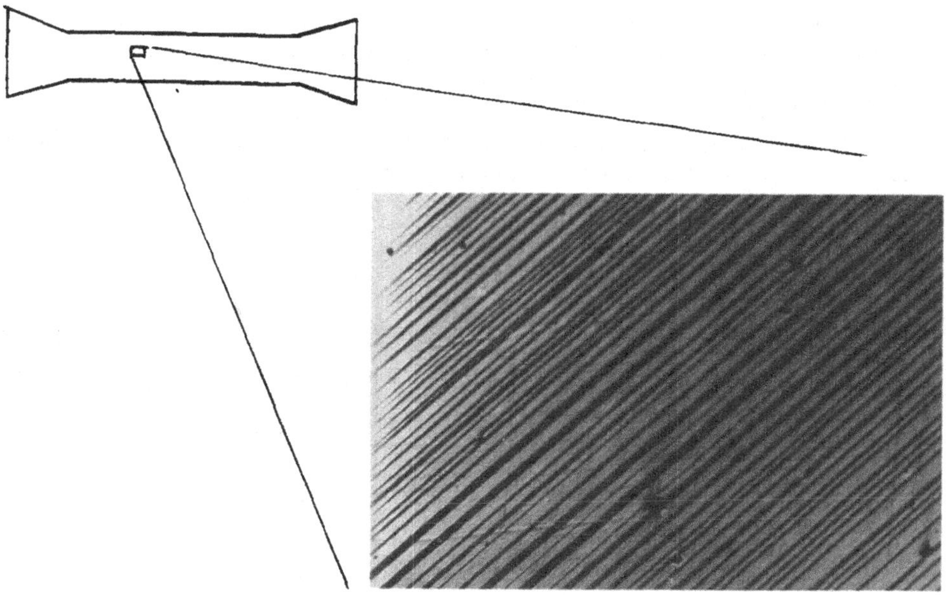

Figure 1.3. Tensile specimen in pseudo-elastic range.

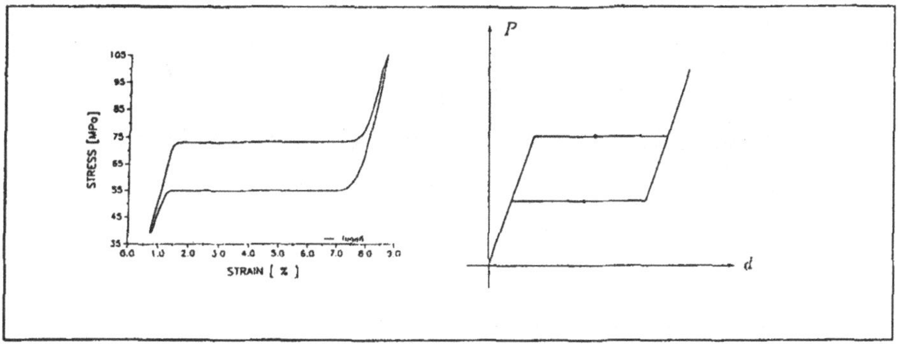

Figure 1.4. Pseudo-elasticity. Left: experimental. Right: ideal.

The shift of the pseudo-elastic hysteresis loop to higher values of **P** with increasing temperatures is documented on the left hand side of Figure 1.5. for four different temperatures. On the right hand side we see how that shift represents itself in ideal pseudo-elasticity.

It may be said that the key to the understanding of hysteresis phenomena does not lie in the loop as such, but in the states inside the loop. Figure 1.6., top left shows what happens when the yield is interrupted and the deformation decreased. At first the state moves steeply into the loop but then - upon reaching a hypothetical diagonal - it turns into a horizontal line which we shall call an *internal recovery line*. Analogous changes occur when the recovery is interrupted, see top right in Figure 1.6. In this case we observe internal yield and, as far as we can tell, this starts on the same line as does the internal recovery. The bottom of Figure 1.6. shows how this behaviour is schematized in ideal pseudo-elasticity. The steep curves are elastic, i.e. we may move up or down along them.

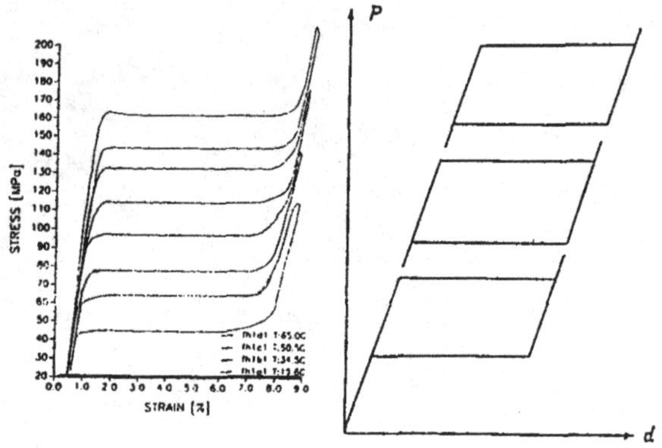

Figure 1.5. Temperature dependence of the hysteresis. Left: experimental. Right: ideal.

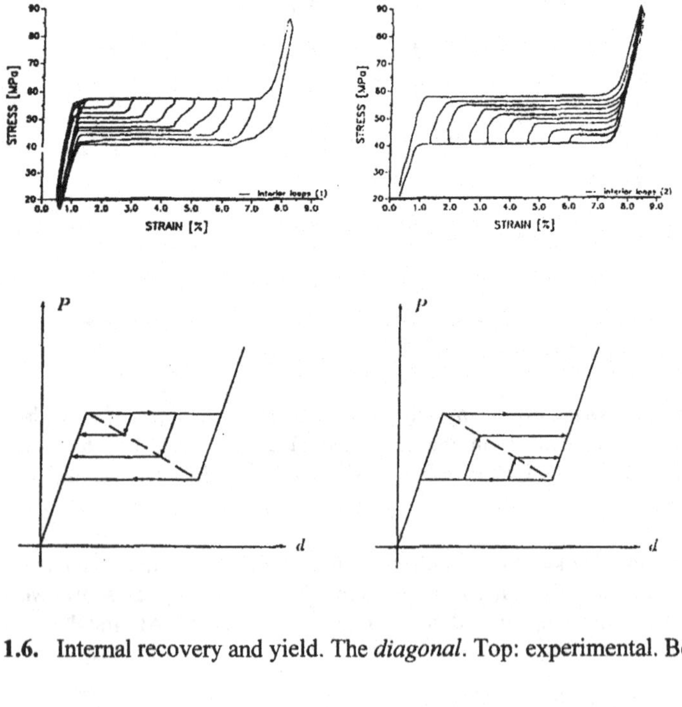

Figure 1.6. Internal recovery and yield. The *diagonal*. Top: experimental. Bottom: ideal

The internal yield may also be interrupted by elastic unloading and reloading as shown in Figure 1.7., left. After reloading the internal yield continues at the same load where it was previously interrupted, see Figure 1.7., right.

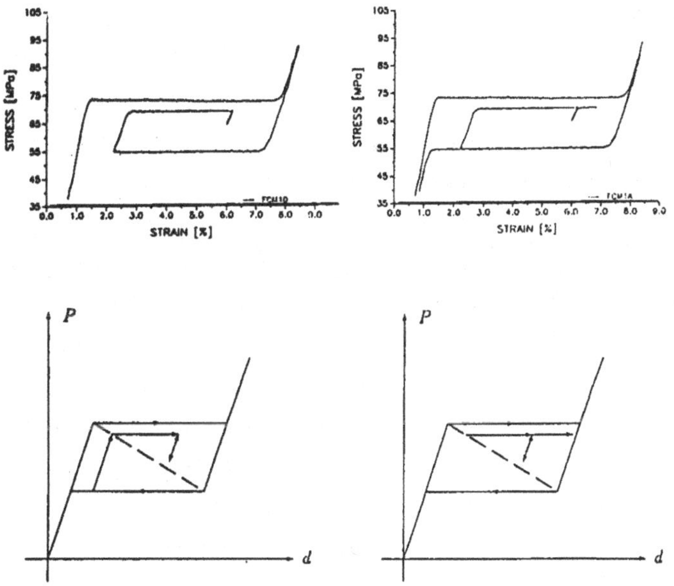

Figure 1.7. Internal elasticity and history dependence. Top: experimental. Bottom: ideal.

1.3. (d,T)-Diagrams and (P,T)-Diagrams

It is not only in the **(P,d)**-diagram that we observe hysteresis loops in shape memory alloys, but also in **(d,T)**-diagrams at fixed loads and in **(P,T)**-diagrams at fixed deformations. Infact these hystereses are the consequence of the loops in the **(P,d)**-diagram. Figure 1.8. illustrates that schematically. That curve refers to a fixed load $\mathbf{P_1}$, the same load which is indicated in Figure 1.2. Also the little circles at temperatures $\mathbf{T_1}$ through $\mathbf{T_4}$ in the two figures correspond to each other and inspection shows that, given **(P,d)**-curves for enough different temperatures, we may construct a **(d,T)**-diagram.

It is diagrams like those of Figure 1.8. which serve as the basis of most applications of shape memory alloys. Note that a simple alternation of temperature triggers contraction and elongation. It is obvious that this phenomenon may be used to construct thermally activated switches and engines for the conversion of thermal into mechanical energy.

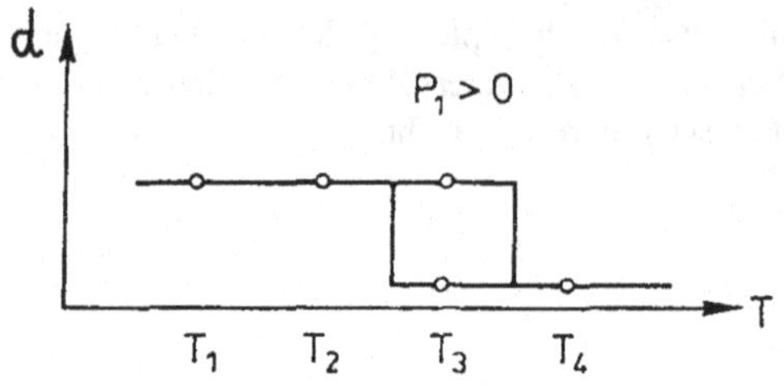

Figure 1.8. **(d,T)-diagram at fixed load $P=P_1>0$.**
(Compare circles with those of Figure 2).

Figure 1.9. shows actually measured **(d,T)**-diagrams (top) and their idealization (bottom). We see that we may move into the hysteresis loops by interrupting the cooling process. Once again, when the diagonal across the loop is touched, a drastic change of deformation occurs inside the loop.

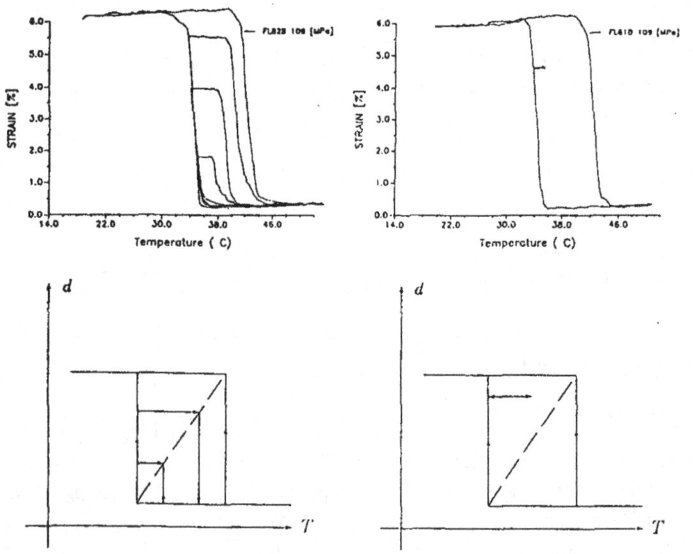

Figure 1.9. Internal jumps and thermal path. Top: experimental. Bottom: ideal.

Figure 1.10. shows **(P,T)**-diagrams at constant deformation. Much the same phenomena occur as in the **(P,d)**- and **(d,T)**-diagrams and they are recorded in the top figures while the ideal behaviour is represented by the bottom figures.

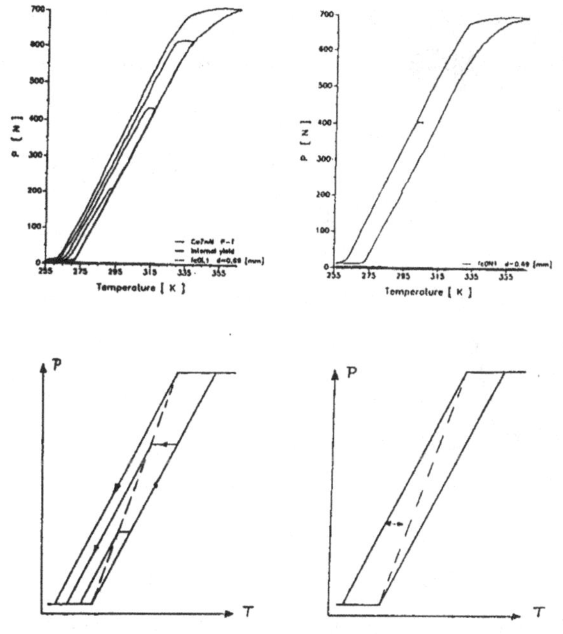

Figure 1.10. The **(P,T)**-diagram. Top: experiment. Bottom: ideal.

1.4. Simultaneous Change of P,d, and T. The Standard Test

Sofar in all figures we have ovserved iso-curves, i.e. curves of constant **T,P** or **d.** Such a restriction helps to fix the ideas and to provide an easy intuitive understanding for the material comportment. But it is also instructive to have two time-dependent input functions and observe the third function as an output. We shall discuss the case where temperature and load are input functions while the deformation is the output.

Specifically we prescribe the load as a triangular tensile load which zig-zags up and down taking about 40 seconds for each period. At the same time the temperature increases in about 10 minutes from a low value, stays constant at a high value for 10 minutes and than falls back to the initial value. Figure 1.11. shows these two input functions and also the deformation as an output. The deformation at first alternates up and down - along with the load - about a large mean value. When the temperature increases, the deformation drops sharply and proceeds to alternate about a small mean value until, upon recooling the former large deformation is reached, see Figure 1.11., left.

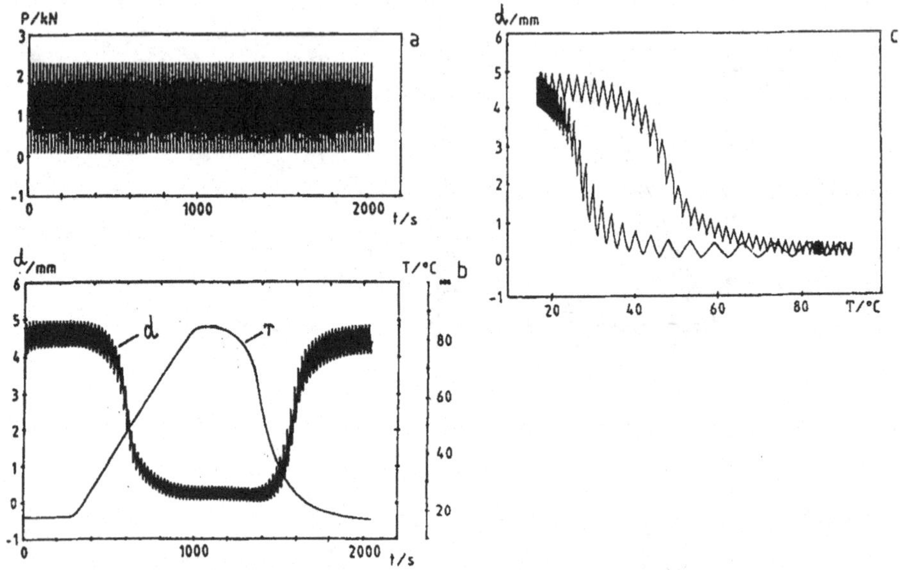

Figure 1.11. Response to triangular tensile loading and change of temperature.

If we eliminate the time between the two functions **d=d(t)** and **T=T(t)** we obtain the **(d,T)**-curve which is shown on the right hand side of Figure 1.11. This hysteretic curve must be compared with the curves of Figures 1.8. and 1.9. and we clearly see the similarity, except that there are no zig-zags in the previous figures since the load was constant.

The reason for choosing this particular process need not concern us here. Ehrenstein has found the process useful since it permitted the determination of many material parameters of the tensile specimen, such as elastic moduli, transition temperatures, etc. For that reason Ehrenstein has called the process the *standard test*.

1.5. Phase Transition

In view of the above description of phenomena we now face two problems

DESCRIBING and EXPLAINING

The "description" of the phenomena is entirely macroscopic. In order to succeed we have to *invent* a free energy, function of **d** and **T**, which provides the observed **(P,d)-**, **(d,T)-** and **(P,T)-diagrams.**
The "explaining" of the phenomena starts from more basic principles. Typically it uses a molecular model and, by virtue of the methods of statistical thermodynamics, it *derives* a free energy with the correct properties.

We shall undertake both, describing and explaining in subsequent chapters. The key to both is the observation - made by metallurgists - that shape memory and pseudo-elasticity are consequences of a martensitic-austenitic phase transition and of twinning in the martensitic phase. The austenite is the high-temperature phase; it has a highly symmetric lattice structure, typically a cubic one. The low temperature phase martensite is less symmetric. Typically lattice cells of the martensitic phase may be considered as having originated from the austenitic cube by a deformation consisting of

a stretch, a shear, and a rotation

as indicated in Figure 1.12. All in all 24 such deformations are possible and accordingly there are 24 possible "twins".

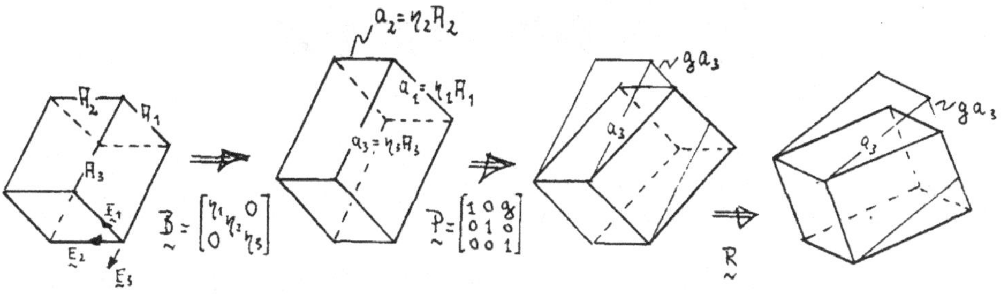

Figure 1.12. Martensitic transformation of an austenitic cubic cell.

The coherency of the lattice is maintained in the phase transition and the twinning. This means that the different lattices of the phases and twins must be carefully accommodated. Such an accommodation is only possible along certain directions and therefore the crystallographic structure of the atomic lattice is a determining factor in the macroscopic distribution of the phases and twins.

Figure 1.13. provides an excellent illustration of this fact. It shows a thin CuAlNi single crystal plate in cooling from 80° C down to room temperature. As the sample cools more and more martensitic "arrows" appear, easily recognizable from their yellow colour on the red austenite.

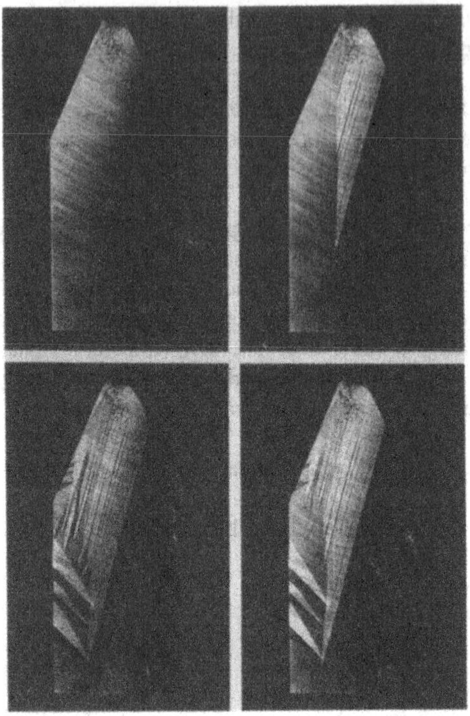

Figure 1.13. Formation of martensitic arrows upon cooling of a $CuAl_{14}Ni_{4.2}$ (at %) single crystal (natural size).

Each arrow has two sides consisting of two different twins. The arrows protrude from the austentic plane in such a way that the ridge between the two sides is elevated by a fraction of a millimeter on one side, while on the opposite side it is indented.

1.6. Transition in Quasi-plasticity and Pseudo-elasticity

While in cooling of unloaded samples usually several different twins appear this is not so in uniaxial tension or compression. Here we observe only one type of twin for tension and one for compression. That, of course will be the one whose direction of shear lies closest to the direction of maximum shear stress which is at 45° C to the direction of the applied load. Therefore in the uniaxial case it suffices to consider austenite and just two twins.

At low temperature, e.g. for $\mathbf{T_1}$ and $\mathbf{T_2}$ in Figure 1.2a,b, there is no austenite, only martensite and the figure is drawn for the case that the two twins - denoted by $\mathbf{M_+}$ and $\mathbf{M_-}$ - are present in equal proportions in the unloaded and undeformed state. The yield in tension is then due to a twinning deformation $\mathbf{M_-} \rightarrow \mathbf{M_+}$.

At high temperatures, e.g. for T_3 and T_4 in Figure 1-2c,d, we have only austenite in the unloaded state. When the load grows in tension and the yield load is reached, the austenite is forced into the martensitic twin M_+. This is a typical case for a load-induced phase transition. During recovery the reverse transition occurs, i.e. $M_+ \rightarrow A$.

In compression we observe the same transition except that **A** is converted into M_- during compressive yield. In Figure 1.14. these transitions are schematically indicated.

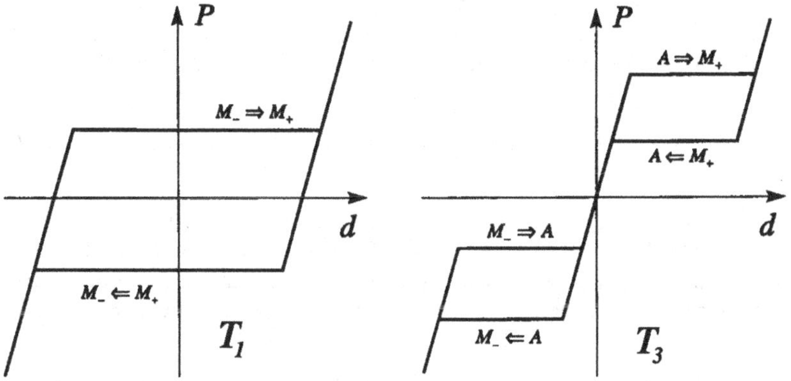

Figure 1.14. Twinning and Phase Transitions in Yield and Recovery.

The striations shown on the tensile specimen in Figure 1.3. represent the martensitic layers appearing during the yield in the $A \rightarrow M_+$ transition in the pseudo-elastic range. In situ observations conducted during the phase transitions always show that the transition proceeds by flipping more and more layers from the austenitic to the martensitic phase. The layers are of unequal width and there are up to 2000 of them on a tensile specimen of 4 cm in length.

2. NON-CONVEX THERMODYNAMICS

2.1. Non-Convexity of Free Energy and Non-Monotonicity of Load

Ever since van der Waals advanced his statistical mechanics of real gases the analytical description of phase transitions has made use of non-convex free energies and non-monotone loads. This is a popular approach among mathematicians. Infact, mathematicians are not concerned with whether a statistical model exists that gives rise to a non-convex free energy, and under what conditions. And certainly they cannot be bothered to *derive* such a free energy from a statistical model. They just *assume* it

In this chapter we do just that: assume a non-convex free energy and exploit its consequences. We shall focus the attention on pseudo-elasticity and assume functions

$$f = f(d,T) \quad \text{and} \quad P = P(d,T) = \left(\frac{\partial f}{\partial d}\right)_T \qquad \textbf{(2.1)}$$

of the form shown in Figure 2.1. On the left hand side of that figure we see smooth functions. Analytically $f(d,T)$ is a bi-cubic function of d. While this is a simple function - certainly much simpler than any function that would follow from statistical mechanics - it is not simple enough for analytical thermodynamic calculations.

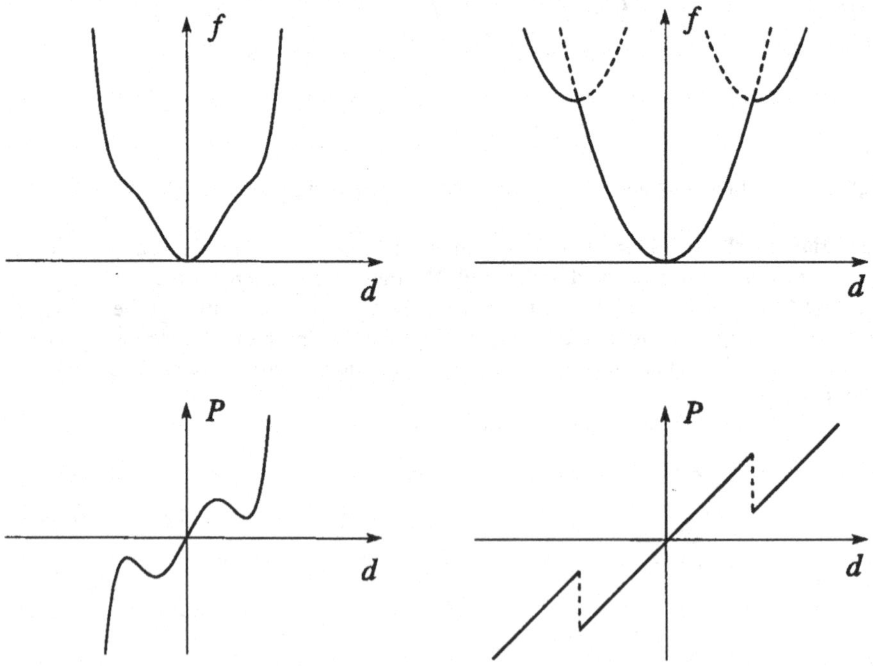

Figure 2.1. Two mathematical models for the description of pseudo-elasticity.

Therefore we shall also consider curves of the type shown on the right hand side of Figure 2.1. Here the free energy consists of a train of parabolae with equal curvatures, and the load-deformation curve consists of straight lines.These types of curves will permit us to model the ideal pseudo-elasticity described in Chapter 1.

In the language of Section 1.5 the functions shown in Figure 2.1. will help us to describe the phenomena, at least qualitatively. An attempt to explain the phenomena will follow in a later chapter.

2.2. Thermodynamic Stability

Thermodynamic stability conditions are consequences of the 1st and 2nd law of thermodynamics. These laws read for a body B with boundary ∂ **B**

$$\frac{d(U+K)}{dt} + \int_{\partial B} q_i n_i dA = \int_{\partial B} t_{ij} v_i n_j dA \quad , \tag{2.2}$$

$$\frac{dS}{dt} + \int_{\partial B} \frac{q_i n_i}{T} dA = \Sigma \geq 0 \quad . \tag{2.3}$$

Here U, K, S are the internal and kinetic energy and the entropy of the body. q_i is the heat flux, t_{ij} the Cauchy stress tensor, v_i the velocity, n_i the outer normal and dA an area element on the boundary ∂ **B.** P is the absolute temperature and Σ the entropy production

If we consider a quasi-static process, the kinetic energy may be ignored. For a uni-axial tensile load P and for a homogeneous temperature T the first and second laws reduce to

$$\frac{dU}{dS} - \mathring{Q} = P\mathring{D} \tag{2.4}$$

$$\frac{dS}{dt} - \frac{\mathring{Q}}{T} = \Sigma \geq 0 \quad . \tag{2.5}$$

Here $\mathring{Q} = -\int_{\partial B} q_i n_i dA$ is the total heat flux or "heat added". $\mathring{Q}$ is positive when the body is heated, negative otherwise. $\mathring{D}$ is the time derivative of the total elongation. The pressure and temperature variations in the surrounding air are neglected.

We eliminate $\mathring{Q}$ from the two equations (2.4), (2.5) and introduce the Helmholtz free energy $F = U - TS$ and the Gibbs free energy $G = F - PD$ and obtain

$$T\Sigma = \quad P\mathring{D} - S\mathring{T} - \frac{dF}{dt} \geq 0 \quad . \tag{2.6}$$

$$T\Sigma = -D\overset{\circ}{P} - S\overset{\circ}{T} - \frac{dG}{dt} \geq 0 \quad . \tag{2.7}$$

It follows that for D and T constant the Helmholtz free energy F approaches a minimum in equilibrium, while for P and T constant it is the Gibbs free energy G that becomes minimal in equilibrium.

2.3. Free Energy and Deformation of a Phase Mixture

We introduce the specific free energy $f = {}^{F}\!/_{m}$ and the specific deformation $d = {}^{D}\!/_{m}$ of the body and assume an $(f,d)-$diagram and a-$(P,d)-$diagram of the qualititative form shown in Figure 2.1. The non-monotonicity of the $(P,d)-$curve gives us the possibility of having two different deformations for the same load: this is a necessary ingredient for a proper description of a phase transition. The third possible deformation in the middle is ignored for being unstable and indeed: An *increase* of the load in the middle range of the $(P,d)-$curve leads to a *decrease* of deformation.

In the present case of a pseudo-elastic body we associate the first ascending branch of the $(P,d)-$curve with the austenitic phase and the second one with the martensitic phase. If the specimen is partly austenitic and partly martensitic we have for the free energy F and for the elongation D

$$F = F_A + F_M + \Delta F, \quad D = D_A + D_M \tag{2.8}$$

or for the specific values

$$f = (1-z) f(d_A) + z\, f(d_M) + \Delta f \tag{2.9}$$

$$d = (1-z) \quad d_A \quad + z \quad d_M \quad . \tag{2.10}$$

Here $f(d_A) = {}^{F_A}\!/_{m_A}$, $f(d_M) = {}^{F_M}\!/_{m_M}$ and $z = \frac{m_M}{m}$, where m_A and m_M are the masses of the body in the A- und M-phase respectively. Thus z is the phase fraction of martensite.

ΔF is a contribution of the free energy due to the mixing of the phases, or better: due to the coherency of phases along the interfaces. Consequently Δf is the specific coherency energy. We assume that each interface contributes the same amount to the coherency energy and that the number of interfaces is proportional to $z(1-z)$.This seems to be the simplest expression that satisfies the necessary requirement that $\Delta f = 0$ holds for either $z = 0$ or $z = 1$.

Thus we come to the assumption

$$\Delta f = \quad A z\,(1-z) \tag{2.11}$$

A will be called the coherency coefficient, it is assumed positive so that there is a penalty for the formation of interfaces.

2.4. Minimizing the Free Energy F

We recall that the free energy approaches a minimum in equilibrium at constant D and T and minimize

$$f(d_A, d_M, z) = (1-z)f(d_A) + z\,f(d_M) + Az(1-z) \tag{2.12}$$

under the constraint

$$d = (1-z)\;d_A + z\;d_M \quad . \tag{2.13}$$

We introduce a Lagrange multiplier λ to take care of the constraint and obtain by differentiation with respect to $d_{A,}d_M$ and z

$$\lambda = \left.\frac{\partial f}{\partial d}\right|_{d_A} = \left.\frac{\partial f}{\partial d}\right|_{d_M} = \frac{f(d_M) - f(d_A) + A(1-2z)}{d_M - d_A} \tag{2.14}$$

These are equilibrium conditions: Two equations for the equilibrium values of $d_{A,}d_M$ and z.

Along with (2.13) there are enough equations for the calculation of the equilibrium values of $d_{A,}d_M$ and z. Unfortunately the equations are not specific, since we have not specified the form of $f(d)$ other than saying that it is a non-convex function. Therefore $d_{A,}d_M$ and z cannot be calculated analytically. However it is instructive to find a graphical solution.

The graphical solution starts out from the observation that λ equals the load on the two phases A and M.

Indeed, since $P = \frac{\partial f}{\partial d}$ holds, we conclude from $(2.14)_{1,2}$ that the load on the two phases is equal in equilibrium and that this common load is equal to λ.

By writing

$$f(d_M) - f(d_A) = \int_{d_A}^{d_M} P(\alpha)\,d\alpha \tag{2.15}$$

we may write the equilibrium condition (2.14) as

$$P(d_M - d_A) - \int_{d_A}^{d_M} P(\alpha)\,d\alpha = A(1-2z), \tag{2.16}$$

and this form lends itself to an easy graphical solution as illustrated in Figure 2.2.

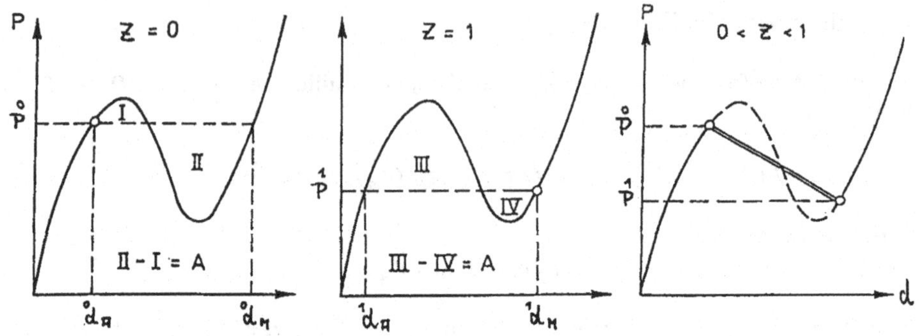

Figure 2.2. Construction of the phase equilibrium line.

According to (2.16) the (common) load on the two phases and the phase deformations d_A and d_M in equilibrium are found by making the difference between the rectangle $P(d_M - d_A)$ and the area $\int_{d_A}^{d_M} P(\alpha)\,d\alpha$ equal to $A(1-2z)$. Obviously $P, d_M,$ and d_A thus depend on z. Figure 2.2 shows the values appropriate for $z = 0$ and $z = 1$.

Intermediate values of z require that the difference of $P(d_M - d_A)$ and $\int_{d_A}^{d_M} P(\alpha)\,d\alpha$ has some value between - A and A. This determines the appropirate equilibrium values of $P(z), d_M(z)$ and $d_A(z)$ and therefore - by (2.13) - the corresponding value $d(z)$. The curve $(P(z), d(z))$ runs downward from $(\overset{0}{P}, \overset{0}{d}_A)$ to $(\overset{1}{P}, \overset{1}{d}_A)$, cf. right hand side of Figure 2.2. Such a curve, along which a decrease of load increases the deformation, raises questions of stability which we shall discuss in some detail later.

Note that the graphical construction described heretofore reduces to the Maxwell construction of "equal areas" in case $A = 0$ holds. In that case there is only one load at which phase equilibrium can occur and the equilibrium deformations are independent of z.

2.5. An Analytical Model

In order to be able to study the question of stability rigorously and so as to apply analytic methods rather than graphical ones we now introduce the mathematical model of ideal pseudo-elasticity. This model is characterized by a free energy that consists of a train of two intersecting parabolae, viz.

$$f(d,T)=\begin{cases} f^A(d,T)=\frac{1}{2}\alpha d^2 & +c(T-T_R)-cT\ln\frac{T}{T_R}+\varepsilon_A-T\eta_A \\ f^M(d,T)=\frac{1}{2}\alpha(d-\Delta_d)^2 & +c(T-T_R)-cT\ln\frac{T}{T_R}+\varepsilon_M-T\eta_M \end{cases} . \quad (2.17)$$

Figure 2.3. shows a plot of these parabolae in the *(f,d)* diagram for one temperature. α is the elastic modulus of both phases and *c* is their common specific heat. T_R is some reference temperature. The form (2.17) implies for the specific values of internal energy and entropy of the phases

$$u^A(d,T)=\frac{1}{2}\alpha\, d^2+c(T-T_R)+\varepsilon_A,\quad u^M(d,T)=\frac{1}{2}\alpha(d-\Delta_d)^2+c(T-T_R)+\varepsilon_M$$
$$s^A(d,T)=c\ln\frac{T}{T_R}+\eta_A,\quad s^M(d,T)=c\ln\frac{T}{T_R}+\eta_M \quad . \quad (2.18)$$

The ε's and η's are the values of *u* and *s* of the phases in the undeformed state at $T=T_R$. For their differences we introduce the notation $\Delta_\varepsilon=\varepsilon_M-\varepsilon_A$, $\Delta_\eta=\eta_M-\eta_A$. Note that s_A and s_M are independent of *d*.

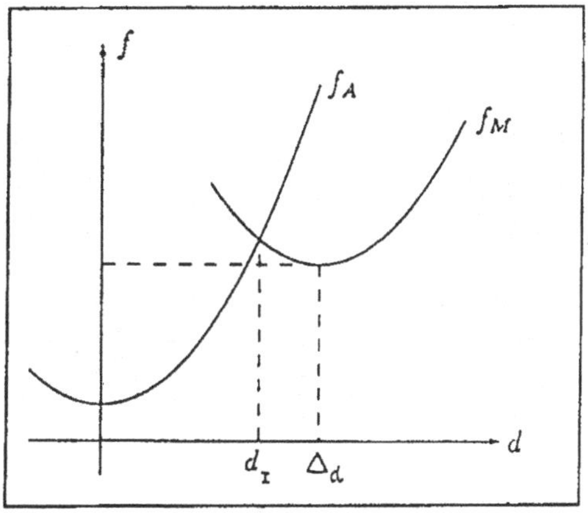

Figure 2.3. Free energy of *A*- and *M*-phases

The point of intersection of the two parabolae is given by

$$d_I=\frac{\Delta_d}{2}\left(1+2\frac{\Delta_\varepsilon-T\Delta_\eta}{\alpha\,\Delta_d^2}\right), \quad (2.19)$$

and the temperature at which that intersection occurs at a prescribed value of *d* is

$$T_I = -\alpha\ \frac{\Delta_d}{\Delta_\eta}\ \left(d - \frac{1}{2}\Delta_d\right) + \frac{\Delta_\varepsilon}{\Delta_\eta}\ . \tag{2.20}$$

The load $P = \left(\frac{\partial f}{\partial d}\right)_T$ comes out as

$$P = \begin{cases} \alpha\ \ d & \textit{for the A phase} \\ \alpha\ (d - \Delta_d) & \textit{for the M phase} \end{cases} \tag{2.21}$$

It is plotted in Figure 2.4. For large d the lower curve is the stable and relevant one, since it corresponds to the lower free energy. The experimental evidence - cf. Figure 1.5 - shows that austenite is stable to higher values of d for increased T. Therefore we expect d_I to grow with T and hence conclude that Δ_η must be negative.

Figure 2.4. also shows the *(d,T)* curves of the two phases and the *(P,T)* curves. None of these depend on *T*.

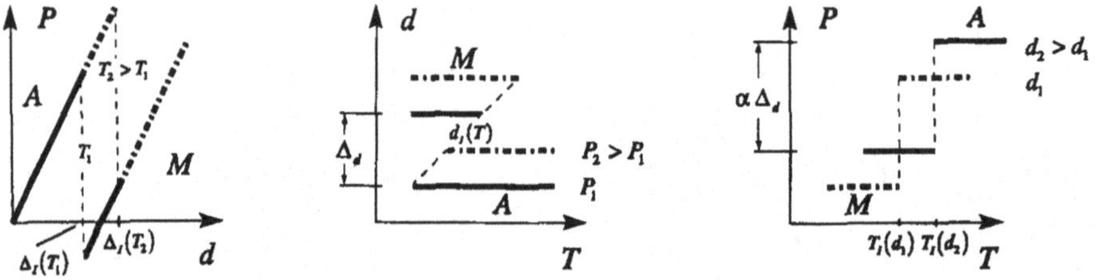

Figure 2.4. ***(P,d), (d,T),*** and ***(P,T)*** relations of pure phases.

2.6. Free Energy and Entropy Production of the Phase Mixture

We recall (2.9) with (2.11) and (2.10) for the phase mixture and write

$$\begin{aligned} f(d_A, d_M, z, T) &= (1-z) f^A(d_A, T) + z\, f^M(d_M, T) + Az(1-z) \\ d &= (1-z)\ d_A + z\ d_M\ . \end{aligned} \tag{2.22}$$

Insertion into (2.6) provides an expression for the specific entropy production σ, viz.

$$T\sigma = -\left(s + \frac{\partial f}{\partial T}\right)\overset{\circ}{T} + (1-z)\left(P - \frac{\partial f^A}{\partial d_A}\right)\overset{\circ}{d}_A + z\left(P - \frac{\partial f^M}{\partial d_M}\right)\overset{\circ}{d}_M + \qquad (2.23)$$

$$+\left[P(d_M - d_A) - (f^M - f^M) - A(1-2z)\right]\overset{\circ}{z} \geq 0 \quad .$$

2.7. Reversible Processes

Reversible processes are those for which the entropy production vanishes for all $\overset{\circ}{T}, \overset{\circ}{d}_A, \overset{\circ}{d}_M, \overset{\circ}{z}$, in particular for the "reversed" process with $-\overset{\circ}{T}, -\overset{\circ}{d}_A, -\overset{\circ}{d}_M, -\overset{\circ}{z}$, hence the name. Hence follow from (2.23) the four necessary conditions for a reversible process

$$s = -\frac{\partial f}{\partial T} \qquad\qquad s = (1-z)s^A(T) + z\, s^M(T)$$

$$P = \frac{\partial f^A}{\partial d_A} = \frac{\partial f^M}{\partial d_M} \qquad \Rightarrow \qquad \begin{array}{l} d_M - d_A = \Delta_d \\ P = \alpha\,(d - z\,\Delta_d) \end{array} \qquad (2.24)$$

$$P(d_M - d_A) - (f^M - f^A) = A(1-2z) \qquad\qquad P\Delta_d - (\Delta_\varepsilon - T\Delta_\eta) = A(1-2z) \quad .$$

Apart from the first one these are the equilibrium conditions which we have derived by minimizing the free energy, cf. (21.4) in Section 2.4. Because of this a reversible process is also often called an equilibrium process. Of course the conditions are now expressed more explicitly than in Section 2.4., since we have simple functions for f^A and f^M.

In particular we may calculate the *phase equilibrium line* - formerly constructed graphically - by elimination of z between $(2.24)_3$ and $(2.24)_4$. We obtain

$$P = P_E(d,T) = -\frac{\alpha\,A}{\frac{1}{2}\alpha\,\Delta_d^2 - A}\,d + \frac{\alpha\,\Delta_d(A + \Delta_\varepsilon - T\Delta_\eta)}{\alpha\,\Delta_d^2 - 2A} \quad . \qquad (2.25)$$

Assuming $A < \frac{1}{2}\alpha\,\Delta_d^2$ we conclude that this is a straight line which has a negative slope as indicated already in Figure 2.2. and, again, in the (P,d)-diagram of Figure 2.3. as part of the thick line.

There are phase equilibrium lines in the ***(d,T)***-diagram and ***(P,T)***-diagram as well. They result from (2.25) by holding ***P = const*** or ***d = const.*** In Figure 2.5. they are represented by the ascending parts of the thick lines.

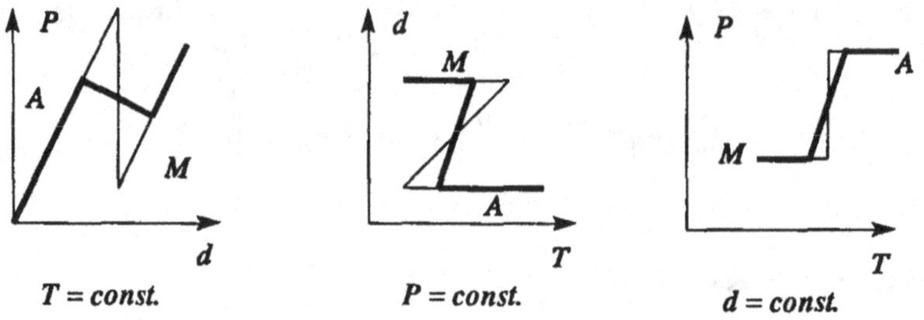

Figure 2.5. Equilibrium States. (The thin lines recall the curves of Figure 2.4.).

2.8. Discussion

Sofar there is no reason to believe that the phase equilibrium lines that connect the pure phases A and M might not be realistic. *But experiments show that infact they are not.* Indeed, instead of a phase change along a single line the experiments represented in Figures 1.4. through 1.11. exhibit hysteretic behaviour.

We must conclude that investigation of equilibriae and reversible processes are not quite appropriate in this case. And we shall proceed and try to link the observed hysteretic phenomena with the suspected instabilities of the phase equilibria.

3. PARTIAL EQUILIBRIUM, STABILITY, HYSTERESIS AND ENERGY DISSIPATION

3.1. Partial Equilibrium

In the previous chapter we have investigated reversible processes with arbitrary values $\overset{\circ}{T}, \overset{\circ}{d}_A, \overset{\circ}{d}_M, \overset{\circ}{z}$ in which the entropy production vanishes. In such processes the four equilibrium conditions (2.24) should hold. We may say therefore that equilibrium has four ingeredients

- thermal equilibrium, where $s = -\frac{\partial f}{\partial T}$,
- dynamic equilibrium, where the loads on the phases are given by $\frac{\partial f^A}{\partial d_A}$ and $\frac{\partial f^M}{\partial d_M}$,
- dynamic phase equilibrium, where the loads on the phases are equal,
- phase equilibrium, where the phase fractions satisfy $(2.24)_4$. This latter condition is sometimes called chemical equilibrium.

The phase equilibrium provides a problem because, according to it, the phase transition ought to occur reversibly along the thick lines of Figure 2.5. Instead in reality there is the hysteresis loop. We conclude that phase equilibrium is not established reversibly - at least not in the experiments reported in Chapter 1, even though these were quasi-static.

We shall now concentrate on this irreversibility and, in order to do so, we shall assume that in our quasi-static process thermal, dynamic and dynamic phase equilibria are established , while phase equilibrium is not. This means that a process with $\overset{\circ}{z} = 0$ and arbitrary values of $\overset{\circ}{T}, \overset{\circ}{d}_A, \overset{\circ}{d}_M$ is now still supposed to be reversible, while a process with $\overset{\circ}{z} \neq 0$ has a non-zero entropy production. Thus we still have

$$s = -\frac{\partial f}{\partial T} \qquad\qquad s = (1-z)s^A(T) + zs^M(T)$$

$$P = \frac{\partial f^A}{\partial d_A} = \frac{\partial f^M}{\partial d_M} \qquad \Rightarrow \qquad \begin{array}{l} d_M - d_A = \Delta_d \\ P = \alpha(d - z\Delta_d) \end{array} \tag{3.1}$$

but now there is an entropy production σ which reads

$$T\sigma = \left[P\Delta_d - (\Delta_\varepsilon - T\Delta_\eta) - A(1-2z)\right]\overset{\circ}{z} \geq 0. \tag{3.2}$$

The idea behind this assumption is that thermal and dynamic equilibrium are established in a very short time on the time scale of a quasi-static process, while phase equilibrium is lagging behind, perhaps far behind. In such a case we speak of a *partial equilibrium*.

The reversible paths experimentally detected inside hysteresis loops in Figures 1.7, 1.9. and 1.10. will be considered as paths of partial equilibrium. In other words we take them to be processes with $\overset{\circ}{z} = 0$ which seems eminently reasonable.

3.2. Stability of Phase Equilibrium

We use the conditions (3.1) and (2.10) of partial equilibrium to calculate

$$d_A = d - z\Delta_d \qquad \text{and} \qquad d_M = d - (z-1)\Delta_d \tag{3.3}$$

With these relations we may write the free energy (2.22) (with (2.17)) as a function of d and z, viz.

$$f(d,T,z) = \frac{1}{2}\alpha(d - z\Delta_d)^2 + c(T - T_R) - cT\,ln\frac{T}{T_R} + \varepsilon_A - T\eta_A + z(\Delta_\varepsilon - T\Delta_\eta) + Az(1-z). \tag{3.4}$$

The conditions for a minimum of f at fixed d and T are

$$\frac{\partial f}{\partial z} = 0 \qquad \Rightarrow \quad -a(d - z\Delta_d)\Delta_d + (\Delta_\varepsilon - T\Delta_\eta) + A(1-2z) = 0 \tag{3.5}$$

$$\frac{\partial^2 f}{\partial z^2} > 0 \qquad \Rightarrow \quad a\,\Delta_d^2 - 2A > 0 \ . \tag{3.6}$$

Equation (3.5) is identical to the condition $(2.24)_4$ of phase equilibrium and (3.6) shows that that equilibrium represents a minimum of the free energy, i.e. the equilibrium is stable - under the reasonable assumption that $A < \frac{1}{2} a\,\Delta_d^2$ holds.

We also investigate the phase equilibrium for fixed values of P and T: From the inequality (2.7) we conclude that in that case the Gibbs free energy g has a minimum in equilibrium. We have $g = f - Pd$ or by (2.22) (with (2.17)) and (3.3), $(3.1)_3$

$$g(P,T,z) = -\frac{P^2}{2\alpha} - Pz\Delta_d + c(T - T_R) - cT\,ln\frac{T}{T_R} + \varepsilon_A - T\eta_A + z(\Delta_\varepsilon - T\Delta_\eta) + Az(1-z). \tag{3.7}$$

The conditions for a minimum of g at fixed P and T are

$$\frac{\partial g}{\partial z} = 0 \qquad \Rightarrow \quad -P\Delta_d + (\Delta_\varepsilon - T\Delta_\eta) + A(1-2z) = 0 \tag{3.8}$$

$$\frac{\partial^2 g}{\partial z^2} > 0 \qquad \Rightarrow \quad -2A > 0. \tag{3.9}$$

Equation (3.8) is, once again, identical to the condition of phase equilibrium and (3.9) shows that that equilibrium represents a maximum - after all, we assume A < 0 - i.e. the equilibrium is *unstable*.

We summarize these results by saying that the points on the phase equilibrium line represent something like a saddle point. Among states with the same ***d*** they are minima while among states with the same ***P*** they are maxima. [The expression "saddle point" is not quite appropriate since it is two different functions, ***f*** and ***g*** respectively, which have a minimum or a maximum.] Figure 3.1. indicates the stability situation on the phase equilibrium line in a schematic, self-explanatory manner.

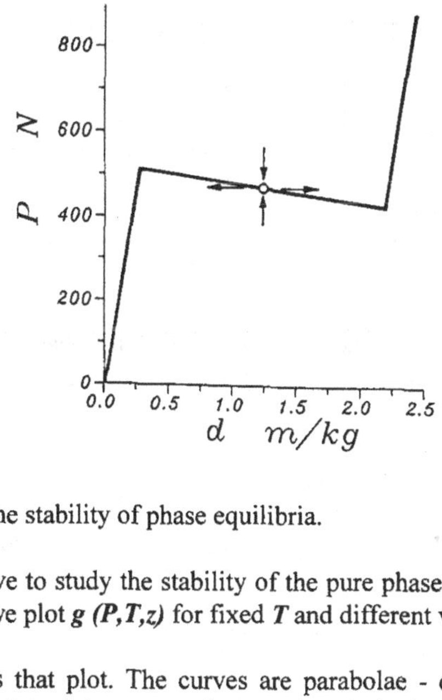

Figure 3.1. On the stability of phase equilibria.

It is also instructive to study the stability of the pure phases ***A*** and ***M*** under load control. For that purpose we plot $g\,(P,T,z)$ for fixed T and different values of P as a function of z.

Figure 3.2. shows that plot. The curves are parabolae - open to the bottom - and the minima are end-point minima at $z = 0$ or $z = 1$, or in both positions. We write (3.7) in the form

$$g+\frac{P^2}{2\alpha}-c(T-T_R)+cT\ln\frac{T}{T_R}-(\varepsilon_A-T\eta_A)=-\left[P\Delta_d-(\Delta_\varepsilon-T\Delta_\eta)\right]z+Az(1-z) \quad (3.10)$$

and denote the left hand side by $g - g_0$ and the square bracket on the right hande side by λ. From Figure 3.2. we conclude that the quantity λ determines the shape of the curves in the range $0 \geq z \geq 1$:

- For $P\Delta_d-(\Delta_\varepsilon-T\Delta_\eta)-A\leq 0$ the only minimum occurs at $z = 0$. Therefore the austenite is stable.

- For $P\Delta_d-(\Delta_\varepsilon-T\Delta_\eta)+A\geq 0$ the only minimum occurs at $z = 1$. Therefore the martensite is stable.

The maxima of the curves of Figure 3.2. occur at

$$z_I(P,T) = \frac{-P\Delta_d - (\Delta_\varepsilon - T\Delta_\eta) + A}{2A} . \tag{3.11}$$

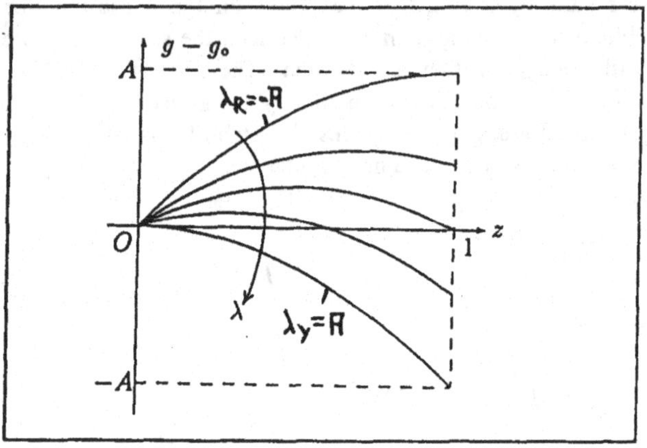

Figure 3.2. ***g*** as a function of *z*.

For given values of ***P*** and ***T*** these are the *z*-values for which the phase equilibrium occurs, cf. $(2.24)_4$. Obviously, being maxima, these equilibria are unstable.

3.3. Conjectures about Yield and Recovery

The fact that for $P\Delta_d - (\Delta_\varepsilon - T\Delta_\eta) = A$ the austenite becomes unstable suggests that the condition determines the yield load.

$$\lambda_y = P_y\Delta_d - (\Delta_\varepsilon - T\Delta_\eta) = A \quad . \tag{3.12}$$

Similarly since $P\Delta_d - (\Delta_\varepsilon - T\Delta_\eta) = -A$ determines the limit of stability of the martensite, we conjecture that this condition determines the recovery load

$$\lambda_R = P_R\Delta_d - (\Delta_\varepsilon - T\Delta_\eta) = -A \ . \tag{3.13}$$

In the ***(P,d)***-diagram of Figure 3.3. we have indicated the hysteresis corresponding to these conjectures by the dashed lines. The area of the hysteresis loop is easily calculated from (3.12), (3.13), viz.

$$(P_y - P_R)\Delta_d = 2A \ . \tag{3.14}$$

By (3.12), (3.13) both values P_y and P_R move up with temperature, as is indeed observed cf. Figure 1.5., but the size of the loop remains constant.

In the *(d,T)*-diagram the situation is similar. Upon heating the transition $M \rightarrow A$ is expected at the temperature T_A while in cooling the reverse transition occurs at T_M. The values of T_M and T_A result from

$$P\Delta_d - (\Delta_\varepsilon - T\Delta_\eta) = \pm A \tag{3.15}$$

by solving for T so that we have

$$T_A = \frac{\Delta_\varepsilon - P\Delta_d + A}{\Delta_d} \quad \text{and} \quad T_M = \frac{\Delta_\varepsilon - P\Delta_d - A}{\Delta_d} \tag{3.16}$$

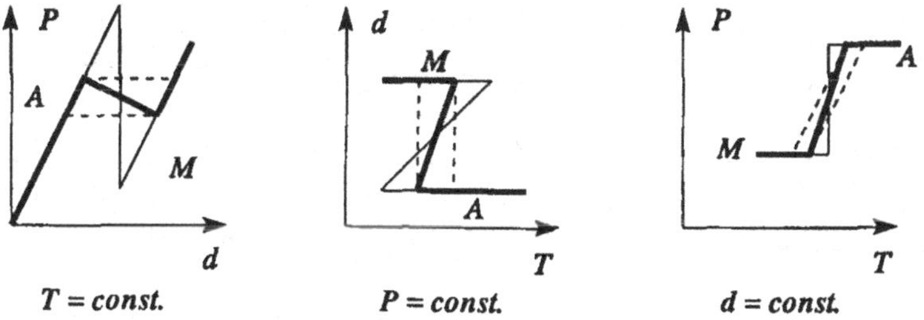

Figure 3.3. Conjectures about hysteresis.

Both temperatures increase with growing load at the same rate. The size of the loop is given by

$$(T_A - T_M)\Delta_d = -\frac{\Delta_d}{\Delta_\eta} A. \tag{3.17}$$

The interpretation of the (P,T)-diagram is less clear. We expect the $M \rightarrow A$ transition to start in the lower corner of the thick curve at

$$P = \alpha(d - \Delta_d), \qquad T = \frac{\Delta_\varepsilon - \alpha(d - \Delta_d) - A}{\Delta_\eta}, \tag{3.18}$$

and the transition should start in the upper corner at

$$P = \alpha d, \qquad T = \frac{\Delta_\varepsilon - \alpha(d - \Delta_d) - A}{\Delta_\eta}, \tag{3.19}$$

The equations (3.18), (3.19), satisfy (3.15) and it is obvious that the transition points move up and to the right for growing d. There is no way, however, of determining the slope of the dashed lines in the (P,T)-diagram of Figure 3.3. in this theory. Therefore we cannot evaluate the size of the loop.

This remark reminds us that even in the *(P,d)*- and *(d,T)*-diagram the theory did not predict horizontal or vertical transitions. They were drawn into the diagrams as conjectures only because experiments suggest these slopes, cf. Figures 1.4. and 1.9.

It is interesting to observe that the size of the hysteresis, which according to (3.14) should equal 2A, does not depend on the present simple mathematical model with parabolae as free energies. Indeed, if our conjectures are applied to the model shown in Figure 2.2. - with an arbitrary non-monotone load curve - we obtain a hysteresis area as in Figure 3.4. and that area also obviously has the value 2A.

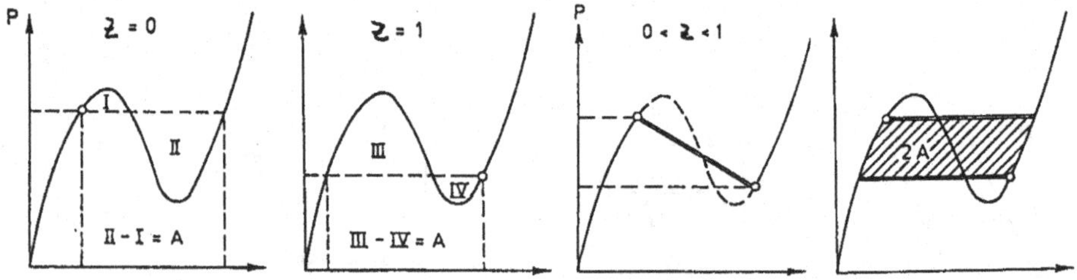

Figure 3.4. Construction of the hysteresis loop and of the phase equilibrium line.

3.4. Conjectures about Internal Yield and Recovery

We recall from the discussion of (3.11) that the points on the phase equilibrium line are unstable. It is tempting to associate this instability with the observed phenomena of internal yield and recovery, see Figure 1.6.: We may say that, even though in these experiments the elongation of the specimen is controlled, an individual layer in the specimen finds itself under load-control; it feels the load which is necessary to maintain the overall elongation and therefore tends to move away from the phase equilibrium line in the manner indicated in Figure 3.1. According to this conjecture the specimen ought to start internal recovery when the phase equilibrium line is reached from above. If it is reached from below, the specimen should start the internal yield.

3.5. Entropy Production and Energy Dissipation

We continue to investigate the case of partial equilibrium and calculate the entropy production during yield and recovery, or internal yield and recovery. By (3.2) we have, using, (3.11)

$$T\sigma = 2A(z - z_I(P,T))\overset{\circ}{z}. \tag{3.20}$$

Integration of this expresssion along an internal yield or recovery line with P and T constant and z running from an initial value z_i to a final value z_e gives

$$\int_{z_i}^{z_e} T\sigma\, dt = A\left[(z_e - z_I)^2 - (z_i - z_I)^2\right]. \tag{3.21}$$

In particular, integration along the main yield and recovery lines starting with $z_i = z_I = 0$ or $z_i = z_I = 1$ respectively gives

$$\int_0^{z_e} T\sigma\, dt = A z_e^2 \quad \text{and} \quad \int_0^{z_e} T\sigma\, dt = A(z_e - 1)^2 . \tag{3.22}$$

Thus in particular, as the yield proceeds the entropy produced increases parabolically with z until, when the yield is complete, it reaches the value $\frac{A}{T}$.

Of course the entropy production is not a quantity that could be measured. What may be measured is the heat added, viz. by (2.5)

$$\overset{\circ}{q} = T\frac{ds}{dt} - T\sigma \qquad \text{or, by } (3.1)_1, (2.18)_2$$

$$\overset{\circ}{q} = c\overset{\circ}{T} + T\Delta_\eta \overset{\circ}{z} - 2A(z - z_I(P,T))\overset{\circ}{z}. \tag{3.23}$$

The first term represents the temperature change that accompanies the heating. In an isothermal process the heat added causes the phase transition. We have

$$\overset{\circ}{q}_{T=const} = T\Delta_\eta \overset{\circ}{z} - 2A(z - z_I(P,T))\overset{\circ}{z}. \tag{3.24}$$

Now the first term is the latent heat of the phase transition. In a reversible process this is the only contribution to the heat added. The second term might be called the *irreversible* heat added.

In a cyclic process like the one shown in Figure 3.5. the heat added is easily calculated in four parts as

$$\int_1^2 \overset{\circ}{q}\, dt = T\Delta_\eta (z_2 - z_I(P_1)) - A(z_2 - z_I(P_1))^2$$

$$\int_2^3 \overset{\circ}{q}\, dt = 0$$

$$\int_3^4 \overset{\circ}{q}\, dt = T\Delta_\eta (z_4 - z_I(P_3)) - A(z_4 - z_I(P_3))^2$$

$$\int_4^1 \overset{\circ}{q}\, dt = 0 \quad .$$

Since $z_2 = z_1(P_3)$ and $z_4 = z_1(P_1)$ the latent heat contributions drop out from the total heat added which reads

$$\oint \overset{\circ}{q}\, dt = -2A(z_1(P_3) - z_1(P_1))^2 \quad , \tag{3.25}$$

i.e. the internal yield and internal recovery contribute equal parts to the irreversible heat added.

In particular, if the cycle runs along the whole hysteresis loop we have

$$\oint \overset{\circ}{q}\, dt = -2A, \tag{3.26}$$

since $z_1(P_3) = 1$ and $z_1(P_1) = 0$. This is equal to the work done $\oint P\overset{\circ}{d}dt$ and we conclude that the work is converted into heat in this irreversible process.

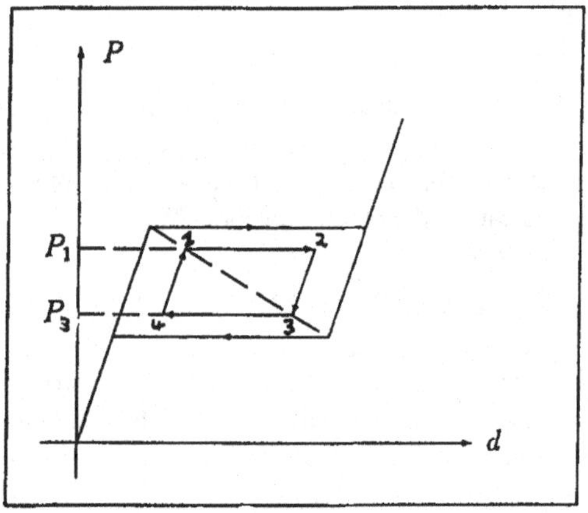

Figure 3.5. A cycle inside the ***(P,d)***-loop.

Another aspect of irreversibility is revealed by a cycle in the ***(d,T)***-diagram as shown in Figure 3.6. The load is constant so that the overall work is zero. By the first law the overall heat added must then also vanish. The heat added on the different steps can easily be calculated from (3.23). We have

$$\int_1^2 \overset{\circ}{q}\, dt = T_1\Delta_\eta(z_2 - z_1(T_1)) - A(z_2 - z_1(T_1))^2$$

$$\int_2^3 \overset{\circ}{q}\, dt(T_3 - T_2)$$

$$\int_3^4 \overset{\circ}{q}\,dt = T_3\Delta_\eta(z_4 - z_1(T_3)) - A(z_4 - z_1(T_3))^2$$

$$\int_4^1 \overset{\circ}{q}\,dt = c(T_1 - T_4)\,. \quad .$$

Since $T_2 = T_1$, $T_4 = T_3$ and $z_2 = z_1(T_3)$, $z_4 = z_1(T_1)$ we obtain for the total heat added

$$\oint \overset{\circ}{q}\,dt = \left[\Delta_\eta(T_3 - T_1) - 2A(z_1(T_1) - z_1(T_3))\right](z_1(T_1) - z_1(T_3)) = 0 \ . \tag{3.27}$$

Thus, by (3.11), the total heat is indeed zero. But at T_3 the process absorbs heat since $\int_3^4 \overset{\circ}{q}\,dt > 0$ and at T_1 it gives off heat - the same amount - since $\int_1^2 \overset{\circ}{q}\,dt < 0$. Therefore the irreversible process in this case consists of a transport of heat from hot to cold.

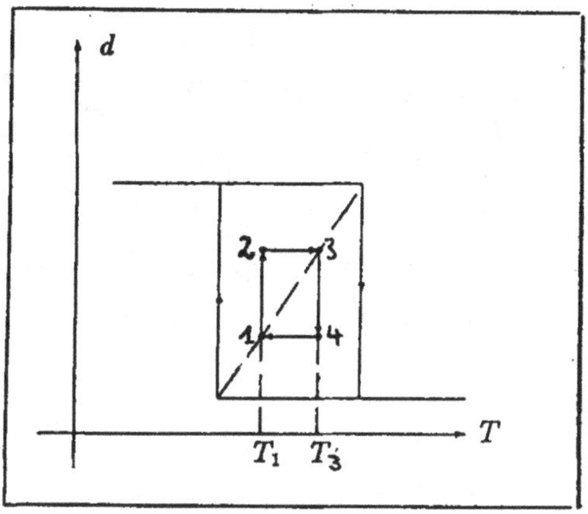

Figure 3.6. A cycle inside the *(d,T)*-loop.

4. MODEL WITHOUT COHERENCY ENERGY

4.1. Basic Element of the Model

The model described here whose properties will be exploited is *not* a microscopic model . We may properly call it mesoscopic, because its basic element is a "lattice particle", i.e. a small part of the metallic lattice that forms the body. Also the model will be constructed so as to simulate only deformations that result from uniaxial loading.

Figure 4.1. shows a lattice particle in three different equilibrium configurations denoted by ***A*** and $M_\pm$. ***A*** stands for austenite, the phase with a highly symmetric lattice structure and $M_\pm$ stands for the martensitic twins. Obviously $M_\pm$ may be considered as sheared versions of the lattice particle and indeed, according to Section 1.5., shear is a primary ingredient of the austenitic-martensitic phase transition.

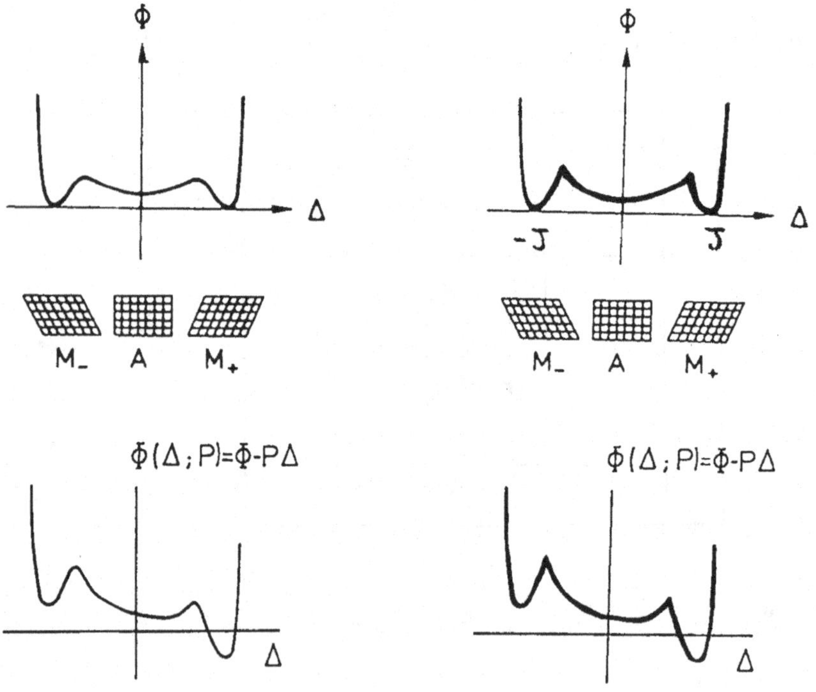

Figure 4.1. Lattice particle and its potential energy without and with load. Left: Smooth. Right: Train of parabolae.

Intermediate positions are also possible, of course, and Figure 4.1. shows the postulated form of the potential energy for different shear length, i.e. the length of the shift of the upper plane with respect to the lower plane. The figure also shows a simplified version of the potential energy, viz. a train of parabolae. This will be the form of the function for which we perform the calculations.

First of all, we apply a load in the vertical direction. In this manner the layers are subject to a shear force; the M_--layers become steeper and the M_+-layers become flatter and the vertical component of the shear lengths of all layers contributes to the elongation D. Thus, cf. Figure 4.2.

$$D = L - L_o = \frac{1}{\sqrt{2}} \sum_{i=1}^{N} \Delta_i \tag{4.1}$$

where the summation extends over all layers. If the load is removed the layers fall back to their previous shape and the elongation vanishes. Thus under a small load *the deformation is elastic.*

On the other hand, if the load is increased, there comes the point at which the M_--layers flip over to become M_+. According to Figure 4.1. this is the point where the left barrier has become so small that the M_--layers can overcome it. Each flipping layer will contribute a big shear length - roughly equal to 2J - to the elongation of the body and therefore the body extends considerably.

Subsequent removal of the load will let all layers drop to the M_+-minimum so that there is a considerable *residual deformation.* Thus we see that the model is capable of simulating

- the initial elastic deformation
- the yield - by flipping $M_- \rightarrow M_+$
- the residual deformation.

We proceed with the process of heating. Just exactly how things happen upon heating will be investigated in detail soon; here it suffices to say that the high temperature phase austenite forms, i.e. the layers straighten up and consequently the body contracts, see Figure 4.2. Now it already has the outward shape of the original martensitic body, except that the internal structure is different and that the surfaces are plane.

Upon cooling the austenitic layers loose their stability, they become martensitic again and - in all likelihood - every other one will be M_+ and M_- as shown in the figure. Now the body has returned to its original configuration, internal structure and all.

Thus we see that the model simulates - in addition to the above points

- recovery of shape upon heating
- conservation of this shape upon cooling.

In other words the model has shape memory.

Before we begin a more formal treatment of the model we continue this qualitative consideration with the discussion of the role of the temperature.

If the particle is under a shear load, the potential energy of the load must be taken into account. This is a linear function of the shear length and the effective potential energy is $\Phi(\Delta,P)=\Phi(\Delta)-P\Delta$. The load deforms the potential energy in the form shown in the lower part of Figure 4.1.: The right minimum becomes deeper and the central and left minimum become shallower. The barriers are affected as shown.

While in the previous two chapters the free energy was postulated as a non-convex function so *as to obtain the desired results,* in the present argument non-convexity of the potential energy *is dictated by the molecular structure of the metallic lattice* which has a highly symmetric phase and a less symmetric phase capable of twinning. In this sense the present model is much closer to first principles; whether or not the free energy is non-convex will have to be seen as a result of the evaluation of the model. Assuming that we will obtain satisfactory results we may feel that the present model *explains* - rather than merely *describes* - the material properties.

4.2. The Model of the Body as a Whole

In order to construct the model of the body as a whole we assemble lattice particles in layers and stack these on top of each other as shown in Figure 4.2. On the left hand side of the figure we see the unloaded body in the martensitic phase with alternating layers of M_- and M_+. We take this to be the reference configuration so that the elongation D is equal to zero.

Before we proceed with the development of the mathematical properties of this model we discuss its qualitative behaviour under a load and during heating and cooling. In this way we shall try to convince ourselves that this is a good model.

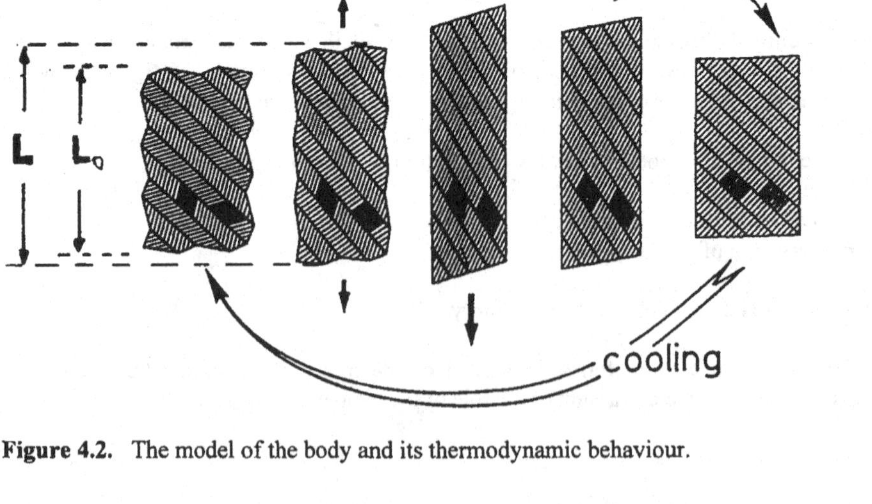

Figure 4.2. The model of the body and its thermodynamic behaviour.

4.3. The Role of Temperature

At zero temperature the layers still lie in their potential wells. For all non-zero temperatures they fluctuate about these minima and the mean kinetic energy of this fluctuation is proportional to the absolute temperature T. In Figure 4.3. this kinetic energy is indicated by the height of the pool of layers in the potential wells.

The figure shows the effect of temperature and load. On its left hand side we have a small temperature - a shallow pool - and it takes a large force to lower the left barrier sufficiently so that the layers can flip from the left potential well M_- to the right potential well M_+. At higher temperatures the pool is less shallow and the flipping will obviously occur at a smaller load.

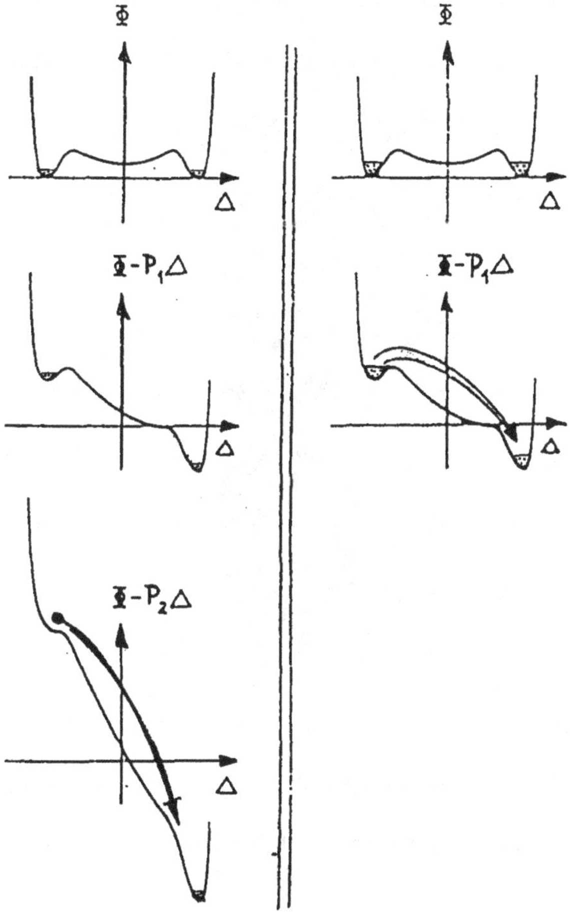

Figure 4.3. Activation of flipping. Left: Cold. Right: Warm.

We conclude that the model is also capable of simulating

- the decrease of the yield load with increasing temperature

which was reported in Figure 1.2. for the quasiplastic range.

One may tend to think that the regularity of the model as shown in Figure 4.2. makes it unnatural. This is not so, however, there are cases in which the $\boldsymbol{M}_{\pm}$ folding is nearly exactly as observed in micrographs. For proof we point to Figure 4.4. which shows a micrograph of a CuAlNi single crystal in the martensitic phase.

Figure 4.4. Martensitic twins in a CuAlNi single crystal. Magnification 200 x.

4.4. Thermal Fluctuation and Free Energy

In statistical thermodynamics the macroscopic behaviour of a model whose elements fluctuate in potential wells is captured in two quantities, viz.

Energy and Entropy .

We illustrate this statement for the present case of the model built of martensitic and austenitic layers as shown in Figure 4.2.

As the layers are fluctuating it becomes incovenient to characterize the shear length of each individual one. Instead we introduce a distribution function N_Δ which denotes the number of layers which have a particular shear length. With N_Δ we have for elongation, (potential) energy and entropy the expressions

$$D = \sum_\Delta \Delta\, N_\Delta \quad , \; E = \sum_\Delta \Phi(\Delta) N_\Delta \quad , \; S = k \ln \frac{N!}{\prod_\Delta N_\Delta !} \quad . \qquad (4.2)$$

The first two of these expressions are self-explanatory and the last one results from Boltzmann's statistical definition of the entropy, viz. $S = k \ln W$, where W is the number of possibilities to arrange the distribution N_Δ.

There is a competition between energy and entropy: In equilibrium the energy attempts to be minimal by assembling all layers in the depths of their potential wells and the entropy attempts to be as big as possible by creating an equal distribution. The outcome of these competing tendencies is that the free energy $F=E-TS$ assumes a minimum in equilibrium. At low temperatures, where the second term has little influence, the free energy becomes a minimum, because E does. But at high temperatures, where the first term is negligible compared to the second one, the free energy becomes a minimum, because S is big. Since the temperature measures the strength of the fluctuation we may say that increased fluctuations provide the entropic tendency toward equal distribution with the chance to make itself felt. The reason is clear: Increased fluctuations permit the layers to overcome the energetic barriers and this is the prerequisite for establishing an equal distribution.

In order to clarify the roles of energy and entropy in the present case of a phase transition we shall split the process of equilibration into three parts.

i.) the establishment of equilibrium inside each potential well irrespective of the numbers $N_{M_\pm}$, N_A,
ii.) the equilibration of the loads on the three phases,
iii.) the establishment of phase equilibrium.

4.5. Step i.) Equilibration in Potential Wells. Entropic Stabilisation of Metastable States.

We assume the potential energy $\Phi(\Delta)$ to have the form

$$\Phi(\Delta) = \begin{cases} \Phi_A = \Phi_o + K_A \Delta^2 & for \;\; |\Delta| < \Delta_s \\ \Phi_{M_\pm} = K_M (\Delta \pm J) & for \;\; |\Delta| > \Delta_s \end{cases} \quad . \qquad (4.3)$$

Figure 4.5. shows this potential energy.

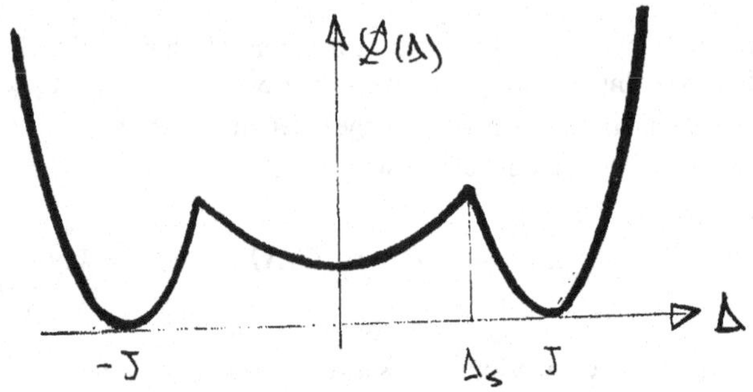

Figure 4.5. Parabolic potentials with curvatures $K_A < K_M$.

Let there be N_A and $N_{M\pm}$ layers in the respective wells and let these layers contribute the elongations D_A and $D_{M\pm}$ to the total elongation. We determine the equilibrium distributions by minimizing the free energies of the three phases.

$$F_A = E_A - TS_A = \sum_{\Delta_A} \left(\Phi(\Delta) + kT \ln \frac{N_\Delta}{N_A} \right) N_\Delta \tag{4.4}$$

under the constraints $N_A = \sum_{\Delta_A} N_\Delta \;,\; D_A = \sum_{\Delta_A} \Delta\, N_\Delta$

$$F_{M_\pm} = E_{M_\pm} - TS_{M_\pm} = \sum_{\Delta_{M_\pm}} \left(\Phi(\Delta) + kT \ln \frac{N_\Delta}{N_{M_\pm}} \right) N_\Delta \tag{4.5}$$

under the constraints $N_{M_\pm} = \sum_{\Delta_{M_\pm}} N_\Delta \;,\; D_{M_\pm} = \sum_{\Delta_{M_\pm}} \Delta\, N_\Delta$.

The result is

$$N_\Delta = N_A \frac{e^{-\frac{\Phi_A(\Delta)-\beta_A\Delta}{kT}}}{\sum_{\Delta_A} e^{-\frac{\Phi_A(\Delta)-\beta_A\Delta}{kT}}} \qquad |\Delta| < \Delta_s$$

$$\tag{4.6}$$

$$N_\Delta = N_{M_\pm} \frac{e^{-\frac{\Phi_{M_\pm}(\Delta)-\beta_{M_\pm}\Delta}{kT}}}{\sum_{\Delta_{M_\pm}} e^{-\frac{\Phi_{M_\pm}(\Delta)-\beta_{M_\pm}\Delta}{kT}}} \qquad |\Delta| > \Delta_s$$

The β's are Lagrange multipliers that take care of the constraints on D_A and $D_{M\pm}$.

Insertion of (4.6) into (4.4), (4.5) and integration provides the energies, entropies and free energies in the forms

$$E_A = N_A\left(\Phi_o + \frac{1}{2}kT + \frac{\beta_A^2}{4K_A}\right), \qquad S_A = kN_A\left(\frac{1}{2} + \ln Y\sqrt{\pi\, kT} - \frac{1}{2}\ln K_A\right), \tag{4.7}$$

$$F_A = N_A\left(\Phi_o + \frac{\beta_A^2}{4K_A} - kT\ln Y\sqrt{\pi\, kT} - \frac{1}{2}kT\ln K_A\right)$$

$$E_{M_\pm} = N_{M_\pm}\left(\Phi_o + \frac{1}{2}kT + \frac{\beta_{M_\pm}^2}{4K_{M_\pm}}\right), \quad S_{M_\pm} = kN_{M_\pm}\left(\frac{1}{2} + \ln Y\sqrt{\pi\, kT} - \frac{1}{2}\ln K_M\right), \tag{4.8}$$

$$F_{M_\pm} = N_{M_\pm}\left(\Phi_o + \frac{\beta_{M_\pm}^2}{4K_{M_\pm}} - kT\ln Y\sqrt{\pi\, kT} - \frac{1}{2}kT\ln K_M\right).$$

The β's may be calculated from the constraints and we obtain

$$\beta_A = 2K_A\frac{D_A}{N_A} \qquad \beta_{M_\pm} = 2K_M\left(\frac{D_{M_\pm}}{N_{M_\pm}} \mp J\right). \tag{4.9}$$

A brief calculation shows that $S = -\frac{\partial F}{\partial T}$ holds, so that - in the language of Section 3.1. - this equilibrium in the potential wells implies thermal equilibrium. Also we see from (4.7) through (4.9) that we have

$$\beta_A = \frac{\partial F_A}{\partial D_A} \qquad \beta_{M\pm} = \frac{\partial F_M}{\partial D_M}. \tag{4.10}$$

Therefore, if we interpret the β's as the loads in the two phases, we have dynamic equilibrium in each phase in the language of Section 3.1. Therefore the present step i) of equilibration corresponds to the first two partial equilibria of Section 3.1. All that remains to be established is the dynamic phase equilibrium and the equilibrium of the phase fractions.

But first we discuss the present step of equilibration. We plot F_A and $F_{M\pm}$ which, according to (4.7) through (4.9) have the forms

$$\frac{F_A}{N_A} = \Psi_o + K_A\left(\frac{D_A}{N_A}\right)^2 + \frac{1}{2}KT\ln K_A - KT\ln Y\sqrt{\pi\ KT}$$

$$\frac{F_{M_I}}{N_{M_I}} = K_M\left(\frac{D_{M_I}}{N_{M_I}} \pm J\right)^2 + \frac{1}{2}KT\ln K_M - KT\ln Y\sqrt{\pi\ KT} \qquad (4.11)$$

Note that - except for the last two terms - the free energy per layer is the same function of the elongation per layer as the potential energy Ψ of the shear length. Among the last two terms the last one is an unimportant term because it is added to both free energies. But the terms with $\ln K$ are interesting and important. Indeed, since $K_M > K_A$ holds according to Figure 4.5., the martensitic well of the free energy is lifted up faster with increasing T than the austenitic well. Figure 4.6. illustrates this behaviour. We conclude that the stable minimum of the free energy - due to the entropic term - is in the austenitic phase at high enough temperatures, although the potential energy of that phase is metastable. One expresses this observation by saying that the austenite is *entropically stabilized.*

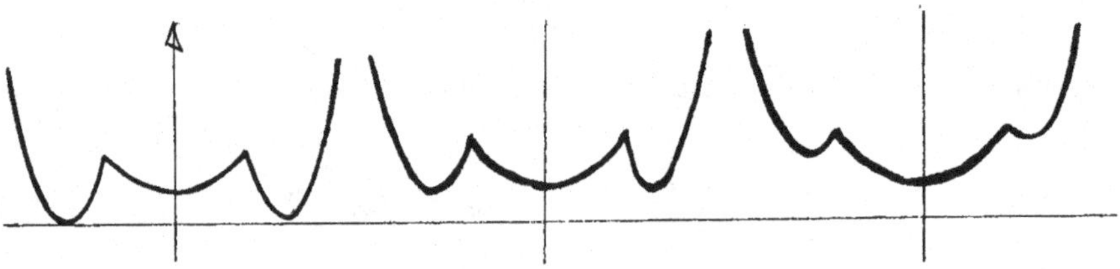

Figure 4.6. Free energy at different temperatures.
Left: T = 0. Free energy = potential energy.
Center and right: T > 0.

In a manner of speaking we have thus *derived* non-convex free energies of the type needed for the description of a phase transition and in particular for a pseudoelastic phase transition. We have obtained the desired form of the free energy from the non-convex shape of the potential energy which is dictated by the molecular lattice structure.

We conclude this section with an intuitive interpretation of entropic stabilisation of metastable minimum of potential energy. This seems appropriate at this point, although it involves a discussion of jumping over barriers which we have otherwise excluded from the considerations of this section.

Let us consider a situation in which we have layers both in the austenitic and in the martensitic potential wells and let the temperature be high enough that the barriers can be overcome, at least occasionally. The rate at which barriers are overcome depends on two conditions:

- the mean kinetic energy of the layers in comparison with the height of the barrier,
- the frequency with which the layers knock against the barrier, and this is determined by the curvature of the potential wells.

In the present case the ***M→A*** barrier is higher than the ***A→M*** barrier and therefore we should expect that the rate of ***A→M*** transitions is bigger than that of ***M→A*** transitions and - consequently - that the density of ***M***-layers is bigger than the density of ***A***-layers, *But* the curvature of M wells is bigger than that of the A wells, so that the ***M***-layers get a chance to jump their high barrier more often than the ***A***-layers have an opportunity to jump across their low barrier. Therefore the expectation about the density of the layers in the potential wells may well be reversed. In other words, when the ***M***-well is narrow enough the density of the layers there may be *less* than the density of ***A***-layers in their shallow wells. This is the nature of entropic stabilisation and this effect finds its formal expression in the form of the free energy function.

4.6. Step ii) Equilibration of Forces on the Three Phases

We consider a phase mixture of ***M±*** and ***A*** and write for the total free energy and total elongation

$$\begin{aligned} F &= N_A\left[\Psi_o + K_A\left(\frac{D_A}{N_A}\right)^2 + \frac{1}{2}kT\ln K_A\right] \\ &+ \sum_{l=\pm} N_{M_l}\left[K_M\left(\frac{D_{M_l}}{N_{M_l}} \pm J\right)^2 + \frac{1}{2}kT\ln K_M\right] - NkT\ln\left(Y\sqrt{\pi kT}\right) \\ D &= D_A + D_{M_-} + D_{M_+} \quad . \end{aligned} \tag{4.12}$$

We consider ***D_A*** and ***D_M±*** as variables and minimize ***F*** under the constraint of a fixed ***D***. We obtain

$$\lambda = 2K_A\frac{D_A}{N_A} = 2K_M\left(\frac{D_{M-}}{N_{M_-}} + J\right) = 2K_M\left(\frac{D_{M_+}}{N_{M_+}} - J\right) \tag{4.13}$$

where λ is the Lagrange multiplier that takes care of the constraint. Obviously λ is equal to the load ***P*** on all three phases.

We conclude that step ii) corresponds to the equilibration of the forces on the three phases. Therefore this step ensures what we have called the ***dynamic phase equilibrium*** in Section 3.1.

4.7. Step iii) Phase Equilibrium of Three Phases

It remains to do the last step in the process of equilibrium, the adjustment of the numbers N_A, $N_{M\pm}$ in such a manner that

$$F = N_A\left[\Psi_o + \frac{P^2}{4K_A} + \frac{1}{2}kT\,ln\,K_A\right] + \left(N_{M_-} N_{M+}\right)\left[\frac{P^2}{4K_M} + \frac{1}{2}kT\,ln\,K_M\right] - NkT\,ln\left(Y\sqrt{\pi\;kT}\right) \tag{4.14}$$

assumes a minimum under the constraint, of course that $N = N_A + N_{M_-} + N_{M_+}$ is constant.

P is the (common) force on the three phases and from (4.12), (4.13) we obtain

$$P = \frac{D + \left(N_{M_-} - N_{M_+}\right)J}{\dfrac{N_A}{2K_A} + \dfrac{N_{M_-} + N_{M+}}{2K_M}} \tag{4.15}$$

We introduce the Lagrange multiplier λ to take care of the constraint and obtain as equilibrium conditions

$$\begin{aligned}
&\left[\Psi_o + \frac{P^2}{4K_A} + \frac{1}{2}kT\,ln\,K_A\right] + \left(\frac{N_A}{2K_A} + \frac{N_{M_-} + N_{M_+}}{2K_M}\right) P\frac{\partial\,P}{\partial\,N_A} = \lambda\\
&\left[\frac{P^2}{4K_M} + \frac{1}{2}kT\,ln\,K_M\right] + \left(\frac{N_A}{2K_A} + \frac{N_{M_-} + N_{M_+}}{2K_M}\right) P\frac{\partial\,P}{\partial\,N_{M_-}} = \lambda\\
&\left[\frac{P^2}{4K_M} + \frac{1}{2}kT\,ln\,K_M\right] + \left(\frac{N_A}{2K_A} + \frac{N_{M_-} + N_{M_+}}{2K_M}\right) P\frac{\partial\,P}{\partial\,N_{M_+}} = \lambda
\end{aligned} \tag{4.16}$$

With

$$\begin{aligned}
&\frac{\partial P}{\partial N_A} = -\frac{D + (N_{M_-} - N_{M_+})}{\left(\dfrac{N_A}{2K_A} + \dfrac{N_{M_-} + N_{M_+}}{2K_M}\right)^2}\frac{1}{2K_A}\,,\\
&\frac{\partial P}{\partial N_{M_\mp}} = -\frac{D + (N_{M_-} - N_{M_+})}{\left(\dfrac{N_A}{2K_A} + \dfrac{N_{M_-} + N_{M_+}}{2K_M}\right)^2}\frac{1}{2K_A} \pm \frac{J}{\left(\dfrac{N_A}{2K_A} + \dfrac{N_{M_-} + N_{M_+}}{2K_M}\right)}
\end{aligned} \tag{4.17}$$

we conclude from (4.16) that we must have

$$P = 0 \qquad \text{and} \qquad \Psi_o - \frac{1}{2} kT \ln \frac{K_M}{K_A} = 0 \tag{4.18}$$

for equilibrium of all three phases to prevail. This means that the load must be zero and that the temperature is such that all three minima of the free energy are equal in height, see (4.11).

Note that the distribution of layers is not fixed in equilibrium. Between the conditions (cf. (4.15), (4.18))

$$N = N_A + N_{M-} + N_{M+} \qquad \text{and} \qquad D = (N_{M-} - N_{M+}) J \tag{4.19}$$

we have one freedom of choice for the values N_A, $N_{M\pm}$

4.8. Step iii) Phase Equilibrium of Two Phases

We retrace the arguments of step iii) under the assumption that $N_{M-} = 0$ holds, so that we are now investigating a phase equilibrium between the austenitic phase and the M+-phase. We minimize

$$F = N_A \left[\Psi_o + \frac{P^2}{4K_A} + \frac{1}{2} kT \ln K_A \right] + N_{M_+} \left[\frac{P^2}{K_M} + \frac{1}{2} kT \ln K_M \right] - NkT \ln\left(Y \sqrt{\pi}\, kT \right) \tag{4.20}$$

under the constraint $N = N_A + N_{M_+} = \boldsymbol{const.}$ λ is the (common) force on the two phases and we have from (4.12), (4.13) with $N_{M_-} = 0$

$$P = \frac{D - N_{M_+} J}{\dfrac{N_A}{2K_A} + \dfrac{N_{M_+}}{2K_M}} \tag{4.21}$$

With the Lagrange multiplier λ we obtain

$$\begin{aligned} &\left[\Psi_o + \frac{P^2}{4K_A} + \frac{1}{2} kT \ln K_A \right] + \left(\frac{N_A}{2K_A} + \frac{N_{M_+}}{2K_M} \right) P \frac{\partial P}{\partial N_A} = \lambda \\ &\left[\qquad \frac{P^2}{4K_M} + \frac{1}{2} kT \ln K_M \right] + \left(\frac{N_A}{2K_A} + \frac{N_{M_+}}{2K_M} \right) P \frac{\partial P}{\partial N_{M_+}} = \lambda \end{aligned} \tag{4.22}$$

We now have

$$\frac{\partial P}{\partial N_A}$$

$$\frac{\partial P}{\partial N_A} = -\frac{D - N_{M_+} J}{\left(\frac{N_A}{2K_A} + \frac{N_{M_+}}{2K_M}\right)^2} \frac{1}{2K_A} ,$$

$$\frac{\partial P}{\partial N_{M_+}} = -\frac{D - N_{M_+} J}{\left(\frac{N_A}{2K_A} + \frac{N_{M_+}}{2K_M}\right)^2} \frac{1}{2K_M} - \frac{J}{\frac{N_A}{2K_A} + \frac{N_{M_+}}{2K_M}} \tag{4.23}$$

and therefore we conclude from (4.22)

$$P^2\left(\frac{1}{4K_A} - \frac{1}{4K_M}\right) - PJ - \left[\Psi_o - \frac{1}{2}kT\ln\frac{K_M}{K_A}\right] = 0$$

$$\Rightarrow P = P(T) = \frac{2J}{\frac{1}{K_A} - \frac{1}{K_M}}\left(1 - \sqrt{1 - \frac{1}{J^2}\left(\frac{1}{K_A} - \frac{1}{K_M}\right)\left(\frac{1}{2}kT\ln\frac{K_M}{K_A} - \Psi_o\right)}\right) . \tag{4.24}$$

This means that for each T there exists a unique load P for which the austenitic phase can be in phase equilibrium with the martensitic phase.

If $P = P(T)$ from (4.24) is inserted into the free energy function (4.20) we observe that F is independent of D in phase equilibrium. Indeed we obtain

$$\frac{F}{N} = (1-z)\left[\Psi_o + \frac{P^2(T)}{4K_A} + \frac{1}{2}kT\ln K_A\right] + z\left[\frac{P^2(T)}{4K_M} + \frac{1}{2}kT\ln K_M\right] - kT\ln\left(Y\sqrt{\pi kT}\right) \tag{4.25}$$

where z is the phase fraction of martensite. This fraction is determined by D, because, according to (4.21) we have

$$\frac{D}{N} = P(T)\left(\frac{1-z}{2K_A} - \frac{z}{2K_M}\right) + jz \tag{4.26}$$

so that the relation between D an z is linear.

A brief calculation shows that $P = P(T)$, calculated in (4.24), is the slope of the common tangent of the branches A and $M+$ of the free energy function on the right hand side of model has provided us with a non-convex free energy, cf. Figure 4.6., much of the argument follows the same lines as in Chapters 2 and 3.

We note that in this chapter we have done without coherency energy in the treatment of the model, and we shall continue to ignore that effect for the moment, postponing it to the next chapter.

4.9. Step iii) Alternative Calculation for Phase Equilibrium of Two Phases

We recall that for equilibrium at a constant load the free enthalpy $G = F - PD$ must be minimal. The free enthalpies of the two phases are given by

$$G_A = N_A\left[\Phi_0 + \frac{P^2}{4K_A} + \frac{1}{2}kT\ln K_A - kT\ln\left(Y\sqrt{\pi kT}\right)\right] - N_A\frac{P^2}{2K_A}$$
$$G_A = N_A\left[\Phi_0 - \frac{P^2}{4K_A} + \frac{1}{2}kT\ln K_A - kT\ln\left(Y\sqrt{\pi kT}\right)\right] \qquad (4.27)$$

$$G_{M_+} = N_{M_+}\left[+ \frac{P^2}{4K_M} + \frac{1}{2}kT\ln K_M - kT\ln\left(Y\sqrt{\pi kT}\right)\right] - P\left(\frac{P}{2K_M} + J\right)N_{M_+}$$
$$G_{M_+} = N_{M_+}\left[- \frac{P^2}{4K_M} + \frac{1}{2}kT\ln K_M - kT\ln\left(Y\sqrt{\pi kT}\right) - PJ\right] \qquad (4.28)$$

We note that for a temperature with

$$\Phi_0 - \frac{1}{2}kT\ln\frac{K_M}{K_A} < 0$$

we have $G_A/N_A < G_{M_+}/N_{M_+}$ for $P = 0$ such that the equilibrium phase is austenitic. As P grows, both free enthalpies decrease, but G_{M_+}/N_{M_+} decreases faster than G_A/N_A, so that for

$$P^2\left(\frac{1}{4K_A} - \frac{1}{4K_M}\right) + PJ - \left[\Phi_0 - \frac{1}{2}kT\ln\frac{K_M}{K_A}\right] = 0 \qquad (4.29)$$

the two free enthalpies are equal and for bigger P's we have $G_{M_+}/N_{M_+} < G_A/N_A$, so that the martensite prevails. Equation (4.29) is the same condition as $(4.24)_1$ and we conclude that with the "free enthalpy argument" we obtain the same phase equilibrium load $P = P(T)$ as with the previous "free energy argument", Figure 4.7. shows the behaviour of the free enthalpies of the phases for a particular, but typical choice of the parameter values.

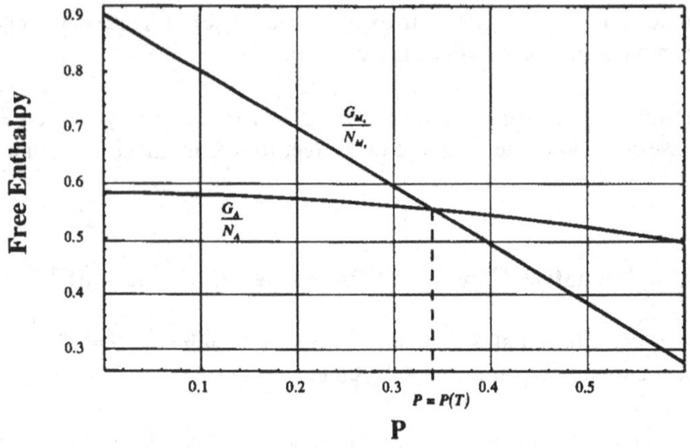

Figure 4.7. Free enthalpies as functions of load.

4.10. Latent Heat

We continue the consideration of the $A \to M$ transition and we shall attempt to calculate the heating that accompanies the transition as it proceeds from a phase fraction z to z+Δz. By (4.26) the corresponding change of deformation is equal to

$$\Delta D = N\left(J - \frac{P(T)}{2}\left(\frac{1}{K_A} - \frac{1}{K_M}\right)\right)\Delta z. \qquad \textbf{(4.30)}$$

The temperature is constant during the transition and so is the load, by (4.24). Given the load ***P(T)*** the potential energy of the layers has the form shown in Figure 4.8.

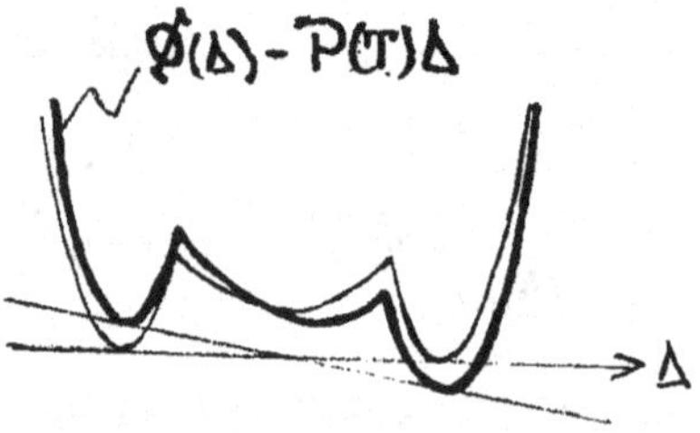

Figure 4.8. The potential energy under the load ***P(T)***.

We recall the consideration of Section 4.5., according to which the potential wells - in the present case only the ***A***- and ***M***+ -ones - are teeming with layers that fluctuate as dictated by the temperature. The two wells are in equilibrium in the sense that the layers moving from ***A*** to ***M***+ and those moving in the opposite direction are equal in numbers. This is so, although the ***M***+ → ***A*** barrier is higher than the the ***A*** → ***M***+ barrier, and even though that difference is increased by the load now.

The latent heat of the phase transition is duc to the fact that as a layer jumps from the ***A***- to the ***M***+ -well it converts potential energy into kinetic energy, i.e. heat. For each layer that conversion is equal to the difference in height of the minima. Thus we have for the "heat added" during an increase from z to z+Δz

$$\mathop{\overset{\circ}{Q}}_{z\to z+\Delta z} dt = -N\left[\Phi_0 - \frac{P^2(T)}{4}\left(\frac{1}{4K_A} - \frac{1}{4K_M}\right) + P(T)J\right]\Delta z \tag{4.31}$$

or, by $(4.24)_1$

$$\mathop{\overset{\circ}{Q}}_{z\to z+\Delta z} dt = -N\frac{1}{2}kT\ln\frac{K_M}{K_A}\Delta z\,. \tag{4.32}$$

We note that - by (4.7), (4.8) - this heat may be expressed in the form

$$\mathop{\overset{\circ}{Q}}_{z\to z+\Delta z} dt = T\Delta S \qquad \text{with} \qquad S = S_{M+} + S_A\,. \tag{4.33}$$

This conforms to thermodynamics, because during a reversible phase transition the heat added equals $\overset{\circ}{Q} = T\overset{\circ}{S}$ according to equilibrium thermodynamics.

5. MODEL WITH COHERENCY ENERGY

5.1. Coherency Energy in Phase Mixture

The model described in the previous chapter remains unchanged and so do all arguments up to Section 4.6., where - for the first time - the coexisting phases are considered. As in the Chapters 2 and 3 we assume that the coherency energy is proportional to $N_A \cdot N_M$ and we write as before, see (2.11)

$$\Delta F = AN\mu\ z(1-z)\ , \tag{5.1}$$

where μ is the mass of a layer.

According to this assumption the coherency energy depends only on z and therefore the arguments of Section 4.6. remain unchanged. The first change occurs in Section 4.7., or - if we restrict the attention to two phases - in Section 4.8., and, specifically, in (4.20) which reads

$$\begin{aligned} F = N_A\left[\Phi_o + \frac{P^2}{4K_A} + \frac{1}{2}kT\ln K_A\right] + N_{M_+}\left[\frac{P^2}{4K_M} + \frac{1}{2}kT\ln K_M\right] - \\ - NkT\ln\left(Y\sqrt{\pi kT}\right) + \frac{A}{N}\mu N_A N_{M_+}\ . \end{aligned} \tag{5.2}$$

We still have

$$P = \frac{D - N_{M_+}J}{\dfrac{N_A}{2K_A} + \dfrac{N_{M_+}}{2K_M}} \tag{5.3}$$

and (4.22) is changed into

$$\begin{aligned} \left[\Phi_o + \frac{P^2}{4K_A} + \frac{1}{2}kT\ln K_A\right] + \left(\frac{N_A}{2K_A} + \frac{N_{M_+}}{2K_M}\right)P\frac{\partial P}{\partial N_A} + \frac{A}{N}\mu N_{M_+} = \Lambda \\ \left[\frac{P^2}{4K_M} + \frac{1}{2}kT\ln K_M\right] + \left(\frac{N_A}{2K_A} + \frac{N_{M_+}}{2K_M}\right)P\frac{\partial P}{\partial N_{M_+}} + \frac{A}{N}\mu N_A = \Lambda \end{aligned} \tag{5.4}$$

Accordingly we obtain, instead of (4.24)

$$P^2\left(\frac{1}{4K_A} + \frac{1}{4K_M}\right) - PJ - \left[\Phi_o - \frac{1}{2}kT\ln\frac{K_M}{K_A}\right] + A\mu(1-2z) = 0 \tag{5.5}$$

$$\Rightarrow P = P(T,z)$$

$$= \frac{2J}{\dfrac{1}{K_A} - \dfrac{1}{K_M}}\left(1 - \sqrt{1 - \frac{1}{J^2}\left(\frac{1}{K_A} - \frac{1}{K_M}\right)\left(\frac{1}{2}kT\ln\frac{K_M}{K_A} - \Phi_o + A\mu(1-2z)\right)}\right).$$

Not unexpectedly the load under which the phase equilibrium can now prevail depends on ***T and z***; it decreases with growing z. This consequence of the coherency energy was discussed at length in Chapters 2 and 3. There the dependence was linear because the mathematical model of ideal pseudo-elasticity was simpler than the present one. Of course, if the coherency energy is considered small compared with Φ_0 (say) we still have a linear relation ***P=P(T,z)*** approximately.

From (5.3) we obtain

$$\frac{D}{N} = P(T,z)\left(\frac{1-z}{2K_A}+\frac{z}{2K_M}\right)+Jz \tag{5.6}$$

so that D and z are no longer linearly dependent. Elimination of z between (5.5) and (5.6) provides an expression for ***P=P(T,D)*** which - for one T - is plotted in Figure 5.1 along with ***F=F(T,D)***. We see here that in phase equilibrium P is a decreasing function of D and F is non-convex. The obvious conclusions about stability have been discussed before.

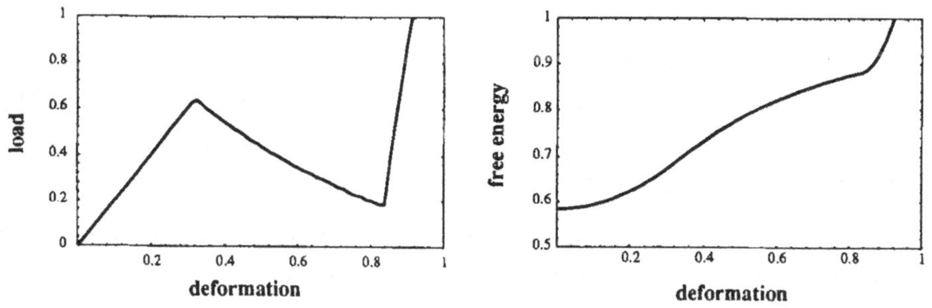

Figure 5.1. Load and free energy as functions of ***D***. $\left(\frac{K_A}{K_M}\right.$ is chosen as $\left.\mathbf{0.3}\right)$

5.2. Latent Heat

It makes eminent sense to assume that the observed pseudo-elastic yield starts out at the load ***P=P(T,0)*** whose value may be read off from (5.5). At this load the potential energy of the layers is slightly more "tilted" than in Figure 4.8 which corresponds to ***P=P(T)*** < ***P(T,0)***. And yet the state ***(P=P(T,0) , z=0)*** is a state of phase equilibrium. To be sure, the ***A→M*** barrier is now lower than at ***P=P(T)*** and the ***M→A*** barrier is higher; but an ***A→M*** jump is made more difficult by the sill of height $A\mu$ which the coherency energy saddles on top of the barrier, whereas the ***M→A*** is made easier because in this reverse transition the energy $A\mu$ is subtracted from the barrier.

Such a compensation of lowering or raising barriers and of adding or subtracting sills to them is valid for all points on the phase equilibrium line. And since for $z=1/2$ there are no sills the phase equilibrium occurs at the same ***P(T)*** as in the non-coherent case. For bigger values of z the ***A→M***-sill is negative and therefore the phase equilibrium load is smaller than ***P(T)***.

Assuming that the phase transition occurs along the phase equilibrium line, we calculate the latent heat. We note that in an ***A→M*** transition the energy difference between the minima is not all converted into heat, part of it is used to create coherency energy. Indeed, if we start at $(P=P(T,0)\,,\ z)$ and increase z by Δz , the difference of the minima has the value

$$\Phi_A^{min}-\Phi_M^{min}=\Phi_o-\frac{P^2(T,z)}{4}\left(\frac{1}{K_A}-\frac{1}{K_M}\right)+P(T,z)J\quad .$$

The heat added is this amount for each of the $N\Delta z$ layers minus the change of coherence energy $A\mu(1-2z)N\Delta z$. We obtain

$$\overset{\circ}{Q}{}^{rev.}_{z\to z+\Delta z}\,dt=-N\left[\Phi_o-\frac{P^2(T,z)}{4}\left(\frac{1}{K_A}-\frac{1}{K_M}\right)+P(T,z)J-A\mu(1-2z)\right]\Delta z \tag{5.7}$$

or, by (5.5)

$$\overset{\circ}{Q}{}^{rev.}_{z\to z+\Delta z}\,dt=-N\frac{1}{2}kT\ln\frac{K_M}{K_A}\,\Delta z \tag{5.8}$$

as before in (4.29). And again as before, cf. (4.30) this is equal to $T\Delta S$.

The index rev. indicates that the "heat added", which is calculated here is reversible heat in the sense that it reverses its sign, if Δz is reversed.

However, as we know from the experiments, the ***A→M*** phase transition does not occur along the equilibrium line; rather it occurs along the horizontal yield line, presumably under the load ***P=P(T,0)***. In that case the heat added is given by (5.7) but with ***P(T,z)*** replaced by ***P(T,0)***. Infact the excess heat added is given by

$$\left(\overset{\circ}{Q}_{z\to z+\Delta z}-\overset{\circ}{Q}{}^{rev.}_{z\to z+\Delta z}\right)dt=\left[\frac{P^2(T,0)-P^2(T,z)}{4}\left(\frac{1}{K_A}-\frac{1}{K_M}\right)+\big(P(T,0)-P(T,z)\big)J\right]\Delta z\,. \tag{5.9}$$

In the case $K_A=K_M$ and with ***P(T,z)*** being linear in z - as was the case in Chapters 2 and 3 - this term represents the irreversible heat, cf. Section 3.5, which increases parabolically with z.

Of course, in order to calculate the irreversible heat added in recovery we must replace ***P(T,0)*** in (5.9) by ***P(T,1)*** thus reversing the sign but also changing the value.

6. PHASE TRANSITION AS AN ACTIVATED PROCESS

6.1. Rate Laws for Phase Fractions

We proceed with the evaluation of the model introduced in Chapter 4 in which the body is composed of layers which have potential energies as shown in Figures 4.1 or 4.3. We shall now assume that the layers may overcome the barriers between the potential wells, at least occasionally, and we determine the rates of change of the phase fractions

$$\mathrm{X}_{M_-} = \frac{N_{M_-}}{N} \qquad \mathrm{X}_A = \frac{N_A}{N} \qquad \mathrm{X}_{M_+} = \frac{N_{M_+}}{N} \quad . \tag{6.1}$$

We write

$$\begin{aligned}
\overset{\circ}{\mathrm{X}}_{M_-} &= -\overset{-A}{p}\,\mathrm{X}_{M_-} + \overset{A-}{p}\,\mathrm{X}_A \\
\overset{\circ}{\mathrm{X}}_A &= \overset{-A}{p}\,\mathrm{X}_{M_-} - \overset{A-}{p}\,\mathrm{X}_A - \overset{A+}{p}\,\mathrm{X}_A + \overset{+A}{p}\,\mathrm{X}_{M_+} \\
\overset{\circ}{\mathrm{X}}_{M_+} &= \overset{A+}{p}\,\mathrm{X}_A - \overset{+A}{p}\,\mathrm{X}_{M_+}
\end{aligned} \tag{6.2}$$

assuming that $\overset{\circ}{X}_{M_-}$ (say) is composed of a loss and a gain. The loss is due to layers that jump across the (MA)-barrier from left to right and it is proportional to the phase fraction X_{M_-}. The gain is due to layers that jump across that barrier from right to left and it is proportional to X_A. The factors of proportionality are the transition probabilities $\overset{-\circ}{p}$ and $\overset{\circ-}{p}$ respectively. The equations for $\overset{\circ}{\mathrm{X}}_A$ and $\overset{\circ}{\mathrm{X}}_{M_+}$ are constructed similarly. Equation $(6.2)_2$ is more complex only, because the central minimum may exchange layers with both adjacent minima.

6. 2. First Law of Thermodynamics

The temperature also obeys a rate law which is none other than the first law of thermodynamics

$$\frac{dU}{dt} = \overset{\circ}{Q} + P\,\overset{\circ}{D} \quad . \tag{6.3}$$

By $(4.7)_1$, $(4.8)_1$ and (4.9) with $\beta_A = \beta_{M_\pm} = P$ we have

$$U = N\mu c(T - T_R) + N\left(\frac{\mathrm{X}_{M_-}}{4K_M} + \frac{\mathrm{X}_A}{4K_A} + \frac{\mathrm{X}_{M_+}}{4K_M}\right)P^2 + N\mathrm{X}_A\Phi_o + N\mu A\mathrm{X}_A(1 - \mathrm{X}_A) \tag{6.4}$$

$$P\overset{\circ}{D} = NP\left[\left(\frac{P}{2K_M} - J\right)\mathrm{X}_{M_-} + \frac{P}{2K_A}\mathrm{X}_A + \left(\frac{P}{2K_M} + J\right)\mathrm{X}_{M_+}\right]^{\bullet}, \tag{6.5}$$

where c is the specific heat of the specimen, N is the total number of layers, as before, and μ is their mass. The term with A stands for the coherency energy; we assume that only interfaces between the A- and the M-phase contribute significantly to that energy.

We consider $\overset{\circ}{Q}$ to be given by a simple Newton law of cooling, viz.

$$\overset{\circ}{Q} = -N\alpha(T - T_E) \tag{6.6}$$

where T_E is the external temperature and α is the coefficient of heat transfer. Thus the temperature equation reads

$$\begin{aligned}\mu c\frac{dT}{dt} = -\alpha(T - T_E) &+ \left[\Phi_o - \frac{P^2}{4K_A} - \left(PJ - \frac{P^2}{4K_M}\right)\right]\overset{\circ}{\mathrm{X}}_{M_-} \\ &+ \left[\Phi_o - \frac{P^2}{4K_A} + \left(PJ + \frac{P^2}{4K_M}\right)\right]\overset{\circ}{\mathrm{X}}_{M_+} - \mu A(1 - 2\mathrm{X}_A)\overset{\circ}{\mathrm{X}}_A \,.\end{aligned} \tag{6.7}$$

Note that the second and third term on the right hand side of (6.7) represent the kinetic energy gained when a layer jumps from the left to the center or from the center to the right respectively. The last term represents the coherency energy gained or expended when the jumps alter the number of interfaces.

6.3. Transition Probabilities

It seems reasonable to assume that the probability for an M_--layer to jump across the barrier at $\Delta = -\Delta_s$ is proportional to the probability of an M_--layer to have the energy $\Phi(-\Delta_s) + P\Delta_s$, i.e. to be on top of the left barrier. By $(4.6)_2$, with $\beta_{M_-} = P$ this probability has the value [1]

$$\frac{e^{-\frac{\Phi_{M_-}(-\Delta_s)+P\Delta_s}{kT}}}{\sum\limits_{\Delta M_-} e^{-\frac{\Phi(\Delta)P\Delta}{kT}}} = \frac{1}{\Upsilon}\sqrt{\frac{K_M}{\pi kT}}\; e^{-\frac{\frac{P^2}{4K_M}+PJ}{kT}}\; e^{-\frac{\Phi(-\Delta_s)+P\Delta_s}{kT}} \tag{6.8}$$

---

[1] In order to obtain the right hand side of (6.8) we have converted the sum into an integral and extended the range of integration from $-\infty$ to $+\infty$.

This is a common assumption in the theory of thermally activated processes and it is based on the expectation that an M-layer whose energy has reached the height of the barrier, will be able to pass that barrier. In our case we have to modify this because the interfacial energy has to be accounted for: Indeed, if an M-layer jumps across the left barrier, it needs to have the energy $\Phi(-\Delta_s)+P\Delta_s+\mu A(1-2X_A)$, because the coherency energy $N\mu AX_A(1-X_A)$ changes by the amount $\mu A(1-2X_A)$.

Therefore we take the transition probability $\overset{-o}{p}$ to be given by the expression

$$\overset{-o}{p}=R\sqrt{\frac{K_M}{\mu}}\sqrt{\frac{\Phi_o}{kT}}\,exp\left\{-\frac{\frac{P^2}{4K_M}-PJ}{kT}\right\}exp\left\{-\frac{\Phi(-\Delta_s)+P\Delta_s+\mu A(1-2X_A}{kT}\right\} \quad . \tag{6.9}$$

The other three transition probabilities are constructed by analogous arguments and we obtain

$$\overset{o-}{p}=R\sqrt{\frac{K_A}{\mu}}\sqrt{\frac{\Phi_o}{kT}}\,exp\left\{-\frac{\frac{P^2}{4K_A}-\Phi_o}{kT}\right\}exp\left\{-\frac{\Phi(-\Delta_s)+P\Delta_s-\mu A(1-2X_A}{kT}\right\}$$

$$\overset{+o}{p}=R\sqrt{\frac{K_M}{\mu}}\sqrt{\frac{\Phi_o}{kT}}\,exp\left\{-\frac{\frac{P^2}{4K_M}+PJ}{kT}\right\}exp\left\{-\frac{\Phi(\Delta_s)-P\Delta_s+\mu A(1-2X_A}{kT}\right\}$$

$$\overset{o+}{p}=R\sqrt{\frac{K_A}{\mu}}\sqrt{\frac{\Phi_o}{kT}}\,exp\left\{-\frac{\frac{P^2}{4K_A}-\Phi_o}{kT}\right\}exp\left\{-\frac{\Phi(\Delta_s)-P\Delta_s-\mu A(1-2X_A}{kT}\right\}. \tag{6.10}$$

R is a dimensionless factor of proportionality.

6.4. Dimensionless Equations and Choice of Parameters

We introduce dimensionless variables for the temperature, load and deformation, viz.

$$a=\frac{kT}{\Phi(\Delta_s)}\ ,\qquad p=\frac{\Delta_s}{\Phi(\Delta_s)}P\ ,\qquad l=\frac{D}{NJ}\ , \tag{6.11}$$

and dimensionless parameters

$$\delta=\frac{\Delta_s}{J},\ \varphi=\frac{\Phi(\Delta_s)}{4K_M\Delta_s^2},\ \varepsilon=\frac{\mu A}{\Phi(\Delta_s)},\ \kappa=\frac{K_A}{K_M},\ \zeta=\frac{c}{k/\mu}\ . \tag{6.12}$$

In terms of these dimensionless quantities the transition probabilities (6.9), (6.10) assume the forms

$$\overset{-o}{p}=R\frac{1}{\tau_X}\frac{1}{\sqrt{\kappa}}\sqrt{1-\frac{1}{4}\frac{\kappa}{\phi}}\ \frac{1}{\sqrt{a}}exp\left\{-\frac{\varphi p^2-\frac{1}{\delta}p}{a}\right\}\quad exp\left\{-\frac{1+p+\varepsilon(1-2X_A)}{a}\right\}$$

$$\overset{o-}{p}=R\frac{1}{\tau_X}\quad\sqrt{1-\frac{1}{4}\frac{\kappa}{\phi}}\ \frac{1}{\sqrt{a}}exp\left\{-\frac{\frac{\varphi}{\kappa}p^2-\left(1-\frac{1}{4}\frac{\kappa}{\varphi}\right)}{a}\right\}exp\left\{-\frac{1+p-\varepsilon(1-2X_A)}{a}\right\}$$

(6.13)

$$\overset{+o}{p}=R\frac{1}{\tau_X}\frac{1}{\sqrt{\kappa}}\sqrt{1-\frac{1}{4}\frac{\kappa}{\phi}}\ \frac{1}{\sqrt{a}}exp\left\{-\frac{\varphi p^2+\frac{1}{\delta}p}{a}\right\}\quad exp\left\{-\frac{1-p+\varepsilon(1-2X_A)}{a}\right\}$$

$$\overset{o+}{p}=R\frac{1}{\tau_X}\quad\sqrt{1-\frac{1}{4}\frac{\kappa}{\phi}}\ \frac{1}{\sqrt{a}}exp\left\{-\frac{\frac{\varphi}{\kappa}p^2-\left(1-\frac{1}{4}\frac{\kappa}{\varphi}\right)}{a}\right\}exp\left\{-\frac{1-p-\varepsilon(1-2X_A)}{a}\right\}\ .$$

The energy balance (6.7) reads

$$\frac{da}{dt}=-\frac{1}{\tau_T}(a-a_E)+\frac{1}{\zeta}\left[1-\frac{1}{4}\frac{\kappa}{\varphi}-\frac{\varphi}{\kappa}p^2-\left(\frac{1}{\delta}p-\varphi p^2\right)\right]\frac{dX_{M-}}{dt}+$$

$$+\frac{1}{\zeta}\left[1-\frac{1}{4}\frac{\kappa}{\varphi}-\frac{\varphi}{\kappa}p^2+\left(\frac{1}{\delta}p+\varphi p^2\right)\right]\frac{dX_{M+}}{dt}-\frac{\varepsilon}{\zeta}(1-2X_A)\frac{dX_A}{dt}\ .$$

(6.14)

$\frac{1}{R}\tau_X$ and τ_T are typical relaxation times of the phase fractions and the temperature respectively

$$\tau_X = \sqrt{\frac{\mu}{K_A}}, \qquad \tau_T = \frac{\mu c}{\alpha} \quad . \tag{6.15}$$

We refer to Figure 4.5. and assume that the potential energy $\Phi_o + K_A\Delta_s^2$ of the barriers is 6 times higher than the potential energy Φ_o of the austenitic minimum. Also as in Figure 5.1. we take K_A/K_M to be equal to 0.3. Previous calculations have shown us that a reasonable choice for the coherency energy lies between 30 and 40 % of the barrier energy. Finally the specific heat c of an oscillator of atomic mass μ_{at} is equal to $3K/\mu_{at}$, we estimate that a layer has 10^3 atoms and obtain with all this

$$\delta = 0.71, \quad \varphi = 0.09, \quad \varepsilon = 0.375, \quad \zeta = 3000 . \tag{6.16}$$

It remains to estimate R and the relaxation times τ_X and τ_T. Obviously τ_T depends on the shape of the specimen - in particular its thickness - and we estimate it to be equal to 1 sec for the specimen that was used for the standard test described in Section 1.4. τ_X is the period of free oscillation of a mass μ in a harmonic potential $K_A\Delta^2$. Its inverse is the eigenfrequency of such a mass which we estimate to be 100 Hz. Roughly speaking that eigenfrequency determines the frequency with which a layer knocks against the barrier and, if on every tenth attempt the barrier will be surpassed, we conclude that $\frac{1}{R}\tau_X$, the relaxation time of the phase fractions, must be equal to 0.1 sec. Therefore we set

$$\frac{1}{R}\tau_X = 0.1\,sec \qquad\qquad \tau_T = 1\,sec . \tag{6.17}$$

6.5. Integration

We observe that the four equations (6.2) with (6.13), and (6.14) constitute a set of ordinary differential equations of the form

$$\overset{\circ}{V} = f(V; P(t), T_E(t)) \tag{6.18}$$

where V is the vector composed of X_{M_-}, X_A, X_{M_+}, T. Therefore, given initial values $V(0)$ and P and T_E as functions of time we may calculate $V(t)$ by a stepwise numerical procedure.

Once done, this will allow us to calculate the specific deformation as a function of time, see (4.15)

$$d(t) = \left(X_{M_-}\frac{1}{2K_M} + X_A\frac{1}{2K_A} + X_{M_+}\frac{1}{2K_M} \right) P - (X_{M-} - X_{M+})J \; . \tag{6.19}$$

which reads in dimensionless form, see (6.11), (6.12)

$$l(t)=(\kappa X_{M_-}+X_A+\kappa X_{M_+})2\frac{\varphi}{\kappa}\delta p-(X_{M_-}-X_{M_+}). \tag{6.20}$$

6.6. Results: Load-Deformation Isotherms

For the choices of the parameter values given in Section 6.4. we obtain results shown in Figures 6.1. and 6.2. In these figures we have calculated the output

$$X_{M_-}(t)\,,X_A(t),X_{M_+}(t)\,,T(t)\,,d(t)$$

that results from an essential input of the form

$P(t)$ - triangular tensile and compressive force
$T_E(t)$ - constant.

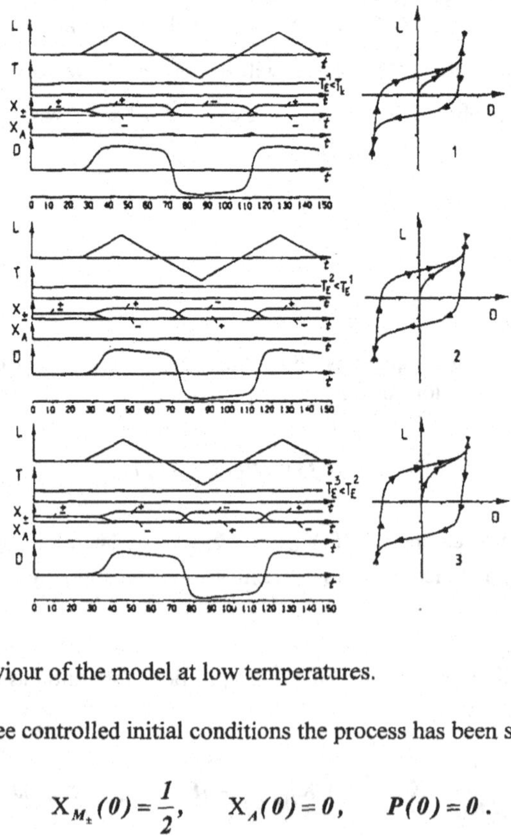

Figure 6.1. Behaviour of the model at low temperatures.

In order to guarantee controlled initial conditions the process has been started in each case with

$$X_{M_\pm}(0)=\frac{1}{2},\qquad X_A(0)=0,\qquad P(0)=0\,.$$

In each case $T_E(0)$ had the same value and it was smaller than the critical temperature T_K at which the unloaded body becomes austenitic.

For the three cases of Figure 6.1. the external temperature was lowered - after 10 time units - to values

$$T_E^3 < T_E^2 < T_E^1 < T_K$$

respectively and kept there throughout the experiment. This was to allow the body any readjustment of initial phase fractions, if needed. None was needed in Figure 6.1. After further 10 time units of readjustment time the triangular load was started first in tension, then in compression.

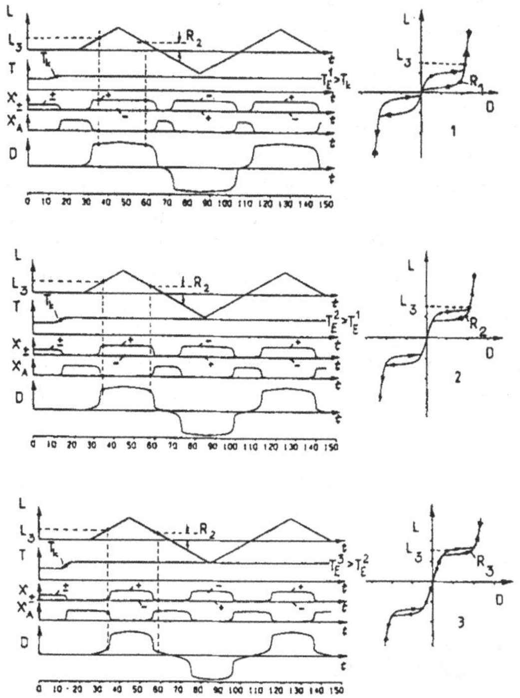

Figure 6.2. Behaviour of the model at high temperatures.

The deformation followed as indicated and the phase fractions alternated between $X_{M_+} = 1$ and $X_{M_-} = 1$, also as indicated. There is no significant difference between the external temperature T_E and the actual temperature T. Most instructive perhaps are the load-deformation diagrams on the right hand side which exhibit a virginal curve on first loading and then hystereses around the origin, much as those shown in the schematic pictures of Figure 1.2. The width of the hysteresis grows with decreasing temperature.

For the three cases of Figure 6.2. the external temperature was raised - after 10 time units - to values

$$T_E^3 > T_E^2 > T_E^1 > T_K$$

respectively and kept constant afterwards. We observe a rapid readjustment of phase fractions: the martensite disappears and X_A assumes the value 1, as is proper for the unloaded body at $T > T_K$. After further 10 time units the triangular tensile and compressive load is started and it is accompanied by a change in deformation and by complex changes in the phase fractions which alternate between $X_{M_-} = 1$, $X_A = 1$, and $X_{M_+} = 1$. No deviation of temperature from the external value can be detected. The load-deformation diagrams on the right hand side of Figure 6.2. exhibit the phenomenon of pseudoelasticity as shown schematically in Figure 1.2. [Note, however, that the pseudoelastic hysteresis grows smaller as T_E increases.]

6.7. Results: The Standard Test

The foregoing calculations and, in particular, the curves of Figures 6.1., 6.2. have given us the confidence that the model is capable - at least qualitatively - to simulate the behaviour of shape memory alloys. On the other hand it is clear that the simple input used in those figures does not really tax the model convincingly. In particular we have not demonstrated how the model behaves under a varying (external) temperature.

In order to investigate that we simulate the standard test which was described in Section 1.4.: A triangular tensile load is often repeated while the external temperature is first raised slowly and then slowly lowered to the original small value, see Figure 6.3.

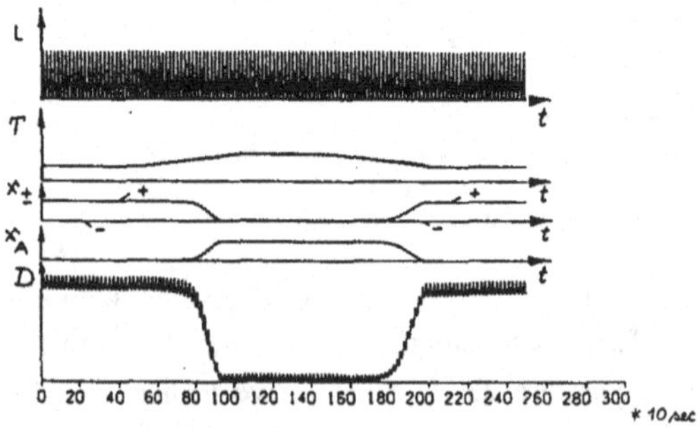

Figure 6.3. The standard test. Simulation.

The initial phase fraction X_{M_+} equals 1 so that X_A and X_{M-} vanish initially. We see that the deformation goes up and down, along with the load, about a large mean value at first. As the temperature is raised the deformation decreases precipitously and then, as long as the temperature is high, the deformation "oscillates" about a small mean value. When the temperature returns to its small value the deformation grows and resumes its oscillation about the original large mean value.

All this is just as observed in the experiment, cf. Figure 1.11. so that we conclude that the model is capable of simulating the standard test. Actually the model provides additional insight in that it shows that the drop in deformation is accompanied, or caused, by the formation of austenite and that the subsequent rise of deformation is due to the reestablishment of the martensitic phase.

In the present case there is also a small observable difference between the external temperature and the temperature of the body. To see this we examine the curve $T(t)$ of Figure 6.3. closely. That curve represents both the external temperature $T_E(t)$ and the temperature of the body. Both nearly coincide, but not entirely. Indeed, as the austenite is formed., the slightly zig-zaggy body temperature falls below $T_E(t)$ and when the martensite is formed the body temperature lies above $T_E(t)$. This is plausible; after all the austenite formation is endothermic and the martensite formation is exothermic.

LITERATURE

Proceedings of international conferences on martensitic transformations and shape memory.

Perkins, J. (ed). Shape Memory Effect in Alloys. Plenum Press New York, London 1976.

Delaey, L., Chandrasekharan, L. (eds). Proc. Int. Conf. on Martensitic Transformation, Leuven (Belgium) 1982. J. de Physique 43 (1982).

Tamura, I. (ed). Proc. Int. Conf. on Martensitic Transformation, Nara (Japan) 1986. The Japan Institute of Metals.

Hornbogen, E., Jost, N. (eds). The Martensitic Transformation in Science and Technology. DGM Informationsgesellschaft Oberursel (Germany) (1989).

Liu, C.T., Kunsmann, H., Otsuka, K., Wuttig, M. Shape Memory Materials and Phenomena - Fundamental Aspects and Applications. Materials Research Society. Fall Meeting Boston (USA) 1991. MRS Symp. Proc. 246.

Perkins, J. (ed.). Proc. Int. Conf. on Martensitic Transformation, Monterey (USA) 1992.

Dissertations

Ehrenstein, H., Formerinnerungsvermögen in NiTi. Dissertation TU Berlin (1985).

Achenbach, M., Ein Modell zur Simulation des Last-Verformungs-Temperaturverhaltens von Legierungen mit Formerinnerungsvermögen. Dissertation TU Berlin (1986).

Huo, Y., On the Thermodynamics of Pseudoelasticity. Dissertation TU Berlin (1992).

Xu, H., Experimentelle und theoretische Untersuchung des Hystereseverhaltens in Formgedächtnislegierungen. Dissertation TU Berlin (1992).

Research papers

Müller, I., Xu, H., On the Pseudoelastic Hysteresis. Acta Metallurgica et Materialia 39 (1991).

Huo, Y., Müller, I., Thermodynamics of Pseudoelasticity - A Graphical Approach. Pitman Research Notes in Mathematics (in press).

Fu, S., Huo, Y., Müller, I., Thermodynamics of Pseudoelasticity - An Analytical Approach. Acta Mechanica (in press).

Huo, Y., Müller, I., Non-Equilibrium Thermodynamics of Pseudoelasticity. Cont. Mech. Thermodyn. 5 (1993).

VARIATIONAL METHODS IN THE STEFAN PROBLEM

José-Francisco Rodrigues

CMAF/Universidade de Lisboa,
Av. Prof. Gama Pinto, 2
1699 LISBOA Codex, PORTUGAL

INTRODUCTION

The Stefan problem is one of the simplest possible macroscopic models for phase changes in a pure material when they occur either by heat conduction or diffusion. Its history, starting in the last century with the works of Lamé and Clapeyron [LC] and, afterwards, Stefan [St], provides a helpful example of the interplay between free boundary problems and the real world (see, for instance, the books [R], [EO], [Cr] or [Me2]).

Here we shall follow the method of weak solutions for elliptic and parabolic equations with discontinuous coefficients, after the works of Kamin [Ka] and Oleinik [O] on the multidimensional Stefan problem, and also the method of variational inequalities, after the works of Baiocchi [B], Duvaut [Du1,2] and Frémond [Fr] (see, for instance, the books [DL], [L3], [BC], [KS], [BL], [EO], [F2] and [R3] or, for a general introduction to the theory, also [Z]).

In this research-expository work we introduce the model problem in several space dimensions and we develop and apply the variational method to the mathematical study of several variants of the Stefan problem, obtaining in many cases new results and extensions of previous works to more general situations. We do not intend to present an overview of the multidimensional Stefan problem nor a survey of the free boundary problems connected with it, although the variational methods used here are applicable to other class of nonlinear problems. Further references can be found also in [Pr], [T3], [M2], [N], [Dan], [D3], [R2,4] or [OPR].

In the first section we describe the mathematical-physics model, under the continuum hypothesis, which leads to an evolution equation of the type

$$\text{(S)} \qquad \frac{d}{dt}(\vartheta + \chi) = \Delta\vartheta ,$$

where a constitutive relation $\chi \in H(\vartheta)$, H being the maximal monotone graph associated with the Heaviside function, is the essential feature of the Stefan problem. We obtain the equivalent variational inequality formulation, following an observation by Damlamian [D1], that showed the natural relation between this approach introduced by Baiocchi, and the generalized solution of Kamin [Ka] and Oleinik [O], also studied by other authors (see [F1,2], [LSU], [L2], [B1], etc.). In the one-phase case we study a nonclassical type of nonlocal boundary condition recently considered in [PR], while in the two-phase case we consider Dirichlet or Neumann conditions on the fixed boundary.

The results on the one-phase Stefan problem with a nonlocal boundary condition are obtained in the Section 2, within a suitable abstract framework to which the variational inequality method is well adapted. This yields the well-posedness of the variational formulation, including the continuous dependence with respect to physical constants covering, as a special case, the results of [PR] and the asymptotic behaviour as $t \to \infty$.

Although the same nonlocal boundary condition could be used in the two-phase problem, in Section 3, we preferred to consider only the more classical Neumann boundary condition. Here we have followed mainly the works of Tarzia [T1,2] and Pawlow [P1,2], but the existence proof follows different ideas for the approximate problem, being the remark used in Lemma 3.6 probably new. The variational inequality approach for the two-phase problem is particularly well adapted to the numerical approximation of solutions and to optimal control problems (see, for instance, [Ba2], [HN], [S1,2] or [P2]). The results on the existence of optimal control of Section 3.2 follow the general approach of [L1] and [P2], but Theorems 3.18 and 3.20 are new and extend previous results of [R4], which were based on a more restrictive non-degeneracy property of [No].

The last section is new and extends to the general case with a convective term, given by a solenoidal velocity field without regularity assumptions, the existence, uniqueness and continuous dependence results for weak solutions to the enthalpy formulation. We consider nonlinear perturbations on the equation and on the Neumann boundary condition and we apply the monotonicity and compactness methods (see [L2] and [V1]) combined with a maximum principle. The uniqueness follows the method of [Ka], extended in [NP] and [RY], but only holds for the evolutionary case. The stabilization when time goes to infinity is considered here only under monotonicity assumptions and still presents several interesting open questions, including the uniqueness of the steady-state problem with convection. For the case of convection only in the liquid phase, we refer to [DO] and [R6] and their bibliographies.

We shall not consider here more complex models taking into account surface tension or hysteresis effects occurring in phase transitions (see [V5], for instance). However, it is interesting to refer that the parabolic equation (S) where the constitutive relation between ϑ and χ is replaced by some hysteresis functional $\chi = X(\vartheta)$ has been considered in [V3,4]. Some uniqueness results were shown in [H] for mh-hysterons and certain type of Preisach operators. Recently some interesting problems on the Stefan model with hysteresis type boundary conditions have been investigated in [FH] and [GHM].

These notes develop the lectures given at the third 1993 CIME Summer Course, where these results were presented. The author wishes to thank to Prof. A. Visintin for his challenging invitation and to Prof.s A. Damlamian and M. Primicerio for many stimulating discussions.

1 – THE MATHEMATICAL–PHYSICS MODELS

1.1 – Balance and state equations

The Stefan problem is included in the number of mathematical problems describing a physical phenomena to which the continuum hypothesis is applied. The model can be obtained with the techniques of continuum mechanics consisting of an appropriate combination of conservation laws (equations of balance) with constitutive relations (equations of state), conducing to partial differential equations with prescribed boundary and initial conditions.

The existence of phase changes give rise to idealized interfaces at which the dependent variables suffer discontinuities, yielding jump conditions across those surfaces whose locations are not "a priori" known (the free boundaries).

We consider a material element with two phases, a solid part $\mathcal{S}$ and a liquid one $\mathcal{L}$, separated by an interface Φ and limited by an external boundary Γ:

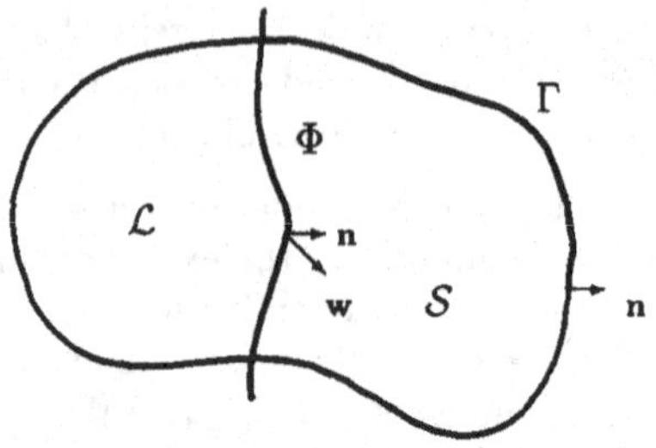

$$\Omega = \mathcal{S} \cup \Phi \cup \mathcal{L} \subset \mathbf{R}^N \quad (N = 2, 3).$$

A general balance-type conservation law has the form

$$\frac{d}{dt} \int_\Omega A \, dx + \int_\Gamma \alpha \, d\sigma = \int_\Omega C \, dx + \int_{\Phi \cap \Omega} \gamma \, d\sigma \,, \tag{1.1}$$

where A in the quantity being conserved per unit volume, α represents its flux across Γ, C and γ are associated with sources of that quantity in the body and on the interface, respectively. From (1.1) we obtain the general balance equation

$$\frac{dA}{dt} + A \nabla \cdot \mathbf{v} = C - \nabla \cdot \mathbf{a} \quad \text{in } \mathcal{S} \cup \mathcal{L} \,, \tag{1.2}$$

where $d/dt = \partial_t + \mathbf{v} \cdot \nabla$ denotes the material derivative ($\partial_t = \partial/\partial t$), $\mathbf{v}$ the velocity field and $\mathbf{a}$ is a vector field such that $\alpha = \mathbf{a} \cdot \mathbf{n}|_\Gamma$. In addition, we obtain also the jump condition at the discontinuity interface Φ:

$$[\![A(\mathbf{w} - \mathbf{v}) \cdot \mathbf{n}]\!] = [\![\mathbf{a} \cdot \mathbf{n}]\!] - \gamma \quad \text{on } \Phi \,, \tag{1.3}$$

where $[\![f]\!] = f_\mathcal{S} - f_\mathcal{L}|_\Phi$ denotes the discontinuity of the quantity f across Φ (the subscripts $\mathcal{S}$ and $\mathcal{L}$ refer to the solid and liquid region, respectively), $\mathbf{n}$ the unit normal pointing to $\mathcal{S}$ and $\mathbf{w}$ denotes the velocity of the interface Φ.

For instance, the balance of mass corresponding to the case $\alpha = \gamma = C \equiv 0$ and $A \equiv \rho = \rho(x,t) > 0$ is the density, yields the continuity equation

$$\frac{d\rho}{dt} + \rho \nabla \cdot \mathbf{v} = 0 \quad \text{in } \mathcal{S} \cup \mathcal{L} \,, \tag{1.4}$$

and, from (1.3), also

$$[\rho_\mathcal{S} \mathbf{v}_\mathcal{S} - \rho_\mathcal{L} \mathbf{v}_\mathcal{L}] \cdot \mathbf{n} = (\rho_\mathcal{S} - \rho_\mathcal{L}) \mathbf{w} \cdot \mathbf{n} \quad \text{on } \Phi \,. \tag{1.5}$$

If we assume there exists no discontinuity of the velocity field ($\mathbf{v}_\mathcal{S} = \mathbf{v}_\mathcal{L}$) across Φ and there is a relative motion of the interface ($(\mathbf{v} - \mathbf{w}) \cdot \mathbf{n} \neq 0$), then we must have $\rho_\mathcal{S} = \rho_\mathcal{L}$.

This is the case of an homogeneous incompressible body with negligible variation of the constant density at each phase, i.e., $\rho \equiv$ constant everywhere, or

$$\nabla \cdot \mathbf{v} = 0 \ \text{ in } \mathcal{S} \cup \mathcal{L} \quad \text{and} \quad [\![\mathbf{v}]\!] = 0 \ \text{ on } \Phi \,. \tag{1.6}$$

For the balance of energy, which is of most interest for the Stefan problem, we take $A = \rho(\frac{v^2}{2} + e)$ where e is the internal energy density and, in general, we must take into account the inherent balance of forces also. In order to simplify, we consider a simplified form of (1.2) corresponding to a normalized constant density $\rho \equiv 1$, $\mathbf{a} = \mathbf{q}$ denoting the energy flux and r the energy source including already the heating term of mechanical origin. The energy equation is, in this case,

$$\frac{de}{dt} = r - \nabla \cdot \mathbf{q} \quad \text{in } \mathcal{S} \cup \mathcal{L} \tag{1.7}$$

and the jump condition (1.3), taking (1.6) into account, is therefore

$$[\![e]\!](\mathbf{v} - \mathbf{w}) \cdot \mathbf{n} = [\![\mathbf{q}]\!] \cdot \mathbf{n} \quad \text{on } \Phi\,. \tag{1.8}$$

In order to obtain the heat equation from (1.7) it is necessary to introduce constitutive relations, which must be consistent with experience and with thermodynamical principles.

The classical Fourier law relates the temperature T with the heat flux,

$$\mathbf{q} = -k(T)\,\nabla T\,, \tag{1.9}$$

where $k > 0$ denotes the thermal conductivity. With the usual Kirchhoff transformation,

$$\vartheta \equiv K(T) = \int_{T_\Phi}^{T} k(\tau)\,d\tau\,, \tag{1.10}$$

T_Φ denoting the constant melting temperature, the relation (1.9) is just $\mathbf{q} = -\nabla\vartheta$ and the interface Φ is at the normalized temperature $\vartheta = 0$, the solid zone is represented by $\mathcal{S} = \{\vartheta < 0\}$ and the liquid one by $\mathcal{L} = \{\vartheta > 0\}$.

A functional relation between the internal energy and the temperature $e = e(T)$ should describe its jump at the interface, which is essentially the latent heat

$$\lambda = [\![e]\!] > 0\,. \tag{1.11}$$

Introducing the Heaviside function h ($h(\vartheta) = 0$ if $\vartheta < 0$, $h(\vartheta) = 1$ if $\vartheta > 0$), we can write

$$e = e(K^{-1}(\vartheta)) = b(\vartheta) + \lambda\,h(\vartheta) \quad \text{for } \vartheta \neq 0\,, \tag{1.12}$$

where b is a given continuous and increasing function.

We observe that assuming the system has constant volume, the internal energy can be identified with the quantity of heat. Since we also neglect the internal stresses and we shall assume a constant pressure, the internal energy can also be identified with the enthalpy, which often we shall refer to.

Consequently, from (1.7) we obtain the heat equation with convection

$$(\partial_t + \mathbf{v} \cdot \nabla)\,b(\vartheta) - \Delta\vartheta = r \quad \text{in } \{\vartheta > 0\} \cup \{\vartheta < 0\} \tag{1.13}$$

and from (1.8) the free boundary Stefan condition ($\partial_n = \partial/\partial n$ denoting the normal derivative)

$$\lambda(\mathbf{w} - \mathbf{v}) \cdot \mathbf{n} = [\![\nabla\vartheta]\!] \cdot \mathbf{n} = [\![\partial_n\vartheta]\!] \quad \text{on } \Phi = \{\vartheta = 0\}\,. \tag{1.14}$$

1.2 – Distributional and enthalpy formulations

Consider now the system (1.13)–(1.14) in the space-time domain $Q = \Omega\times\,]0,T[$, which is bisected by the smooth interface $\Phi = \{\vartheta = 0\} = \bigcup_{0<t<T} \Phi(t)$ separating the solid and the liquid regions, respectively

$$\mathcal{S} = \{\vartheta < 0\} = \bigcup_{0<t<T} \mathcal{S}(t) \quad \text{and} \quad \mathcal{L} = \{\vartheta > 0\} = \bigcup_{0<t<T} \mathcal{L}(t)\,.$$

Considering a smooth test function $\psi \in \mathcal{D}(Q)$ and integrating by parts separately in $\mathcal{S}$ and in $\mathcal{L}$, we have

$$\int_{\mathcal{S}\cup\mathcal{L}} \psi\,\Delta\vartheta - \int_{\Phi} [\partial_n\vartheta]\psi = -\int_Q \nabla\vartheta\cdot\nabla\psi = \int_Q \vartheta\,\Delta\psi = \langle\Delta\vartheta,\psi\rangle_{\mathcal{D}'}\,, \tag{1.15}$$

where $\langle\ ,\ \rangle_{\mathcal{D}'}$ means "in the sense of distributions in Q". If $\chi_{\mathcal{L}}$ denotes the characteristic function of $\mathcal{L}$ ($\chi_{\mathcal{L}} = 1$ in $\mathcal{S}$ and $\chi_{\mathcal{L}} = 0$ in $Q\backslash\mathcal{L}$), we obtain

$$\begin{aligned}\Big\langle\lambda(\partial_t + \mathbf{v}\cdot\nabla)\,\chi_{\mathcal{L}},\,\psi\Big\rangle_{\mathcal{D}'} &= -\lambda\int_{\mathcal{L}}(\partial_t\psi + \nabla\cdot(\mathbf{v}\psi)) = -\lambda\int_{\Phi}\psi(N_t + \mathbf{v}\cdot\mathbf{n})\\ &= \lambda\int_{\Phi}\psi(\mathbf{w}-\mathbf{v})\cdot\mathbf{n} = \int_{\Phi}\psi[\partial_n\vartheta]\end{aligned} \tag{1.16}$$

by (1.14), where $-N_t = \mathbf{w}\cdot\mathbf{n}$ denotes the normal velocity of Φ.

Since $b(\vartheta)$ is continuous across Φ and (1.6) is assumed, we have

$$\Big\langle(\partial_t + \mathbf{v}\cdot\nabla)\,b(\vartheta),\,\psi\Big\rangle_{\mathcal{D}'} = -\int_Q b(\vartheta)(\partial_t\psi + \nabla\cdot(\mathbf{v}\psi)) = \int_{\mathcal{S}\cup\mathcal{L}}\psi(\partial_t + \mathbf{v}\cdot\nabla)\,b(\vartheta)\,. \tag{1.17}$$

From (1.15), (1.16) and (1.17) using (1.14) we deduce

$$\Big\langle(\partial_t + \mathbf{v}\cdot\nabla)\,(b(\vartheta) + \lambda\,\chi_{\mathcal{L}}) - \Delta\vartheta,\,\psi\Big\rangle_{\mathcal{D}'} = \int_{\mathcal{S}\cup\mathcal{L}} r\,\psi, \qquad \forall\psi\in\mathcal{D}(Q)\,. \tag{1.18}$$

When Φ is a smooth surface, $\chi_{\mathcal{L}}$ and $h(\vartheta)$ define the same distribution, and therefore (1.18) condensates into a single equation ($d/dt = \partial_t + \mathbf{v}\cdot\nabla$)

$$\frac{d}{dt}e(\vartheta) - \Delta\vartheta = r \quad \text{in } \mathcal{D}'(Q) \tag{1.19}$$

both conditions (1.13) and (1.14), where $e(\vartheta) = b(\vartheta) + \lambda\,h(\vartheta)$ is exactly the constitutive relation (1.12).

We note, however, that the distributional formulation (1.19) is more general than the corresponding pointwise equation with the jump condition, since all references to the free boundary Φ have disappeared. In particular, (1.19) includes the possibility of the degeneration of the level set $\{\vartheta = 0\}$ into a region without any requirement of smoothness except measurability.

These considerations lead to the weak or enthalpy formulation of the corresponding (fixed) boundary and initial value problems for the Stefan equation (1.19). Consider the maximal monotone graph H associated with the Heaviside function, i.e.

$$H(\vartheta) = \begin{cases} 0 & \vartheta < 0,\\ [0,1] & \vartheta = 0,\\ 1 & \vartheta > 0.\end{cases} \tag{1.20}$$

For a given temperature $\vartheta = \vartheta(x,t)$, define the enthalpy η by the pointwise inclusion

$$\eta \in \gamma(\vartheta) \equiv b(\vartheta) + \lambda H(\vartheta) \tag{1.21}$$

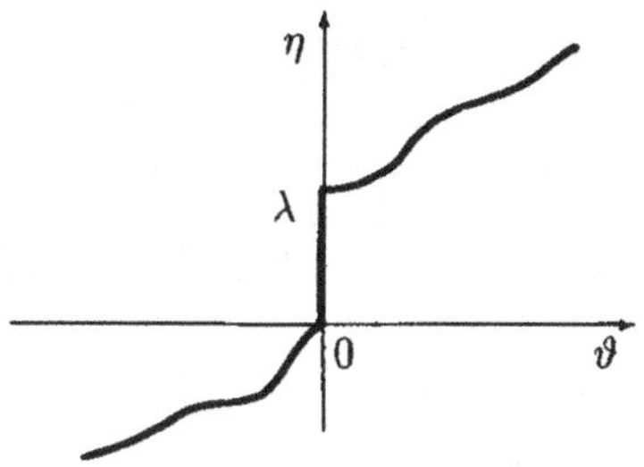

Consider now in the prescribed domain $Q = \Omega \times]0,T[$ with lateral boundary $\Sigma = \partial\Omega \times]0,T[$, for instance, a Dirichlet boundary condition for the temperature on $\Sigma_{\mathrm{D}} = \Sigma$

$$\vartheta = \vartheta_{\mathrm{D}} \quad \text{on } \Sigma_{\mathrm{D}} \tag{1.22}$$

and an initial condition for the enthalpy

$$\eta = \eta_0 \quad \text{on } \Omega \text{ at } t = 0\,. \tag{1.23}$$

If φ denotes a smooth test function such that $\varphi(T) = 0$ and $\varphi = 0$ on Σ_{D}, under the incompressibility assumption ($\nabla \cdot \mathbf{v} = 0$), we obtain from (1.19) and integration by parts the variational condition

$$-\int_Q \eta(\partial_t + \mathbf{v}\cdot\nabla)\varphi + \int_Q \nabla\vartheta\cdot\nabla\varphi = \int_Q r\varphi + \int_\Omega \eta_0\,\varphi(0), \quad \forall\varphi:\ \varphi|_{\Sigma_{\mathrm{D}}}=0,\ \varphi(T)=0\,, \tag{1.24}$$

which is valid under weak regularity conditions on η and on ϑ.

The problem of finding a pair (ϑ, η) verifying the conditions (1.21), (1.22) and (1.24) is called here the Stefan problem in its enthalpy formulation (with prescribed convection). This formulation is easily extended to the case of a Neumann boundary condition on part Σ_{N} of Σ (possibly the whole Σ). In fact if (1.22) holds only on $\Sigma\backslash\Sigma_{\mathrm{N}}$ and we have

$$-\partial_n\vartheta = g(x,t,\vartheta) \quad \text{on } \Sigma_N\,, \tag{1.25}$$

with the assumption $\mathbf{v}\cdot\mathbf{n} = 0$ on Σ_{N}, we should replace (1.24) by

$$-\int_Q \eta(\partial_t + \mathbf{v}\cdot\nabla)\varphi + \int_Q \nabla\vartheta\cdot\nabla\varphi + \int_{\Sigma_{\mathrm{N}}} g(\vartheta)\,\varphi = \int_Q r\varphi + \int_\Omega \eta_0\,\varphi(0), \tag{1.26}$$
$$\forall\varphi:\ \varphi|_{\Sigma_{\mathrm{D}}}=0,\ \varphi(T)=0\,,$$

i.e., φ vanishes only on $\Sigma_{\mathrm{N}} = \Sigma\backslash\Sigma_{\mathrm{D}}$, if this part of the boundary is not empty as in the case of Dirichlet–Neumann mixed boundary value problem.

We note that (1.25) without the restriction $\mathbf{v}\cdot\mathbf{n} = 0$ on Σ_{N} will yield (1.26) with $g(\vartheta)$ replaced by $g(\vartheta) + \eta\,\mathbf{v}\cdot\mathbf{n}$, which is a more delicate problem not considered here.

Finally we should remark that if (η, ϑ) is a weak solution of the Stefan problem such that the level set $\{\vartheta = 0\}$ is a smooth surface, which would imply enough regularity on ϑ, then by reversing the above arguments, it is easy to conclude that the equation (1.13) and the jump condition (1.14) are both satisfied in the classical sense.

1.3 – Variational inequality formulation: two-phase problem

In this section we assume $\mathbf{v} \equiv 0$ and $r = 0$. While the absence of convection is a crucial restriction, the second condition stands only for simplification of the presentation.

The essence of the enthalpy formulation (1.24) can be easily seen in the distributional equation in the form

$$\partial_t[b(\vartheta) + \lambda\chi] = \Delta\vartheta \, , \tag{1.27}$$

where $\chi = \frac{1}{\lambda}(\eta - b(\vartheta))$ is a function such that,

$$\chi \in H(\vartheta) \quad \text{a.e. in } Q \, , \tag{1.28}$$

or equivalently (recall (1.20))

$$0 \leq \chi_{\{\vartheta>0\}} \leq \chi \leq 1 - \chi_{\{\vartheta<0\}} \leq 1 \quad \text{a.e. in } Q \, . \tag{1.29}$$

Since, in general, the equation (1.27) contains a measure concentrated on the discontinuity set $\{\vartheta = 0\}$, a natural way to overcome this difficulty consists of integrating (1.27) with respect to time. In fact this procedure leads to consider the new unknown function

$$u(x,t) = \int_0^t \vartheta(x,\tau)\,d\tau \, , \quad x \in \overline{\Omega}, \;\; t \geq 0 \, . \tag{1.30}$$

Hence

$$u_t \equiv \partial_t u = \vartheta \, , \tag{1.31}$$

and integrating (1.27) in time, using the initial condition (1.23), we obtain

$$b(u_t) + \lambda\chi = \Delta u + \eta_0 \, . \tag{1.32}$$

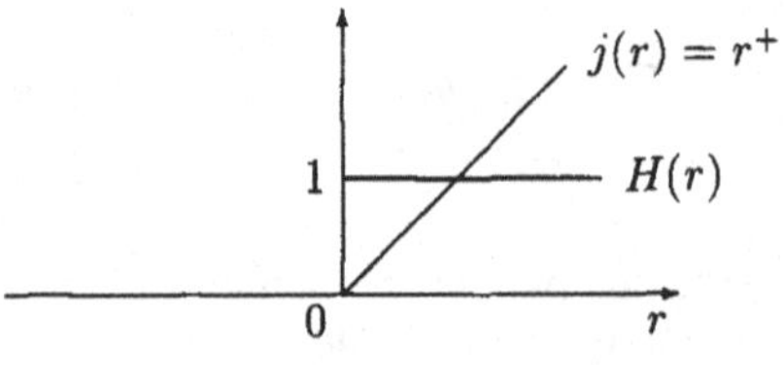

The graph $H = \partial j$

Now we observe the pointwise inequality ($r^+ = \max(r,0)$)

$$r^+ - s^+ \geq h(r-s), \qquad \forall h \in H(s), \ \forall r, s \in \mathbf{R}, \tag{1.33}$$

with the relations (1.28) and (1.31), implies

$$v^+ - u_t^+ \geq \chi(v - u_t) \quad \text{a.e. in } Q, \tag{1.34}$$

for any integrable function v. Introducing the notation

$$J(v) = \lambda \int_\Omega v^+ \, dx, \qquad \forall v \in L^2(\Omega), \tag{1.35}$$

and the integrating (1.34) in Ω, we have

$$J(v) - J(u_t) \geq \lambda \int_\Omega \chi(v - u_t), \qquad \forall v \in L^2(\Omega), \tag{1.36}$$

which means exactly that, $\chi = \chi(t)$ satisfies, as an element of $L^2(\Omega)$,

$$\lambda\chi \in \partial J(u_t) \quad \text{(at each fixed time } t\text{)}, \tag{1.37}$$

where ∂J denotes the subdifferential(*) of the convex functional J defined by (1.35).

Then combining (1.36) with (1.32) we obtain the following variational inequality for $u = u(t)$, for each fixed $t > 0$:

$$\int_\Omega b(u_t)(v - u_t) - \int_\Omega \Delta u(v - u_t) + J(v) - J(u_t) \geq \int_\Omega \eta_0(v - u_t), \qquad \forall v \in L^2(\Omega). \tag{1.38}$$

We observe that (1.30) yields the initial condition

$$u(0) = 0. \tag{1.39}$$

The variational inequality (1.38) has not yet included the boundary conditions for the temperature on the fixed boundary $\partial\Omega$. If they are of Dirichlet type, by (1.30) they remain the same type for the new unknown u. However if they are of Neumann type (1.25), the corresponding condition after the integration in time may, in general, involve the value of u_t on the boundary. In order to avoid this situation, we shall limit ourselves to the linearized case

$$g(x,t,\vartheta) = \sigma(x)(\vartheta - h'(x,t)), \qquad x \in \partial\Omega, \ t > 0, \tag{1.40}$$

where $\sigma \geq 0$ and h' are give functions. Then we obtain

$$-\partial_n u = \sigma(u - h) \quad \text{on } \partial\Omega, \ t > 0, \tag{1.41}$$

where $h(x,t) = \int_0^t h'(x,\tau)\, d\tau$.

(*) We recall from convex analysis, that if $J\colon V \to \mathbf{R}$ is a convex l.s.c. functional, its subdifferential at the point $u \in V$ is defined by $\partial J(u) = \{\xi \in V'\colon J(v) - J(u) \geq \langle \xi, v - u\rangle, \forall v \in V\}$. If J is differentiable at u, then $\partial J(u) = \{DJ(u)\}$.

In order to incorporate this boundary condition in (1.38) we observe that for any $w \in H^1(\Omega)$,

$$-\int_\Omega w\,\Delta u = \int_\Omega \nabla u \cdot \nabla w - \int_{\partial\Omega} w\,\partial_n u \tag{1.42}$$

and we introduce the following notations

$$a(u,v) = \int_\Omega \nabla u \cdot \nabla v + \int_{\partial\Omega} \sigma\, u\, v\,, \qquad \forall\, u,v \in H^1(\Omega)\,, \tag{1.43}$$

$$(u,v) = \int_\Omega u\, v,\quad |u| = \Big(\int_\Omega |u|^2\Big)^{1/2}, \qquad \forall\, u,v \in L^2(\Omega)\,, \tag{1.44}$$

$$\langle F, v\rangle = \int_\Omega \eta_0\, v + \int_{\partial\Omega} \sigma\, h\, v\,, \qquad \forall\, v \in H^1(\Omega)\,. \tag{1.45}$$

The variational formulation for the two-phase Stefan problem is now given by the following conditions:

$$u, u_t \equiv \partial_t u \in L^2(0,T;H^1(\Omega)), \qquad \text{such that } u(0)=0 \text{ and for a.e. } t>0\,, \tag{1.46}$$

$$(b(u_t), v-u_t) + a(u, v-u_t) + J(v) - J(u_t) \geq \langle F, v-u_t\rangle\,, \qquad \forall\, v \in H^1(\Omega)\,. \tag{1.47}$$

This is a parabolic variational inequality of second kind introduced and considered by Duvaut and Lions [DL], which has a unilateral condition on the time derivative u_t of the solution.

This formulation is quite useful for the theoretical and numerical resolution of the underlying two-phase Stefan problem and has the advantage of giving almost immediately the uniqueness of its solution. Indeed, for instance in the typical example where

$$b(s) = \gamma\, s^+ - \delta\, s^-\,, \qquad s \in \mathbf{R},\quad \gamma,\delta > 0\,, \tag{1.48}$$

if we have two solutions u and $\widehat{u}$ to (1.46)–(1.47), by letting $v = \widehat{u}(t)$ and $v = u(t)$ respectively as test functions in the corresponding inequalities (1.47), we obtain, after subtraction,

$$\inf(\gamma,\delta)\, |u-\widehat{u}|^2 + \frac{1}{2}\frac{d}{dt} a(u-\widehat{u}, u-\widehat{u}) \leq 0$$

which, integrating in time, implies $u = \widehat{u}$.

Consequently the uniqueness of the variational solution implies the existence of at most one solution of the original problem for the temperature $\vartheta = u_t$, the free boundary being *a posteriori* the level set $\{\vartheta = 0\}$.

1.4 – Variational inequality formulation: one-phase problem

An important simplification, often used in practice, consists of neglecting the variation of the temperature in one of the phases, say, for instance, in the solid one. A concrete example is the melting of ice at uniform zero temperature. This situation can be deduced from the general case and yields a parabolic variational inequality of first kind in the terminology of [DL].

Assume that a solution of (1.38) or, equivalently, of (1.32)–(1.28), is such that

$$u_t = \vartheta \geq 0 \quad \text{a.e. in } Q = \Omega \times]0, T[\tag{1.49}$$

and we are in the case (1.48), for simplicity. Then the function χ, given by

$$\chi = \frac{1}{\lambda}(\Delta u + \eta_0 - \gamma\, u_t) \tag{1.50}$$

is such that

$$0 \leq \chi_{\{u_t > 0\}} \leq \chi \leq 1 \quad \text{a.e. in } Q\ . \tag{1.51}$$

Note that we have assumed that the set $\{u_t < 0\}$ is empty but this does not imply that the level set $\{u_t = 0\}$ is neglectable in Q. In fact we are assuming that this set has positive measure in Q.

Now we suppose also that, up to a set of null measure, $\{u_t = 0\} \subset \{u = 0\}$, or equivalently

$$\{u > 0\} \subset \{u_t > 0\}, \quad \text{i.e.} \quad \chi_{\{u>0\}} \leq \chi_{\{u_t>0\}}\ . \tag{1.52}$$

Hence, since $u \geq 0$, we conclude from (1.51) that also

$$\chi \in H(u) \quad \text{or} \quad J(v) - J(u) \geq \lambda \int_\Omega \chi(v - u)\,, \qquad \forall v \in L^2(\Omega)\ ,$$

and, from (1.50), we obtain, for each fixed $t > 0$, the following inequality for $u = u(t)$:

$$\int_\Omega \gamma\, u_t(v-u) - \int_\Omega \Delta u(v-u) + J(v) - J(u) \geq \int_\Omega \eta_0(v-u)\,, \qquad \forall v \in L^2(\Omega)\ . \tag{1.53}$$

However, this formulation is more general, and of course not all solutions to (1.53) must satisfy the natural constrain (note that no assumption on η_0 nor on the fixed boundary conditions have yet been made)

$$u \geq 0 \quad \text{a.e. in } Q\ . \tag{1.54}$$

If this holds for the solution u of (1.53), then $u = u^+$ and it also solves

$$\int_\Omega \gamma\, u_t(v-u) - \int_\Omega \Delta u(v-u) \geq \int_\Omega (\eta_0 - \lambda)\,(v-u)\,, \qquad \forall v \in L^2(\Omega),\ \ v \geq 0\ , \tag{1.55}$$

which is a variational inequality of obstacle type (see [R3]), still without the fixed boundary conditions, which may also be of Dirichlet, Neumann or mixed type. Note that, compatible assumptions on the initial and boundary data must be imposed in order to recover the one-phase condition $u_t = \partial_t u \geq 0$.

To fix the mathematical model, we shall consider now a different type of boundary conditions corresponding to the following physical situation

$$\partial\Omega = \Gamma_0 \cup \Gamma_1 \,,$$
$$\Phi(t) = \partial\{\vartheta(t) > 0\} \subset \Omega \,,$$
$$\Gamma_1 = \partial\omega \,.$$

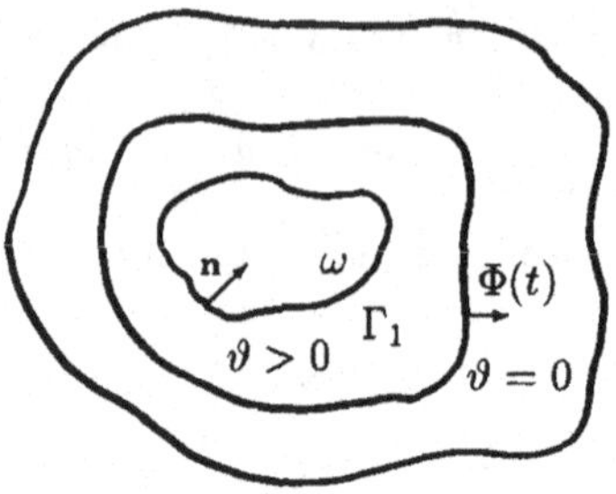

We assume that, as indicated in the figure above, the inner fixed boundary Γ_1 is a heat transfer interface between the exterior melting ice and an interior domain ω ($\partial\omega = \Gamma_1$) containing a hot liquid at temperature $\Theta = \Theta(x,t)$. Then the transmission boundary conditions are given by ($\mathbf{n}$ denotes the normal vector directed to the interior of ω)

(1.56) $$\vartheta = \Theta \,,$$

(1.57) $$\partial_n\vartheta = k\,\partial_n\Theta \quad \text{on } \Gamma_1 \,,$$

and Θ satisfies, for instance, a linear heat equation

(1.58) $$\sigma\,\partial_t\Theta = k\,\Delta\Theta + f \quad \text{on } \omega \,.$$

If we assume the interior liquid well stirred, Θ may be considered homogeneous in the space variables and therefore

(1.59) $$\Theta = \Theta(t) \quad \text{(function of } t \text{ only).}$$

This implies that the boundary condition (1.57) must be replaced by the information on the total heat flux across Γ_1. In fact, using the equation (1.58) we have

$$-k\int_{\Gamma_1} \partial_n\Theta = k\int_\omega \Delta\Theta = \int_\omega (\sigma\,\partial_t\Theta - f) = \sigma\,|\omega|\,\dot{\Theta} - |\omega|\,f(t) \,,$$

where the heat source f is also assumed space independent, $|\omega| = \int_\omega 1$ and $\dot{\Theta} = \frac{d\Theta}{dt}$. Accordingly we shall replace (1.57) by the natural condition ($\alpha = \sigma|\omega| \geq 0$)

(1.60) $$\alpha\,\dot{\Theta} + \int_{\Gamma_1} \partial_n\vartheta = |\omega|\,f(t) \quad \text{on } \Gamma_1, \ \ t > 0 \,,$$

to which we add a initial condition (necessary if $\alpha > 0$)

(1.61) $$\Theta(0) = \Theta_0 > 0 \,.$$

Applying the transformation (1.30), and taking this initial condition into account, the boundary conditions on Γ_1 for u are easily found:

(1.62) $$u = U(t) \quad (U(t) \text{ unknown constant at each } t > 0),$$

(1.63) $$\alpha\dot{U} + \int_{\Gamma_1} \partial_n u = g + \alpha\,\Theta_0 \ \text{ on } \Gamma_1, \quad \text{for } \ t > 0 \,,$$

where $g = g(t) = |\omega| \int_0^t f(\tau)\,d\tau \geq 0$ is a given function. In addition, since we shall assume that the outer boundary $\Gamma_0 = \partial\Omega \backslash \Gamma_1$ is in contact with the ice at zero temperature, it is natural to impose

$$(1.64) \qquad u = 0 \quad \text{on } \Gamma_0, \qquad \text{for } t > 0 \,.$$

The natural functional space to solve the variational inequality (1.53) with the boundary conditions (1.62)–(1.64) is then

$$(1.65) \qquad H^1_{\#} = \left\{ v \in H^1(\Omega) \colon \; v|_{\Gamma_0} = 0 \text{ and } v|_{\Gamma_1} = V = \text{constant} \right\} .$$

Using integration by parts with $(v-u) \in H^1_{\#}$, we have, for fixed t,

$$(1.66) \qquad \begin{aligned} -\int_\Omega \Delta u\,(v-u) &= \int_\Omega \nabla u \cdot \nabla (v-u) - (V-U) \int_{\Gamma_1} \partial_n u \\ &= \int_\Omega \nabla u \cdot \nabla (v-u) + (V-U)\,[\alpha \dot{U} - g - \alpha \Theta_0] \,. \end{aligned}$$

We introduce the notations

$$(1.67) \qquad (u,v)_{\alpha\gamma} = \gamma \int_\Omega uv + \alpha\, UV \,, \qquad \forall\, u,v \in H^1_{\#} \,,$$

$$(1.68) \qquad ((u,v)) = \int_\Omega \nabla u \cdot \nabla v \,, \qquad \forall\, u,v \in H^1_{\#} \,,$$

$$(1.69) \qquad \langle L(t), v \rangle = \int_\Omega \eta_0\, v + (g(t) + \alpha \Theta_0)\, V \,, \qquad \forall\, v \in H^1_{\#} \,,$$

and we observe that each function $v \in H^1_{\#}$ can be associated with the pair $(v,V) \in L^2(\Omega) \times \mathbf{R}$, where $V = v|_{\Gamma_1}$, and (1.67) can be regarded as a inner product in $H^1_{\#}$. The closure $\mathbf{H}$ of $H^1_{\#}$ with respect to $(\cdot,\cdot)_{\alpha\gamma}$ (provided $\alpha, \gamma > 0$) is isomorphic to $L^2(\Omega) \times \mathbf{R}$ (it is easy to see that $\mathbf{H}^\perp = \{(0,0)\}$).

Hence, from (1.66) and (1.53), with these notations we obtain the following variational inequality for this one-phase Stefan problem:

$$(1.70) \qquad u \in L^2(0,T; H^1_{\#}) \,, \qquad u' \equiv (u_t, \dot{U}) \in L^2(0,T; L^2(\Omega) \times \mathbf{R}), \quad u(0) = 0 \,,$$

$$(1.71) \qquad (u', v-u)_{\alpha\gamma} + ((u, v-u)) + J(v) - J(u) \geq \langle L, v-u \rangle \,, \qquad \forall\, v \in H^1_{\#} \,,$$

for a.e. $t > 0$.

Analogously, if we use (1.55) instead of (1.53) and we impose the constrain (1.54), i.e., in addition to (1.70) we require, for a.e. $t > 0$,

$$(1.72) \qquad u = u(t) \in \mathbf{K} \equiv \left\{ v \in H^1_{\#} \colon \; v \geq 0 \text{ in } \Omega \right\} ,$$

then one must replace (1.71) by

$$(1.73) \qquad (u', v-u)_{\alpha\gamma} + ((u, v-u)) \geq \langle L - \lambda, v-u \rangle \,, \qquad \forall\, v \in \mathbf{K} \,,$$

where we set

$$(1.74) \qquad \langle L - \lambda, v \rangle = \langle L, v \rangle - \lambda \int_\Omega v \,, \qquad \forall\, v \in H^1_{\#} \,.$$

The uniqueness of the solution to both problems can be easily shown, but the equivalence between the formulations (1.71) and (1.73) follows only if the solution of (1.71) also satisfies (1.72). In the next sections we shall discuss the problems of existence of solutions, their properties and their consequences to the existence of a solution to the original problem for the temperature.

2 – THE ONE-PHASE PROBLEM VIA VARIATIONAL INEQUALITIES

2.1 – Resolution of the variational inequality

Using the notations (1.35) and (1.67)–(1.69) we shall assume here the following hypothesis on the data:

$$\alpha, \gamma, \lambda > 0 \text{ and } 0 \le \Theta_0 \le M \quad \text{are given constants;} \tag{2.1}$$

$$\eta_{0\gamma} = \gamma\vartheta_0 + \lambda\chi_{\{\vartheta_0 > 0\}}, \quad \text{for } \vartheta_0 \in L^\infty(\Omega),\ 0 \le \vartheta_0 \le M; \tag{2.2}$$

$$g \in W^{1,\infty}(0,T),\ g(0) = 0,\ \dot{g} \ge 0 \quad \text{a.e. } t \in [0,T]\ . \tag{2.3}$$

We suppose $\Omega \subset \mathbf{R}^N$ is an open bounded domain with a smooth boundary $\partial\Omega = \Gamma_0 \cup \Gamma_1$. Hence, by the Poincaré inequality, in the Hilbert subspace $\mathbf{V} \equiv H^1_\# \subset H^1(\Omega)$, the inner product (1.68) induces an equivalent topology to the one defined by $H^1(\Omega)$. Also by the trace theorem, there exists a constant $C_\# = C(\Gamma_0, \Gamma_1) > 0$, such that

$$|U| \le C_\# \|\nabla u\| = C_\# \Big(\int_\Omega |\nabla u|^2\Big)^{1/2}, \quad \forall u \in H^1_\#,\ U = u|_{\Gamma_1}\ . \tag{2.4}$$

Consequently, the topology associated with the inner product $(\cdot,\cdot)_{\alpha\gamma}$ defined by (1.67) in $\mathbf{V}$ is weaker than the $H^1(\Omega)$-topology, since

$$|u|_{\alpha\gamma} \equiv (u,u)^{1/2}_{\alpha\gamma} \le C\,\|\nabla u\|, \quad \forall u \in H^1_\#\ . \tag{2.5}$$

Therefore, if we denote by $\mathbf{H}$ the completion of $\mathbf{V}$ for the topology of $|\cdot|_{\alpha\gamma}$, we have $\mathbf{V} \subsetneq \mathbf{H}$ and, with the identification of $v \in H^1_\#$ with $(v, V) \in L^2(\Omega) \times \mathbf{R}$, $V \equiv v|_{\Gamma_1}$, we easily obtain a natural isomorphism between $\mathbf{H}$ and $L^2(\Omega) \times \mathbf{R}$, allowing the identification of these two spaces.

In addition to the usual notation for the Sobolev spaces $W^{m,p}(\Omega)$ ($m \in \mathbf{N}$, $1 \le p \le \infty$) with $L^p(\Omega) = W^{0,p}(\Omega)$ and $H^1(\Omega) = W^{1,2}(\Omega)$, we shall also use the same notation for vector valued space functions $W^{m,p}(0,T;\mathbf{V})$, with $H^1(0,T;\mathbf{V}) = W^{1,2}(0,T;\mathbf{V})$ and $W^{1,\infty}(0,T;\mathbf{H}) = C^{0,1}[0,T;\mathbf{H}]$, if $\mathbf{H}$ is an Hilbert space.

The following theorem summarizes the most relevant results in the resolution of the variational inequality one-phase Stefan problem, in the Hilbertian framework $\mathbf{V} = H^1_\#$ and $\mathbf{H} \simeq L^2(\Omega) \times \mathbf{R}$.

Theorem 2.1. *Let the assumptions (2.1)–(2.3) hold. Then, there exists a unique solution to the evolution variational problem (1.70)– (1.71)*

$$u \in H^1(0,T;\mathbf{V}) \cap W^{1,\infty}(0,T;\mathbf{H})\ . \tag{2.6}$$

In addition, the solution u and its time derivative u' (we write $u' = u_t$ in Ω and $u' = \dot{U}$ on Γ_1) have the following properties:

(2.7) $0 \leq u'(t) \leq M + \frac{1}{\alpha} g(t)$, *fot* $t > 0$ *and a.e. on* Ω *and on* Γ_1 ;

(2.8) $0 \leq u(x,t) \leq tM + \frac{1}{\alpha}\int_0^t g(s)\,ds$ *for* $(x,t) \in \overline{\Omega} \times [0,t]$;

(2.9) $u \in L^\infty(0,T; W^{2,p}(\Omega) \cap C^{1,\alpha}(\overline{\Omega})) \cap C^{0,1}(\overline{\Omega} \times [0,T])$, $\forall p < \infty$, $0 \leq \alpha < 1$;

(2.10) $U = u|_{\Gamma_1} \in C^{0,1}[0,T]$. ∎

Remark 2.2. The uniqueness follows immediately from (1.71): if u_1 and u_2 are two solutions, their difference $w = u_1 - u_2$, $W = w|_{\Gamma_1}$ verifies

$$\frac{\gamma}{2}\frac{d}{dt}\int_\Omega w^2 + \frac{\alpha}{2}\frac{d}{dt}W^2 + \int_\Omega |\nabla w|^2 = (w', w)_{\alpha\gamma} + ((w,w)) \leq 0$$

and, integrating in t, we find $w \equiv 0$ since $w(0) = W(0) = 0$. □

Remark 2.3. The property (2.8) follows from (2.7) by integration in time. In particular, it implies that $u(t) \in \mathbf{K}$, as defined in (1.72) and if we restrict in (1.71) v to $\mathbf{K}$, we obtain that u also satisfies (1.73). Since this later variational inequality also admits at most one solution, we conclude the equivalence of the two formulations, under the above assumptions. □

Remark 2.4. The regularity property (2.10) is an immediate consequence of (2.6) and the isomorphism $\mathbf{H} \simeq L^2(\Omega) \times \mathbf{R}$, while the first regularity property of (2.9) is a consequence of the estimate (2.7) on the time derivative and the well know regularity properties for the elliptic obstacle problem (see [R3], for instance). Here, as well as in all cases where $W^{2,p}$-regularity up to the boundary is used, a $C^{1,1}$-smoothness of boundary $\partial\Omega$ is required, while in the rest of this paper the Lipschitz regularity of the fixed boundary is sufficient. Finally, since all derivatives of u are bounded, it follows that $u \in C^{0,1}(\overline{\Omega} \times [0,T])$. □

The existence result may be obtained as an application of different mathematical techniques, such as, for instance, nonlinear semi-groups associated with maximal monotone operators in Hilbert Spaces (see, for instance, [B2] or [Ba1]) or the semi-discretization in time, also called the Rothe's method (see, for instance, [L2] or [Na1]). However we shall employ here the variational approach, also called the energy method (see [L2], [DL] or [Z]), which is based on the Faedo–Galerkin method for the resolution of an approximated (by regularization) problem, from which the *a priori* estimates leading to (2.6) and (2.7) can be easily obtained

For each $\varepsilon > 0$, we consider a C^∞-approximation H_ε of the Heaviside function (1.20), such that

$$(2.11)\qquad \begin{cases} H_\varepsilon(\tau) = 0 \ \text{ if } \tau \leq 0, \ \ H_\varepsilon(\tau) = 1 \ \text{ if } \tau \geq \varepsilon, \ \ H'_\varepsilon(\tau) \geq 0, \ \ \tau \in \mathbf{R}, \\ \text{and } H_\varepsilon \to H \text{ uniformly in the compact subsets of } \mathbf{R}\backslash\{0\} \text{ as } \varepsilon \to 0, \end{cases}$$

and we consider the following auxiliar semi-linear approximating problem for $u_\varepsilon = u_\varepsilon(x,t)$:

$$(2.12)\qquad \gamma u_{\varepsilon t} - \Delta u_\varepsilon + \lambda H_\varepsilon(u_\varepsilon) = \eta_{0\gamma} \quad \text{in } \Omega, \ \ t > 0 \ ,$$

$$(2.13)\qquad u_\varepsilon = 0 \quad \text{on } \Gamma_0,\ t>0\,,$$

$$\left.\begin{aligned}&(2.14)\qquad u_\varepsilon(t) = U_\varepsilon(t)\ \text{(unknown constant)}\\ &(2.15)\qquad \alpha\dot{U}_\varepsilon + \int_{\Gamma_1}\partial_n u_\varepsilon = g(t) + \alpha\Theta_0\end{aligned}\right\}\ \text{on } \Gamma_1,\ t>0\,,$$

$$(2.16)\qquad u_\varepsilon(0) = z_\varepsilon \ \text{in } \Omega\,,\quad U^\varepsilon(0) = Z_\varepsilon\,.$$

Here $Z_\varepsilon = z_\varepsilon|_{\Gamma_1}$ is a constant and z_ε is the solution of the elliptic problem (where $\chi_0 \equiv \chi_{\{\vartheta_0>0\}}$ is as given in (2.2))

$$(2.17)\qquad \begin{cases} -\Delta z_\varepsilon + \lambda H_\varepsilon(z_\varepsilon) = \lambda\chi_0 \quad \text{in } \Omega\,,\\ z_\varepsilon|_{\Gamma_0} = 0,\ \ z_\varepsilon|_{\Gamma_1} = Z_\varepsilon\ \text{(unknown) and } \displaystyle\int_{\Gamma_1}\partial_n z_\varepsilon = 0\,. \end{cases}$$

Lemma 2.6. *The unique solution z_ε to (2.17) is such that*

$$(2.18)\qquad 0\le z_\varepsilon\le\varepsilon \ \text{ and } \int_\Omega |\nabla z_\varepsilon|^2 \le \varepsilon C_0\,,\quad \forall\varepsilon>0\,.$$

Proof: If we write (2.17) in variational form

$$(2.19)\qquad z_\varepsilon\in H^1_\#:\ \int_\Omega \nabla z_\varepsilon\cdot\nabla\zeta + \lambda\int_\Omega H_\varepsilon(z_\varepsilon)\zeta = \lambda\int_\Omega \chi_0\zeta\,,\quad \forall\zeta\in H^1_\#\,,$$

since H_ε is a monotone increasing bounded nonlinearity, the existence and uniqueness of its solution z_ε follows easily (see, for instance, [R3], §4.3). Taking $\zeta = \min(z_\varepsilon,0) = -z_\varepsilon^- \in H^1_\#$ in (2,19), we have

$$\int_\Omega |\nabla z_\varepsilon^-|^2 = -\lambda\int_\Omega \chi_0\, z_\varepsilon^- \le 0$$

and $z_\varepsilon^- = 0$, yielding $z_\varepsilon\ge 0$ in Ω. Analogously, taking $\zeta = (z_\varepsilon-\varepsilon)^+\in H^1_\#$ in (2.19), we conclude $z_\varepsilon\le\varepsilon$ from

$$\int_\Omega |\nabla(z_\varepsilon-\varepsilon)^+|^2 = \lambda\int_{\{z_\varepsilon>\varepsilon\}} (\chi_0-1)(z_\varepsilon-\varepsilon)\le 0\,.$$

Finally, the last estimate of (2.18) is an immediate consequence of the first one, by taking $\zeta = z_\varepsilon$ in (2.19) and letting $C_0 = \lambda\int_\Omega \chi_0$. ∎

We consider now the hilbertian framework defined above

$$H^1_\# \equiv \mathbf{V}\subset\mathbf{H}\simeq L^2(\Omega)\times\mathbb{R}\subset\mathbf{V}'$$

and we observe that these imbeddings are continuous and compact. We introduce the parabolic monotone equation for u_ε corresponding to (2.12)–(2.16), for each $\varepsilon>0$,

$$(2.20)\qquad (u_\varepsilon', v)_{\alpha\gamma} + \langle A_\varepsilon u_\varepsilon, v\rangle = \langle L(t), v\rangle\,,\quad \forall v\in\mathbf{V},\ t>0\,,$$

$$(2.21)\qquad u_\varepsilon(0) = z_\varepsilon \quad \text{in } \mathbf{V}\ (\subset\mathbf{H})\,,$$

where L is given by (1.69) and the monotone operator $A_\varepsilon\colon \mathbf{V} \to \mathbf{V}'$ is defined by

$$\langle A_\varepsilon u, v\rangle = \int_\Omega \nabla u \cdot \nabla v + \lambda \int_\Omega H_\varepsilon(u)\, v\,, \quad \forall\, u, v \in \mathbf{V}\,. \tag{2.22}$$

Proposition 2.7. *Under the preceding assumptions, there exists a unique solution to (2.20)–(2.21), which is such that*

$$u_\varepsilon \in H^1(0,T;\mathbf{V}) \cap W^{1,\infty}(0,T;\mathbf{H}), \quad \text{uniformly in } 0 < \varepsilon \le 1\,. \tag{2.23}$$

Due to the special structure of our problem, we shall use a particular choice of the Galerkin approximations in the Hilbert space $\mathbf{V} = H^1_\#$, by considering the finite dimensional subspace $\mathbf{V}_m$, for arbitrary $m \in \mathbf{N}$, spanned by the first m vectors of some basis of $\mathbf{V}$. In particular we also consider the Galerkin approximations of the solutions of the auxiliary problem (2.17) or (2.19) for each fixed $\varepsilon > 0$:

$$z_m \in \mathbf{V}_m\colon \int_\Omega \nabla z_m \cdot \nabla \varphi + \lambda \int_\Omega H_\varepsilon(z_m)\, \varphi = \lambda \int_\Omega \chi_0\, \varphi\,, \quad \forall \varphi \in \mathbf{V}_m\,. \tag{2.24}$$

Lemma 2.8. *The unique solution z_m to (2.24) converges in $\mathbf{V} = H^1_\#$-strong, as $m \to +\infty$, to the solution z_ε of (2.19).*

Proof: Letting $\varphi = z_m$ in (2.24), we easily obtain the uniform estimate $\|z_m\|_V \le C_0|\chi_0|$. Consequently, for a subsequence $m \to +\infty$ and for some element $z_\varepsilon \in \mathbf{V}$, we have

$$z_m \rightharpoonup z_\varepsilon \ \text{ in } \mathbf{V}\text{-weak} \ \text{ and in } \mathbf{H}\text{-strong.}$$

Let $\varphi \in \mathbf{V}_{m_0}$, for $m_0 \in \mathbf{N}$ arbitrary, and take the limit $m \to \infty$ in (2.24), we find that

$$\langle A_\varepsilon z_\varepsilon, \varphi\rangle \equiv \int_\Omega \nabla z_\varepsilon \cdot \nabla \varphi + \lambda \int_\Omega H_\varepsilon(z_\varepsilon)\, \varphi = \lambda \int_\Omega \chi_0\, \varphi\,, \quad \forall \varphi \in \mathbf{V}_{m_0}\,.$$

Since m_0 is arbitrary and $\bigcup_{m \in \mathbf{N}} \mathbf{V}_m$ is dense in $\mathbf{V}$, we conclude that z_ε also solves (2.19). The strong convergence $z_m \to z_\varepsilon$ in $\mathbf{V}$ holds from the following application of the weak convergence with the equalities (2.24) and (2.19), respectively for $\varphi = z_m$ and $\zeta = z_\varepsilon$:

$$\begin{aligned}\int_\Omega |\nabla(z_\varepsilon - z_m)|^2 &= \lambda \int_\Omega [\chi_0 - H_\varepsilon(z_\varepsilon)]\,(z_\varepsilon - z_m) - \int_\Omega \nabla z_m \cdot \nabla z_\varepsilon + \lambda \int_\Omega [\chi_0 - H_\varepsilon(z_m)] z_m \\ &\underset{m \to +\infty}{\longrightarrow} 0 - \int_\Omega |\nabla z_\varepsilon|^2 + \lambda \int_\Omega [\chi_0 - H_\varepsilon(z_\varepsilon)]\, z_\varepsilon = 0\,. \ \blacksquare\end{aligned}$$

Proof of Proposition 2.7.: Let $v_1, \ldots, v_n, \ldots$ be a basis of $\mathbf{V}$ and $\mathbf{V}_m$ the subspace spanned by the first m elements. Since $z_\varepsilon \in \mathbf{V}$ we may consider the sequence $z_m \in \mathbf{V}_m$ of solutions to (2.24) and we know that $z_m \to z_\varepsilon$ in $\mathbf{V}$, as $m \to \infty$. We look for a finite dimensional approximation of u_ε in the form

$$u_m(t) = \sum_{i=1}^{m} \xi_{mi}(t)\, v_i$$

and such that the vector $\{\xi_{mi}\}_{i=1}^{m}$ solves the system of ordinary differential equations in t:

$$\text{(2.25)} \qquad \begin{cases} (u'_m(t), v_j)_{\alpha\gamma} + \langle A_\varepsilon u_m(t), v_j\rangle = \langle L(t), v_j\rangle \quad \forall j = 1, ..., m \ , \\ u_m(0) = z_m \ . \end{cases}$$

Multiplying (2.25) by $\xi_{mj}(t)$, summing on $j = 1, ..., m$, we obtain

$$\text{(2.26)} \qquad \frac{\gamma}{2}\frac{d}{dt}|u_m|^2 + \frac{\alpha}{2}\frac{d}{dt}U_m^2 + \int_\Omega |\nabla u_m|^2 + \lambda \int_\Omega H_\varepsilon(u_m)\, u_m = \langle L(t), u_m\rangle \ .$$

Integrating in time, since $H_\varepsilon(u_m)\, u_m \geq 0$, by recalling the definition (1.69), Poincaré inequality and (2.4), we obtain easily with the aid of Schwartz inequality, the following energy estimate ($Z_m = z_m|_{\Gamma_1}$):

$$\text{(2.27)} \qquad \gamma |u_m(t)|^2 + \alpha\, U_m^2(t) + \int_0^t \int_\Omega |\nabla u_m|^2 \leq \gamma |z_m|^2 + \alpha\, Z_m^2 + \int_0^t C_L^2 \ ,$$

where $C_L = C_L(t) = C_0(|\eta_{0\gamma}| + g(t) + \alpha\, \Theta_0) > 0$ is such that

$$\text{(2.28)} \qquad \begin{aligned} \langle L(t), v\rangle &\leq |\eta_{0\gamma}|\, |v| + (g(t) + \alpha\Theta_0)\, |V| \leq C_0(|\eta_{0\gamma}| + g(t) + \alpha\Theta_0)\, \|\nabla v\| \\ &\leq \frac{1}{2}\int_\Omega |\nabla v|^2 + \frac{1}{2} C_L^2(t) \ , \qquad \forall v \in \mathbf{V} = H^1_{\#} \ . \end{aligned}$$

Since the right hand side of (2.27) is independent of m (and of ε), we can extract a subsequence such that as $m \to \infty$

$$\text{(2.29)} \qquad u_m \rightharpoonup u_\varepsilon \quad \text{in } L^2(0,T;\mathbf{V})\text{-weak and } L^\infty(0,T;\mathbf{H})\text{-weak*},$$

$$\text{(2.30)} \qquad u'_m \rightharpoonup u'_\varepsilon \text{ and } A_\varepsilon u_m \rightharpoonup X \quad \text{in } L^2(0,T;\mathbf{V}')\text{-weak},$$

$$\text{(2.31)} \qquad u_m \to u_\varepsilon \quad \text{in } C^0[0,T;\mathbf{H}] \text{ strongly } (^*).$$

In particular, (2.31) implies that $X = A_\varepsilon u_\varepsilon$ (recall (2.22)) and from (2.24) we easily conclude that u_ε solves

$$\text{(2.32)} \qquad u'_\varepsilon + A_\varepsilon u_\varepsilon = L \quad \text{in } L^2(0,T;\mathbf{V}') \ ,$$

with the initial condition (2.21) in $\mathbf{H}$. In order, to show that u_ε has the regularity (2.23) and consequently also solves (2.20)–(2.21) we need an additional estimate on u'_ε.

Let $w_h(t) = \frac{1}{h}[u_\varepsilon(t+h) - u_\varepsilon(t)]$, for $h > 0$ such that $t + h \leq T$, and multiply it by the difference in $\mathbf{V}'$ of the equations (2.32) at time $t + h$ and t. We obtain for $t \in [0, T-h]$

$$\Big\langle u'_\varepsilon(t+h) - u'_\varepsilon(t) + A_\varepsilon u_\varepsilon(t+h) - A_\varepsilon u_\varepsilon(t), w_h(t)\Big\rangle = \Big\langle L(t+h) - L(t), w_h(t)\Big\rangle$$

and, recalling that H_ε is monotone increasing, we have

$$\frac{1}{2}\frac{d}{dt}(w_h, w_h)_{\alpha\gamma} + \int_\Omega |\nabla w_h|^2 \leq g_h(t)\, W_h(t) \qquad \text{for a.e. } t > 0 \ ,$$

(*) Here we use the compactness of the imbedding $\mathbf{V} \subset \mathbf{H}$, which implies by the Aubin' lemma (see [Si], for instance) that $\{v \in L^2(0,T;\mathbf{V}) \colon v' = \frac{dv}{dt} \in L^2(0,T;\mathbf{V}')\}$ is compactly imbedded in $C^0[0,T;\mathbf{H}]$.

where $g_h(t) = \frac{1}{h}[g(t+h) - g(t)]$. Integrating in time we obtain (using (2.4))(**)

$$\text{(2.33)} \qquad \gamma|w_h(t)|^2 + \alpha W_h^2(t) + \int_0^t \int_\Omega |\nabla w_h|^2 \le \gamma|w_h(0)|^2 + \alpha W_h^2(0) + C_\#^2 \int_0^t |\dot{g}|^2 .$$

Now we observe that, from the finite dimensional approximations (2.25), since the $\zeta_{mi}(t) \in C^1[0,T]$, we have, recalling (1.69), (2.2) and (2.24):

$$\text{(2.34)} \qquad (u_m'(0), v_j)_{\alpha\gamma} = \langle L(0) - A_\varepsilon z_m, v_j \rangle = \gamma \int_\Omega \vartheta_0 \, v_j + \alpha\, \Theta_0 V_j \le |\widetilde{\vartheta}_0|_{\alpha\gamma} \, |v_j|_{\alpha\gamma}$$

for $j = 1, \ldots, m$ and $\widetilde{\vartheta}_0 = (\vartheta_0, \Theta_0) \in \mathbf{H} \simeq L^2(\Omega) \times \mathbf{R}$. Since $\{v_j\}$ form a basis of $\mathbf{V}$, which is dense in $\mathbf{H}$ for the norm of $|\cdot|_{\alpha\gamma}$, we conclude

$$|u_\varepsilon'(0)|_{\alpha\gamma} \le \liminf_{m \to \infty} |u_m'(0)|_{\alpha\gamma} \le |\widetilde{\vartheta}_0|_{\alpha\gamma} .$$

But this means also that, as $h \to 0$, $w_h(0) \to u_\varepsilon'(0)$ in $\mathbf{H}$ and in (2.33), the right hand side is bounded independently of h. Therefore, by letting also $h \to 0$ in (2.33), we obtain the estimate

$$\text{(2.35)} \qquad \gamma|u_{\varepsilon t}(t)|^2 + \alpha \dot{U}_\varepsilon^2(t) + \int_0^t \int_\Omega |\nabla u_{\varepsilon t}|^2 \le \gamma|\vartheta_0|^2 + \alpha\Theta_0^2 + C_\#^2 \int_0^t |\dot{g}|^2 ,$$

which, in particular, implies the regularity

$$u_\varepsilon' \in L^\infty(0,T;\mathbf{H}) \cap L^2(0,T;\mathbf{V}), \qquad \text{uniformly in } \varepsilon > 0 .$$

The uniqueness follows easily by an argument similar to the one in Remark 2.2. ∎

Remark 2.9. From the proof of Proposition 2.7, we easily conclude that $\{\zeta_{mi}'(t)\}_{i=1}^m$ also solves the following system of ordinary differential equations in t (recall (2.25) and (2.34)):

$$\begin{cases} (u_m''(t), v_j)_{\alpha\gamma} + \langle A_\varepsilon'(u_m)\, u_m'(t), v_j \rangle = \langle L'(t), v_j \rangle \\ (u_m'(0), v_j)_{\alpha\gamma} = (\widetilde{\vartheta}_0, v_j)_{\alpha\gamma} , \qquad j = 1, \ldots, m , \end{cases}$$

where $\langle L'(t), v \rangle = \dot{g}(t)\, V$, $\forall\, v \in H_\#^1$ and

$$\text{(2.37)} \qquad \langle A_\varepsilon'(u)w, v \rangle = \int_\Omega \nabla w \cdot \nabla v + \lambda \int_\Omega H_\varepsilon'(u)\, wv , \qquad \forall\, v, w \in H_\#^1, \ \forall\, u \in L^2(\Omega) .$$

Hence, we have $u_m' \rightharpoonup w = u_\varepsilon'$, when $m \to +\infty$, in $L^2(0,T;\mathbf{V})$-weak and in $L^\infty(0,T;\mathbf{H})$-weak* and, in the limit, due to (2.31), we have

$$\text{(2.38)} \qquad \begin{cases} w' + A_\varepsilon'(u_\varepsilon)\, w = L' & \text{in } L^2(0,T;\mathbf{V}') \\ w(0) = \widetilde{\vartheta}_0 \quad \text{in } \mathbf{H} & (\widetilde{\vartheta}_0 = (\vartheta_0, \Theta_0)). \end{cases}$$

Notice that also that $w = u_\varepsilon' \in H^1(0,T;\mathbf{V}^1) \cap C^0[0,T;\mathbf{H}]$, but this property is not uniform with respect to $\varepsilon > 0$, as well as the further regularity $w \in L^2(\delta,T;H^2(\Omega)) \cap H^1(\delta,T;\mathbf{H})$, $\forall\, \delta \in]0,T[$ (see [B2]). □

(**) Here we use the fact that $g \in H^1(0,T)$ and if, we set $g_h \equiv 0$ for $T - h \le t \le T$, we have $\|g_h\|_{L^2(0,T)} \le \|\dot{g}\|_{L^2(0,T)}$ and $g_h \to \dot{g}$ in $L^2(0,T)$, as $h \to 0$.

Proposition 2.10. *Under the previous assumptions, we have*

$$0 \le w = u'_\varepsilon \le M + \frac{1}{\alpha} g(t), \qquad \text{for } t > 0, \text{ a.e in } \Omega \text{ and on } \Gamma_1 . \tag{2.39}$$

Proof: From (2.38), after multiplication by $v(t) = -[w(t)]^- \in \mathbf{V}$, and using the assumption (2.3) and the regularity of Remark 2.9 we have for a.e. $t > 0$

$$-(w', w^-)_{\alpha\gamma} - \int_\Omega \nabla w \cdot \nabla w^- - \lambda \int_\Omega H'_\varepsilon(u_\varepsilon)\, w\, w^- = -\dot{g}\, W^- \le 0 ;$$

recalling that $H'_\varepsilon \ge 0$ and $-ww^- = (w^-)^2$, integrating in time between $\delta > 0$ and $t > \delta$, we have

$$\frac{1}{2}\int_\delta^t \frac{d}{dt}(w^-, w^-)_{\alpha\gamma} + \int_\delta^t \int_\Omega |\nabla w^-|^2 \le 0 ,$$

whence, letting $\delta \to 0$ and setting $v = -w^-$, we find

$$\frac{1}{2}|v(t)|^2_{\alpha\gamma} + \int_0^t \int_\Omega |\nabla v|^2 \le 0 , \tag{2.40}$$

since $-v(0) = w^-(0) = (\vartheta_0^-, \Theta_0^-) = (0,0)$ by the assumptions (2.1) and (2.2). Therefore $-v = w^- = 0$ and we conclude $w \ge 0$.

Also from (2.38), multiplying by $v(t) \equiv (w(t) - M - \frac{1}{\alpha} g(t))^+ \in \mathbf{V}$ we easily get, for a.e. t

$$\left(\left[w(t) - M - \frac{1}{\alpha} g(t)\right]', v\right)_{\alpha\gamma} + \int_\Omega |\nabla v|^2 = -\lambda \int_\Omega H'_\varepsilon(u_\varepsilon)\, w\, v - \frac{\gamma}{\alpha}\dot{g}(t) \int_\Omega v \le 0 \tag{2.41}$$

since

$$\frac{1}{\alpha}\left(\dot{g}(t), v(t)\right)_{\alpha\gamma} = \frac{\gamma}{\alpha}\dot{g}(t) \int_\Omega v(t) + \dot{g}(t)\, V(t) .$$

Hence, integrating (2.41) in time, we also get (2.40) and therefore also $v(t) \equiv 0$, because $M \ge \max(\vartheta_0, \Theta_0)$ by (2.1) and (2.2). ∎

Proof of the Theorem 2.1: From the proof of the Proposition 2.7, there exists a subsequence $\varepsilon \to 0$ and a function $u \in H^1(0,T;\mathbf{V}) \cap W^{1,\infty}(0,T;\mathbf{H})$ such that

$$u_\varepsilon \to u \quad \text{in } H^1(0,T;\mathbf{V})\text{-weak and } C^0(0,T;\mathbf{H})\text{-strong}, \tag{2.42}$$

$$U_\varepsilon \to U \quad \text{in } C^{0,\alpha}[0,T], \ \forall 0 < \alpha < 1 . \tag{2.43}$$

Now if we set

$$J_\varepsilon(v) = \lambda \int_\Omega \Big(\int_0^v H_\varepsilon(\tau)\, d\tau\Big) dx ,$$

since $\int_0^v H_\varepsilon(\tau)\, d\tau \to v^+$ uniformly in the compact subsets of $\mathbf{R}$, by (2.42), we have

$$J_\varepsilon(u_\varepsilon) \to J_\varepsilon(u) \quad \text{in } C^0[0,T] .$$

Using the following property of the convex function J_ε

$$J_\varepsilon(v) - J_\varepsilon(u_\varepsilon) \ge \lambda \int_\Omega H_\varepsilon(u_\varepsilon)\,(v - u_\varepsilon), \qquad \forall v \in L^2(\Omega) ,$$

in the equation (2.20), by recalling (2.22) we obtain for a.e. $t > 0$:

$$(2.44) \qquad (u'_\varepsilon, v - u_\varepsilon)_{\alpha\gamma} + ((u_\varepsilon, v - u_\varepsilon)) + J_\varepsilon(v) - J_\varepsilon(u_\varepsilon) \geq \langle L(t), v - u_\varepsilon \rangle, \qquad \forall v \in \mathbf{V}.$$

Since (2.42), in particular, implies $\liminf_{\varepsilon\to 0} \|\nabla u_\varepsilon(t)\|^2 \geq \|\nabla u(t)\|^2$, for a.e. $t > 0$, we can take the limit in (2.44) and we conclude that u solves the variational inequality (1.71) for a.e. $t > 0$.

On the other hand, $u_\varepsilon(0) = z_\varepsilon \to 0$ in $\mathbf{V}$-strong, by the Lemma 2.6, so that the initial condition $u(0) = 0$ is satisfied.

The property (2.7) is a consequence of the estimate (2.39).

Finally, if we take $v = \varphi \in \mathcal{D}(\Omega)$ in (2.20) we obtain

$$\gamma \int_\Omega u_{\varepsilon t}\varphi + \int_\Omega \nabla u_\varepsilon \cdot \nabla\varphi + \lambda \int_\Omega H_\varepsilon(u_\varepsilon)\varphi = \int_\Omega \eta_{0\gamma}\varphi, \qquad \forall \varphi \in \mathcal{D}(\Omega),$$

and consequently, recalling (2.39) and $0 \leq H_\varepsilon \leq 1$, we have a.e. in $\Omega\times]0,T[$

$$(2.45) \qquad -\eta_{0\gamma} \leq \Delta u_\varepsilon(t) = \gamma\, u_{\varepsilon t}(t) + \lambda\, H_\varepsilon(u_\varepsilon(t)) - \eta_{0\gamma} \leq \gamma M + \frac{\gamma}{\alpha}\, g(t) + \lambda.$$

Therefore by elliptic regularity $u_\varepsilon(t)$ having constant trace on $\partial\Omega$, is such that $u_\varepsilon \in L^\infty(0,T; W^{2,p}(\Omega))$, $\forall p < \infty$, uniformly in $\varepsilon > 0$, which implies the additional property (2.9) and completes the proof of the Theorem 2.1. ∎

Remark 2.11. The variational inequational (1.70)–(1.71) can be shown to have a solution under weaker assumptions on the data, but the conditions of this section are natural for the Stefan problem. However, we have not required that the initial conditions ϑ_0 and Θ_0 have any relation. It would be natural to assume $\vartheta_0|_{\Gamma_1} = \Theta_0$, but this is not necessary. Of course, in this case we cannot expect to pass to the limit in ε the property $u'_\varepsilon(0) = \tilde{\vartheta}_0$ from (2.38). □

Remark 2.12. The free boundary representing the ice-water interface can be recovered *a posteriori* as the boundary of the open set (see [F2] or [R4], for a more detailed discussion)

$$(2.46) \qquad \Lambda(t) = \left\{x \in \Omega:\ u(x,t) > 0\right\} = \left\{x \in \Omega:\ u_t(x,t) > 0\right\},$$

which complement $I(t) = \Omega\backslash\Lambda(t)$ represents the ice region. It can be shown (see [R4]) that u solves the nonlinear equation

$$(2.47) \qquad \gamma\, u_t - \Delta u = \eta_{0\gamma}\, \chi_{\{u>0\}} = \eta_{0\gamma} - \lambda\, \chi_{\{u=0\}} \qquad \text{a.e. in } Q_T = \Omega\times]0,T[,$$

and $\vartheta = u_t$, $\eta = \gamma\vartheta + \lambda\chi_{\{\vartheta>0\}}$ also solve (1.24) with $\mathbf{v} \equiv 0$ and $r = 0$. □

2.2 – Stability with respect to α, γ, $\alpha \to 0$ and $\gamma \to 0$

In this section we discuss the dependence of the variational solution $u = u_{\alpha\gamma}$ with respect to the parameters α and γ, which is denoted, respectively, by $u_{\alpha o}$ and $u_{o\gamma}$ or by

u_{oo}, if both variations are considered. The subscripts are also used with the same purpose for L, defined for $\alpha, \gamma \geq 0$, by

$$\langle L_{\alpha\gamma}(t), v\rangle = \int_\Omega (\gamma\vartheta_0 + \lambda\chi_{\vartheta_0>0})\, v + (g(t) + \alpha\,\Theta_0)\, V\,, \quad \forall v \in \mathbf{V} = H^1_\# \,.$$

A simple stability result can be obtained with the standard energy estimates. From the proof of the Proposition 2.3 we easily obtain the following auxiliary result.

Lemma 2.13. *Under the assumptions of Theorem 2.1 if $u = u_{\alpha\gamma}$ denotes the solution to (1.70)–(1.71) we have the following estimates for all $t \in [0, T]$*

$$\gamma|u(t)|^2 + \alpha U^2(t) + \int_0^t\int_\Omega |\nabla u|^2 \leq C(T)\,(1 + \alpha^2 + \gamma^2)\,, \tag{2.48}$$

$$\gamma|u_t(t)|^2 + \alpha \dot{U}^2(t) + \int_0^t\int_\Omega |\nabla u_t|^2 \leq C'(T)\,(1 + \alpha + \gamma)\,, \tag{2.49}$$

where the constants $C(T), C'(T) > 0$ are independent of α, γ.

Proof: The estimate (2.27) as $m \to \infty$ yields

$$\gamma|u_\varepsilon(t)|^2 + \alpha U_\varepsilon^2(t) + \int_0^t\int_\Omega |\nabla u_\varepsilon|^2 \leq (\gamma + \alpha)\,\varepsilon + \int_0^t C_L^2\,, \quad \forall \varepsilon > 0\,,$$

where $|C_L| \leq C(t)\,(1 + \alpha + \gamma)$, and (2.48) follows by letting $\varepsilon \to 0$. In a similar way (2.49) follows immediately from (2.35). ∎

The following continuous dependence estimate holds.

Theorem 2.14. *Denoting $u = u_{\alpha\gamma}$ and $\widehat{u} = u_{\widehat{\alpha}\widehat{\gamma}}$ for $0 < \alpha, \widehat{\alpha}, \gamma, \widehat{\gamma} \leq \mu$, we have for some constant $C = C(\mu, T) > 0$:*

$$\sqrt{\alpha}\,\|U - \widehat{U}\|_{L^\infty(0,T)} \leq C\Big\{|\alpha - \widehat{\alpha}| + |\gamma - \widehat{\gamma}|\Big\}\,, \tag{2.50}$$

$$\sqrt{\gamma}\,\|u - \widehat{u}\|_{L^\infty(0,T;L^2(\Omega))} \leq C\Big\{|\alpha - \widehat{\alpha}| + |\gamma - \widehat{\gamma}|\Big\}\,, \tag{2.51}$$

$$\|u - \widehat{u}\|_{L^2(0,T;H^1_\#)} \leq C\Big\{|\alpha - \widehat{\alpha}| + |\gamma - \widehat{\gamma}|\Big\}\,. \tag{2.52}$$

Proof: Recalling (1.71) for u with $v = \widehat{u}$,

$$\begin{aligned}\gamma\int_\Omega u_t(\widehat{u} - u) + \alpha\dot{U}(\widehat{U} - U) + \int_\Omega \nabla u\cdot\nabla(\widehat{u} - u) + J(\widehat{u}) - J(u) \geq \\ \geq \int_\Omega (\gamma\vartheta_0 + \lambda\chi_{\{\vartheta_0>0\}})\,(\widehat{u} - u) + (\alpha\Theta_0 + g(t))\,(\widehat{U} - U)\,,\end{aligned}$$

and, changing the role of u and $\widehat{u}$, we find for $w = u - \widehat{u}$:

$$\begin{aligned}\frac{\gamma}{2}\frac{d}{dt}|w|^2 + \frac{\dot{\alpha}}{2}\frac{d}{dt}W^2 + \int_\Omega |\nabla w|^2 &\leq (\gamma - \widehat{\gamma})\,(\vartheta_0 - \widehat{u}_t, w) + (\alpha - \widehat{\alpha})\,(\Theta_0 - \dot{\widehat{U}}, W) \\ &\leq \frac{1}{2}\int_\Omega |\nabla w|^2 + (\alpha - \widehat{\alpha})^2\, C_\#^2\, |\Theta_0 - \dot{\widehat{U}}|^2 + (\gamma - \widehat{\gamma})^2\, C_\#^2\, |\vartheta_0 - \widehat{u}_t|^2 \\ &\leq \frac{1}{2}\int_\Omega |\nabla w|^2 + 2C_\#^2\Big(|\vartheta_0|^2 + |\Theta_0|^2 + 2C_\#^2\int_\Omega |\widehat{u}_t|^2\Big)\,\Big\{(\alpha - \widehat{\alpha})^2 + (\gamma - \widehat{\gamma})^2\Big\}\,,\end{aligned} \tag{2.53}$$

where $C_\#$ is a constant satisfying (2.4) and $|v| \leq C_\# \|v\|$, $\forall v \in H^1_\#$.
Hence, integrating (2.53) in time and using (2.49) for $\widehat{u}_t$, we find

$$\gamma|w(t)|^2 + \alpha|W(t)|^2 + \int_0^t \int_\Omega |\nabla w(t)|^2 \leq C^2(\mu, T)\left\{(\alpha-\widehat{\alpha})^2 + (\gamma-\widehat{\gamma})^2\right\}$$

and the estimates (2.50),(2.51) and (2.52) follow. ∎

We observe that formally the estimates (2.50)–(2.52) still hold for the limit cases $\widehat{\alpha}=0$ or $\widehat{\gamma}=0$, provided the corresponding limit problems have solutions sufficiently regular, which, in fact, is the case. Rewriting (1.71) for $\widehat{\alpha} \geq 0$ and $\widehat{\gamma} \geq 0$ and a.e. $t \in]0,T[$

$$\text{(2.54)} \quad \widehat{\gamma}\int_\Omega \widehat{u}_t(v-\widehat{u}) + \widehat{\alpha}\overset{\bullet}{\widehat{U}}(v-\widehat{U}) + \int_\Omega \nabla\widehat{u}\cdot\nabla(v-\widehat{u}) + J(v) - J(\widehat{u}) \geq$$
$$\geq \int_\Omega (\widehat{\gamma}\vartheta_0 + \lambda\chi_{\{\vartheta_0>0\}})(v-\widehat{u}) + (\widehat{\alpha}\Theta_0 + g(t))(V-\widehat{U}), \quad \forall v \in H^1_\#\ ,$$

we see that, this variational inequality still makes sense for $\widehat{\gamma}=0$ or $\widehat{\alpha}=0$ and, in each case does not require any conditions on $\widehat{u}_t$ and ϑ_0 or on $\overset{\bullet}{\widehat{U}}$ and Θ_0, respectively.

Theorem 2.15. *In each one of the three cases $\widehat{\alpha}>0$ and $\widehat{\gamma}=0$, $\widehat{\alpha}=0$ and $\widehat{\gamma}>0$ or $\widehat{\alpha}=\widehat{\gamma}=0$, there exists a unique*

$$\text{(2.55)} \qquad \widehat{u} = u_{\widehat{\alpha}\widehat{\gamma}} \in H^1(0,T;\mathbf{H}^1_\#)\,, \qquad \widehat{u}(0)=0 \text{ solution of (2.54).}$$

Moreover, if $u = u_{\alpha\gamma}$ solves (1.70)–(1.71) for $\alpha>0$, $\gamma>0$, then $u_{\alpha\gamma} \rightharpoonup u_{\widehat{\alpha}\widehat{\gamma}}$ weakly in $H^1(0,T;\mathbf{H}^1_\#)$ and strongly in $L^2(0,T;H^1_\#) \cap C^0[0,T;L^2(\Omega)\times\mathbb{R}]$ with the order of convergence given by the estimates (2.50)–(2.52).

Proof: It is sufficient to recall from (2.48) and (2.49) that

$$\int_0^T\int_\Omega \left(|\nabla u_{\alpha\gamma}|^2 + |\nabla(u_{\alpha\gamma})_t|^2\right) \leq C(T)\ ,$$

where $C(T)>0$ does not depend on α and γ, for $\alpha,\gamma \in]0,\mu]$. Therefore, for some subsequence we have

$$u_{\alpha\gamma} \rightharpoonup u_{\widehat{\alpha}\widehat{\gamma}} \quad \text{in } H^1(0,T;\mathbf{V})\text{-weak and } C^0[0,T;\mathbf{H}]\text{-strong}$$

and by integrating (1.71) between s and t, for arbitrary $0<s<t<T$, we easily obtain (2.54), first also integrated in time, and by the arbitrariness of s and t, also a.e. $t \in]0,T[$.
Since the uniqueness of the solution of (2.54) with the initial condition (2.55) is immediate, we can repeat the proof of Theorem 2.14 in each one of the three cases, to conclude the respective order of convergences. ∎

Remark 2.16. When $(\widehat{\alpha},\widehat{\gamma}) = (0,\gamma)$, $\gamma>0$, we have the one-phase Stefan problem with total heat flux prescribed and spacely homogeneous unknown temperature on the boundary Γ_1. Also in this case, the estimates (2.50)–(2.52) yield for $\alpha \to 0$

$$\|U_{\alpha\gamma} - U_{o\gamma}\|^2_{L^\infty(0,T)} + \|u_{\alpha\gamma} - u_{o\gamma}\|_{L^\infty(0,T;L^2(\Omega))} + \|u_{\alpha\gamma} - u_{o\gamma}\|_{L^2(0,T;H^1_\#)} = O(\alpha)\ . \ \square$$

Remark 2.17. When $(\widehat{\alpha},\widehat{\gamma}) = (\alpha, 0)$, $\alpha > 0$, this case corresponds to the Stefan problem with zero specific heat or to the model of the Hele–Shaw Problem with nonlocal injection condition in [PR]. In particular for $\gamma \to 0$, we have

$$\|U_{\alpha\gamma} - U_{\alpha o}\|_{L^\infty(0,T)} + \|u_{\alpha\gamma} - u_{\alpha o}\|^2_{L^\infty(0,T;L^2(\Omega))} + \|u_{\alpha\gamma} - u_{\alpha o}\|_{L^2(0,T;H^1_\#)} = O(\gamma)\ . \square$$

Remark 2.18. The limit case $(\widehat{\alpha},\widehat{\gamma}) = (0,0)$, corresponds to an elliptic problem where the time has a role of a parameter through the nonhomogeneous term g in (2.54) and where $\chi_{\{\vartheta_0>0\}}$ may be replaced by the characteristic function of any measurable subset of Ω, since no information on ϑ_0, is needed. Of course, this observation is also valid for the case of Remark 2.17.

In this case, only (2.52) is directly applicable, yielding

$$\|u_{\alpha\gamma} - u_{oo}\|_{L^2(0,T;H^1_\#)} = O(\alpha+\gamma) \quad \text{as } \alpha+\gamma \to 0\ . \square \tag{2.58}$$

Remark 2.19. In addition to (2.58), we can also compare the rates of convergences of $u_{\alpha o} \to u_{oo}$ as $\alpha \to 0$ or $u_{o\gamma} \to u_{oo}$ as $\gamma \to 0$ respectively from (2.50) and (2.51)

$$\|U_{\alpha o} - U_{oo}\|_{L^\infty(0,T)} = O(\sqrt{\alpha})\ , \tag{2.59}$$

$$\|u_{o\gamma} - u_{oo}\|_{L^\infty(0,T;L^2(\Omega))} = O(\sqrt{\gamma})\ . \tag{2.60}$$

Consequently, from the Theorem 2.15 we can draw the following diagram with the respective rates of convergences

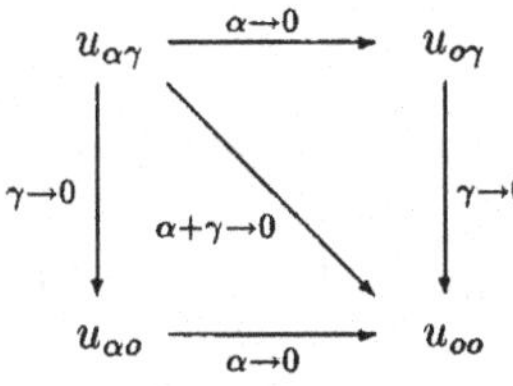

For the case $\alpha > 0$ when $\gamma \to 0$, we may take advantage of the estimate (2.7) and regard, for each fixed $t > 0$, the solution $u_\gamma \equiv u_{\alpha\gamma}$ as an elliptic perturbation of the solution $u_o \equiv u_{\alpha o}$. In fact, recalling (2.47) and the definition (2.2), $u_\gamma = u_\gamma(t)$ solves, for fixed $t > 0$, the nonlinear elliptic Dirichlet problem

$$-\Delta u_\gamma = f_\gamma\, \chi_{\{u_\gamma>0\}} \quad \text{in } \Omega\ , \tag{2.60}$$

$$u_\gamma = U_\gamma \ \text{ on } \Gamma_1 \quad \text{and} \quad u_\gamma = 0 \ \text{ on } \Gamma_0\ , \tag{2.61}$$

where

$$f_\gamma = \gamma(\vartheta_0 - u_{\gamma t}) - \lambda\, \chi_{\{\vartheta_0=0\}}, \quad \text{for } \gamma \geq 0\ . \tag{2.62}$$

Remark 2.20. We note that, in particular, Theorem 2.5 gives a different proof of the existence result of [PR] for the case $\alpha > 0$ and $\gamma = 0$. Since $u'_\gamma \rightharpoonup u'_o$ in $L^2(0,T;H^1_\#)$-weak, from (2.7), where we may take without loss of generality $M = \Theta_0 \geq \vartheta_0$, we also obtain the Proposition 2 of [PR]:

$$0 \leq u_o' \leq \Theta_0 + \frac{1}{\alpha}\dot{g} \quad \text{on } \Omega \cup \Gamma_1, \ \text{for } t > 0 ,$$

which was proved directly using a different method. □

Theorem 2.21. *Assume that $d(\Gamma_1, \{\vartheta_0 = 0\}) > 0$. Then for each $t > 0$, if $u_\gamma = u_{\alpha\gamma}$ and $u_0 = u_{\alpha 0}$ denote the variational solutions for fixed $\alpha > 0$, then as $\gamma \to 0$ we have*

$$\|u_\gamma(t) - u_o(t)\|_{H^1(\Omega)} = O(\gamma) , \tag{2.63}$$

$$|I_\gamma(t) \div I_o(t)| = O(\gamma) , \tag{2.64}$$

$$\|u_\gamma(t) - u_o(t)\|_{W^{2,p}(\Omega)} = O(\gamma^{1/p}) , \quad 1 < p < \infty , \tag{2.65}$$

where $\div$ denotes the symmetric difference between the coincidence sets $I_\gamma(t) = \{x \in \Omega : u_\gamma(x,t) = 0\}$, $\gamma \geq 0$, and $|A| = \text{meas}(A)$ denotes the Lebesgue measure of the set A.

Proof: By (2.60)–(2.61) we note that $u_\gamma = u_\gamma(t)$ $(\gamma \geq 0)$ solves the elliptic variational inequality of obstacle type (see [R3], for instance)

$$u_\gamma \in \mathbb{K}_\gamma : \int_\Omega \nabla u_\gamma \cdot \nabla(v - u_\gamma) \geq \int_\Omega f_\gamma (v - u_\gamma) , \quad \forall v \in \mathbb{K}_\gamma ,$$

where f_γ is given by (2.62) and $\mathbb{K}_\gamma = \{v \in H^1(\Omega) : v \geq 0 \text{ in } \Omega, \ v|_{\Gamma_0} = 0 \text{ and } v|_{\Gamma_1} = U_\gamma\}$, being U_γ the constant value of $u_\gamma(t)$ on Γ_1 at fixed time $t > 0$.

From Proposition 3 of [PR] we know that $U_o > 0$ for every $t > 0$. Hence, using (2.57) we have $|U_\gamma - U_o| = O(\gamma)$ and we can apply the argument of Theorem 1 of [LR] to obtain first (2.63). Then taking a function $\varphi \in C^\infty(\Omega)$ such that $\varphi \equiv 1$ on $\{\vartheta_0 = 0\}$ and $\varphi = 0$ on a neighbourhood of Γ_1, we localize the solutions u_γ by taking $w_\gamma = \varphi u_\gamma \in H_0^1(\Omega)$ and we apply Theorems 4.7 and 4.8 of [R3], Chap. 5, to conclude, respectively, (2.64) and (2.65). ∎

Remark 2.22. Note that (2.64) gives a stability result, in the sense of the Lebesgue measure, of the coincidence sets, which represent the zone occupied by the "ice". This results may be found in a similar (but simpler) case in [LR]. □

2.3 – Asymptotic convergence as time $t \to \infty$

In this section we shall assume $\alpha, \gamma > 0$ and

$$g(t) \to g_\infty \quad \text{as } t \to \infty , \tag{2.66}$$

where g_∞ is a finite number. Let us define $L_\infty \in \mathbf{V}'$ by

$$\langle L_\infty, v \rangle = \int_\Omega (\gamma \vartheta_0 + \lambda \chi_{\{\vartheta_0 > 0\}}) v + (\alpha \Theta_0 + g_\infty) V , \quad \forall v \in \mathbf{V} , \tag{2.67}$$

which is the strong limit in $\mathbf{V}'$ of $L(t)$, defined by (1.69), as $t \to \infty$. The steady-state variational inequality corresponding to (1.71) is now

$$u_\infty \in \mathbf{V} : \ ((u_\infty, v - u_\infty)) + J(v) - J(u_\infty) \geq \langle L_\infty, v - u_\infty \rangle , \quad \forall v \in \mathbf{V} . \tag{2.68}$$

Proposition 2.13. *The unique solution of (2.68) satisfies*

(2.69) $$u_\infty \in W^{2,p}(\Omega) \cap C^{1,\alpha}(\overline{\Omega}), \quad \forall 1 \le p < \infty, \ 0 \le \alpha < 1 ,$$

(2.70) $$0 \le u(t) \le u_\infty , \quad \forall 0 < t < \infty ,$$

where $u(t)$ is the solution of (1.70)–(1.71).

Proof: The existence and uniqueness of u_∞ is a direct consequence of well-known results on the minimization of the strictly convex and coercive functional on the Hilbert space $\mathbf{V} = H^1_\#$:

$$F(v) = \frac{1}{2}((v,v)) + J(v) - \langle L_\infty, v\rangle .$$

Since (2.3) implies $L_\infty \ge L(t)$, $\forall t > 0$, we apply the same argument of the Proposition 2.10 to prove (2.70) by showing that $w(t) = (u(t) - u_\infty)^+$ is zero: take $v = u_\infty + w(t)$ in (2.68) and $v = u(t) - w(t)$ in (1.71); integrating in time their difference we obtain

$$|w(t)|^2_{\alpha\gamma} + 2\int_0^t \|w(t)\|^2 \le |(-u_\infty)^+|^2_{\alpha\gamma} = 0 , \quad \forall t > 0 ,$$

since $u_\infty \ge 0$, as it can be seen by taking $v = u^+_\infty$ in (2.68).

But u_∞ is also the solution of the obstacle problem (recall (1.72) for the definition of $\mathbb{K}$)

(2.71) $$u_\infty \in \mathbb{K}: \ ((u_\infty, v - u_\infty)) \ge \langle L_\infty - \lambda, v - u_\infty\rangle , \quad \forall v \in \mathbb{K} ,$$

and (2.69) follows by standard regularity results (see [R3], for instance). ∎

Let us defined the "liquid" zones by

(2.72)
$$\Lambda_0 = \{x \in \Omega: \ \vartheta_0(x) > 0\}, \ \Lambda(t) = \{x \in \Omega: \ u(x,t) > 0\} \text{ and } \Lambda_\infty = \{x \in \Omega: \ u_\infty(x) > 0\}$$

and assume that Λ_0 is also an open subset. We denote the coincidence sets by $I(t) = \Omega\backslash\Lambda(t)$ and $I_\infty = \Omega\backslash\Lambda_\infty$ and the free boundaries by

(2.73) $$\Phi(t) = \partial\Lambda(t) \cap \Omega = \partial I(t) \cap \Omega \quad \text{and} \quad \Phi_\infty = \partial\Lambda_\infty \cap \Omega = \partial I_\infty \cap \Omega .$$

Proposition 2.24. *We have, for every $t > 0$*

(2.74) $$\Lambda_0 \subset \Lambda(t) \subset \Lambda(t+\tau) \subset \Lambda_\infty , \quad \forall \tau > 0 ,$$

and denoting $\omega_\gamma \equiv |\Lambda_0| + (\alpha\Theta_0 + g_\infty + \gamma \int_\Omega \vartheta_0)/\lambda > 0$, we have

(2.75) $$|\Omega\backslash\Lambda_\infty| = |I_\infty| \ge |\Omega| - \omega_\gamma \quad \textit{if } |\Omega| > \omega_\gamma ,$$

(2.76) $$|\Lambda_\infty| = |\Omega| \ \textit{(no free boundary)} \quad \textit{if } |\Omega| \le \omega_\gamma .$$

Proof: The first inclusion in (2.74) is an easy consequence of (2.60) and the strong maximum principle, while the others two are immediate consequences of (2.70) and (2.7).

Next, it is easy to conclude that $u_\infty|_{\Gamma_1} = U_\infty > 0$ and $\int_{\Gamma_1} \partial_n u_\infty = \alpha\Theta_0 + g_\infty > 0$, arguing as in [PR]. Since we have (compare with (2.60))

$$\Delta u_\infty = (-\gamma\vartheta_0 + \lambda\chi_{\Omega\setminus\Lambda_0})\chi_{\Lambda_\infty} = -\gamma\vartheta_0 + \lambda\chi_{\Lambda_\infty\setminus\Lambda_0} \quad \text{in } \Omega$$

by integrating in Ω, since $\partial_n u_\infty|_{\Gamma_0} \le 0$, we find

$$|\Lambda_\infty\setminus\Lambda_0| = \int_\Omega \chi_{\Lambda_\infty\setminus\Lambda_0} = \frac{1}{\lambda}\Big(\int_\Omega \Delta u_\infty + \gamma\int_\Omega \vartheta_0\Big) \le \frac{1}{\lambda}\Big(\alpha\Theta_0 + g_\infty + \gamma\int_\Omega \vartheta_0\Big)$$

and since $|\Omega\setminus\Lambda_\infty| = |\Omega| - |\Lambda_0| - |\Lambda_\infty\setminus\Lambda_0|$ the conclusions (2.75) and (2.76) follow. ∎

Remark 2.25. This criteria for the existence of the free boundary was given in [PR] for the case $\gamma = 0$. It is clear that if $\partial\Lambda_\infty \cap \Gamma_0 = \emptyset$ then $\partial_n u_\infty|_{\Gamma_0} = 0$ and the number ω_γ, $\gamma \ge 0$, is exactly equal to $|\Lambda_\infty|$, i.e. the measure (area or volume) of the limit liquid zone. For sufficiently large Ω the existence of the free boundary can be guaranteed since we may estimate the support of the solution of (2.71) by comparison with supersolutions that vanish in a neighbourhood of Γ_0, as in [LR], [PR] or [R3] section 6:7, for instance. □

We shall apply an useful lemma on differential inequalities, which proof can be found, for instance, in [Na2] and was used in [R1].

Lemma 2.26. *Let $\varphi(t)$ be a non-negative absolutely continuous function on every compact interval of $[0,\infty[$, $l(t) \in L^1_{\text{loc}}(0,\infty)$ and $\omega > 0$ a constant such that*

(2.77) $$\dot\varphi(t) + \omega\,\varphi(t) \le l(t) \quad \text{a.e. } t > 0\,.$$

a) *If $\sup_{t>\sigma}\int_t^{t+1} l(s)\,ds = C(\sigma)$, then $\varphi(\sigma+t) \le e^{-\omega t}\varphi(\sigma) + C(\sigma)\,(1-e^{-\omega})^{-1}$, $\forall t,\sigma \ge 0$;*

b) *If $\int_t^{t+1} l(s)\,ds = O(t^{-\beta})$, $\beta > 0$, then $\varphi(t) = O(t^{-\beta})$ as $t \to \infty$;*

c) *If $\int_t^{t+1} l(s)\,ds = O(e^{-\nu t})$, $\nu > 0$, then $\varphi(t) = O(e^{-\mu t})$ as $t \to \infty$, where $\mu = \min(\nu,\omega)$.* ∎

Theorem 2.27. *Let (2.66) hold, $u(t) = u_{\alpha\gamma}(t)$ denote the solution to (1.70)–(1.71) with $\alpha,\gamma > 0$ and u_∞ the solution to (2.68). Then the following convergences hold*

(2.78) $$u(t) \to u_\infty \quad \text{in } W^{2,p}(\Omega)\cap C^{1,\alpha}(\overline{\Omega}),\ \ \forall p<\infty,\ \ \alpha<1\,,$$

(2.79) $$u_t(t) \to 0 \quad \text{in } L^p(\Omega),\ \ \forall p<\infty,\ \ \dot U(t) \to 0\,,$$

(2.80) $$|\Lambda_\infty\setminus\Lambda(t)| \to 0, \quad \text{as } t\to\infty\,.$$

Proof: We note that by our assumptions

$$\eta(t) \equiv L_\infty - L(t) = g_\infty - g(t) \to 0 \quad \text{as } t\to\infty$$

and $\dot\eta = -\dot g \le 0$ is bounded in $[0,\infty)$. Hence letting $v = u(t)$ in (2.68) and $v = u_\infty$ in (1.71), we see that $w(t) = u(t) - u_\infty$ satisfies (2.77) for $\varphi(t) = |w(t)|^2_{\alpha\gamma}$, with $l(t) = C\eta^2(t)$. Therefore (2.78) holds in $\mathbf{H} \simeq L^2(\Omega)\times\mathbb{R}$, by Lemma 2.26 a).

We also note that (2.66) implies that

$$\zeta(t) = \int_t^{t+1} \|L'(s)\|_*^2 \, ds \le C\big(g(t+1) - g(t)\big) \le C\,\eta(t) \to 0 \quad \text{as } t \to \infty \tag{2.81}$$

and we can apply Theorems 2 and 3 of [R1] to conclude, respectively (2.78) in $\mathbf{V} = H^1_\#$ strong and (2.79), first in $\mathbf{H} \simeq \mathbf{L}^2(\Omega) \times \mathbf{R}$ strong, and by (2.7) also in every $L^p(\Omega)$, $\forall p < \infty$.

Now we can repeat the argument of Theorem 2.21 and consider the evolution problem as an elliptic perturbation of the steady-state one and we can conclude analogously the convergences (2.78) and (2.80) (note that $\Lambda_\infty \backslash \Lambda(t) = I(t) \div I_\infty$) with a straightforward application of the strong continuous dependence results for obstacle type problems (see [R3] section 5:4). ∎

Remark 2.28. In the case $\gamma = 0$, $\alpha > 0$, we have similar results to the convergences of Theorem 2.27, of course, with the exception of the convergence $u_t \to 0$ as $t \to \infty$, which still holds in $L^p(\Omega)$-weak, $\forall p < \infty$. See Theorem 5 of [PR] for a direct proof of this result. □

Remark 2.29. In the case $\alpha = 0$ and $\gamma > 0$, we cannot apply the estimate (2.7) to obtain the *a priori* L^∞-bound of the time derivative u'. Nevertheless all the results of Theorem 2.27 still hold, by restricting ourselves to the case $p = 2$ and replacing $C^{1,\alpha}(\overline{\Omega})$ in (2.78) by $C^{0,\beta}(\overline{\Omega})$ with $0 \le \beta < 2 - N/2$ (for $N = 2, 3$, by the Rellich–Kondrachov theorem). □

Remark 2.30. In the doubly degenerated case $\alpha = \gamma = 0$ the time is just a parameter and the problem is reduced to a simple continuous dependence result with respect to t, with the lost of information on the time derivatives. In particular, if $g(t) = g_\infty$ for all $t \ge T_*$ for a finite time T_* then also $u_{oo}(t) = u_\infty$ for all $t \ge T_*$. □

Exploiting parts b) and c) of Lemma 2.26, we can improve the asymptotic convergences (2.78), (2.79) and then also (2.80). In fact, if we assume, for some $\beta > 0$ and $\nu > 0$

$$\eta(t) \equiv g_\infty - g(t) = O(t^{-\beta}) \quad \text{as } t \to \infty\,, \tag{2.82}$$

or

$$\eta(t) = O(e^{-\nu t}) \quad \text{as } t \to \infty\,, \tag{2.83}$$

then we can apply the second parts of Theorems 2 and 3 of [R1], to conclude the following results (see also [R2,3]).

Corollary 2.31. *Under the assumptions of Theorem 2.27, if (2.82) or (2.83) hold, we have*

$$\begin{cases} \|u(t) - u_\infty\|_{H^1_\#} + \|u_t(t)\|_{L^p(\Omega)} + |\dot{U}(t)| + |\Lambda_\infty \backslash \Lambda(t)| = O(t^{-\beta/2})\,, \\ \|u(t) - u_\infty\|_{W^{2,p}(\Omega)} = O(t^{-\beta/2p})\,, \quad \forall p < \infty\,, \end{cases} \tag{2.84}$$

or

$$\begin{cases} \|u(t) - u_\infty\|_{H^1_\#} + \|u_t(t)\|_{L^p(\Omega)} + |\dot{U}(t)| + |\Lambda_\infty \backslash \Lambda(t)| = O(e^{-\mu t})\,, \\ \|u(t) - u_\infty\|_{W^{2,p}(\Omega)} = O(e^{-\mu t/p})\,, \quad \forall p < \infty\,, \end{cases} \tag{2.85}$$

respectively, where $\mu > 0$ is some number, $\mu \le \nu$. ∎

In some cases with special geometries its is possible to explicite the asymptotic stabilization in terms of the convergence of the free boundaries defined by (2.73).

Let us consider the case of the starshaped configuration suggested by the picture of section 1.4 and defined in the following way: we suppose the sets ω and $\Omega \cup \overline{\omega}$ are starshaped with respect to a ball B_δ of radius $\delta > 0$ and center at the origin; we assume that, in polar coordinates $x = (\rho, \sigma)$, $\rho = |x|$ and

$$\sigma = \arg x \in \Sigma \equiv \left\{\sigma \colon -\pi \leq \sigma_i \leq \pi,\ 1 \leq i < N-1,\ 0 \leq \sigma_{N-1} \leq \pi\right\}$$

the initial temperature ϑ_0 is such that

$$\frac{\partial}{\partial \rho}(\rho^2 \vartheta_0(x)) < 0 \quad \text{in } \{\vartheta_0 > 0\} \neq \emptyset, \quad \vartheta_0|_{\Gamma_1} = \Theta_0 > 0$$

and $\Phi_\infty \cap \Gamma_0 = \emptyset$ (if this holds than also $\Phi(t) \cap \Gamma_0 = \emptyset$, $\forall t > 0$). We shall call these assumptions the starshaped geometry since it is well known (see [FK] and [R3] section 6:4) that, in this case the free boundaries $\Phi(t)$ and Φ_∞ admit a graphic representation in the form

$$\Phi(t) \colon \rho = \rho^*(\sigma, t), \quad \sigma \in \Sigma, \quad t > 0 \quad \text{and} \quad \Phi_\infty \colon \rho = \rho^*_\infty(\sigma), \quad \sigma \in \Sigma\,,$$

where $\rho^*(t)$ and ρ^*_∞ are Lipschitz functions in σ, being ρ^* also Hölder continuous in t. As in [PR] (see also [R2] and [R3] Chap. 6), for each $t > 0$ and $\Phi(t)$ and Φ_∞ are analytic surfaces and, as a consequence, of the general stability of smooth free boundaries and Theorem 2.27 and Corollary 2.31 (see also [R4]) we may state the following result, which holds for $\alpha > 0$ and $\gamma \geq 0$.

Theorem 2.32. *Under the starshaped geometry assumptions if (2.66) holds, the graphs of the respective free boundaries satisfy $\rho^*(t) \to \rho^*_\infty$ as $t \to \infty$, uniformly in $\sigma \in \Sigma$. Moreover, as $t \to \infty$, they also satisfy*

$$\|\rho^*(t) - \rho^*_\infty\|_{C^{0,\lambda}(\Sigma)} = O(t^{-(1-\lambda)\beta t/2N}) \quad \textit{if (2.82) holds}$$

and

$$\|\rho^*(t) - \rho^*_\infty\|_{C^{0,\lambda}(\Sigma)} = O(e^{-(1-\lambda)\mu t/N}) \quad \textit{if (2.83) holds,}$$

were for any λ, $0 \leq \lambda < 1$, $\beta > 0$ and $\mu > 0$ are given, respectively, by (2.84) and (2.85) and N is the space dimension. ∎

3 – THE TWO-PHASE PROBLEM VIA VARIATIONAL INEQUALITIES

3.1 – Study of the variational inequality

In order to solve the two phase Stefan problem in the variational form (1.46)–(1.47) with the notations introduced in (1.35) and (1.43-45), we shall use the following assumptions:

(3.1) $$b \in C^{0,1}(\mathbb{R}), \quad b(0) = 0 \quad \text{and} \quad 0 < b_* \leq b'(s) \leq b^* \ \text{ a.e. } s \in \mathbb{R}\,;$$

(3.2) $$\sigma = \sigma(x) \in L^\infty(\Gamma) \quad \text{and} \quad 0 < \nu \leq \sigma(x) \ \text{ a.e. } x \in \Gamma = \partial\Omega\,;$$

(3.3) $$\eta_0 = b(\vartheta_0) + \lambda\chi_{\{\vartheta_0>0\}} \quad \text{for some } \vartheta_0 \in H^1(\Omega) \text{ and } \lambda > 0\,;$$

(3.4) $$h = h(x,t) \in H^2(0,T;L^2(\Gamma))\,, \qquad h(0) = 0\,.$$

We note that (3.1) implies that b is bilipschitz continuous and strictly increasing, in particular, we have

(3.5) $$b_*|u-v|^2 \leq \big(b(u)-b(v), u-v\big) \leq b^*|u-v|^2\,, \qquad \forall\, u,v \in L^2(\Omega)\,.$$

The assumption (3.2) implies that the bilinear form $a(\cdot,\cdot)$ introduced in (1.43) defines a inner product in $H^1(\Omega)$, whose topology is equivalent to the usual one given by the $H^1(\Omega)$-norm $\|v\|_1^2 = |v|^2 + |\nabla v|^2$. In fact, when $\Gamma = \partial\Omega$ is of class $C^{0,1}$, which has been assumed here, we have

$$c_\sigma\|v\|_1^2 \leq a(v,v) = \int_\Omega |\nabla v|^2 + \int_\Gamma \sigma\, v^2 \leq C_\sigma\|v\|_1^2\,, \qquad \forall\, v \in H^1(\Omega)\,,$$

for positive constants $C_\sigma = C_\sigma(\Gamma, \|\sigma\|_{L^\infty(\Gamma)})$ and $c_\sigma(\Gamma,\nu)$, by well-known results in the trace theorem and on the extended Poincaré inequality (see, for instance, [R3] pg.75). In fact, we only need to assume $\sigma(x) \geq \nu > 0$ on an open subset $\Gamma \subset \partial\Omega$ such that the $(n-1)$ dimensional Lebesgue measure of Γ is positive.

The assumption (3.3), in particular, implies that $\eta_0 \in \gamma(\vartheta_0)$ where γ is defined by (1.21), and, as it will be seen in the proof below, is used to obtain enough regularity on the time derivative $u_t = \vartheta$ of the variational solution to (1.47), which can be rewritten in the form, for a.e. $t > 0$

(3.6) $$\int_\Omega b(u_t)\,(v-u_t) + \int_\Omega \nabla u \cdot \nabla(v-u_t) + \int_\Gamma \sigma\, u(v-u_t) + \lambda\int_\Omega v^+ - \lambda\int_\Omega u_t^+ \geq$$
$$\geq \int_\Omega \eta_0(v-u_t) + \int_\Gamma \sigma\, h\,(v-u_t)\,, \qquad \forall\, v \in H^1(\Omega)\,.$$

In fact, we need $u_t = u_t(t) \in H^1(\Omega)$ for a.e. $t > 0$, and this is given in the following existence theorem.

Theorem 3.1. *Under the assumptions (3.1)–(3.4) there exists a unique solution to (1.46)–(1.47), i.e. a function*

(3.7) $$u \in W^{1,\infty}(0,T;H^1(\Omega)) \cap H^2(0,T;L^2(\Omega)) \cap L^\infty(0,T;H^2_{\text{loc}}(\Omega))$$

satisfying (3.6) for a.e. $t > 0$, and $u(0) = 0$. ∎

Remark 3.2. The uniqueness follows immediately from (3.6): if u and $\widehat{u}$ are two solutions, their difference $w = u - \widehat{u}$ satisfies

$$b_*|w_t|^2 + \frac{1}{2}\frac{d}{dt}a(w,w) \le \int_\Omega [b(u_t) - b(\widehat{u}_t)]\, w_t + a(w,w_t) \le 0 \quad \text{a.e. } t > 0\,;$$

integrating in time and using $w(0) = 0$, we find $w(t) = 0$ for $t > 0$. □

Remark 3.3. The regularity of u in $L^\infty(0,T;H^2_{\rm loc}(\Omega))$ can be improved up to the boundary if $\partial\Omega \in C^2$, $\sigma \in C^{0,1}(\partial\Omega)$ and $h \in W_2^{1/2,1/4}(\partial\Omega\times\,]0,T[)$ (see [LSU] for the definition of this space, that contains, in particular, $H^1(0,T;H^{1/2}(\partial\Omega))$). In this case we have

$$u \in L^\infty(0,T;H^2(\Omega)) \cap H^2(0,T;L^2(\Omega)) \subset H^2(Q)\,. \ \square$$

Remark 3.4. From the assumption (3.1), the function $\widetilde{b}(s) = \int_0^s b(\tau)\,d\tau$ is a $C^{1,1}$ strictly convex function. Introducing the convex functional

$$B(v) = \int_\Omega \widetilde{b}(v) + \lambda \int_\Omega v^+$$

the variational inequality (1.47) (or (3.6)) can be written, for a.e. $t > 0$, in the form

$$a(u(t), v - u_t(t)) + B(v) - B(u_t(t)) \ge \langle F(t), v - u_t(t)\rangle\,, \qquad \forall v \in H^1(\Omega)\,, \tag{3.8}$$

which belongs to a type considered by Brézis [B2] (see also [Bal], §3.3). Notice that, taking $t = 0$ in (3.8), we obtain that $w_0 = u_t(0) \in H^1(\Omega)$ is the unique solution of

$$\int_\Omega b(w_0)\,(v - w_0) + \lambda\int_\Omega v^+ - \lambda\int_\Omega w_0^+ \ge \int_\Omega \eta_0(v - w_0)\,, \qquad \forall v \in H^1(\Omega)\,.$$

But this means $\eta_0 \in \gamma(w_0) = b(w_0) + \lambda H(w_0)$ and, by the assumption (3.3), $w_0 = \vartheta_0$. □

In order to prove the existence and regularity in the Theorem 3.1 we shall consider, for each $\varepsilon > 0$, the following parabolic problem:

$$b(u_{\varepsilon t}) - \Delta u_\varepsilon + \lambda H_\varepsilon(u_{\varepsilon t}) = \eta_\varepsilon \quad \text{in } \Omega,\ t > 0\,, \tag{3.9}$$

$$\partial_n u_\varepsilon = \sigma(h - u_\varepsilon) \quad \text{on } \Gamma,\ t > 0\,, \tag{3.10}$$

$$u_\varepsilon(0) = 0 \quad \text{on } \Omega\,, \tag{3.11}$$

where H_ε is defined by (2.11) and $\eta_\varepsilon = \gamma_\varepsilon(\vartheta_0) \in H^1(\Omega)$, where, by (3.1),

$$\gamma_\varepsilon(s) = b(s) + \lambda H_\varepsilon(s) \quad \text{satisfies } 0 < b_* \le \gamma_\varepsilon'(s) \le b_\varepsilon^* \ \text{ a.e. } s \in \mathbf{R}\,, \tag{3.12}$$

being $b_\varepsilon^* = b^* + \lambda\, l_\varepsilon$ (l_ε is the Lipschitz constant of H_ε: $l_\varepsilon = O(\frac{1}{\varepsilon})$).

We note that, by the assumption (3.3) and (2.11) we have, using Lebesgue's theorem,

$$\eta_\varepsilon \to \eta_0 \quad \text{in } L^2(\Omega) \ \text{ as } \varepsilon \to 0\,. \tag{3.13}$$

Let us denote

$$\langle F_\varepsilon(t), v\rangle = \int_\Omega \eta_\varepsilon\, v + \int_\Gamma \sigma\, h(t)\, v\,, \qquad \forall\, v \in H^1(\Omega), \quad t > 0\,.$$

Proposition 3.5. *Under the preceding assumptions, there exists a unique solution to (3.9), (3.11), such that,* $u_\varepsilon(0) = 0$

$$u_\varepsilon \in W^{1,\infty}(0,T; H^1(\Omega)) \cap H^2(0,T; L^2(\Omega)) \cap L^\infty(0,T; H^2_{\mathrm{loc}}(\Omega)), \quad \text{uniformly in } 0 < \varepsilon \le 1\,, \tag{3.14}$$

$$(\gamma_\varepsilon(u_{\varepsilon t}(t)), v) + a(u_\varepsilon(t), v) = \langle F_\varepsilon(t), v\rangle\,, \qquad \forall\, v \in H^1(\Omega), \quad t > 0\,. \tag{3.15}$$ ∎

We shall use a fixed point argument in the Hilbert space $\mathbf{Z} = \{\zeta \in H^1(0,T; L^2(\Omega)) : \zeta(0) = 0\}$, with norm

$$|||\zeta||| = \Big(\int_0^T \int_\Omega \zeta_t^2\Big)^{1/2} = \|\zeta_t\|_{L^2(Q)}\,,$$

and we shall study first the following linear parabolic problem

$$\rho(z_t, v) + a(z, v) = \langle F_\varepsilon, v\rangle - (\rho\zeta_t + \gamma_\varepsilon(\zeta_t), v)\,, \qquad \forall\, v \in H^1(\Omega), \quad t > 0\,, \tag{3.16}$$

with $z(0) = 0$, for given $\zeta \in \mathbf{Z}$ and $\rho > 0$.

Lemma 3.6. *Let $\rho > 0$ and $\zeta \in \mathbf{Z}$. Then there exists a unique solution $z \in \mathbf{Z} \cap L^\infty(0,T; H^1(\Omega))$ to (3.16). Moreover, if z and $\hat{z}$ denote solutions corresponding to ζ and $\hat{\zeta}$, respectively, then, if $\rho > (b_\varepsilon^*)^2/2b_*$, there exists $\delta < 1$ such that*

$$|||z - \hat{z}||| \le \delta |||\zeta - \hat{\zeta}|||\,. \tag{3.17}$$

Proof: As in the proof of Proposition 2.7 we shall use the Faedo–Galerkin method: let $\{v_i\}$ be a basis of $H^1(\Omega)$ and let $V_m = \operatorname{span}\{v_1, \ldots, v_m\}$; the finite dimensional approximation of z will be of the form $z_m(x,t) = \sum_{i=1}^m \xi_{mi}(t)\, v_i(x)$, where the coefficients ξ_{mi} will be defined by the system

$$\rho(z_m'(t), v_j) + a(z_m(t), v_j) = \langle L_\zeta(t), v_j\rangle\,, \qquad j = 1, \ldots, m\,, \tag{3.18}$$

with the initial condition $z_m(0) = 0$, where we have set

$$\langle L_\zeta, v\rangle = \langle F_\varepsilon, v\rangle - (\rho\zeta_t + \gamma_\varepsilon(\zeta_t), v)\,, \qquad \forall\, v \in H^1(\Omega)\,.$$

Multiplying (3.18) by $\xi_{mj}'(t)$ and summing on j, we obtain

$$\rho|z_m'|^2 + \frac{1}{2}\frac{d}{dt} a(z_m, z_m) \le |\eta_\varepsilon - \rho\zeta_t - \gamma_\varepsilon(\zeta_t)|\, |z_m'| + \int_\Gamma \sigma\, h\, z_m'\,. \tag{3.19}$$

Remarking that

$$\begin{aligned} \int_0^t \int_\Gamma \sigma\, h\, z_m' &= \int_\Gamma \sigma\, h(t)\, z_m(t) - \int_0^t \int_\Gamma \sigma\, h'\, z_m \\ &\le \frac{1}{4} a(z_m(t), z_m(t)) + c\|h\|^2_{L^\infty(0,t;L^2(\Gamma))} + \int_0^t a(z_m, z_m) + c'\|h'\|^2_{L^2(0,t;L^2(\Gamma))} \end{aligned}$$

and, by integration in t, from (3.19), we obtain

$$\rho \int_0^t |z_m'|^2 + \frac{1}{2} a(z_m(t), z_m(t)) \leq C(t) + 2 \int_0^t a(z_m, z_m) ,$$

where $C(t) = \int_0^t |\eta_\varepsilon - \rho\zeta_t - \gamma_\varepsilon(\zeta_t)|^2 + 2(c+c') \|h\|^2_{H^1(0,t;L^2(\Gamma))}$. Hence, by Gronwall inequality, we conclude that

$$\rho |||z_m|||^2 + \sup_{0<t<T} \|z_m(t)\|^2_{H^1(\Omega)} \leq C \text{ (independent of } m), \tag{3.20}$$

and letting $m \to \infty$ we obtain $z \in \mathbf{Z} \cap L^\infty(0,T;H^1(\Omega))$ solution of (3.16).

Let $\hat{z}_m(t) = \sum_{i=1}^m \hat{\xi}_{mi}(t)\, v_i$ denotes the finite dimensional approximation corresponding to $\hat{L}_{\hat{\zeta}}(t)$ associated with $\hat{\zeta}$. Multiplying (3.18) by $\xi'_{mj}(t) - \hat{\xi}'_{mj}(t)$, as well as the system corresponding to $\hat{z}_m$, summing on j and taking their respective difference, we obtain for $\overline{z}_m(t) = z_m(t) - \hat{z}_m(t)$, and a.e. $t > 0$:

$$\rho |\overline{z}_m'|^2 + \frac{1}{2}\frac{d}{dt} a(\overline{z}_m, \overline{z}_m) \leq |\rho\overline{\zeta}_t + \gamma_\varepsilon(\zeta_t) - \gamma_\varepsilon(\hat{\zeta}_t)|\, |\overline{z}_m'| , \tag{3.21}$$

where we set $\overline{\zeta} = \zeta - \hat{\zeta}$. Using the property (3.12) we have

$$\begin{aligned} |\rho\overline{\zeta}_t + \gamma_\varepsilon(\zeta_t) - \gamma_\varepsilon(\hat{\zeta}_t)|^2 &= \rho^2 |\overline{\zeta}_t|^2 - 2\rho\,\overline{\zeta}_t(\gamma_\varepsilon(\zeta_t) - \gamma_\varepsilon(\hat{\zeta}_t)) + |\gamma_\varepsilon(\zeta_t) - \gamma_\varepsilon(\hat{\zeta}_t)|^2 \\ &\leq (\rho^2 - 2\rho b_* + (b_\varepsilon^*)^2)\, |\overline{\zeta}_t|^2 = \rho^2 \delta^2 |\overline{\zeta}_t|^2 , \end{aligned}$$

where $\delta = \sqrt{1 - 2b_*/\rho + (b_\varepsilon^*)^2/\rho^2} < 1$ if $\rho > (b_\varepsilon^*)^2/2b_*$. Hence, from (3.21) we obtain, by integrating in time,

$$\rho |||\overline{z}_m|||^2 = \rho \int_0^T |\overline{z}_m'|^2 \leq \rho\,\delta \int_0^T |\overline{\zeta}_t|\, |\overline{z}_m'| \leq \rho\,\delta\, |||\overline{\zeta}|||\, |||\overline{z}_m||| ,$$

and consequently also

$$|||z_m - \hat{z}_m||| \leq \delta |||\zeta - \hat{\zeta}||| ,$$

which, by taking the "$\liminf_{m\to\infty}$", yields (3.17). ∎

Proof of Proposition 3.5: Let $S : \mathbf{Z} \to \mathbf{Z}$ be the mapping defined by $z = S(\zeta)$, where z is the solution of (3.16) for some $\rho > (b_\varepsilon^*)^2/2b_*$. By (3.17) S is a strict contraction in $\mathbf{Z}$ and by the Banach fixed point theorem there exists a unique u_ε

$$u_\varepsilon = S(u_\varepsilon)$$

which solves (3.16) with $z = \zeta = u_\varepsilon$, i.e., (3.15).

Consider the solution of the nonlinear parabolic problem

$$\big\langle [\gamma_\varepsilon(w)]_t, v \big\rangle + a(w,v) = \int_\Gamma \sigma\, h_t(t)\, v , \qquad \forall v \in H^1(\Omega) , \tag{3.22}$$

with the initial condition $w(0) = \vartheta_0$. Using the Faedo–Galerkin method we can show the existence of at least one solution $w \in L^2(0,T;H^1(\Omega)) \cap H^1(0,T;(H^1(\Omega))')$. Hence $w \in C^0([0,T];L^2(\Omega))$ and defining $\tilde{w}(t) = \int_0^t w(s)\,ds$, by integrating (3.22) in time, we conclude that $\tilde{w}$ solves (3.15), which, by uniqueness, implies $\tilde{w} = u_\varepsilon$ and $u_{\varepsilon t} = w$.

Taking $v = w_t$ in (3.22) (this can be done in the Galerkin approximation as before) we have

$$\int_0^t \int_\Omega \gamma_\varepsilon'(w)\, w_t^2 + \int_0^t a(w, w_t) = \int_0^t \int_\Gamma \sigma\, h_t\, w_t = \int_\Gamma \sigma\, h_t(t)\, w(t) - \int_\Gamma \sigma\, h_t(0)\, \vartheta_0 - \int_0^t \int_\Gamma \sigma\, h_{tt}\, w \,.$$

Recalling (3.12) and using Gronwall inequality, we obtain

$$b_* \int_0^t \int_\Omega w_t^2 + \sup_{0 \le t \le T} \|w(t)\|^2_{H^1(\Omega)} \le C(\|\vartheta_0\|_{H^1(\Omega)}, \|h\|_{H^2(0,T;L^2(\Omega))}) \,,$$

which implies $u_{\varepsilon t} \in H^1(0,T; L^2(\Omega)) \cap L^\infty(0,T; H^1(\Omega))$ uniformly in ε.

On the other hand taking $v = \varphi \in \mathcal{D}(\Omega)$ in (3.15), we find a.e. in Q:

$$-\Delta u_\varepsilon = \eta_\varepsilon - \gamma_\varepsilon(u_{\varepsilon t}) \in L^\infty(0,T; L^2(\Omega)) \quad \text{uniformly in } \varepsilon \,; \tag{3.23}$$

hence, due to (3.13) and (3.12), what implies

$$|\gamma_\varepsilon(u_{\varepsilon t})| \le b^* \, |u_{\varepsilon t}| + \lambda$$

and knowing $u_{\varepsilon t} \in L^\infty(0,T; L^2(\Omega))$ independently of ε, by the elliptic theory, we finally conclude $u_\varepsilon \in L^\infty(0,T; H^2_{\text{loc}}(\Omega))$ also uniformly in ε. ∎

Proof of Theorem 3.1: From the estimates (3.14), we can select a subsequence as $\varepsilon \to 0$ such that

$$u_\varepsilon \rightharpoonup u \quad \text{in } W^{1,\infty}(0,T; H^1(\Omega)) \cap L^\infty(0,T; H^2_{\text{loc}}(\Omega))\text{-weak}^* \,, \tag{3.24}$$

$$u_{\varepsilon t} \rightharpoonup u_t \quad \text{in } H^1(0,T; L^2(\Omega))\text{-weak and } L^2(Q_T)\text{-strong} \,, \tag{3.25}$$

$$u_\varepsilon(t) \rightharpoonup u(t) \quad \text{in } H^1(\Omega)\text{-weak}, \ \forall t \in [0,T] \,. \tag{3.26}$$

We note that, as $\varepsilon \to 0$, we have $\int_0^r H_\varepsilon(s)\, ds \to r^+$ uniformly in the compacts subsets of $\mathbb{R}$ and therefore, (3.25) yields

$$\int_0^T J_\varepsilon(u_{\varepsilon t}) = \lambda \int_0^T \int_\Omega \int_0^{u_{\varepsilon t}} H_\varepsilon(s)\, ds \to \int_0^T J(u_t) \quad \text{as } \varepsilon \to 0 \,.$$

Since we have

$$J_\varepsilon(v) - J_\varepsilon(u_{\varepsilon t}) \ge \lambda(H_\varepsilon(u_{\varepsilon t}), v - u_{\varepsilon t}) \,, \qquad \forall v \in L^2(\Omega) \,,$$

taking any $\tilde v \in L^2(0,T; H^1(\Omega))$, from (3.15) we obtain, for a.e. $t > 0$:

$$(b(u_{\varepsilon t}), \tilde v - u_{\varepsilon t}) + a(u_\varepsilon, \tilde v - u_{\varepsilon t}) + J_\varepsilon(\tilde v) - J_\varepsilon(u_{\varepsilon t}) \ge \langle F_\varepsilon(t), \tilde v - u_{\varepsilon t} \rangle \,.$$

Integrating in time, we have

$$\int_0^T \left\{ \langle b(u_{\varepsilon t}) - F_\varepsilon, \tilde v - u_{\varepsilon t} \rangle + a(u_\varepsilon, \tilde v) + J_\varepsilon(\tilde v) - J_\varepsilon(u_{\varepsilon t}) \right\} \ge \frac{1}{2} a(u_\varepsilon(T), u_\varepsilon(T))$$

and recalling (3.13) and the convergences (3.24)–(3.26), which in particular imply

$$\liminf_{\varepsilon \to 0} a(u_\varepsilon(T), u_\varepsilon(T)) \ge a(u(T), u(T)) = 2 \int_0^T a(u, u_t)$$

we obtain in the limit

$$(3.27)\quad \int_0^T \left\{\langle b(u_t)-F,\tilde{v}-u_t\rangle + a(u,\tilde{v}-u_t) + J(\tilde{v}) - J(u_t)\right\} \geq 0\,, \quad \forall \tilde{v} \in L^2(0,T;H^1(\Omega))\,.$$

For arbitrary $v \in H^1(\Omega)$, we take in (3.27) for $t \in]\delta, T-\delta[$

$$\tilde{v}(\tau) = \begin{cases} u_t(\tau) & \text{if } \tau \notin]t-\delta, t+\delta[, \\ v & \text{if } \tau \in]t-\delta, t+\delta[\,, \end{cases}$$

with $\delta > 0$. Dividing by 2δ and letting $\delta \to 0$ we finally obtain (3.6) or (1.47) a.e. $t \in]0,T[$, i.e., u has the required properties. ∎

In order to apply the weak maximum principle, we shall assume that, for some given numbers $\mu \leq 0 \leq M$,

$$(3.28)\qquad \mu \leq \vartheta_0 \leq M \ \text{ in } \Omega \quad \text{and} \quad \mu \leq h_t \leq M \ \text{ on } \Gamma \times (0,T)\,.$$

Proposition 3.7. *Under the preceding assumptions, namely (3.28), we have*

$$(3.29)\qquad \mu \leq u_t \leq M \quad \text{a.e. in } Q = \Omega \times (0,T)\,.$$

Proof: It is sufficient to prove this property for the approximating function $w = u_{\varepsilon t}$. Choosing $v = (w(t)-M)^+$ in (3.22) we have

$$(3.30)\qquad \int_{\Omega\cap\{w>M\}} [\gamma_\varepsilon(w)]_t\,(w-M) + \int_{\Omega\cap\{w>M\}} \nabla w\cdot\nabla(w-M) = $$
$$= \int_{\Gamma\cap\{w>M\}} \sigma(h_t - w)\,(w-M) \leq 0\,,$$

denoting $z = \gamma_\varepsilon(w)$ and $w = \beta_\varepsilon(z)$, we set

$$\tilde{\beta}_M(r) = \begin{cases} \displaystyle\int_0^r (\beta_\varepsilon(s) - M)\,ds & \text{if } r > \gamma_\varepsilon(M), \\ 0 & \text{if } r \leq \gamma_\varepsilon(M)\,, \end{cases}$$

and, remarking that $\{w > M\} = \{z > \gamma_\varepsilon(M)\}$, we have

$$\frac{d}{dt}\int_\Omega \tilde{\beta}_M(\gamma_\varepsilon(w)) = \int_{\Omega\cap\{z>\gamma_\varepsilon(M)\}} (\beta_\varepsilon(z) - M)\,z_t = \int_{\Omega\cap\{w>M\}} (w-M)\,[\gamma_\varepsilon(w)]_t$$

and, integrating (3.30) in time, since $\vartheta_0 \leq M$, we find

$$\int_\Omega \tilde{\beta}_M(\gamma_\varepsilon(w(t))) + \int_0^t\int_\Omega |\nabla(w-M)^+|^2 \leq \int_\Omega \tilde{\beta}_M(\gamma_\varepsilon(\vartheta_0)) = 0\,,$$

which clearly implies $(w-M)^+ = 0$ a.e. in $\Omega\times(0,T)$, i.e. $w = u_{\varepsilon t} \leq M$. In a similar way we prove that $u_{\varepsilon t} \geq \mu$. ∎

Remark 3.8. Under the additional assumption (3.28) we have, in particular, $u_{\varepsilon t}(t) \in L^\infty(0,T;L^\infty(\Omega))$, and from (3.23) we may conclude $\Delta u_\varepsilon \in L^\infty(0,T;L^\infty(\Omega))$, uniformly in ε. Hence, by the elliptic theory, we have the additional regularity

$$u_\varepsilon \ \text{ and } \ u \in W^{1,\infty}(0,T;L^\infty(\Omega)) \cap L^\infty(0,T;W^{2,p}_{\text{loc}}(\Omega))\,, \quad \forall p < \infty\,,$$

and, in particular, by the Sobolev imbedding, we also have

$$u_\varepsilon \text{ and } u \in C^{0,\alpha}(Q), \quad \forall 0 \leq \alpha < 1 . \square$$

Remark 3.9. Note that if $\vartheta_0 \geq 0$ and $h_t \geq 0$ then also $u_t \geq 0$ and $u \geq 0$ for all $t > 0$ and we are reduced to a one-phase Stefan problem. □

Remark 3.10. Since $\Delta u_\varepsilon \in L^\infty(0,T;L^2(\Omega))$ and $\Gamma \in C^{0,1}$ by the generalized Green formula

$$\langle \partial_n u_\varepsilon, v\rangle_\Gamma = \int_\Omega \Delta u_\varepsilon\, v + \int_\Omega \nabla u_\varepsilon \cdot \nabla v , \qquad \forall v \in H^1(\Omega)$$

we have $\partial_n u_\varepsilon \rightharpoonup \partial_n u$ in $L^\infty(0,T;H^{-1/2}(\Gamma))$-weak* and from (3.10) we conclude also

$$\partial_n u = \sigma(h-u) \quad \text{a.e. on } \Gamma, \ t > 0 . \square$$

3.2 – Application to a control problem

We are interested in the control problem with respect to the boundary cooling coefficient σ, so that we shall be first concerned with a simple continuous dependence result.

Proposition 3.11. *Under the assumptions of Theorem 3.1, if u and $\widehat{u}$ denote, respectively, the solutions of (1.46)–(1.47) corresponding to σ and $\widehat{\sigma}$, then there exists a constant $C > 0$ depending only on ν, $\|\sigma\|_{L^\infty(\Gamma)}$, b_*, ϑ_0 and h:*

$$\|u_t - \widehat{u}_t\|_{L^2(Q)} + \|u - \widehat{u}\|_{L^\infty(0,T;H^1(\Omega))} \leq C\|\sigma - \widehat{\sigma}\|_{L^\infty(\Gamma)} . \tag{3.30}$$

Moreover,

$$\sigma \rightharpoonup \widehat{\sigma} \ \textit{in } L^\infty(\Gamma)\textit{-weak*} \quad \textit{then } u \rightharpoonup \widehat{u} \tag{3.31}$$

in $W^{1,\infty}(0,T;H^1(\Omega)) \cap L^\infty(0,T;H^2_{\text{loc}}(\Omega))$-weak, in $H^2(0,T;L^2(\Omega))$-weak and also strongly in $H^1(0,T;L^2(\Omega)) \cap L^\infty(0,T;H^1(\Omega))$ and in $H^1(0,T;L^2(\Gamma))$ for the traces.*

Proof: We take $v = \widehat{u}_t$ in (3.6) and $v = u_t$ in the corresponding inequality for $\widehat{u}$. We have for $\overline{u} = u - \widehat{u}$

$$\begin{aligned} b_* \int_0^t \int_\Omega |\overline{u}_t|^2 + \frac{1}{2} a(\overline{u}(t),\overline{u}(t)) &\leq \int_0^t \int_\Gamma (\sigma - \widehat{\sigma})(h - \widehat{u})\,\overline{u}_t \leq \\ &\leq \int_\Gamma (\sigma-\widehat{\sigma})(h(t)-\widehat{u}(t))\,\overline{u}(t) - \int_0^t\int_\Gamma (\sigma-\widehat{\sigma})(h_t - \widehat{u}_t)\,\overline{u} \equiv \mathcal{F}(\sigma-\widehat{\sigma}) . \end{aligned} \tag{3.32}$$

Since we have

$$\mathcal{F}(\sigma-\widehat{\sigma}) \leq \frac{1}{4} a(\overline{u}(t),\overline{u}(t)) + \int_0^t a(\overline{u},\overline{u}) + C\|\sigma-\widehat{\sigma}\|_{L^\infty(\Gamma)} , \tag{3.33}$$

now, recalling (3.32), it is clear that (3.30) follows easily by the Gronwall lemma.

Let $\sigma \rightharpoonup \widehat{\sigma}$ in $L^\infty(\Gamma)$-weak*. Then σ is uniformly bounded in $L^\infty(\Gamma)$ and, from the proof of the Theorem 3.1 we easily see that the regularity (3.7) holds also uniformly in σ and therefore (3.31) follows by compactness. In fact since u_t is bounded in $H^1(0,T;L^2(\Omega))\cap L^\infty(0,T;H^1(\Omega))$ its trace on Γ is compact in the space $L^2(0,T;L^2(\Gamma)) = L^2(\Gamma\times]0,T[)$ (see Lemma 3.12, below) and the passage to the limit in (3.6), as $\sigma \rightharpoonup \widehat{\sigma}$, can be done as in the proof of the Theorem 3.1. Finally the convergence $\overline{u} \to 0$ in $L^\infty(0,T;H^1(\Omega))\cap H^1(0,T;L^2(\Omega))$ is a consequence of (3.32) since $\overline{u} \to 0$ in $L^\infty(0,T;L^2(\Gamma))$ implies $\mathcal{F}(\sigma-\widehat{\sigma}) \to 0$. ∎

Lemma 3.12. *The trace mapping* $\tau\colon v \mapsto v|_\Gamma$ *(with* $\Gamma \in C^{0,1}$*) in the spaces*

$$\tau\colon\ L^2(0,T;H^1(\Omega))\cap H^1(0,T;H^{-1}(\Omega)) \to L^2(0,T;L^2(\Gamma))$$

is compact.

Proof: Since $H^1(\Omega)\hookrightarrow H^s(\Omega)$ is compact, for any $0 \le s < 1$, then the embedding

$$I_s\colon\ L^2(0,T;H^1(\Omega))\cap H^1(0,T;H^{-1}(\Omega)) \hookrightarrow L^2(0,T;H^s(\Omega)),\qquad \forall s<1\ ,$$

is also compact, by Aubin type results (see [Si], for instance). On the other hand, by the trace theorem, $\tau_s\colon L^2(0,T;H^s(\Omega)) \to L^2(0,T;L^2(\Gamma))$ is a continuous mapping for any $1/2 < s \le 1$. Hence, we can write τ as a composition of a compact mapping I_s with a continuous one τ_s, $\forall\, 1/2 < s < 1$, and the conclusion follows. ∎

Proposition 3.13. *Under the conditions of Proposition 3.11 if also (3.28) holds, then*

$$\|u_t-\widehat{u}_t\|_{L^2(Q)} + \|u-\widehat{u}\|_{L^\infty(0,T;H^1(\Omega))} \le C_1\|\sigma-\widehat{\sigma}\|_{L^1(\Gamma)}\ , \tag{3.34}$$

where the constant $C_1>0$ *depends only on* ν, $\|\sigma\|_{L^\infty(\Gamma)}$, b_*, T *and* μ, M. *Moreover if*

$$\sigma \rightharpoonup \widehat{\sigma}\ \text{ in } L^2(\Omega)\text{-weak}\qquad \text{then } u \to \widehat{u} \tag{3.35}$$

in $H^1(0,T;L^2(\Omega))\cap L^\infty(0,T;H^1(\Omega))$*-strong, their traces in* $H^1(0,T;L^2(\Gamma))$*-strong and also in* $W^{1,\infty}(0,T;H^1(\Omega))\cap L^\infty(0,T;H^2_{\text{loc}}(\Omega))$*-weak* and in* $H^2(0,T;L^2(\Omega))$*-weak.*

Proof: We recall that now $\mu \le \widehat{u}_t \le M$ and we repeat the proof of Proposition 3.12, replacing in (3.33) the $L^\infty(\Gamma)$-norm of $\sigma-\widehat{\sigma}$ by its L^1-norm. The convergence (3.35) may be shown also in a similar way. ∎

We have already observed in Remark 3.10 that the Neumann boundary condition (1.41) for the variational solution u is satisfied a.e.. It is natural also to re-obtain the equation (1.32) by passing to the limit in (3.9) as $\varepsilon \to 0$. In fact, since $0 \le H_\varepsilon \le 1$, there exists $\chi \in L^2(Q)$ such that

$$H_\varepsilon(u_{\varepsilon t}) \rightharpoonup \chi\quad \text{in } L^2(Q)\text{-weak}\ , \tag{3.36}$$

with $0 \le \chi \le 1$ a.e. in $Q = \Omega\times]0,T[$. Then, the estimate (3.14) allow us to pass to the limit in the sense of distributions in (3.9), and it implies the equation (1.32):

$$b(u_t) - \Delta u + \lambda\chi = \eta_0 \quad \text{a.e. in } \Omega,\ t>0\ . \tag{1.32}$$

Theorem 3.14. *The function* χ *satisfying (1.32) is such that* $\chi \in H(u_t)$ *a.e. in* Q, *or equivalently*

$$0 \le \chi_{\{u_t>0\}} \le \chi \le 1-\chi_{\{u_t<0\}} \le 1 \quad \text{a.e. in } Q\ . \tag{3.37}$$

Moreover, if χ is associated with σ and $\hat{\chi}$ with $\hat{\sigma}$, as in Propositions 3.11 or 3.13, respectively, if $\sigma \to \hat{\sigma}$ in $L^\infty(\Gamma)$-weak or in $L^2(\Gamma)$-weak then*

$$\chi \rightharpoonup \hat{\chi} \text{ in } L^p(Q)\text{-weak}, \quad \forall 1 \le p < \infty . \tag{3.38}$$

Proof: By (3.25) we know that $u_{\varepsilon t} \to u_t$ in $L^2(Q_T)$-strong. Recalling the definition of H_ε and (3.36), we have as $\varepsilon \to 0$

$$0 = \int_Q H_\varepsilon(u_{\varepsilon t})\, u_{\varepsilon t}^- \;\to\; \int_Q \chi\, u_t^- = 0 ,$$

which implies $\chi = 0$ if $u_t < 0$, i.e., $\chi \le 1 - \chi_{\{u_t<0\}}$. Similarly we conclude $\chi \ge \chi_{\{u_t>0\}}$ from

$$0 = \int_Q (1 - H_\varepsilon(u_{\varepsilon t}))\,(u_{\varepsilon t} - \varepsilon)^+ \;\to\; \int_Q (1-\chi)\, u_t^+ = 0 \quad \text{as } \varepsilon \to 0 .$$

Let $\sigma \to \hat{\sigma}$ in $L^\infty(\Gamma)$-weak* (resp. in $L^2(\Gamma)$-weak). The continuous dependence results of Proposition 3.11 (resp. Proposition 3.13) imply, in particular, that

$$b(u_t) \to b(\hat{u}_t) \text{ in } L^2(Q) \quad \text{and} \quad \Delta u \to \Delta \hat{u} \text{ in } L^\infty(0,T;L^2(\Omega))\text{-weak*},$$

and taking the limit in the equation (1.32) the conclusion follows. ■

Remark 3.15. If we define the temperature $\vartheta = u_t$ and the enthalpy $\eta = b(u_t) + \lambda\,\chi$, from (3.37), we conclude that the relation $\eta \in \gamma(\vartheta)$ (see (1.21)) holds a.e. in Q. Then as a consequence of (3.38) we have

$$\sigma \rightharpoonup \hat{\sigma} \text{ in } L^\infty(\Gamma)\text{-weak*} \quad \text{implies } \eta \rightharpoonup \hat{\eta} \text{ in } L^2(Q)\text{-weak} . \; \square$$

Remark 3.16. If $\text{meas}_Q\{u_t = 0\} = 0$, then from (3.37) it follows that $\chi = \chi_{\{u_t>0\}} = 1 - \chi_{\{u_t<0\}}$ a.e. in Q. In this case we have $\chi = \chi^p$, $\forall 1 < p < \infty$, and

$$\int_Q \chi = \int_Q \chi^p \le \liminf_{\varepsilon\to 0} \int_Q [H_\varepsilon(u_{\varepsilon t})]^p \le \lim_{\varepsilon\to 0} \int_Q H_\varepsilon(u_{\varepsilon t}) = \int_Q \chi$$

implies also

$$H_\varepsilon(u_{\varepsilon t}) \to \chi_{\{u_t>0\}} \quad \text{in } L^p(Q)\text{-strong}, \; 1 \le p < \infty .$$

A sufficient condition for the property $\text{meas}_Q\{u_t = 0\} = 0$ for the case $u_t \in L^\infty(Q)$ is $\text{meas}_\Omega\{\vartheta_0 = 0\} = 0$ and its proof can be found in [GZ]. □

As a consequence of this remark and the second part of (3.38), we have the following interesting continuous dependence result.

Corollary 3.17. *If (3.28) holds and* $\text{meas}_\Omega\{\vartheta_0 = 0\}$, *then*

$$\sigma \rightharpoonup \hat{\sigma} \text{ in } L^2(\Gamma)\text{-weak} \quad \text{implies } \eta \to \hat{\eta} \text{ in } L^p(Q)\text{-strong}, \quad \forall 1 \le p < \infty . \; ■$$

We can now consider an application to the boundary control of the two-phase Stefan problem, by taking as cost functional $\mathcal{F}\colon L^\infty(\Gamma) \to \mathbf{R}^+$ defined by

$$\mathcal{F}(\sigma) = \rho_1 \int_Q |\eta - \eta_*|^2 + \rho_2 \int_\Sigma |\vartheta - \vartheta_*|^2 + \rho_3 \int_\Omega |\vartheta(T) - \vartheta_{*T}|^2 , \tag{3.41}$$

where $Q = \Omega\times\,]0,T[$, $\Sigma = \Gamma\times\,]0,T[$, $\rho_1, \rho_2, \rho_3 \geq 0$ are constants ($\rho_1+\rho_2+\rho_3 > 0$) and $\eta_* \in L^2(Q)$, $\vartheta_* \in L^2(\Sigma)$ and $\vartheta_{*T} \in L^2(\Omega)$ are given functions corresponding to the ideal state and $\eta = b(\vartheta) + \lambda\,\chi$, $\vartheta = u_t$ is the solution of (1.46)-(1.47) corresponding to σ.

For arbitrarily fixed $\nu \in\,]0,1[$ and $C \geq 1$ we shall consider the set of control variables defined by

$$S_\infty = \Big\{\sigma \in L^\infty(\Gamma):\ \nu \leq \sigma(x) \leq C \ \text{ a.e. } x \in \Gamma\Big\}. \tag{3.42}$$

Theorem 3.18. *Under the assumptions of Theorem 3.1, then there exists an optimal control $\sigma^* \in S_\infty$, i.e., a solution to the minimization problem*

$$\mathcal{F}(\sigma^*) = \inf_{\sigma\in S_\infty} \mathcal{F}(\sigma)\,. \tag{3.43}$$

Proof: Take a minimization sequence for the functional (3.41), i.e. $\{\sigma_n\} \subset S_\infty$ such that

$$\inf_{\sigma\in S_\infty} \mathcal{F}(\sigma) \equiv \mathcal{F}_0 \leq \mathcal{F}(\sigma_n) \leq \mathcal{F}_0 + \frac{1}{n}\,. \tag{3.44}$$

Then by the definition (3.42), there exists a subsequence, still denoted by σ_n and an element $\sigma^* \in S_\infty$, such that, $\sigma_n \rightharpoonup \sigma^*$ in $L^\infty(\Gamma)$-weak*. Then by the Proposition 3.11 and Remark 3.15 we have

$$\begin{aligned} \eta_n &\rightharpoonup \eta^* \quad \text{in } L^2(Q)\text{-weak}\,,\\ \vartheta_n &\rightharpoonup \vartheta^* \quad \text{in } L^2(\Sigma)\text{-weak}\,,\\ \vartheta_n(T) &\rightharpoonup \vartheta^*(T) \quad \text{in } L^2(\Omega)\,, \end{aligned}$$

where η^*, ϑ^* is the solution corresponding to σ^*. By the lower semi-continuity of $\mathcal{F}$ and recalling (3.44), we have

$$\mathcal{F}_0 \leq \mathcal{F}(\sigma^*) \leq \liminf_{\sigma_n \rightharpoonup \sigma^*} \mathcal{F}(\sigma_n) \leq \mathcal{F}_0$$

and the conclusion follows. ∎

Remark 3.19. In [P2] similar control problems were considered by using only the temperature in the definition of $\mathcal{F}$ and h as the control variable instead of σ. Also in [P2], optimality conditions were given for the regularized problem (3.15), as well as, numerical results for the discrete approximations to the corresponding optimal control problems. □

We may define the liquid zone as being the subset $\{u_t > 0\}$, up to a set of null measure in Q, and introduce a different cost functional $\mathcal{G}\colon L^2(\Gamma) \to \mathbf{R}^+$ defined by

$$\mathcal{G}(\sigma) = \rho_0 \int_Q |\chi - \chi_\Lambda|^2 + \rho \int_\Gamma \sigma^2 \tag{3.45}$$

for $\rho_0, \rho > 0$, where χ_Λ is the characteristic function of some desired subset $\Lambda \subset Q$ and $\chi \in H(u_t)$ is associated with the solution u corresponding to the coefficient σ, that will be taken in the set of control variables (ν fixed)

$$S_2 = \Big\{\sigma \in L^2(\Gamma):\ \sigma(x) \geq \nu > 0 \ \text{ a.e. in } \Gamma\Big\}. \tag{3.46}$$

Theorem 3.20. *Under the assumptions of the Corollary 3.17, namely (3.28) and* $\text{meas}_\Omega\{\vartheta_0 = 0\}$*, then there exists* $\sigma^* \in S_2$*:*

$$\mathcal{G}(\sigma^*) = \inf_{\sigma \in S_2} \mathcal{G}(\sigma) . \tag{3.47}$$

Proof: If σ_n denotes a minimizing sequence in S_2, then $\mathcal{G}(\sigma_n)$ is bounded and, consequently $\{\sigma_n\}$ is also bounded in $L^2(\Gamma)$. Then for a subsequence $\sigma_n \rightharpoonup \sigma_*$ in $L^2(\Gamma)$-weak, we have, as a consequence of (3.40) and Remark 3.17, that

$$\chi_n = \chi_{\{u_t^n > 0\}} \to \chi_{\{u_t^* > 0\}} \quad \text{in } L^2(Q)\text{-strong} .$$

Therefore we may conclude

$$\inf_{\sigma \in S_2} \mathcal{G}(\sigma) \leq \mathcal{G}(\sigma^*) \leq \liminf_{\sigma_n \rightharpoonup \sigma^*} \mathcal{G}(\sigma_n) \leq \inf_{\sigma \in S_2} \mathcal{G}(\sigma) ,$$

what ends the proof of Theorem 3.20. ∎

Remark 3.21. In the multidimensional two-phase Stefan problem, an existence result for the boundary optimal control with direct observation of the non-degenerate free boundary was given in [R4]. Some extensions to cost functional defined in terms of characteristic functions in spaces of bounded variation have been proposed recently in [RZ]. □

3.3 – Asymptotic stabilization as time $t \to \infty$

In this section we shall assume

$$h_t(t) \to h_\infty \quad \text{in } L^2(\Gamma) \text{ as } t \to \infty \tag{3.48}$$

for some $h_\infty \in L^2(\Gamma)$, under the hypothesis

$$h - t\, h_\infty \in L^\infty(0,\infty; L^2(\Gamma)) , \tag{3.49}$$

$$h_t - h_\infty \in L^1(0,\infty; L^2(\Gamma)) \cap L^\infty(0,\infty; L^2(\Gamma)) , \tag{3.50}$$

$$h_{tt} \in L^1(0,\infty; L^2(\Gamma)) . \tag{3.51}$$

By the assumption (3.4), the condition (3.49) means that the convergence (3.48) in time average has the order of convergence

$$\left\| \frac{1}{t} \int_0^t h_t - h_\infty \right\|_{L^2(\Gamma)} = O\left(\frac{1}{t}\right) \quad \text{as } t \to \infty .$$

We also note that (3.50) and (3.51) are sufficient to imply (3.48), as a simple application of the following lemma, which proof is an easy adaptation of well-known results.

Lemma 3.22. *Let* H *be a Hilbert space, and for any* p', $1 \leq p' \leq \infty$, *and let* $v \in W_p(a,b;H) = \{w \in L^p(a,b;H)\colon w' = \frac{dw}{dt} \in L^{p'}(a,b;H)\}$ *with* $1 \leq p \leq \infty$, $1/p + 1/p' = 1$. *Then, for any* $\delta > 0$, $b - a = \tau > 0$, *we have*

$$\|v\|_{L^\infty(a,b;H)} \leq (\delta^{-1} + \tau^{-1/p}) \|v\|_{L^p(a,b;H)} + \delta \|v'\|_{L^{p'}(a,b;H)} , \tag{3.52}$$

with the convention $-1/\infty = 0$. In particular, any $v \in W_p(t_0, \infty; H)$ satisfies $v(t) \to 0$ in H as $t \to \infty$.

Proof: Define $w(t) = v(t) - \frac{1}{\tau}\int_a^b v$ and if $t_0 \in [a,b]$ is such that $w(t_0) = 0$, we have ((v,w) denotes the inner product in H)

$$\frac{1}{2}|w(t)|^2 = \int_{t_0}^t (w, w') \le \|w\|_{L^p(a,b;H)} \|w'\|_{L^{p'}(a,b;H)} \le 2\|v\|_{L^p(a,b;H)} \|v'\|_{L^{p'}(a,b;H)} .$$

Hence, using the Cauchy inequality with $\sqrt{\delta}$, for any $t \in [a,b]$ we find

$$\begin{aligned} |v(t)| &\le |w(t)| + \left|\frac{1}{\tau}\int_a^b v\right|_H \le 2\|v\|^{1/2}_{L^p(a,b;H)} \|v'\|^{1/2}_{L^{p'}(a,b;H)} + \tau^{-1/p}\|v\|_{L^p(a,b;H)} \\ &\le (\delta^{-1} + \tau^{-1/p})\|v\|_{L^p(a,b;H)} + \delta\|v'\|_{L^{p'}(a,b;H)} . \blacksquare \end{aligned}$$

Let us consider the corresponding linear stationary problem

$$(3.53) \qquad -\Delta\vartheta_\infty = 0 \ \text{ in } \Omega , \qquad \partial_n\vartheta_\infty = \sigma(h_\infty - \vartheta_\infty) \ \text{ on } \Gamma ,$$

which has a unique variational solution given by

$$(3.54) \qquad \vartheta_\infty \in H^1(\Omega): \quad a(\vartheta_\infty, v) = \int_\Gamma \sigma\, h_\infty\, v, \quad \forall v \in H^1(\Omega) .$$

Defining $\eta_\infty = b(\vartheta_\infty) + \lambda\,\chi_{\{\vartheta_\infty > 0\}}$, $u_\infty = t\,\vartheta_\infty$, $t > 0$, and recalling the notations of Section 1.3, from (3.54) we obtain ($u_{\infty t} = \vartheta_\infty$), for any $v \in H^1(\Omega)$,

$$(3.54) \qquad (b(u_{\infty t}), v - u_{\infty t}) + a(u_\infty, v - u_{\infty t}) + J(v) - J(u_{\infty t}) \ge \langle F_\infty, v - u_{\infty t}\rangle$$

where, following (1.45), we set

$$(3.55) \qquad \langle F_\infty - F, v\rangle = \int_\Omega (\eta_\infty - \eta_0)\, v + \int_\Gamma \sigma(t\,h_\infty - h)\, v , \qquad \forall v \in H .$$

By comparing (3.54) with (1.47), we have the following asymptotic stabilization result for the temperature $\vartheta = u_t$.

Theorem 3.23. *Let (3.49)–(3.51) hold. Then*

$$(3.56) \qquad \left\|\frac{1}{t}\int_0^t \vartheta - \vartheta_\infty\right\|_{H^1(\Omega)} = O\Big(\frac{1}{t}\Big) \quad \text{as } t \to \infty ,$$

$$(3.57) \qquad \vartheta(t) \to \vartheta_\infty \quad \text{in } H^1(\Omega)\text{-weak and in } L^2(\Omega)\text{-strong as } t \to \infty .$$

Proof: For fixed time $t > 0$ we take $v = u_t(t)$ in (3.54) and $v = u_{\infty t} = \vartheta_\infty$ in (1.47). Denoting $z(t) = u(t) - u_\infty$, we obtain

$$b_* \int_\Omega |z_t|^2 + \frac{1}{2}\frac{d}{dt}a(z,z) \le \int_\Omega (\eta_0 - \eta_\infty)\, z_t + \int_\Gamma \sigma(h - t\,h_\infty)\, z_t ,$$

and integrating in time

$$b_* \int_0^t \int_\Omega |z_t|^2 + \frac{1}{2} a(z(t), z(t)) \leq \int_\Omega (\eta_0 - \eta_\infty)\, z(t) - \int_0^t \int_\Gamma \sigma(h_\infty - h_t)\, z + \int_\Gamma \sigma(h(t) - t\, h_\infty)\, z(t)\,.$$

Hence, we deduce for some constants $C_\sigma, C_0, C_1 > 0$ independent of t,

$$\int_0^t \int_\Omega |z_t|^2 + C_\sigma \|z(t)\|^2_{H^1(\Omega)} \leq C_0 \Big(\|\eta_0 - \eta_\infty\|^2_{L^2(\Omega)} + \|h(t) - t h_\infty\|^2_{L^2(\Gamma)} \Big) \\ + C_1 \int_0^t \|h_\infty - h_t\|_{L^2(\Gamma)}\, \|z\|_{H^1(\Omega)}\,,$$

and applying Lemma 3.24 below to $y(t) = \|z(t)\|_{H^1(\Omega)}$, with the assumptions (3.49) and (3.50) we conclude that

$$\int_0^t \int_\Omega |\vartheta - \vartheta_\infty|^2 + \Big\| \int_0^t \vartheta - t\, \vartheta_\infty \Big\|_{H^1(\Omega)} \leq C \quad \text{independent of } t > 0\,, \tag{3.58}$$

and in particular (3.56) follows.

In order to obtain $\vartheta(t) \to \vartheta_\infty$ in $L^2(\Omega)$ by Lemma 3.22, it will be sufficient to prove that $\vartheta - \vartheta_\infty \in W_2(0, \infty, L^2(\Omega))$. By (3.58), it is enough to show that $\vartheta_t \in L^2(0, \infty; L^2(\Omega))$. As in the proof of the Proposition 3.5, from (3.54) and (3.22), denoting $\zeta = u_{\varepsilon t} - \vartheta_\infty$, we find

$$b_* \int_0^t \int_\Omega \zeta_t^2 + \frac{1}{2} a(\zeta(t), \zeta(t)) \leq \frac{1}{2} a(\vartheta_0 - \vartheta_\infty, \vartheta_0 - \vartheta_\infty) - \int_\Gamma \sigma(h_t(0) - h_\infty)\, \vartheta_0 \\ + \int_\Gamma \sigma(h_t(t) - h_\infty)\, \zeta(t) - \int_0^t \int_{\Gamma \sigma h_{tt}} \zeta \tag{3.59}$$

and hence, for some constants C'_σ, C'_0 and $C'_1 > 0$ independent of $t > 0$

$$\int_0^t \int_\Omega \zeta_t^2 + C'_\sigma \|\zeta(t)\|^2_{H^1(\Omega)} \leq C'_0 \Big(\|\vartheta_0 - \vartheta_\infty\|^2_{H^1(\Omega)} + \|h_t - h_\infty\|^2_{L^\infty(0,T;L^2(\Omega))} \Big) \\ + C'_1 \int_0^t \|h_{tt}\|_{L^2(\Omega)}\, \|\zeta\|_{H^1(\Omega)}\,.$$

Again by Lemma 3.24 below and now by the assumptions (3.50) and (3.51) we obtain

$$\int_0^t \int_\Omega |u_{\varepsilon tt}|^2 + \|u_{\varepsilon t}(t) - \vartheta_\infty\|_{H^1(\Omega)} \leq C' \quad \text{(independent of } t > 0 \text{ and } \varepsilon > 0). \tag{3.60}$$

Recalling that $u_{\varepsilon t} \to u_t = \vartheta$ as $\varepsilon \to 0$, from (3.60) we obtain the required estimates to conclude the convergence of (3.57). ∎

Lemma 3.24. *Let $A \in L^1(0,T)$ and $y \in L^\infty(0,T)$ be nonnegative functions and $B > 0$ a constant such that*

$$y^2(t) \leq \int_0^t A(s)\, y(s)\, ds + B \quad \textit{for } t \in (0,T)\,.$$

Then we have

$$y(t) \leq \frac{1}{2} \int_0^t A(s)\, ds + \sqrt{B} \quad \textit{for all } t \in [0,T]\,. \tag{3.61}$$

Proof: Let $x(t) = \int_0^t A(s)\, y(s)\, ds + B$. We have $x' = Ay \le A\sqrt{x}$, and integrating in time $\frac{d}{dt}(\sqrt{x}) = \frac{x'}{2\sqrt{x}} \le \frac{A}{2}$ we find (3.61). ∎

Remark 3.25. The results of Theorem 3.23 for the variational inequality (with $h(t) = t\, h_\infty$) can be found in [T2] with some additional variants and are similar to the previous results of [F1] for the Dirichlet boundary condition. For the mixed boundary value problem, using a direct approach based on the theory of nonlinear evolution equations governed by time-dependent subdifferential operators in Hilbert spaces, some extensions were given in [DK2]. □

4 – THE GENERAL PROBLEM WITH PRESCRIBED CONVECTION

4.1 – Existence of the weak solution

We consider now the direct weak formulation (1.26) of the two-phase Stefan problem in general form, first with Neumann boundary condition on $\Sigma = \Sigma_N$, where $\Sigma = \partial\Omega \times]0,T[$ and $\partial\Omega$ is the Lipschitz boundary of the bounded domain Ω. The precise weak formulation consists of finding the pair temperature-enthalpy

$$(4.1)\quad (\vartheta, \eta) \in L^2(0,T;H^1(\Omega)) \times L^2(Q)\colon\ \eta \in \gamma(\vartheta) \ \text{ a.e. in } Q = \Omega \times]0,T[\,,$$

$$(4.2)\quad -\int_Q \eta(\partial_t \varphi + \mathbf{v} \cdot \nabla\varphi) + \int_Q \nabla\vartheta \cdot \nabla\varphi + \int_Q f(\vartheta)\,\varphi + \int_{\Sigma_N} g(\vartheta)\,\varphi = \int_\Omega \eta_0\, \varphi(0)\,,$$

for all $\varphi \in H^1(Q)\colon \varphi(T) = 0$. Here we shall assume γ given by (1.21) with b satisfying the assumption (3.1), and for some constants μ, M:

$$(4.3)\qquad \eta_0 \in \gamma(\vartheta_0) \text{ for some } \vartheta_0 \in L^\infty(\Omega)\colon\quad \mu \le \vartheta_0 \le M \quad \text{a.e. in } \Omega\,,$$

$$(4.4)\qquad f(\vartheta) \ge 0 \text{ and } g(\vartheta) \ge 0 \quad \text{for } \vartheta \ge M \ge 0\,,$$

$$(4.5)\qquad f(\vartheta) \le 0 \text{ and } g(\vartheta) \le 0 \quad \text{for } \vartheta \le \mu \le 0\,,$$

for prescribed Carathéodory functions $f(\vartheta) = f(x,t,\vartheta) : Q \times \mathbf{R} \to \mathbf{R}$ and $g(\vartheta) = g(x,t,\vartheta)\colon \Sigma \times \mathbf{R} \to \mathbf{R}$, i.e., f (resp. g) is continuous in ϑ for a.e. $(x,t) \in Q$ (resp. a.e. $(x,t) \in \Sigma$) and measurable in (x,t) for every $\vartheta \in \mathbf{R}$. In addition we shall require the existence of functions f^* and g^* such that

$$(4.6)\qquad f^* \in L^2(Q)\colon\quad \sup_{\mu \le \vartheta \le M} |f(x,t,\vartheta)| \le f^*(x,t) \quad \text{a.e. in } Q\,,$$

$$(4.7)\qquad g^* \in L^2(\Sigma)\colon\quad \sup_{\mu \le \vartheta \le M} |g(x,t,\vartheta)| \le g^*(x,t) \quad \text{a.e. on } \Sigma\,.$$

Finally we prescribe the velocity $\mathbf{v}$ in the space

$$(4.8)\qquad \mathbf{v} \in \mathbf{L}^2_\sigma(Q) \cap [L^\infty(Q)]^N\,,$$

where we set

$$\mathbf{L}^2_\sigma(Q) = \left\{\mathbf{w} \in [L^2(Q)]^N\colon \int_Q \mathbf{w} \cdot \nabla\varphi = 0,\ \forall \varphi \in L^2(0,T;H^1(\Omega))\right\}\,,$$

which is easily seen to be the closure in $[L^2(Q)]^N$ of

$$S_\sigma \equiv \left\{ \vec{\psi} \in [\mathcal{D}(Q)]^N : \nabla \cdot \vec{\psi} = 0 \text{ in } Q \right\} \tag{4.9}$$

and corresponds to functions that satisfy $\nabla \cdot \mathbf{w} = 0$ in Q and $\mathbf{w} \cdot \mathbf{n} = 0$ on Σ in a weak sense (actually, these constraints are satisfied a.e. if $\mathbf{w} \in \mathbf{L}^2_\sigma(Q) \cap L^2(0,T;[H^1(\Omega)]^N)$, for instance).

Theorem 4.1. *Under the preceding assumptions, namely (4.3)–(4.8) there exists at least one solution to (4.1)–(4.2), which in addition satisfies* $\partial_t \vartheta \in L^2_{\mathrm{loc}}(Q)$ *and*

$$\mu \le \vartheta(x,t) \le M \quad \text{a.e. } (x,t) \in Q\,. \blacksquare \tag{4.10}$$

Remark 4.2. In (4.1) we have considered $\eta \in L^2(Q)$ because in (4.2) we have prescribed $\mathbf{v} \in \mathbf{L}^2_\sigma(Q) \cap [L^\infty(\Omega)]^N$ and $\varphi \in H^1(Q)$. This is coherent also with the assumptions of f^* and g^* in L^2. If instead we choose test functions φ, for instance, in $W^{1,\infty}(Q)$, we could take f^* and g^* in L^1 and only $\mathbf{v} \in \mathbf{L}^1_\sigma(Q)$, the closure of S_σ in $[L^1(Q)]^N$ with $\eta \in L^\infty(Q)$ in (4.1), or else $\mathbf{v} \in \mathbf{L}^2_\sigma(Q)$ and then take again $\eta \in L^2(Q)$ in (4.1). However an existence result to (4.1)–(4.2) for unbounded velocities $\mathbf{v} \in \mathbf{L}^2_\sigma(Q)$ can be obtain only under an additional assumption on the monotonicity of f and g (see Theorem 4.8). □

Remark 4.3. Since we have assumed the conditions (4.4) and (4.5) and a prescribed solenoidal velocity field, that are compatible with the maximum principle, we have not required any growth nor monotonicity conditions on $f(\vartheta)$ or on $g(\vartheta)$ and the existence result of Theorem 4.1 is new. □

Remark 4.4. The assumptions (4.4)–(4.5) are satisfied, for instance, for the case

$$f(x,t,\vartheta) = \alpha(x,t)\,\psi(\vartheta) + \beta(x,t)\,, \tag{4.11}$$

with $\alpha, \beta \in L^\infty(Q)$, $\alpha \ge \alpha_* > 0$ a.e. in Q and $\psi \in C^0(\mathbf{R})$ is such that $\psi(\vartheta) \to +\infty$ (resp. $-\infty$) as $\vartheta \to +\infty$ (resp. $\vartheta \to -\infty$). Analogous remark holds for g and, for instance, in the case (1.40)–(3.2), the assumption (3.28) implies (4.4)–(4.5) for g. More general nonlinear fluxes like the Stefan–Boltzmann's radiation law can also be covered (see [V1], for instance). □

In order to prove the existence result we shall use also the Faedo–Galerkin method for an approximate problem, which is obtained by a simultaneous regularization and truncation. Let $\varepsilon > 0$ and recall the bilipschitz function γ_ε defined in (3.12). Its inverse γ_ε^{-1} satisfies $\gamma_\varepsilon^{-1}(\gamma_\varepsilon(u)) = u$ and

$$0 < 1/b^*_\varepsilon \le (\gamma_\varepsilon^{-1})'(s) \le 1/b_* \quad \text{a.e. } s \in \mathbf{R}\,. \tag{4.12}$$

We consider the truncations $\overline{f}(\vartheta)$ and $\overline{g}(\vartheta)$ of $f(\vartheta)$ and $g(\vartheta)$, respectively, in the form

$$\overline{f}(x,t,\vartheta) = \begin{cases} f(x,t,M) & \text{if } \vartheta \ge M, \\ f(x,t,\vartheta) & \text{if } \mu \le \vartheta \le M, \\ f(x,t,\mu) & \text{if } \vartheta \le \mu\,, \end{cases} \tag{4.13}$$

with analogous definition for $\overline{g}$, and we consider regularizations $\overline{f}_\varepsilon$, $\overline{g}_\varepsilon$ with respect to the time variable, such that, for each $\varepsilon > 0$

$$\sup_{\vartheta} |\partial_t \overline{f}_\varepsilon| \in L^2(Q) \quad \text{and} \quad \sup_{\vartheta} |\partial_t \overline{g}_\varepsilon| \in L^2(\Sigma)\ , \tag{4.14}$$

(4.4)–(4.5) still hold for $\overline{f}_\varepsilon$ and $\overline{g}_\varepsilon$, and they must also satisfy as $\varepsilon \to 0$

$$\sup_{\mu\le\vartheta\le M} |\overline{f}_\varepsilon - \overline{f}| \to 0 \text{ in } L^2(Q) \quad \text{and} \quad \sup_{\mu\le\vartheta\le M} |\overline{g}_\varepsilon - \overline{g}| \to 0 \text{ in } L^2(\Sigma)\ . \tag{4.15}$$

Setting $f_\varepsilon(u) = \overline{f}_\varepsilon(\gamma_\varepsilon^{-1}(u))$, $g_\varepsilon(u) = \overline{g}_\varepsilon(\gamma_\varepsilon^{-1}(u))$ and considering as $\varepsilon \to 0$ sequences $H^1(\Omega) \ni \eta_{0\varepsilon} \to \eta_0$ in $L^2(\Omega)$ and

$$\mathbf{w}_\varepsilon \in \mathcal{S}_\sigma \colon \mathbf{w}_\varepsilon \to \mathbf{v} \text{ in } \mathbb{L}^2_\sigma(Q) \quad \text{with } \|\mathbf{w}_\varepsilon\|_{L^\infty(Q)} \le \|\mathbf{v}\|_{L^\infty(Q)}\ , \tag{4.16}$$

we shall study the approximate problem of finding for each $\varepsilon > 0$ a function $u_\varepsilon \in H^1(Q)$ which satisfies

$$-\int_Q u_\varepsilon(\partial_t\varphi + \mathbf{w}_\varepsilon\cdot\nabla\varphi) + \int_Q \nabla\gamma_\varepsilon^{-1}(u_\varepsilon)\cdot\nabla\varphi + \int_Q f_\varepsilon(u_\varepsilon)\,\varphi + \int_\Sigma g_\varepsilon(u_\varepsilon)\,\varphi = \int_\Omega \eta_{0\varepsilon}\,\varphi(0) \tag{4.17}$$

for all $\varphi \in H^1(Q) \colon \varphi(T) = 0$. We note that this is the variational formulation of the following parabolic Neumann problem:

$$\partial_t u_\varepsilon + \mathbf{w}_\varepsilon \cdot \nabla u_\varepsilon - \Delta\gamma_\varepsilon^{-1}(u_\varepsilon) + f_\varepsilon(u_\varepsilon) = 0 \quad \text{in } Q\ , \tag{4.18}$$

$$\mathbf{n}\cdot\nabla\gamma_\varepsilon^{-1}(u_\varepsilon) + g_\varepsilon(u_\varepsilon) = 0 \quad \text{on } \Sigma\ , \tag{4.19}$$

$$u_\varepsilon(0) = \eta_{0\varepsilon} \quad \text{on } \Omega\ . \tag{4.20}$$

Proposition 4.5. *Under the preceding assumptions there exists a solution $u_\varepsilon \in H^1(Q)$ of (4.10) for each $\varepsilon > 0$.*

Proof: As in the previous constructions of solutions with the Faedo-Galerkin method (see Lemma 3.6), we let $\{v_i\}$ be a basis of $H^1(\Omega)$, assumed orthonormal with respect to the $L^2(\Omega)$-inner product and $V_m = \text{span}\{v_1, ..., v_m\}$.

We look for a finite dimensional approximation of u_ε in the form

$$u_m(x,t) = \sum_{i=1}^m \xi_{mi}(t)\, v_i(x)\ ,$$

where the ξ_{mi} solve the following system of ordinary differential equations

$$\xi'_{mj}(t) = \int_\Omega u'_m(t)\, v_j = \int_\Omega \Big[\mathbf{w}_\varepsilon u_m(t) - \nabla\gamma_\varepsilon^{-1}(u_m(t))\Big]\cdot\nabla v_j - \int_\Omega f_\varepsilon(u_m(t))\, v_j - \int_{\partial\Omega} g_\varepsilon(u_m(t))\, v_j\ , \tag{4.21}$$

for $j = 1, ..., m$, with initial condition $\xi_{mj}(0) = \eta_j$, where η_j is the jth coefficient of $P_m\eta_{0\varepsilon} = \sum_{i=1}^m \eta_i v_i$, where $P_m \colon L^2(\Omega) \to V_m$ denotes the usual orthogonal projection.

Multiplying (4.21) by $\xi_{mj}(t)$ and summing on j, we obtain by recalling (4.12) and the assumptions (4.6)–(4.7)

$$\begin{aligned}\frac{1}{2}\frac{d}{dt}\int_\Omega u_m^2 + \frac{1}{b_\varepsilon^*}\int_\Omega |\nabla u_m|^2 &\le \frac{1}{2}\int_\Omega \mathbf{w}_\varepsilon\cdot\nabla(u_m)^2 + \int_\Omega |f^*|\,|u_m| + \int_{\partial\Omega} |g^*|\,|u_m| \\ &\le C_\varepsilon\Big(|f^*(t)|^2_{L^2(\Omega)} + |g^*(t)|^2_{L^2(\partial\Omega)}\Big) + \frac{1}{2b_\varepsilon^*}\int_\Omega\Big(|u_m|^2 + |\nabla u_m|^2\Big)\end{aligned}$$

and, applying Gronwall inequality, we find that

$$\sup_{0\le t\le T}\int_\Omega |u_m(t)|^2 + \int_Q |\nabla u_m|^2 \le C_{\varepsilon,T} \quad \text{independently of } m\,. \tag{4.22}$$

Setting $\vartheta_m(t) = P_m[\gamma_\varepsilon^{-1}(u_m)] = \sum_{i=1}^m \zeta_{mi}(t)\, v_i \in V_m$, we multiply (4.21) by $\zeta'_{mj}(t)$ and, summing on j, we obtain, for fixed $t>0$,

$$\int_\Omega u'_m\vartheta'_m + \frac{1}{2}\frac{d}{dt}\int_\Omega |\nabla\vartheta_m|^2 = \int_\Omega u_m\,\mathbf{w}_\varepsilon\cdot\nabla\vartheta'_m - \int_\Omega f_\varepsilon(u_m)\,\vartheta'_m - \int_{\partial\Omega} g_\varepsilon(u_m)\,\vartheta'_m\,. \tag{4.23}$$

Noting that $\vartheta'_m = P_m\{[\gamma_\varepsilon^{-1}]'(u_m)\,u'_m\}$, we have, taking into account (4.12) and (3.12),

$$\int_\Omega u'_m\,\vartheta'_m \ge \frac{1}{b_\varepsilon^*}\int_\Omega |u'_m|^2 \ge \frac{b_*^2}{b_\varepsilon^*}\int_\Omega |\vartheta'_m|^2\,. \tag{4.24}$$

Since $\mathbf{w}_\varepsilon \in \mathcal{S}_\sigma$, the first term of the right side of (4.23) yields, after integration in time and using (4.16)

$$\int_0^t\int_\Omega u_m\mathbf{w}_\varepsilon\cdot\nabla\vartheta'_m = -\int_0^t\int_\Omega \vartheta'_m\mathbf{w}_\varepsilon\cdot\nabla u_m \le \frac{b_*^2}{4b_\varepsilon^*}\int_0^t\int_\Omega |\vartheta'_m|^2 + C_\varepsilon\, C_{\varepsilon,T}\,. \tag{4.25}$$

Using the assumption (4.6), we also have by Cauchy's inequality

$$-\int_\Omega f_\varepsilon(u_m)\,\vartheta'_m \le \frac{b_*^2}{4b_\varepsilon^*}\int_\Omega |\vartheta'_m|^2 + \frac{b_\varepsilon^*}{b_*^2}\int_\Omega |f^*|^2\,. \tag{4.26}$$

Setting $\tilde g_\varepsilon(\vartheta_m) = g_\varepsilon(u_m)$ and $G_\varepsilon(x,t,\vartheta) = \int_0^\vartheta \tilde g_\varepsilon(x,t,\tau)\,d\tau$, the third term of the right side of (4.23) can be estimated as follows

$$\begin{aligned}
-\int_0^t\int_{\partial\Omega} g_\varepsilon(u_m)\,\vartheta'_m &= \int_0^t\int_{\partial\Omega} \partial_t G_\varepsilon(\vartheta_m) - \int_{\partial\Omega} G_\varepsilon(\vartheta_m)(t) + \int_{\partial\Omega} G_\varepsilon(\vartheta_m(0))\\
&\le C_\varepsilon^\# + \int_0^t\int_{\partial\Omega} |\vartheta_m|^2 + \int_{\partial\Omega} |\vartheta_m(t)|^2 + \int_{\partial\Omega} |\vartheta_m(0)|^2\\
&\le C_\varepsilon^\# + C_{\varepsilon,T}^\# + \frac{1}{2}\int_\Omega |\nabla\vartheta_m(t)|^2\,,
\end{aligned} \tag{4.27}$$

where the constant $C_\varepsilon^\#$ depends on the L^2-norms of $\partial_t\tilde g_\varepsilon$ and on g^* and $C_{\varepsilon,T}^\#$ depends on the constant $C_{\varepsilon,T}$ of (4.22) and on $\|\eta_{0\varepsilon}\|_{H^1(\Omega)}$, but both constants are independent of m. In the last inequality of (4.27) we have used the Lemma 4.6 below.

Now combining the estimates (4.25)–(4.26) and (4.27) with (4.24) in (4.23), we obtain

$$\frac{b_*^2}{b_\varepsilon^*}\int_0^t\int_\Omega |\vartheta'_m| + \int_\Omega |\nabla\vartheta_m(t)|^2 \le K_\varepsilon \quad (\text{independent of } m)\,. \tag{4.28}$$

In particular, both ϑ_m and $u_m = \gamma_\varepsilon(\vartheta_m)$ belong to a bounded subset of $H^1(Q)$ independently of m. Hence for some subsequence, still denoted by $m\to\infty$, and for some $\vartheta_\varepsilon, u_\varepsilon \in H^1(Q)$ we have

$$\vartheta_m \rightharpoonup \vartheta_\varepsilon \ \text{ and } \ u_m \rightharpoonup u_\varepsilon \quad \text{in } H^1(Q)\text{-weak}\,, \tag{4.29}$$

$$\vartheta_m \to \vartheta_\varepsilon \ \text{ and } \ u_m \to u_\varepsilon \quad \text{in } L^2(Q)\text{-strong and } L^2(\Sigma)\text{-strong}\,. \tag{4.30}$$

For $\varphi_n = \sum_{i=1}^n \psi_i(t)\, v_i(x)$, with $\psi_i \in C^1[0,T]$: $\psi_i(T) = 0$, $i = 1, ..., n, ...$, we obtain from (4.21), for $m > n$, that u_m satisfies

$$-\int_Q u_m(\partial_t\varphi_n + \mathbf{w}_\varepsilon\cdot\nabla\varphi_n) + \int_Q \nabla\gamma_\varepsilon^{-1}(u_m)\cdot\nabla\varphi_n + \int_Q f_\varepsilon(u_m)\,\varphi_n + \int_\Sigma g_\varepsilon(u_m)\,\varphi_n = \int_\Omega P_m\,\eta_{0\varepsilon}\,\varphi_m(0)$$

and due to (4.30) we can pass to the limit as $m \to \infty$, obtaining (4.17) first for $\varphi = \varphi_n$ and since the φ_n are dense in $H^1(Q)$, it follows that (4.17) also holds for all $\varphi \in H^1(Q)$: $\varphi(T) = 0$. Notice that (4.12) and the convergences (4.29) and (4.30) imply, in particular, that $\gamma_\varepsilon^{-1}(u_m) \rightharpoonup \gamma_\varepsilon^{-1}(u_\varepsilon) = \vartheta_\varepsilon$ in $H^1(Q)$-weak. ∎

Lemma 4.6. *For any $\delta > 0$ there exists $C_\delta > 0$, such that*

$$\|v\|_{L^2(\partial\Omega)} \leq \delta\,\|\nabla v\|_{L^2(\Omega)} + C_\delta\|v\|_{L^2(\Omega)}, \qquad \forall\, v \in H^1(\Omega)\,. \tag{4.31}$$

Proof: Since $\partial\Omega$ is a Lipschitz boundary of a bounded domain we know that the embedding $H^1(\Omega) \subset L^2(\Omega)$ and the trace mapping $H^1(\Omega) \to L^2(\partial\Omega)$ are compact. Let us suppose, by absurd, that (4.31) does not hold. Then, for any $\delta > 0$, there exists $v_n \in H^1(\Omega)$ and $c_n \to +\infty$ such that

$$\|v_n\|_{L^2(\partial\Omega)} \geq \delta\,\|\nabla v_n\|_{L^2(\Omega)} + c_n\,\|v_n\|_{L^2(\Omega)}\,.$$

Setting $w_n = v_n/\|\nabla v_n\|_{L^2(\Omega)}$, we obtain

$$\|w_n\|_{L^2(\partial\Omega)} \geq \delta + c_n\,\|w_n\|_{L^2(\Omega)} \geq \delta > 0 \tag{4.32}$$

and, for all n sufficiently large $c_n > 2C_0$, we find by the trace inequality

$$\|w_n\|_{L^2(\partial\Omega)} \leq C_0\big(\|w_n\|_{L^2(\Omega)} + 1\big) \leq \frac{1}{2}\,\|w_n\|_{L^2(\partial\Omega)} + C_0\,,$$

since $\|\nabla w_n\|_{L^2(\Omega)} = 1$.

Hence w_n is bounded in $H^1(\Omega)$, and we can find a subsequence such that $w_n \rightharpoonup w$ in $H^1(\Omega)$-weak. Then, by compactness, $w_n \to w$ in $L^2(\Omega)$-strong and their traces also converge strongly in $L^2(\partial\Omega)$. Consequently, from (4.32) we conclude that $\|w_n\|_{L^2(\Omega)} \to 0$ and $w \equiv 0$ and, in particular, $w = 0$ on $\partial\Omega$ which is a contradiction with the strict inequality of (4.32). ∎

Proposition 4.7. *Let $u_\varepsilon \in H^1(Q)$ be a solution to (4.17) and set $\vartheta_\varepsilon = \gamma_\varepsilon^{-1}(u_\varepsilon)$. Then we have,*

$$\mu \leq \vartheta_\varepsilon \leq M \quad \text{a.e. in } Q\,, \tag{4.33}$$

$$\|\vartheta_\varepsilon\|_{L^2(0,T;H^1(\Omega))} \leq C\,, \tag{4.34}$$

$$\|\partial_t\vartheta_\varepsilon\|_{L^2(A)} \leq C_A \quad \text{for any } A \subset\subset Q\,, \tag{4.35}$$

where $C, C_A > 0$ are constants independent of $\varepsilon > 0$.

Proof: Since $u_\varepsilon \in H^1(Q) \subset C^0([0,T]; L^2(\Omega))$, from (4.17) it is easy to conclude that $u_\varepsilon(0) = \eta_{0\varepsilon}$ and, for a.e. $t > 0$, $u_\varepsilon = \gamma_\varepsilon(\vartheta_\varepsilon)$ satisfies

$$\int_\Omega(\partial_t u_\varepsilon + \mathbf{w}_\varepsilon\cdot\nabla u_\varepsilon)\,\psi + \int_\Omega \nabla\vartheta_\varepsilon\cdot\nabla\psi + \int_\Omega \overline{f}_\varepsilon(\vartheta_\varepsilon)\,\psi + \int_{\partial\Omega} \overline{g}_\varepsilon(\vartheta_\varepsilon)\,\psi = 0\,, \quad \forall\,\psi \in H^1(\Omega)\,. \tag{4.36}$$

We have also $\vartheta_\varepsilon \in H^1(Q)$ and we can assume, without loss of generality, that $\vartheta_\varepsilon(0) = \gamma_\varepsilon^{-1}(\eta_{0\varepsilon})$ satisfies $\mu \leq \vartheta_\varepsilon(0) \leq M$ by the assumption (4.3). Taking $\psi = [\vartheta_\varepsilon(t) - M]^+$ in (4.36), the construction of $\overline{f}_\varepsilon$ and $\overline{g}_\varepsilon$ with the assumption (4.4) implies

$$\int_\Omega \left[\partial_t \widetilde{\beta}_M(\gamma_\varepsilon(\vartheta_\varepsilon)) + \mathbf{w}_\varepsilon \cdot \nabla \widetilde{\beta}_M(\gamma_\varepsilon(\vartheta_\varepsilon))\right] + \int_\Omega \left|\nabla(\vartheta_\varepsilon - M)^+\right|^2 \leq 0\,, \tag{4.37}$$

where $\widetilde{\beta}_M$ is defined as in Proposition 3.7. Hence, integrating (4.37) in time, and noting that

$$\int_\Omega \mathbf{w}_\varepsilon \cdot \nabla \widetilde{\beta}_M(\gamma_\varepsilon(\vartheta_\varepsilon)) = 0 \quad \text{since } \mathbf{w}_\varepsilon \in \mathcal{S}_\sigma\,,$$

we conclude that $(\vartheta_\varepsilon - M)^+ = 0$, i.e. $\vartheta_\varepsilon \leq M$. Analogously by taking $\psi = [\mu - \vartheta_\varepsilon(t)]^+$ in (4.36) we conclude $\vartheta_\varepsilon \geq \mu$.

Taking $\psi = \vartheta_\varepsilon(t)$ in (4.36) and integrating in time, setting $\widetilde{\gamma}_\varepsilon(\tau) = \int_0^\tau \gamma_\varepsilon(\sigma)\, d\sigma$, we deduce

$$\int_\Omega \widetilde{\gamma}_\varepsilon(\vartheta_\varepsilon(T)) + \int_Q |\nabla\vartheta_\varepsilon|^2 \leq \int_\Omega \widetilde{\gamma}_\varepsilon(\vartheta_\varepsilon(0)) + \int_Q |f^*|\,|\vartheta_\varepsilon| + \int_\Sigma |g^*|\,|\vartheta_\varepsilon|\,,$$

which, by (4.33), implies the estimate (4.34), because

$$\int_Q |\nabla\vartheta_\varepsilon|^2 \leq \widetilde{C} \quad (\text{independent of } \varepsilon)\,. \tag{4.38}$$

In a similar way, taking $\psi = u_\varepsilon(t) = \gamma_\varepsilon(\vartheta_\varepsilon(t))$ in (4.36) and setting $\widehat{\gamma}_\varepsilon(\tau) = \int_0^\tau \sqrt{\gamma_\varepsilon'(\sigma)}\, d\sigma$, we deduce

$$\frac{1}{2}\int_\Omega \gamma_\varepsilon^2(\vartheta_\varepsilon(T)) + \int_Q |\nabla\widehat{\gamma}_\varepsilon(\vartheta_\varepsilon)|^2 \leq \frac{1}{2}\int_\Omega \eta_{\varepsilon 0}^2 + \int_Q |f^*|\,|u_\varepsilon| + \int_\Sigma |g^*|\,|u_\varepsilon|\,,$$

since $\int_\Omega \mathbf{w}_\varepsilon \cdot \nabla(u_\varepsilon)^2 = 0$ because $\mathbf{w}_\varepsilon \in \mathcal{S}_\sigma$ and $u_\varepsilon^2 \in H^1(Q) \cap L^\infty(Q)$. Hence

$$\int_Q |\nabla\widehat{\gamma}_\varepsilon(\vartheta_\varepsilon)|^2 \leq \widehat{C} \quad (\text{independent of } \varepsilon)\,. \tag{4.39}$$

In particular, (4.36) implies that u_ε satisfies (4.18) a.e. in Q, or equivalently

$$\partial_t\gamma_\varepsilon(\vartheta_\varepsilon) + \mathbf{w}_\varepsilon \cdot \nabla\gamma_\varepsilon(\vartheta_\varepsilon) - \Delta\vartheta_\varepsilon + \overline{f}_\varepsilon(\vartheta_\varepsilon) = 0 \quad \text{a.e. in } Q\,.$$

For any open set $A \subset Q$, such that $\overline{A} \subset Q$, we consider a cut-off function $\varphi \in \mathcal{D}(Q)$: $0 \leq \chi_A \leq \varphi \leq 1$. Multiplying (4.40), by $\varphi^2\,\partial_t\vartheta_\varepsilon$ and integrating by parts in Q, since φ has compact support in Q, recalling (3.12), we have

$$\frac{\gamma_*}{2}\int_Q \varphi^2\,|\partial_t\vartheta_\varepsilon|^2 + \frac{1}{2}\int_Q \varphi^2\,|\partial_t\widehat{\gamma}_\varepsilon(\vartheta_\varepsilon)|^2 \leq \int_Q \varphi^2\,\partial_t\gamma_\varepsilon(\vartheta_\varepsilon)\,\partial_t\vartheta_\varepsilon = \sum_{j=1}^3 I_j\,, \tag{4.41}$$

where each I_j can be estimated as follow:

$$\begin{aligned}
I_1 &\equiv -\int_Q \varphi^2\mathbf{w}_\varepsilon \cdot \sqrt{\gamma_\varepsilon'}\,\nabla\vartheta_\varepsilon\,\sqrt{\gamma_\varepsilon'}\,\partial_t\vartheta_\varepsilon = -\int_Q \varphi^2\,\mathbf{w}_\varepsilon \cdot \nabla\widehat{\gamma}_\varepsilon(\vartheta_\varepsilon)\,\partial_t\widehat{\gamma}(\vartheta_\varepsilon)\\
&\leq \frac{1}{4}\int_Q \varphi^2|\partial_t\,\widehat{\gamma}_\varepsilon(\vartheta_\varepsilon)|^2 + \|\mathbf{w}_\varepsilon\|^2_{L^\infty(Q)}\int_Q |\nabla\widehat{\gamma}_\varepsilon(\vartheta_\varepsilon)|^2\\
&\leq \frac{1}{4}\int_Q \varphi^2|\partial_t\widehat{\gamma}_\varepsilon(\vartheta_\varepsilon)|^2 + \widehat{C}\|\mathbf{v}\|^2_{L^\infty(Q)}\,,
\end{aligned}$$

where we have used (4.39)and (4.16);

$$I_2 \equiv \int_Q \varphi^2 \Delta\vartheta_\varepsilon \, \partial_t\vartheta_\varepsilon = \int_Q |\nabla\vartheta_\varepsilon|^2 \varphi \, \partial_t\varphi - 2\int_Q \nabla\vartheta_\varepsilon \cdot \nabla\varphi(\varphi\, \partial_t\vartheta_\varepsilon)$$
$$\le C_\varphi \int_Q |\nabla\vartheta_\varepsilon|^2 + \frac{\gamma_*}{8}\int_Q \varphi^2|\partial_t\vartheta_\varepsilon|^2 \le C_\varphi \tilde{C} + \frac{\gamma_*}{8}\int_Q \varphi^2|\partial_t\vartheta_\varepsilon|^2 \,,$$

where $\tilde{C}$ satisfies (4.38) and C_φ depends only on φ and γ_*;

$$I_3 \equiv -\int_Q \overline{f}_\varepsilon(\vartheta_\varepsilon)\varphi^2 \, \partial_t\vartheta_\varepsilon \le \frac{\gamma_*}{8}\int_Q \varphi^2|\partial_t\vartheta_\varepsilon|^2 + \frac{2}{\gamma_*}\int_Q |f^*|^2 \,.$$

Consequently, from (4.41) we obtain (4.35), since we have

$$\int_A |\partial_t\vartheta_\varepsilon|^2 \le \int_Q \varphi^2|\partial_t\vartheta_\varepsilon|^2 \le \frac{4}{\gamma_*}\Big\{\hat{C}\|\mathbf{v}\|^2_{L^\infty(Q)} + C_\varphi\tilde{C} + \frac{2}{\gamma_*}\|f^*\|^2_{L^2(Q)}\Big\} \equiv C_A^2 \,. \blacksquare$$

Proof of Theorem 4.1: From the estimates (4.33)–(4.35) we can extract a subsequence, still denoted by $\varepsilon \to 0$, such that

(4.42) $\qquad \vartheta_\varepsilon \to \vartheta$ in $L^\infty(Q)$-weak* and $L^p(Q)$-strong, $\forall p < \infty$,

(4.43) $\qquad \nabla\vartheta_\varepsilon \rightharpoonup \nabla\vartheta$ in $L^2(Q)$-weak and $\gamma_\varepsilon(\vartheta_\varepsilon) \rightharpoonup \eta$ in $L^\infty(Q)$-weak* ,

where $\vartheta \in L^2(0,T;H^1(\Omega)) \cap H^1_{\text{loc}}(Q)$ satisfies (4.10) and $\eta \in L^\infty(Q)$. The L^2-strong convergence of ϑ_ε follows first locally in $A \subset\subset Q$, by compactness of $H^1(A) \subset L^2(A)$, then a.e. in Q and afterwards in all $L^p(Q)$, $\forall p < \infty$, by (4.33) and the Lebesgue theorem. To prove that $\eta \in \gamma(\vartheta)$, or equivalently that $\vartheta = \gamma^{-1}(\eta)$ (note that $\gamma_\varepsilon^{-1} \xrightarrow{\varepsilon\to 0} \gamma^{-1}$ uniformly in the compact subsets of $\mathbf{R}$), we take the limit in

(4.44)
$$\int_Q [\gamma_\varepsilon(\vartheta_\varepsilon) - \zeta]\,(\vartheta_\varepsilon - \gamma_\varepsilon^{-1}(\zeta)) \ge 0\,, \quad \forall \zeta \in L^\infty(Q)\,.$$

We obtain, due to the convergences (4.42) and (4.43),

$$\int_Q (\eta - \zeta)\,(\vartheta - \gamma^{-1}(\zeta)) \ge 0\,, \quad \forall \zeta \in L^\infty(Q)\,,$$

and substituting $\zeta = \eta + \lambda\xi$, for $\lambda \in \mathbf{R}$ and $\zeta \in L^\infty(Q)$, we find

$$\int_Q \xi(\vartheta - \gamma^{-1}(\eta)) = 0\,, \quad \forall \xi \in L^\infty(Q)\,,$$

which yields the conclusion $\vartheta = \gamma^{-1}(\eta)$ or $\eta \in \gamma(\vartheta)$.

Using (4.42)–(4.43) and the Lemma 4.6, we conclude that $\vartheta_\varepsilon \to \vartheta$ strongly in $L^p(\Sigma)$, first for $p = 2$ and then for all $p < \infty$, by (4.26). Consequently, by (4.15) it follows

$$\overline{g}_\varepsilon(\vartheta_\varepsilon) \to \overline{g}(\vartheta) = g(\vartheta) \quad \text{in } L^2(\Sigma)$$

and similarly also $\overline{f}_\varepsilon(\vartheta_\varepsilon) \to \overline{f}(\vartheta) = f(\vartheta)$ in $L^2(Q)$, as $\varepsilon \to 0$.

Taking the limit in (4.17), and recalling (4.16) we finally conclude that (ϑ, η) satisfies (4.2) ∎

In the preceding proof we have used the strong L^p-convergence of $\vartheta_\varepsilon \to \vartheta$ and of their traces. This was a consequence of the local estimate (4.35), which is the only step where the assumption $\mathbf{v} \in [L^\infty(Q)]^N$ has been used. In fact this restriction can be droped by using monotonicity methods (see [L2] and [V1]) and assuming in addition that a.e. in (x,t), the functions

$$(4.45) \qquad f = f(x,t,\vartheta) \text{ and } g = g(x,t,\vartheta) \quad \text{are monotone non-decreasing in } \vartheta .$$

Theorem 4.8. *Under the assumptions (4.3)–(4.7) and (4.45), for any* $\mathbf{v} \in \mathbf{L}^2_\sigma(Q)$ *there exists at least one solution to (4.1)–(4.2), which also satisfies (4.10).*

Proof: The estimates (4.33) and (4.34) being still valid, we can extract a subsequence $\varepsilon \to 0$, such that

$$(4.46) \qquad \vartheta_\varepsilon \rightharpoonup \vartheta \text{ in } L^\infty\text{-weak} \quad \text{and in} \quad L^2(0,T;H^1(\Omega))\text{-weak} ,$$

$$(4.47) \qquad \gamma_\varepsilon(\vartheta_\varepsilon) \rightharpoonup \eta \quad \text{in } L^\infty(Q)\text{-weak*} ,$$

$$(4.48) \qquad \overline{f}_\varepsilon(\vartheta_\varepsilon) \rightharpoonup f \quad \text{in } L^2(Q)\text{-weak} ,$$

$$(4.49) \qquad \overline{g}_\varepsilon(\vartheta_\varepsilon) \rightharpoonup g \quad \text{in } L^2(\Sigma)\text{-weak} ,$$

for some functions $\vartheta \in L^2(0,T;H^1(\Omega))$ satisfying (4.10), $\eta \in L^\infty(Q)$, $f \in L^2(Q)$ and $g \in L^2(\Sigma)$. Passing to the limit in (4.17) we obtain, for all $\varphi \in H^1(Q)$: $\varphi(T) = 0$,

$$(4.50) \qquad -\int_Q \eta(\partial_t\varphi + \mathbf{v}\cdot\nabla\varphi) + \int_Q \nabla\vartheta\cdot\nabla\varphi + \int_Q f\,\varphi + \int_\Sigma g\,\varphi = \int_\Omega \eta_0\,\varphi(0) ,$$

and it remains to identify $f = f(\vartheta)$, $g = g(\vartheta)$ and to show $\eta \in \gamma(\vartheta)$.

This last property also follows as before, by passing to the limit in (4.44), provided we can guarantee that

$$(4.51) \qquad \int_Q \gamma_\varepsilon(\vartheta_\varepsilon)\,\vartheta_\varepsilon = \langle u_\varepsilon, \vartheta_\varepsilon\rangle_Q \to \langle \eta, \vartheta\rangle_Q = \int_Q \eta\,\vartheta \quad \text{as } \varepsilon \to 0 .$$

Since now we only have $\vartheta_\varepsilon \rightharpoonup \vartheta$ in $L^2(0,T;H^1(\Omega))$-weak, it is sufficient to show that $u_\varepsilon = \gamma_\varepsilon(\vartheta_\varepsilon) \to \eta$ in $L^2(0,T;[H^1(\Omega)]')$-strong. We note that from (4.36), recalling $\mathbf{w}_\varepsilon \in S_\sigma$ and the estimates (4.33) and (4.34) we conclude that $\partial_t u_\varepsilon \in L^2(0,T;[H^1(\Omega)]')$ uniformly with respect to $\varepsilon > 0$; hence, since $u_\varepsilon \in L^\infty(Q)$, also uniformly in ε, and $L^\infty(\Omega) \hookrightarrow [H^1(\Omega)]'$ is a compact imbedding (by compactness of $H^1(\Omega) \hookrightarrow L^1(\Omega)$), by Aubin type results (see [Si], for instance) we have the compactness of $H^1(0,T;[H^1(\Omega)]') \cap L^\infty(Q) \hookrightarrow C^0([0,T];[H^1(\Omega)]')$ and (4.51) follows. In particular, this argument shows that from (4.50) we can write $\partial_t\eta \in L^2(0,T;[H^1(\Omega)]')$ in the form

$$(4.52) \quad \langle\partial_t\eta,\psi\rangle_Q = \int_Q \eta\mathbf{v}\cdot\nabla\psi - \int_Q \nabla\vartheta\cdot\nabla\psi - \int_Q f\,\psi - \int_\Sigma g\,\psi , \quad \forall\psi \in L^2(0,T;H^1(\Omega)) .$$

In order to use the monotonicity argument, we set

$$X_\varepsilon = \int_Q \overline{f}_\varepsilon(\vartheta_\varepsilon)\,\vartheta_\varepsilon + \int_\Sigma \overline{g}_\varepsilon(\vartheta_\varepsilon)\,\vartheta_\varepsilon$$

and will show that

$$\limsup_{\varepsilon\to 0} X_\varepsilon \le \int_Q f\,\vartheta + \int_\Sigma g\,\vartheta\,. \tag{4.53}$$

In fact, if (4.53) holds, we can take "lim sup" in

$$\int_Q \left[\overline{f}_\varepsilon(\vartheta_\varepsilon) - \overline{f}_\varepsilon(\zeta)\right](\vartheta_\varepsilon - \zeta) + \int_\Sigma \left[\overline{g}_\varepsilon(\vartheta_\varepsilon) - \overline{g}_\varepsilon(\zeta)\right](\vartheta_\varepsilon - \zeta) \ge 0$$

for any fixed $\zeta \in L^2(0,T;H^1(\Omega))$, in order to obtain

$$\int_Q [f - \overline{f}(\zeta)](\vartheta - \zeta) + \int_\Sigma [g - \overline{g}(\zeta)](\vartheta - \zeta) \ge 0\,,$$

whence we get first $\overline{f}(\vartheta) = f(\vartheta) = f$ and afterwards also $\overline{g}(\vartheta) = g(\vartheta) = g$.

Noting

$$\widetilde{\beta}_\varepsilon(z) = \int_0^z \gamma_\varepsilon^{-1}(\tau)\,d\tau \to \int_0^z \gamma^{-1}(\tau)\,d\tau = \widetilde{\beta}(z)\,,$$

uniformly in the compact subsets of $\mathbf{R}$, as $\varepsilon \to 0$, we remark that

$$\int_Q u_\varepsilon \mathbf{w}_\varepsilon \cdot \nabla\vartheta_\varepsilon = \int_Q \mathbf{w}_\varepsilon \cdot \nabla\widetilde{\beta}_\varepsilon(u_\varepsilon) = 0 \tag{4.54}$$

and

$$\int_Q \eta\mathbf{v} \cdot \nabla\vartheta = \int_Q \mathbf{v} \cdot \nabla\widetilde{\beta}(\eta) = 0\,, \tag{4.55}$$

since $\mathbf{w}_\varepsilon$ and $\mathbf{v}$ are in $\mathbf{L}^2_\sigma(Q)$, and $\vartheta = \gamma^{-1}(\eta) \in L^2(0,T;H^1(\Omega))$ and $\eta \in L^\infty(Q)$ imply $\widetilde{\beta}(\eta) \in L^2(0,T;H^1(\Omega))$.

Multiplying (4.18) by ϑ_ε we obtain, using (4.54)

$$X_\varepsilon = -\int_Q \vartheta_\varepsilon\,\partial_t u_\varepsilon - \int_Q |\nabla\vartheta_\varepsilon|^2 = \int_\Omega \widetilde{\beta}_\varepsilon(\eta_{0\varepsilon}) - \int_\Omega \widetilde{\beta}_\varepsilon(u_\varepsilon(T)) - \int_Q |\nabla\vartheta_\varepsilon|^2\,.$$

Since, we know that $\eta_{0\varepsilon} \to \eta_0$ in $L^2(\Omega)$ and $u_\varepsilon(T) \rightharpoonup \eta(T)$ weakly-* in $L^\infty(\Omega)$ and strongly in $[H^1(\Omega)]'$, by the lower semi-continuity of convex functionals, we obtain (4.53) from

$$\begin{aligned}\liminf_{\varepsilon\to 0}(-X_\varepsilon) &\ge \int_\Omega \widetilde{\beta}(\eta(T)) - \int_\Omega \widetilde{\beta}(\eta_0) + \int_Q |\nabla\vartheta|^2 \\ &= \langle \partial_t\eta, \vartheta\rangle_Q + \int_Q |\nabla\vartheta|^2 = -\int_Q f\,\vartheta - \int_\Sigma g\,\vartheta\,,\end{aligned}$$

where we have used (4.52) with $\psi = \vartheta$ and taken (4.55) into account.

The proof of Theorem 4.8 is now complete. ∎

Remark 4.9. In the proof of Theorem 4.8 we have combined the techniques of monotonicity with compactness already used by Visintin [V1] in the case without convection (i.e. $\mathbf{v} \equiv 0$). In the proof of Theorem 4.7, with the additional local estimate (4.35) the compactness argument does not require the assumption (4.45). This improves earlier results, in particular in the special case of the continuous casting problem by [RY]. In the case without convection of [V1] (see also [D1,2]) under additional assumptions on the partial time derivates $\partial_t f$ and $\partial_t g$ a global estimate on $\partial_t\vartheta$ in $L^2(Q)$ can also be obtained. □

Remark 4.10. The general case considered in [CD] treats only a convective nonlinear term depending on the temperature, instead of the enthalpy. On the other hand, the cases considered in [Ru] and recently in [YQ] deal with a prescribed velocity field, respectively, $\mathbf{v} = \mathbf{v}(x) \in [C^1(\overline{\Omega})]^N$ and $\mathbf{v} \in L^\infty(0,T;[W^{1,\infty}(\Omega)]^N)$, without the incompressibility condition ($\nabla \cdot \mathbf{v} = 0$), in the non-physical form $\partial_t \eta + \nabla \cdot [\mathbf{v}\eta - \nabla\vartheta] = f$. □

4.2 – Continuous dependence of the solution

We extend now to the case of the multidimensional Stefan problem with prescribed convection a continuous dependence result which is based on the concept of weak solution introduced by Kamin [Ka]: if in (4.2) the test function $\varphi \in H^1(Q)$, $\varphi(T) = 0$ has the additional regularity $\Delta\vartheta \in L^2(Q)$ and $\partial_n \varphi \in L^2(\Sigma)$, we also have

$$(4.56)\quad -\int_Q \eta(\partial_t\varphi + \mathbf{v}\cdot\nabla\varphi) - \int_Q \vartheta\,\Delta\varphi + \int_Q f(\vartheta)\,\varphi + \int_{\Sigma_N}\left[\vartheta\,\partial_n\varphi + g(\vartheta)\,\varphi\right] = \int_\Omega \eta_0\,\vartheta(0)\,.$$

In this section we shall assume the nonlinearities f and g locally Lipschitz continuous, i.e., we assume there exist positive constants L_f and L_g, such that, for all $\vartheta, \widehat{\vartheta} \in [\mu, M]$,

$$(4.57)\qquad \left|f(x,t,\vartheta) - f(x,t,\widehat{\vartheta})\right| \le L_f\,|\vartheta - \widehat{\vartheta}|,\qquad \text{a.e. } (x,t)\in Q\,,$$

$$(4.58)\qquad \left|g(x,t,\vartheta) - g(x,t,\widehat{\vartheta})\right| \le L(g)|\vartheta - \widehat{\vartheta}|,\qquad \text{a.e. } (x,t)\in \Sigma\,.$$

Let us denote by $(\widehat{\vartheta},\widehat{\eta})$ a weak solution to (4.1) (4.2) corresponding to a perturbation of the nonlinear data $\widehat{f} = \widehat{f}(x,t,\vartheta)$ and $\widehat{g} = \widehat{g}(x,t,\vartheta)$ and the initial condition $\widehat{\eta}_0 = \widehat{\eta}_0(x)$, under the existence conditions of Theorems 4.1 or 4.8, and we set

$$(4.59)\qquad |f-\widehat{f}|(x,t) = \sup_{\mu\le\vartheta\le M}\left|f(x,t,\vartheta) - \widehat{f}(x,t,\vartheta)\right|,\qquad \text{a.e. } (x,t)\in Q\,,$$

$$(4.60)\qquad |g-\widehat{g}|(x,t) = \sup_{\mu\le\vartheta\le M}|g(x,t,\vartheta) - \widehat{g}(x,t,\vartheta)|,\qquad \text{a.e. } (x,t)\in \Sigma\,.$$

Theorem 4.11. *If (ϑ,η) and $(\widehat{\vartheta},\widehat{\eta})$ denote bounded solutions to the problem (4.1)–(4.2), i.e., satisfying (4.10), corresponding to different sets of data, respectively, (f,g,η_0) and $(\widehat{f},\widehat{g},\widehat{\eta}_0)$, for any p, $1\le p<\infty$, there exists a constant $K_p>0$, independent of the data, such that*

$$(4.61)\qquad \int_Q |\eta-\widehat{\eta}|^p \le K_p\Big\{\int_Q |f-\widehat{f}| + \int_\Sigma |g-\widehat{g}| + \int_\Omega |\eta_0-\widehat{\eta}_0|\Big\}\,,$$

where f, g satisfy (4.57)–(4.58) and $\widehat{f}$, $\widehat{g}$ (4.59)–(4.60).

Proof: Writing for $(\widehat{\vartheta},\widehat{\eta})$ the corresponding variational identity (4.56), by subtraction and rearrangement, we have

$$(4.62)\qquad \int_Q (\eta-\widehat{\eta})\,(\partial_t\varphi + \mathbf{v}\cdot\nabla\varphi + \alpha\,\Delta\varphi - \beta\,\varphi) - \int_\Sigma (\vartheta-\widehat{\vartheta})\,(\partial_n\varphi + \delta\,\varphi) =$$
$$= \int_Q \left[\widehat{f}(\widehat{\vartheta}) - f(\widehat{\vartheta})\right]\varphi + \int_\Sigma \left[\widehat{g}(\widehat{\vartheta}) - g(\widehat{\vartheta})\right]\varphi + \int_\Omega (\widehat{\eta}_0 - \eta_0)\,\varphi(0)\,,$$

for all $\varphi \in H^1(Q)$: $\varphi(T) = 0$, $\Delta\varphi \in L^2(Q)$ and $\partial_n\varphi \in L^2(\Sigma)$, where

$$\alpha = \alpha(x,t) = \begin{cases} \dfrac{\vartheta - \widehat{\vartheta}}{\eta - \widehat{\eta}} & \text{if } \eta \neq \widehat{\eta}, \\ 0 & \text{if } \eta = \widehat{\eta}\,, \end{cases} \tag{4.63}$$

$$\beta = \beta(x,t) = \begin{cases} \dfrac{f(\vartheta) - f(\widehat{\vartheta})}{\eta - \widehat{\eta}} & \text{if } \eta \neq \widehat{\eta}, \\ 0 & \text{if } \eta = \widehat{\eta}\,, \end{cases} \tag{4.64}$$

$$\delta = \delta(x,t) = \begin{cases} \dfrac{g(\vartheta) - g(\widehat{\vartheta})}{\vartheta - \widehat{\vartheta}} & \text{if } \vartheta \neq \widehat{\vartheta}, \\ 0 & \text{if } \vartheta = \widehat{\vartheta}\,. \end{cases} \tag{4.65}$$

By the assumption (3.12) we have, in particular,

$$b_*|\vartheta - \widehat{\vartheta}| \leq |\eta - \widehat{\eta}| \quad \text{a.e. in } Q \tag{4.66}$$

and therefore, also $0 \leq \alpha \leq 1/b_*$. By (4.57) we have also $|\beta| \leq L_f/b_*$ and, by (4.58), $|\delta| \leq L_g$.

We introduce the following sequence of regularizations $\alpha_m, \beta_m \in C^\infty(\overline{Q})$, $\delta_m \in C^\infty(\overline{\Sigma})$ and $\mathbf{v}_m \in \mathcal{S}_\sigma$, such that for $m \in \mathbb{N}$ and for some constant $C > 0$ independent of m:

$$0 < \frac{1}{m} \leq \alpha_m \leq 1/b_* \quad \text{and} \quad \|\alpha - \alpha_m\|_{L^2(Q)} + \left\| \frac{\alpha - \alpha_m}{\sqrt{\alpha_m}} \right\|^2_{L^2(Q)} \leq \frac{C}{m}\,, \tag{4.67}$$

$$\|\beta - \beta_m\|_{L^2(Q)} + \|\delta - \delta_m\|_{L^1(\Sigma)} + \|\mathbf{v} - \mathbf{v}_m\|_{L^2(Q)} \leq \frac{C}{\sqrt{m}}\,, \tag{4.68}$$

$$\|\nabla\beta_m\|^2_{L^2(Q)} + \|\partial_t\delta_m\|_{L^1(\Sigma)} \leq C\, m^{1/4} \quad \text{and} \quad \|\nabla\mathbf{v}_m\|_{L^\infty(Q)} \leq C \log(m^{1/8})\,. \tag{4.69}$$

For any $\xi \in \mathcal{D}(\Omega)$ we consider the unique $\vartheta_m \in H^1(Q) \cap C^0(\overline{Q})$ solution of the linear parabolic problem

$$\begin{cases} \partial_t\vartheta_m + \mathbf{v}_m \cdot \nabla\varphi_m + \alpha_m\,\Delta\vartheta_m - \beta_m\varphi_m = \xi & \text{in } Q\,, \\ \partial_n\varphi_m + \delta_m\varphi_m = 0 \ \text{ on } \Sigma_{\mathrm{N}} \quad \text{and} \quad \varphi_m(T) = 0 & \text{in } \Omega\,. \end{cases} \tag{4.70}$$

We shall show in Lemma 4.12 below, that

$$\|\varphi_m\|_{L^\infty(Q)} \leq C_\xi\,, \tag{4.71}$$

$$\|\nabla\varphi_m\|_{L^\infty(0,T;L^2(\Omega))} \leq C_1 + C_2\, m^{1/4}\,, \tag{4.72}$$

$$\|\sqrt{\alpha_m}\,\Delta\varphi_m\|_{L^2(Q)} \leq C_1 + C_2\, m^{1/4}\,, \tag{4.73}$$

where the constants $C_\xi, C_1, C_2 > 0$ may depend on $\|\xi\|_{L^\infty(Q)}$, but are independent of $m \in \mathbb{N}$.

Taking $\varphi = \varphi_m$ in (4.62) we obtain

$$\int_Q (\eta - \widehat{\eta})\,\xi \leq \sum_{i=1}^{3} I_i^m + C_\xi\,\widehat{I}\,, \tag{4.74}$$

where C_ξ is given by (4.71), each $I_i^m \to 0$ as $m \to \infty$ according to the estimates below, and

$$\hat{I} = \int_Q |f - \hat{f}| + \int_\Sigma |g - \hat{g}| + \int_\Omega |\eta_0 - \hat{\eta}_0| \,.$$

In fact, we have

$$I_1^m \equiv \left| \int_Q (\eta - \hat{\eta})\,(\alpha_m - \alpha)\,\Delta\vartheta_m \right| \le C \int_Q |\sqrt{\alpha_m}\,\Delta\varphi_m| \left|\frac{\alpha - \alpha_m}{\sqrt{\alpha_m}}\right|$$
$$\le C \|\sqrt{\alpha_m}\,\Delta\varphi_m\|_{L^2(Q)} \left\|\frac{\alpha - \alpha_m}{\sqrt{\alpha_m}}\right\|_{L^2(Q)} \le C'\, m^{-1/4}$$

by (4.67) and (4.73);

$$I_2^m \equiv \left| \int_Q (\eta - \hat{\eta})\,(\mathbf{v}_m - \mathbf{v}) \cdot \nabla\varphi_m \right| \le C \|\mathbf{v}_m - \mathbf{v}\|_{L^2(Q)}\, \|\nabla\varphi_m\|_{L^2(Q)} \le C'\, m^{-1/4}$$

by (4.68) and (4.72), and using (4.68) and (4.71)

$$I_3^m \equiv \left| \int_Q (\eta - \hat{\eta})\,(\beta_m - \beta)\,\varphi_m + \int_\Sigma (\vartheta - \hat{\vartheta})\,(\delta_m - \delta)\,\varphi_m \right| \le C\, C_\xi\, m^{-1/2} \,.$$

Consequently letting $m \to \infty$ in (4.74) we obtain

$$\int_Q (\eta - \hat{\eta})\,\xi \le C_\xi\, \hat{I} \,, \tag{4.75}$$

first for any $\xi \in \mathcal{D}(Q)$ and, by approximation, also for any $\xi \in L^\infty(Q)$, since the constant C_ξ in (4.71) only depends on the L^∞-norm of ξ. Finally, taking in (4.75) $\xi = |\eta - \hat{\eta}|^{p-1} \operatorname{sign}(\eta - \hat{\eta}) \in L^\infty(Q)$ we conclude (4.61). ∎

Lemma 4.12. *Under the conditions (4.67)–(4.69) there exists a unique solution φ_m to (4.70), which satisfies the estimates (4.71), (4.72) and (4.73).*

Proof: Assume first the monotonicity condition (4.45). Then we can choose $\beta_m \ge 0$ and $\delta_m \ge 0$ and we may take $\Lambda(t) = A(T - t)$ with $A = \|\xi\|_{L^\infty(Q)}$ as a supersolution to the problem (4.70). Hence by the maximum principle we obtain $-\Lambda \le \varphi_m \le \Lambda$ and (4.71) holds with $C_\xi = T\|\xi\|_{L^\infty(Q)}$.

Notice that making the change of variables $\tau = T - t$ in the problem (4.70) we reduce it to a standard form and the existence, uniqueness and maximum principles apply to φ_m (see [LSU], for instance). Multiplying (4.70) by $\Delta\varphi_m$ and integrating by parts in Ω, we have for $\varphi(\tau) = \varphi_m(T - \tau)$ (dropping the subscript m for simplicity)

$$\frac{1}{2}\frac{d}{d\tau}\int_\Omega |\nabla\varphi|^2 + \int_\Omega \alpha_m |\Delta\varphi|^2 = -\int_{\partial\Omega} \delta\,\varphi\,\partial_\tau\varphi + \int_\Omega (\xi - \mathbf{v}_m \cdot \nabla\varphi + \beta_m\,\varphi)\,\Delta\varphi \tag{4.76}$$

and, integrating in time, each term on the left-hand side may be estimated as follows:

$$\left|\int_0^t \int_{\partial\Omega} \delta\,\varphi\,\partial_\tau\varphi\right| \le \frac{1}{2}\int_0^t \int_{\partial\Omega} |\partial_\tau \delta\,\varphi^2| + \int_{\partial\Omega} |\delta(t)\,\varphi^2| \le C\, m^{1/4} \,,$$

$$\left|\int_\Omega \xi\,\Delta\varphi\right| = \left|\int_\Omega \varphi\,\Delta\xi\right| \le C \,,$$

$$\left|\int_\Omega (\mathbf{v}_m \cdot \nabla\varphi)\,\Delta\varphi\right| = \left|\int_\Omega \nabla(\mathbf{v}_m \cdot \nabla\varphi)\cdot\nabla\varphi\right| = \left|\int_\Omega (\nabla\mathbf{v}_m \cdot \nabla\varphi)\cdot\nabla\varphi\right| + \frac{1}{2}\left|\int_\Omega \mathbf{v}_m \cdot \nabla|\nabla\varphi|^2\right|$$
$$\leq C\|\nabla\mathbf{v}_m\|_{L^\infty(Q)} \int_\Omega |\nabla\varphi|^2 \leq C\log(m^{1/8}) \int_\Omega |\nabla\varphi|^2 ,$$

$$\left|\int_\Omega \beta_m\,\varphi\,\Delta\varphi\right| = \left|\int_\Omega (\varphi\,\nabla\beta_m\cdot\nabla\varphi + \beta_m|\nabla\varphi|^2) + \int_{\partial\Omega} \beta_m\,\delta_m\,\varphi^2\right|$$
$$\leq C + \int_\Omega |\nabla\beta_m|^2 + C\int_\Omega |\nabla\varphi|^2 \leq C\Big(1 + m^{1/4} + \int_\Omega |\nabla\varphi|^2\Big) .$$

Hence, from (4.76) we have, with C denoting different constants independent of m,

$$\int_\Omega |\nabla\varphi(t)|^2 + \int_0^t\int_\Omega \alpha_m|\Delta\varphi|^2 \leq C(1 + m^{1/4}) + C\Big(1 + \log(m^{1/8}) \int_0^t\int_\Omega |\nabla\varphi|^2\Big) ,$$

and, by Gronwall inequality, we find first

$$\int_\Omega |\nabla\varphi(t)|^2 \leq C(1 + m^{1/4})\, m^{1/8} \leq C\, m^{3/8}$$

and afterwards also (4.73).

In the general case, we replace φ_m by $\widetilde{\varphi}_m(x,t) = e^{-\lambda t}\,\nu_m(x)\,\varphi_m(x,t)$ where ν_m is the mollification of the solution $\nu \in H^1(\Omega)$ of the Neumann problem

$$-\Delta\nu + \nu = 1 \;\text{ in } \Omega \qquad \text{and} \qquad \partial_n\nu + L_g\nu = 0 \;\text{ on } \partial\Omega ,$$

which is such that $\nu \geq 1$ in Ω, and therefore we can suppose that $\nu_m \in C^\infty(\overline{\Omega})$ satisfies also $\nu_m \geq 1$ and $-\frac{1}{\nu_m}\partial_n\nu_m = L_g \geq |\delta_m|$ on $\partial\Omega$. Hence, for $\lambda > 0$ sufficiently large, $\widetilde{\varphi}_m$ satisfies an equation similar to (4.70) with modified coefficients $\widetilde{\alpha}_m$, $\widetilde{\beta}_m$, $\widetilde{\mathbf{v}}_m$, $\widetilde{\delta}_m$ and $\widetilde{\xi}$ satisfying similar properties, in particular, with $\widetilde{\beta}_m \geq 0$ and $\widetilde{\delta}_m \geq 0$. ∎

As an immediate consequence of (4.66), from (4.61) we have the following uniqueness result.

Corollary 4.13. *Under the assumptions of Theorem 4.11, we have for* $1 \leq p < \infty$

$$\int_Q |\vartheta - \widehat{\vartheta}|^p \leq \frac{K_p}{b_*^p}\Big\{\int_Q |f - \widehat{f}| + \int_\Sigma |g - \widehat{g}| + \int_\Omega |\eta_0 - \widehat{\eta}_0|\Big\} ,$$

and, in particular, there exist at most one bounded solution to (4.1)–(4.2). ∎

We can obtain the Dirichlet problem, or the mixed Dirichlet-Neumann as a consequence of a weaker continuous dependence result. Consider from the Section 1.2 the case where the lateral boundary $\partial\Omega = \Gamma_D \cup \overline{\Gamma}_N$ has a non-empty component Γ_D where a Dirichlet boundary condition is prescribed (in particular, we can take $\Gamma_D = \partial\Omega$ and $\Gamma_N = \emptyset$)

(4.77) $$\vartheta = \vartheta_D \quad \text{on } \; \Sigma_D = \Gamma_D \times]0,T[.$$

We shall assume there exist an extension $\widetilde{\vartheta}_D$ such that

(4.78) $\widetilde{\vartheta}_D \in L^2(0,T;H^1(\Omega))$: $\;\widetilde{\vartheta}_D = \vartheta_D$ on Σ_D, $\;\partial_t\widetilde{\vartheta}_D \in L^1(Q)$ and $\mu \leq \widetilde{\vartheta}_D \leq M$ in Q .

The corresponding weak formulation is still given by (4.1) (4.2) with the restrictions $\vartheta - \tilde{\vartheta}_D \in L^2(0,T;H^1_D(\Omega))$ and the test functions φ in (4.2) must also belong to $L^2(0,T;H^1_D(\Omega))$, where

$$H^1_D(\Omega) = \Big\{ v \in H^1(\Omega):\ v = 0 \ \text{ on } \Gamma_D \Big\} \quad (H^1_D(\Omega) = H^1_0(\Omega) \ \text{ if } \Gamma_D = \partial\Omega) \ .$$

For fixed $\sigma \geq \sigma_* > 0$ we shall denote by $(\vartheta_\sigma, \eta_\sigma)$ the solution to (4.1) (4.2) corresponding to the case

$$g(x,t,\vartheta) = \sigma\Big[\vartheta - \vartheta_D(x,t)\Big] \quad \text{for a.e. } (x,t) \in \Sigma_D \ , \tag{4.79}$$

which, by (4.78), is compatible with (4.4) and (4.5), as well as with the other assumptions of Theorems 4.8 and 4.11. For simplicity we shall consider only the case of monotone nonlinearities.

Theorem 4.14. *Under the hypothesis of Theorem 4.8 and 4.11, if (4.78) and (4.79) hold, then as* $\sigma \to +\infty$

$$(\vartheta_\sigma, \eta_\sigma) \rightharpoonup (\vartheta, \eta) \quad \text{in } [L^\infty(Q)]^2\text{-weak}^* \ , \tag{4.80}$$

$$\vartheta_\sigma \rightharpoonup \vartheta \quad \text{in } L^2(0,T;H^1(\Omega))\text{-weak} \ , \tag{4.81}$$

where (ϑ, η) *is the unique solution satisfying (4.10) to the problem (4.1) (4.2) with Dirichlet data (4.77).*

Proof: Letting $\psi = \vartheta_{\sigma\varepsilon} - \tilde{\vartheta}_D$ in (4.36), where $\vartheta_{\sigma\varepsilon}$ denotes the ε-approximation of ϑ_σ, we obtain first for $\vartheta_{\varepsilon\sigma}$ and then for ϑ_σ, the following estimate independently of σ:

$$\int_Q |\nabla \vartheta_\sigma|^2 + \sigma \int_{\Sigma_D} |\vartheta_\sigma - \tilde{\vartheta}_D|^2 \leq C \ . \tag{4.82}$$

This follows as in the proof of Proposition 4.7, taking into account the assumption (4.78), in particular, in the term

$$\int_0^t\!\!\int_\Omega \tilde{\vartheta}_D\, \partial_t u_{\sigma\varepsilon} = \int_\Omega u_{\sigma\varepsilon}(t)\, \tilde{\vartheta}_D(t) - \int_\Omega \eta_{0\varepsilon}\, \tilde{\vartheta}_D(0) - \int_0^t\!\!\int_\Omega u_{\sigma\varepsilon}\, \partial_t \tilde{\vartheta}_\varepsilon \ ,$$

since the estimate $\mu \leq \vartheta_{\sigma\varepsilon} \leq M$ still holds in this case.

Therefore we can extract a subsequence as $\sigma \to \infty$ satisfying (4.80) and (4.81) and it remains to identify the limit (ϑ, η) as a solution to (4.1) (4.2)– (4.77). The Dirichlet condition follows easily from

$$\int_{\Sigma_D} |\vartheta - \tilde{\vartheta}_D|^2 \leq \liminf_{\sigma\to\infty} \int_{\Sigma_D} |\vartheta_\sigma - \tilde{\vartheta}_D|^2 = 0 \ ,$$

while the rest of the proof can be done using the monotonicity arguments of the proof of Theorem 4.8.

Indeed, we know that now $\partial_t \eta_\sigma \in L^2(0,T;[H^1_D(\Omega)]')$ uniformly in σ, by taking $\psi \in L^2(0,T;H^1_D(\Psi))$ in

(4.83)

$$\langle \partial_t \eta_\sigma, \psi \rangle_Q = \int_Q \eta_\sigma \mathbf{v} \cdot \nabla\psi - \int_Q \nabla\vartheta_\sigma \cdot \nabla\psi - \int_Q f(\vartheta_\sigma)\,\psi - \int_{\Sigma_N} g(\vartheta_\sigma)\,\psi - \sigma \int_{\Sigma_D} (\vartheta_\sigma - \tilde{\vartheta}_D)\,\psi \ .$$

Therefore, since $[H^1(\Omega)]' \subset [H^1_{\rm D}(\Omega)]'$, we still have

$$\eta_\sigma \to \eta \quad \text{in } L^2(0,T;[H^1(\Omega)]')\text{-strong}, \quad \text{as } \sigma \to \infty ,$$

and, consequently $\langle \eta_\sigma, \vartheta_\sigma \rangle \to \langle \eta, \vartheta \rangle$ and we may conclude that $\vartheta = \gamma^{-1}(\eta)$ or $\eta \in \gamma(\vartheta)$ a.e. in Q. Analogously, we can identify $f = f(\vartheta)$ in Q and $g = g(\vartheta)$ on $\Sigma_{\rm N}$, by showing

$$\limsup_{\sigma\to\infty} X_\sigma \le \int_Q f\,\vartheta + \int_{\Sigma_{\rm N}} g\,\vartheta$$

where, taking $\psi = \vartheta_\sigma - \tilde{\vartheta}_{\rm D}$ in (4.83), we have

$$X_\sigma \equiv \int_Q f(\vartheta_\sigma)\,\vartheta_\sigma + \int_{\Sigma_{\rm N}} g(\vartheta_\sigma)\,\vartheta_\sigma \le$$

$$\le \int_Q f(\vartheta_\sigma)\,\tilde{\vartheta}_{\rm D} + \int_{\Sigma_{\rm N}} g(\vartheta_\sigma)\,\tilde{\vartheta}_{\rm D} + \langle \partial_t \eta_\sigma, \tilde{\vartheta}_{\rm D} - \vartheta_\sigma \rangle_Q - \int_Q \eta_\sigma\,\mathbf{v}\cdot\nabla\tilde{\vartheta}_{\rm D} - \int_Q \nabla\vartheta_\sigma\cdot\nabla(\vartheta_\sigma - \tilde{\vartheta}_{\rm D}) .$$

The passage to the limit can be easily done by using $\eta_\sigma(T) \to \eta(T)$ in $L^p(\Omega)$ in the formula

$$\langle \partial_t \eta_\sigma, \vartheta_\sigma \rangle_Q = \int_0^T \langle \partial_t \eta_\sigma, \gamma^{-1}(\eta_\sigma) \rangle_\Omega = \int_\Omega \tilde{\beta}(\eta_\sigma(T)) - \int_\Omega \tilde{\beta}(\eta_0) ,$$

with $\tilde{\beta}'(\tau) = \gamma^{-1}(\tau)$, exactly as in the proof of Theorem 4.8, where now we need also to use the identity

$$\langle \partial_t \eta, \tilde{\vartheta}_{\rm D} - \vartheta \rangle_Q = \int_Q \eta\,\mathbf{v}\cdot\nabla\tilde{\vartheta}_{\rm D} - \int_Q \nabla\vartheta\cdot\nabla(\tilde{\vartheta}_{\rm D} - \vartheta) - \int_Q f(\tilde{\vartheta}_{\rm D} - \vartheta) - \int_{\Sigma_{\rm N}} g(\tilde{\vartheta}_{\rm D} - \vartheta) .$$

Finally, the uniqueness of the solution may be proved as in the Theorem 4.11, where now the right-hand side of (4.62) is zero and the test function φ is taken to be the unique solution φ_m of the parabolic problem (4.70) with the additional Dirichlet condition $\varphi_m = 0$ on $\Sigma_{\rm D}$, which does not change the Lemma 4.12. Consequently all the sequence $\sigma \to \infty$ converges and the proof is complete. ∎

Remark 4.15. The uniqueness of the Dirichlet (or the mixed) problem in the case of f and g not necessarily monotone, of course, still holds. However, for the proof of the existence with the approximation $\sigma \to \infty$ by Neumann problem requires additional assumptions on $\partial_t f$ and on $\partial_t g$ on $\Sigma_{\rm N}$, as well as the compatibility condition $\vartheta_0|_{\Gamma_{\rm D}} = \vartheta_{\rm D}(0)$, in order to obtain the global estimate $\partial_t\vartheta_\sigma \in L^2(Q)$ independently of σ. In fact, we can easily check the local estimate (4.35) as done in Proposition 4.7 depends also on σ. □

Remark 4.16. The continuous dependence estimate (4.61) corresponding to the additional Dirichlet boundary condition would be naturally of the form

$$\int_Q |\eta - \hat{\eta}|^p \le C_p\Big\{\int_Q |f - \hat{f}| + \int_{\Sigma_{\rm N}} |g - \hat{g}| + \int_{\Sigma_{\rm D}} |\vartheta_{\rm D} - \hat{\vartheta}_{\rm D}| + \int_\Omega |\eta_0 - \hat{\eta}_0|\Big\} .$$

However, this would require the additional estimate

$$\|\partial_n \varphi_m\|_{L^\infty(\Sigma_{\rm D})} \le C, \quad \text{independently of } m , \tag{4.84}$$

for the solution to (4.70), with $\varphi_m|_{\Sigma_{\rm D}} = 0$, which is a very delicate problem. In fact, if one imposes that $|\vartheta_{\rm D}| \ge a > 0$ on $\Sigma_{\rm D}$ and it is possible to guarantee that the coefficient

α_m is bounded below by a positive constant uniformly in a neighbourhood of Σ_D, by local estimates near Σ_D, the estimate (4.84) can be proved in certain cases. For instance, this is done in [F1] for the Dirichlet case without convection and in [RY] for the continuous casting problem with constant velocity and cylindrical geometry. □

4.3 – Stabilization towards a steady-state

The asymptotic behaviour as $t \to \infty$ of the weak solution to the general Stefan problem with prescribed convection is a complex and delicate problem, in particular, the uniqueness of the general steady-state problem is an open question. In this section we shall use a comparison principle and solutions that are monotone in time. For simplicity we consider only the stabilization towards a steady-state solution to the Dirichlet case ($\Sigma = \Sigma_D$), with monotone nonlinearities. We start by a comparison principle for the general case.

Proposition 4.17. *Under the assumptions of Theorem 4.14, suppose that* $\widehat{\vartheta}_D \geq \vartheta_D$ a.e. *on* Σ_D, $\widehat{\eta}_0 \geq \eta_0$ a.e. *in* Ω, $\widehat{f}(\vartheta) \leq f(\vartheta)$ a.e. *in* Q *and* $\widehat{g}(\vartheta) \leq g(\vartheta)$ a.e. *on* Σ_N, *for all* $\vartheta \in \mathbb{R}$. *Then, we have for the corresponding solutions*

$$\widehat{\eta} \geq \eta \quad \text{and} \quad \widehat{\vartheta} \geq \vartheta \qquad \text{a.e. in } Q\ . \tag{4.85}$$

Proof: Recalling from (2.11) the approximation H_δ of the Heaviside function, we consider in the ε-regularized Neumann problems for ϑ_ε, $u_\varepsilon = \gamma_\varepsilon(\vartheta_\varepsilon)$ and $\widehat{\vartheta}_\varepsilon$, $\widehat{u}_\varepsilon = \gamma_\varepsilon(\widehat{\vartheta}_\varepsilon)$, $\psi = H_\delta(\vartheta_\varepsilon - \widehat{\vartheta}_\varepsilon)$ in (4.36). For their differences $\overline{\vartheta} = \vartheta_\varepsilon - \widehat{\vartheta}_\varepsilon$ and $\overline{u} = u_\varepsilon - \widehat{u}_\varepsilon$, we obtain

$$\int_\Omega (\partial_t \overline{u} + \mathbf{w}_\varepsilon \cdot \nabla \overline{u})\, H_\delta(\overline{\vartheta}) \leq \int_\Omega \left[\widehat{f}_\varepsilon(\widehat{\vartheta}_\varepsilon) - f_\varepsilon(\vartheta_\varepsilon)\right] H_\delta(\overline{\vartheta}) + \int_{\partial\Omega} \left[\widehat{g}_\varepsilon(\widehat{\vartheta}_\varepsilon) - g_\varepsilon(\vartheta_\varepsilon)\right] H_\delta(\overline{\vartheta})\ ,$$

since $\int_\Omega |\nabla\overline{\vartheta}|^2 H_\delta'(\overline{\vartheta}) \geq 0$. Letting $\delta \to 0$, since $H_\delta(\overline{\vartheta}) \to \operatorname{sign}_0^+(\overline{\vartheta})$ where $\operatorname{sign}_0^+(\overline{\vartheta}) = \operatorname{sign}_0^+(\overline{u})$ takes the value 1 if $\vartheta_\varepsilon > \widehat{\vartheta}_\varepsilon$ or 0 if $\vartheta_\varepsilon \leq \widehat{\vartheta}_\varepsilon$, we obtain

$$\int_\Omega \partial_t(\overline{u})^+ = \int_\Omega (\partial_t \overline{u} + \mathbf{w}_\varepsilon \cdot \nabla\overline{u}) \operatorname{sign}_0^+(\overline{u}) \leq 0\ , \tag{4.86}$$

since $\int_\Omega \mathbf{w}_\varepsilon \cdot \nabla(\overline{u})^+ = 0$ and by monotonicity we have

$$\int_{\Omega\cap\{\vartheta_\varepsilon > \widehat{\vartheta}_\varepsilon\}} \left\{[\widehat{f}_\varepsilon(\widehat{\vartheta}_\varepsilon) - \widehat{f}_\varepsilon(\vartheta_\varepsilon)] + [\widehat{f}_\varepsilon(\vartheta_\varepsilon) - f_\varepsilon(\vartheta_\varepsilon)]\right\} \leq 0\ ,$$

and similarly for the term on $\partial\Omega$. Integrating (4.85) with respect to t and taking, without loss of generality, $\widehat{\eta}_{0\varepsilon} \geq \eta_{0\varepsilon}$ we obtain

$$\int_\Omega (u_\varepsilon - \widehat{u}_\varepsilon)^+(t) \leq \int_\Omega (\eta_{0\varepsilon} - \widehat{\eta}_{0\varepsilon})^+ = 0\,, \qquad \forall t > 0\ .$$

Consequently we conclude $u_\varepsilon \leq \widehat{u}_\varepsilon$ and $\vartheta_\varepsilon \leq \widehat{\vartheta}_\varepsilon$. Letting $\varepsilon \to 0$, by uniqueness of the solutions, we conclude first $\widehat{\eta}_\sigma \geq \eta_\sigma$ and $\widehat{\vartheta}_\sigma \geq \vartheta_\sigma$ for all finite σ. Notice that by (4.78), we have $\widehat{g}(\vartheta) - g(\vartheta) = \sigma(\vartheta_D - \widehat{\vartheta}_D) \leq 0$ a.e. on Σ_D for all ϑ. Now letting $\sigma \to \infty$, by (4.80) we conclude (4.85). ∎

In the rest of this section we shall consider only the Dirichlet problem by setting $\Sigma_N = \emptyset$, and we shall assume the velocity field time independent, i.e.

$$(4.87) \qquad \mathbf{v} = \mathbf{v}(x) \in \mathbb{L}^2_\sigma(\Omega) = \left\{ \mathbf{w} \in [L^2(\Omega)]^N : \int_\Omega \mathbf{w} \cdot \nabla\varphi = 0,\ \forall \varphi \in H^1(\Omega) \right\} .$$

In this case, the comparison principle allows us to consider solutions which are monotone in time.

Proposition 4.18. *Under the assumptions of Theorem 4.14 with $\Sigma = \Sigma_D$ and (4.87), let the initial data be also such that $\vartheta_0 = \gamma^{-1}(\eta_0) \in H^1(\Omega)$. If $t \mapsto f(t,\cdot,\cdot)$ is non-increasing (resp. non-decreasing), $t \mapsto \vartheta_D(t)$ is non-decreasing and $\vartheta_D(0) \geq \vartheta_0$ on $\partial\Omega$ (resp. non-increasing and $\vartheta_D(0) \leq \vartheta_0$) and if*

$$\mathbf{v} \cdot \nabla\eta_0 - \Delta\vartheta_0 + f(0, \vartheta_0) \leq 0 \ \ (\textit{resp.} \geq 0) \quad \textit{in } \mathcal{D}'(\Omega) ,$$

then the weak solution $(\vartheta(t), \eta(t))$ is non-decreasing (resp. non- increasing) with respect to time.

Proof: We apply Proposition 4.17 firstly to conclude $\eta_0 \leq \eta(t)$ (resp. $\eta_0 \geq \eta(t)$) for all $t > 0$. Then, we apply again the Proposition 4.17 to $\eta(t)$ and $\widehat{\eta}(t) = \eta(t+h)$, $h > 0$ on the interval $[0, T-h]$. Using the assumptions and the previous remarks, it is easy to check that $\eta(t) \leq \eta(t+h)$ (resp. $\eta(t) \geq \eta(t+h)$) and by the monotonicity of γ^{-1} the same conclusions hold for ϑ. ∎

Consider now the associated steady-state problem in the variational form:

$$(4.88) \quad (\vartheta_\infty, \eta_\infty) \in H^1(\Omega) \times L^\infty(\Omega): \quad \eta_\infty \in \gamma(\vartheta_\infty) \ \text{ a.e. in } \Omega , \qquad \vartheta = \vartheta_D^\infty \ \text{ on } \partial\Omega ,$$

$$(4.89) \quad \int_\Omega (\nabla\vartheta_\infty - \eta_\infty \mathbf{v}) \cdot \nabla\varphi + \int_\Omega f_\infty(\vartheta_\infty)\, \varphi = 0 , \qquad \forall \varphi \in H^1_0(\Omega) .$$

Here we assume $\mathbf{v}$ given by (4.87), $f_\infty = f_\infty(x, \vartheta)$ is a Carathéodory function satisfying (4.4)–(4.6), (4.45) and (4.57), and

$$(4.90) \qquad \exists\, \widetilde{\vartheta}_D^\infty \in H^1(\Omega): \quad \widetilde{\vartheta}_D^\infty = \vartheta_D^\infty \ \text{ on } \partial\Omega \quad \text{and} \quad \mu \leq \widetilde{\vartheta}_D^\infty \leq M \ \text{ in } \Omega .$$

Theorem 4.19. *Under the preceding assumptions on f_∞, (4.90) on ϑ_D^∞ and (4.87) on $\mathbf{v}$, there exists at least a solution to (4.88)–(4.89), satisfying*

$$(4.91) \qquad \mu \leq \vartheta_\infty \leq M \quad \text{on } \Omega .$$

Proof: As in the Section 4.1, we consider a family of approximating solutions $\vartheta_\varepsilon \in H^1(\Omega)$ of the regularized elliptic problem

$$(4.92) \quad \vartheta_\varepsilon - \vartheta_D^\infty \in H^1_0(\Omega): \quad \int_\Omega (\nabla\vartheta_\varepsilon - \gamma_\varepsilon(\vartheta_\varepsilon)\, \mathbf{v}) \cdot \nabla\varphi + \int_\Omega f_\infty(\vartheta_\varepsilon)\, \varphi = 0, \quad \forall \varphi \in H^1_0(\Omega) ,$$

which has a unique solution, satisfying the uniform estimates

$$\mu \leq \vartheta_\varepsilon \leq M \ \text{ in } \Omega \quad \text{and} \quad \|\vartheta_\varepsilon\|_{H^1(\Omega)} \leq C .$$

The a priori L^∞-bound may be obtained by taking $\varphi = (\vartheta_\varepsilon - M)^+$ and $\varphi = (\mu - \vartheta_\varepsilon)^+$ in (4.92), while the estimate in $H^1(\Omega)$ follows by taking $\varphi = \vartheta_\varepsilon - \vartheta_D^\infty$ also in (4.92). The existence of a solution can be obtained with the Schauder fixed point theorem and the uniqueness by standard comparison arguments for elliptic problems (see [R6], for instance).

Hence we can extract a subsequence as $\varepsilon \to 0$, such that

$$\vartheta_\varepsilon \to \vartheta_\infty \quad \text{in } H^1(\Omega)\text{- weak and } L^p(\Omega)\text{-strong}, \ \forall p < \infty ,$$
$$\gamma_\varepsilon(\vartheta_\varepsilon) \rightharpoonup \eta_\infty \quad \text{in } L^\infty(\Omega)\text{-weak}^* ,$$

and we find $\eta_\infty \in \gamma(\vartheta_\infty)$ by the usual monotonicity argument. Taking the limit in (4.92), we obtain (4.89) and the existence of a solution is proved. ∎

Remark 4.20. If $(\vartheta_\infty, \eta_\infty)$ solves (4.88)–(4.89) it is also a solution in a weaker sense, i.e., a solution to

$$\int_\Omega \Big[\vartheta_\infty \Delta\varphi + \eta_\infty \mathbf{v}\cdot\nabla\varphi - f_\infty(\vartheta_\infty)\varphi\Big] = \int_{\partial\Omega} \vartheta_D^\infty \,\partial_n\varphi , \tag{4.93}$$

for all $\varphi \in H_0^1(\Omega)$: $\Delta\varphi \in L^2(\Omega)$ and $\partial_n\varphi \in L^2(\partial\Omega)$. In particular, this definition does not require $\vartheta_\infty \in H^1(\Omega)$ but only $\vartheta_\infty \in L^2(\Omega)$. However the uniqueness with the general convective case is an open problem. Partial results have been obtain for the continuous casting problem in [RY], in which $\mathbf{v}$ is a constant vector, and in [YQ] under the additional restrictions on $f_\infty \equiv 0$ and on the velocity $\mathbf{v}$, namely $\mathbf{v} \in [C^1(\overline{\Omega})]^N$, $\mathbf{v}\cdot\mathbf{n} \geq 0$ on $\partial\Omega$ and $(\nabla\cdot\mathbf{v})I - 2(\nabla\mathbf{v}) = [(\partial v_k/\partial x_k)\delta_{ij} - 2(\partial v_i/\partial x_j)] \geq 0$, i.e., is a positive definite matrix in Ω. □

In order to consider the stabilization towards a steady-state we shall assume

$$\sup_{\mu\leq\vartheta\leq M} \Big|f(x,t,\vartheta) - f_\infty(x,\vartheta)\Big| \to 0 \quad \text{a.e. } x \in \Omega, \ \text{ as } t \to \infty , \tag{4.94}$$

$$\vartheta_D(x,t) \to \vartheta_D^\infty(x) \quad \text{a.e. } x \in \partial\Omega, \ \text{ as } t \to \infty . \tag{4.95}$$

Theorem 4.21. *Let (4.87), (4.94) and (4.95) hold. If $t \mapsto (\vartheta(t), \eta(t))$ is a non-decreasing (resp. non-increasing) weak solution to the Stefan problem, then as $t \to \infty$*

$$\vartheta(t) \to \vartheta_\infty \ \text{ and } \ \eta(t) \to \eta_\infty \quad \text{in } L^p(\Omega)\text{-strong}, \ \forall p < \infty , \tag{4.96}$$

where ϑ_∞ and $\eta_\infty \in \gamma(\vartheta_\infty)$ are a bounded solution to the steady-state problem in the sense of (4.93).

Proof: Since ϑ and η are bounded monotone non-decreasing (non-increasing) functions, we can define

$$\vartheta_\infty(x) = \lim_{t\to\infty} \vartheta(x,t) \ \text{ and } \ \eta_\infty(x) = \lim_{t\to\infty} \eta(x,t) \quad \text{for a.a. } x \in \Omega , \tag{4.97}$$

where ϑ_∞ and η_∞ are measurable functions such that

$$\mu \leq \vartheta_\infty \leq M \ \text{ and } \ \gamma(\mu) \leq \eta_\infty \leq \gamma(M) \quad \text{a.e. in } \Omega .$$

By Lebesgue's theorem, it follows that (4.96) holds and, by the monotonicity argument, it is clear that $\eta_\infty \in \gamma(\vartheta_\infty)$ a.e. in Ω. It remains to show that $(\vartheta_\infty, \eta_\infty)$ satisfies (4.93).

Fix $\varphi \in H_0^1(\Omega)$ with $\Delta\varphi \in L^2(\Omega)$, $\partial_n\varphi \in L^2(\partial\Omega)$ and let $\zeta = \zeta(s)$ be an arbitrary function in $\mathcal{D}(0,1)$ such that $\int_0^1 \zeta(s)\,ds \neq 0$. From (4.56) it is clear that for any $t > 0$ we have

$$\int_t^{t+1}\int_\Omega \varphi\,\eta(\tau)\,\zeta'(\tau - t) =$$
$$= \int_t^{t+1} \zeta(\tau - t)\Big\{\int_\Omega \big[f(\tau, \vartheta(\tau))\,\varphi - \vartheta(\tau)\,\Delta\varphi + \eta(\tau)\,\mathbf{v}\cdot\nabla\varphi\big] + \int_{\partial\Omega} \vartheta_{\mathrm{D}}(\tau)\,\partial_n\varphi\Big\}.$$

Using the notation $u^t(x,s) = u(x, s+t)$ for $(x,s) \in \Omega\times\,]0,1[$ and changing variables by $s = \tau - t$, we have

$$\int_0^1\int_\Omega \varphi\,\eta^t\,\zeta'(s) = \int_0^1 \zeta(s)\Big\{\int_\Omega \big[f^t(\vartheta^t)\,\varphi - \vartheta^t\,\Delta\varphi + \eta^t\,\mathbf{v}\cdot\nabla\varphi\big] + \int_{\partial\Omega} \vartheta_{\mathrm{D}}^t\,\partial_n\varphi\Big\}\,. \tag{4.98}$$

From (4.97) and again Lebesgue's theorem, we also have

$$\vartheta^t \to \vartheta_\infty \text{ and } \eta^t \to \eta_\infty \quad \text{in } L^p(\Omega\times\,]0,1[)\text{-strong } (\forall p < \infty) \text{ as } t \to \infty\,,$$

and letting $t \to \infty$ in (4.98), since (4.94) also implies $f^t(\vartheta^t) \to f_\infty(\vartheta_\infty)$ in $L^2(\Omega\times\,]0,1[)$, we obtain (4.93) from

$$0 = \Big(\int_0^1 \zeta(s)\,ds\Big)\Big\{\int_\Omega \big[f_\infty(\vartheta_\infty)\,\varphi - \vartheta_\infty\,\Delta\varphi + \eta_\infty\,\mathbf{v}\cdot\nabla\varphi\big] + \int_{\partial\Omega} \vartheta_{\mathrm{D}}^\infty\,\partial_n\varphi\Big\}\,,$$

since

$$\int_0^1\int_\Omega \varphi\,\eta^t\,\zeta'(s) \to \int_0^1\int_\Omega \varphi\,\eta_\infty\,\zeta'(s) = \big(\zeta(1) - \zeta(0)\big)\int_\Omega \varphi\,\eta_\infty = 0\,. \blacksquare$$

This result can be used to discuss the stabilization of some special solutions (non-monotone in time) by comparison with monotone ones. Suppose there exists two sets of evolutionary data $f^\downarrow(t)$, $\vartheta_{\mathrm{D}}^\downarrow(t)$ and $f^\uparrow(t)$, $\vartheta_{\mathrm{D}}^\uparrow(t)$, monotone in time, such that

$$\begin{cases} (f^\downarrow, \vartheta_{\mathrm{D}}^\downarrow) \text{ and } (f^\uparrow, \vartheta_{\mathrm{D}}^\uparrow) \text{ satisfy, as } t \to \infty,\ (4.94)\text{–}(4.95),\ f^\downarrow \text{ and } \vartheta_{\mathrm{D}}^\uparrow \\ \text{are non-decreasing, } f^\uparrow \text{ and } \vartheta_{\mathrm{D}}^\downarrow \text{ are non-increasing and } f^\downarrow(0),\ f^\uparrow(0), \\ \vartheta_{\mathrm{D}}^\downarrow(0) \text{ and } \vartheta_{\mathrm{D}}^\uparrow(0) \text{ satisfy the assumptions of Theorem 4.19.} \end{cases} \tag{4.99}$$

Then, there exist steady-state solutions (ϑ^*, η^*) and (ϑ_*, η_*) by solving the problem (4.88)–(4.89), respectively, with data $(f^\downarrow(0), \vartheta_{\mathrm{D}}^\downarrow(0))$ and $(f^\uparrow(0), \vartheta_{\mathrm{D}}^\uparrow(0))$, such that $\vartheta^* \geq \vartheta_*$ in Ω and we may assume that

$$\Theta_*^* \equiv \Big\{\vartheta \in L^\infty(\Omega) \cap H^1(\Omega):\ \vartheta_* \leq \vartheta \leq \vartheta^* \text{ a.e. in } \Omega\Big\} \neq \emptyset\,.$$

Assuming non-monotone data $(f, \vartheta_{\mathrm{D}})$ such that

$$f^\downarrow(x,t,\vartheta) \leq f(x,t,\vartheta) \leq f^\uparrow(x,t,\vartheta) \quad \text{a.e. } x \in \Omega,\ \ t > 0,\ \ \vartheta \in [\mu, M]\,, \tag{4.100}$$

$$\vartheta_{\mathrm{D}}^\uparrow(x,t) \leq \vartheta_{\mathrm{D}}(x,t) \leq \vartheta_{\mathrm{D}}^\downarrow(x,t) \quad \text{a.e. } x \in \partial\Omega,\ \ t > 0\,, \tag{4.101}$$

we can consider the solution $(\vartheta^{\downarrow},\eta^{\downarrow})$ (resp. $(\vartheta^{\uparrow},\eta^{\uparrow})$) corresponding to the data $(f^{\downarrow},\vartheta_D^{\downarrow},\vartheta^*)$ (resp. $(f^{\uparrow},\vartheta_D^{\uparrow},\vartheta_*)$), which is monotone non-increasing (resp. non-decreasing) in time and, by the Theorem 4.21 are such that

$$\vartheta^{\downarrow}(t) \searrow \vartheta^{\downarrow}_{\infty} \text{ and } \vartheta^{\uparrow}(t) \nearrow \vartheta^{\uparrow}_{\infty} \quad \text{as } t \to \infty\,, \tag{4.102}$$

for some steady-state solutions $\vartheta^{\downarrow}_{\infty}$, $\vartheta^{\uparrow}_{\infty}$ with $\vartheta^{\downarrow}_{\infty} \geq \vartheta^{\uparrow}_{\infty}$. If (ϑ,η) denotes the solution corresponding to the data f, ϑ_D and on initial condition $\vartheta_0 = \gamma^{-1}(\eta_0)$ in Θ^*_*, by comparison, we have

$$\vartheta^{\uparrow} \leq \vartheta \leq \vartheta^{\downarrow} \text{ and } \eta^{\uparrow} \leq \eta \leq \eta^{\downarrow} \quad \text{a.e. in } \Omega,\ t>0\,.$$

Therefore if we are able to show that $\vartheta^{\downarrow}_{\infty} = \vartheta^{\uparrow}_{\infty} = \vartheta_{\infty}$ from the convergence (4.102) it is clear that (4.96) still holds for the non-monotone evolutionary solution (ϑ,η) and we have proved the following result:

Corollary 4.22. *Assume the steady-state problem has a unique weak solution $(\vartheta_\infty,\eta_\infty)$ in the sense of Remark 4.20. Then if (ϑ,η) is the weak solution under the preceding assumptions, namely (4.99)–(4.101) and with initial condition $\vartheta_0 \in \Theta^*_*$, then the asymptotic stabilization (4.96) still holds.* ∎

Remark 4.23. As it was observed in Remark 4.20, the uniqueness of steady-state solutions is known only in very special cases, besides the case without convection, which was studied in [F1] for the Dirichlet problem and improved in [DK2] for the mixed Dirichlet–Neumann problem with the approach of the theory of nonlinear evolution equations governed by time-dependent subdifferential operators in Hilbert spaces. For the continuous casting problem, with the same framing technique, in [RY] it was proved the asymptotic stabilization (4.96) with the order of convergence $O(t^{-1/p})$ as $t\to\infty$. Finally the periodicity and almost-periodicity of solutions has been investigated in [DK3] for the case $\mathbf{v}\equiv 0$ and in [Sh] for the continuous casting problem. □

REFERENCES

[B] Baiocchi, C. – Sur un problème à frontière libre traduisant le filtrage de liquides à travers des milieux poreux, *C.R. Acad. Sci. Paris*, 273 (1971), 1215–1217.

[BC] Baiocchi, C. & Capelo, A. – *Disequazioni variazionali e quasi variazionali. Applicazioni a problemi di frontiera libera*, Vol. I, II, Quaderni dell'U. Mat. Ital., Pitagora, Bologna, 1978; English transl. J. Wiley, Chichester-New York, 1984.

[Ba1] Barbu, V. – *Nonlinear semigroups and differential equations in Banach Spaces*, Noordhoof Int. Publ., Leyden, 1976.

[Ba2] Barbu, V. – *Optimal Control of Variational Inequalities*, Research Notes in Math. No. 100, Pitman, Boston, London, 1984.

[BL] Bensoussan, A. & Lions J.L. – *Applications des Inéquations Variationelles en Contrôle Stochastique*, Dunod, Paris, 1978; English transl., North-Holland, Amsterdam, 1982.

[BDF] Bossavit, A., Damlamian, A. & Frémond, M. – *Free Boundary problems: applications and theory*, Vol. III, IV, Research Notes in Math. Nos. 120/121, Pitman, Boston, London, 1985.

[BFN] Brauner, C.M., Frémond, M. & Nicolaenko, B. – *A new homographic approximation to multiphase Stefan problem*, in: "Free Boundary Problems – Theory and Applications" (A. Fasano, M. Primicerio Eds.), Vol. II, Pitman, Boston, 1983, 365–379."

[B1] BRÉZIS, H. – On some degenerate nonlinear parabolic equations, (in Browder F.E., ed.) *Nonlinear functional analysis, Part I, Symp. Pure Math.*, 18 (1970), 28–38.

[B2] BRÉZIS, H. – Problèmes unilatéraux, *J. Math. Pures et Appl.*, 51 (1972), 1–168.

[C1] CAFFARELLI, L.A. – The Regularity of Free Boundaries in Higer Dimensionsm, *Acta Math.*, 139 (1977), 155–184.

[C2] CAFFARELLI, L.A. – Some Aspects of the One-Phase Stefan Problem, *Indiana Univ. Math. J.*, 27 (1978), 73–77.

[CE] CAFFARELLI, L. & EVANS, L. – Continuity of the temperature in the two-phase Stefan problems, *Arch. Rational Mech. Anal.*, 81, 3 (1983), 199–220.

[CF] CAFFARELLI, L.A. & FRIEDMAN, A. – Continuity of the Temperature in the Stefan Problem, *Indiana Univ. Math. J.*, 28 (1979), 53–70.

[CD] CANNON, J.R. & DIBENEDETTO, E. – On the existence of weak solutions to an n-dimensional Stefan problem with nonlinear boundary conditions, *SIAM J. Math. Anal.*, 11 (1980), 632–645.

[CR] CHIPOT, M. & RODRIGUES, J.F. – On the Steady-State Continuous Casting Stefan Problem with Non-Linear Cooling, *Quart. Appl. Math.*, 40 (1983), 476–491.

[Cr] CRANK, J. – *Free and moving boundary problems*, Oxford, Univ. Press, Oxford, 1984.

[D1] DAMLAMIAN, A. – *Résolution de certaines inéquations variationnelles stationnaires et d'évolution*, Thése Doct., Univ., Paris VI, 1976.

[D2] DAMLAMIAN, A. – Some results on the multiphase Stefan problem, *Comm. in P.D.E.*, 2(10) (1977), 1017–1044.

[D3] DAMLAMIAN, A. – *On the Stefan problem: the variational approach and some applications*, in "Math. Models Meth. in Mech.", Banach Cent. Publ. (Warsaw), Vol. 15 (1985), 253–275.

[DK1] DAMLAMIAN, A. & KENMOCHI, N. – Le problème de Stefan avec conditions latérales variables, *Hiroshima Math. J.*, 10 (1980), 271–293.

[DK2] DAMLAMIAN, A. & KENMOCHI, N. – Asymptotic behaviour of the solution to a multiphase Stefan problem, *Japan J. Appl. Math.*, 3 (1986), 15–36.

[DK3] DAMLAMIAN, A. & KENMOCHI, N. – Periodicity and almost-periodicity of solutions to a multiphase Stefan problem in several space variables. *Nonlinear analysis, Th. Math. and Appl.*, 12 (1988), 921–934.

[DK4] DAMLAMIAN, A. & KENMOCHI, N. – Uniqueness of the solution of a Stefan problem with variable lateral boundary conditions, *Adv. Math. Sci. Appl.*, 1 (1992), 175–194.

[Dan] DANILYUK, I.I. – On the Stefan problem, *Russian Math. Surveys*, 40:5 (1985), 157–223.

[Db] DIBENEDETTO, E. – Continuity of weak solutions to certain singular parabolic equations, *Ann. Mat. Pura Appl.*, (4), 130 (1982), 131–176.

[DF] DIBENEDETTO, E. & FRIEDMAN, A. – Periodic Behaviour for the evolutionary dam problem and related free boundary problems, *Comm. P.D.E.*, 11 (1986), 1297–1377.

[DO] DIBENEDETTO, E. & O'LEARY, M. – 3-D conduction-convection with change of phase, *Arch. Rational Mech. Anal.* (in press).

[Du1] DUVAUT, G. – Résolution d'un problème de Stefan, *C. R. Ac. Sci. Paris*, 276-A (1973), 1461–1463.

[Du2] DUVAUT, G. – *The solution of a two-phase Stefan problem by a variational inequality*, in: J.R. Ockendon, A.R. Hodgkins Eds., "Moving Boundary Problems in Heat Flow and Diffusion", Clarendon Press, Oxford, 1975, 173–181.

[DL] Duvaut, G. & Lions, J.L. – *Les inéquations en mécanique et en physique*, Dunod, Paris, 1972; English transl. Springer, Berlin, 1976.

[EO] Elliott, C.M. & Ockendon, J.R. – *Weak and Variational Methods for Moving Boundary Problems*, Research Notes in Maths. 59, Pitman, Boston, London, 1982.

[FP] Fasano, A. & Primicerio, M. (eds.) – *Free Boundary Problems: Th. & Appl.*, Research Notes in Math. 78/79, Pitman, Boston, 1983.

[Fr] Frémond, M. – *Variational formulation of the Stefan problem – Coupled Stefan problem – Frost propagation in porous media*, in: "Computational Methods in Nonlinear Mechanics" (J.T. Oden, Ed.), The University of Texas, Austin, 1974.

[F1] Friedman, A. – The Stefan problem in several space variables, *Trans. Amer. Math. Soc.*, 133 (1968), 51–87.

[F2] Friedman, A. – *Variational Principles and Free-Boundary Problems*, Wiley, New York, 1982.

[FH] Friedman, A. & Hoffmann, K.-H. – Control of free boundary problems with hysteresis, *SIAM J. Control and Optim.*, 26 (1988), 42–55.

[FHY] Friedman, A., Huang, S. & Yong, J. – Optimal periodic control for the two-phase Stefan problem, *SIAM J. Control and Optim.*, 26 (1988), 23–41.

[FK] Friedman, A. & Kinderlehrer, D. – A one phase Stefan problem, *Indiana Univ. Math. J.*, 24 (1975), 1005–1035.

[GHM] Götz, I.G., Hoffmann, H.-H. & Meirmanov, A.M. – *Periodic solutions of the Stefan problem with hysteresis-type boundary conditions*, Rep. 399, T.U. München, 1992.

[GZ] Götz, I.G. & Zaltzman, B.B. – Nonincrease of mushy region in a nonhomogeneous Stefan problem, *Quart. Appl. Math.*, 49 (1991), 741–746.

[H] Hilpert, M. – On uniqueness for evolution problems with hysteresis, *Int. Ser. Numer. Math.*, 88 (1989), 377–388.

[HN] Hoffmann, K.H. & Niezgodka, M. – *Control of Parabolic Systems Involving Free Boundaries*, in [FP], Vol. II (1983), 431–453.

[HS] Hoffmann, K.H. & Sprekels, J. – *Free Boundary Problems, Theory and Applications*, Pitman Res. Notes in Math., 185/186, Longman, 1990.

[Ka] Kamenomostkaya, S. – On the Stefan problem (in Russian), *Naučnye Dokl. Vysšei Školy*, 1 (1958), 60–62; *Mat. Sb.*, 53(95) (1961), 489–514.

[KS] Kinderlehrer, D. & Stampacchia, G. – *An introduction to variational inequalities and their applications*, Acad. Press, New York, 1980.

[LSU] Ladyzenskaya, O.A., Solonnikov, V.A. & Ural'eva, N.N. – *Linear and Quasilinear Equations of Parabolic Type*, A.M.S. Transl. Monog. 23, Providence, 1968.

[LC] Lamé, G. & Clapeyron, B.P. – Mémoire sur la solidification par refroidissement d'un globe solid, *Ann. Chem. Phys.*, 47 (1831), 250–256.

[L1] Lions, J.L. – *Sur le contrôle optimale de systèmes governés par des équations aux dérivées partielles*, Dunod, Paris, 1968.

[L2] Lions, J.L. – *Quelques Méthodes de Résolution des Problèmes aux Limites Non Linéaires*, Dunod, Paris, 1969.

[L3] Lions, J.L. – *Sur quelques questions d'Analyse, de Mécanique et de Contrôle Optimal*, Presses Univ. Montreal, 1976.

[LR] Louro, B. & Rodrigues, J.F. – Remarks on the quasi-steady one phase Stefan problem, *Proc. Royal Soc. Edinburg*, 102-A (1986), 263–275.

[M1] Magenes, E. (ed.) – *Free Boundary Problems*, Proc. Pavia Seminar Sept.–Oct. 1979, Instituto Nazionale di Alta Matematica Francesco Severi, Vol. I and II, Roma, 1980.

[M2] MAGENES, E. – Problemi di Stefan Bifase in più Variabili Spaziali, *Le Matematiche*, (Catania) 36 (1983), 65–108.

[MVV1] MAGENES, E., VERDI, C. & VISINTIN, A. – Semigroup approach to the Stefan problem with nonlinear flux, *Rend. Arad. Naz. Lincei*, 75(2) (1983), 24–33.

[MVV2] MAGENES, E., VERDI, C. & VISINTIN, A. – Theoretical and numerical results on the two-phase Stefan Problem, *SIAM J. Numer. Anal.*, 26 (1989), 1425–1438.

[Me1] MEIRMANOV, A.M. – On the Classical Solution of the Multidimensional Stefan Problem for Quasilinear Parabolic Equations, *Mat. Sbornik*, 112 (1980), 170–192; *Math. USSR-Sb.*, 40, 2 (1981), 157–179).

[Me2] MEIRMANOV, A.M. – *The Stefan Problem*, Nauka, Novosibirsk, 1986; English transl., De Gruyter, Berlin, 1992.

[Na1] NAGASE, H. – On an application of Rothe's method to nonlinear parabolic variational inequalities, *Funkcialaj Ekvacioj*, 32 (1989), 273–299.

[Na2] NAGASE, H. – On an asymptotic behaviour of solutions of nonlinear parabolic variational inequalities, *Japan J. Math.*, 15 (1989), 169–189.

[N] NIEZGODKA, M. – *Stefan-like problems*, in [FP] (1983), 321–347.

[NP] NIEZGODKA, N. & PAWLOW, I. – A generalized Stefan problem in several space variables, *Appl. Math. Optim.*, 9 (1983), 193–224.

[No] NOCHETTO, R. – A class of non-degenerate two-phase Stefan problems in several space variables, *Comm. P.D.E.*, 12 (1987), 21–45.

[O] OLEINIK, O.A. – A method of solution of the general Stefan problem, *Soviet Math. Dokl.*, 1 (1960), 1350–1354.

[OPR] OLEINIK, O.A., PRIMICERIO, M. & RADKEVICH, E.V. – Stefan-like problems, *Meccanica*, 28 (1993), 129–143.

[P1] PAWLOW, I. – A variational inequality approach to generalized two-phase Stefan problem in several space variables, *Annali Mat. Pura Appl.*, 131 (1982), 333–373.

[P2] PAWLOW, I. – *Analysis and control of evolution multiphase problems with free boundaries*, Polska Akad. Nauk., Warszawa, 1987.

[Pr] PRIMICERIO, M. – Problemi di diffusione a frontiera libera, *Boll. U.M.I.*, (5) 18-A (1981), 11–68.

[PR] PRIMICERIO, M. & RODRIGUES, J.F. – The Hele–Shaw Problem with nonlocal injection condition, *Gakuto Int. Ser., Math. Sci. Appl.*, Vol. 2 (1993), 375–390.

[R1] RODRIGUES, J.F. – *Some remarks on the asymptotic behaviour of strong solutions to monotone parabolic variational inequalities*, Rendiconti di Mat., 3 (1984), 457–470.

[R2] RODRIGUES, J.F. – On the variational inequality approach to the one-phase Stefan problem, *Acta Appl. Math.*, 8 (1987), 1–35.

[R3] RODRIGUES, J.F. – *Obstacle Problems in Mathematical Physics*, North-Holland, Amsterdam, 1987.

[R4] RODRIGUES, J.F. – The Stefan problem revisited, *Int. Ser. Num. Math.*, 88 (1989), 129–190.

[R5] RODRIGUES, J.F. (Editor) – *Mathematical models for phase change problems*, ISNM no. 88, Birkhäuser, Basel, 1989.

[R6] RODRIGUES, J.F. – Weak solutions for thermoconvective flows of Boussinesq–Stefan type, *Pitman Res. Notes in Math. Ser.*, 279 (1992), 93–116.

[RY] RODRIGUES, J.F. & YI FAHUAI – On a two-phase continuous casting Stefan problem with nonlinear flux, *Euro. J. Appl. Math.*, 1 (1990), 259–278.

[RZ] RODRIGUES, J.F. & ZALTZMAN, B. – *Free boundary optimal control in the multidimensional Stefan problem*, in: Intern. Conf. on "Free Boundary Problems: Theory and Applications", Toledo, June, 1993 (to appear).

[R] RUBINSTEIN, L.I. – *The Stefan problem*, Amer. Math. Soc. Transl. Monogr. 27, Providence, 1971.

[Ru] RULLA, J. – Weak Solutions to Stefan Problems with Prescribed Convection, *SIAM J. Math. Anal.*, 18 (1987), 1784–1800.

[S1] SAGUEZ, C. – Contrôle optimal d'inéquations variationnelles avec observations de domains, Rapport no. 286, I.R.I.A. (1978); see also *C. R. Acad. Sc. Paris*, 287-A (1978), 957–959.

[S2] SAGUEZ, C. – *Contrôle optimal de systèmes à frontière libre*, Thèse d'État, Université Technologie de Compiègne, 1980.

[Si] SIMON, L. – Compact sets in the space $L^p(0,T;B)$, *Annali Mat. pura ed Appl.*, 146 (1987), 65–96.

[Sh] SHINODA, J. – On a continuous casting problem with periodicity in time, *Gakuto Int. Ser., Math. Sci. Appl.*, Vol. 2 (1993), 655–669.

[St] STEFAN, J. – Über einige Probleme der Theorie der Wärmeleitung, *Sitzungsber, Wien, Akad. Mat. Natur.*, 98 (1889), 473–484; see also pp. 614–634; 965–983; 1418–1442.

[T1] TARZIA, D.A. – Sur le problème de Stefan à deux phase, *C. R. Acad. Sci. Paris*, 288 (1980), 941–944.

[T2] TARZIA, D.A. – Etude de l'inéquation variationnelle proposée par Duvaut pour le problème de Stefan à deux phases, I, *Boll. UMI*, 1-B (1982), 865–883; II, *Boll. UMI*, 2-B (1983), 589–603.

[T3] TARZIA, D.A. – *A bibliography on moving-free boundary problems for the heat diffusion equations*, Rep. Math. Dept. Univ. Firenze, 1988.

[V1] VISINTIN, A. – Sur le problème de Stefan avec flux non linéaire, *Boll. UMI C.*, 18 (1981), 63–86.

[V2] VISINTIN, A. – General free boundary evolution problems in several space dimensions, *J. Math. Anal. Appl.*, 95 (1983), 117–143.

[V3] VISINTIN, A. – A phase transition problem with delay, *Control and Cybernetics*, 11 (1982), 5–18.

[V4] VISINTIN, A. – A model for hysteresis of distributed systems, *Annal. Mat. Pura ed Appl.*, 131 (1982), 203–231.

[V5] VISINTIN, A. – *Mathematical Models of Hysteresis*, in: "Topics in Nonsmooth Mechanics" (J.J. Moreau, P.D. Panagiotopoulos, G. Strang, Eds.),. Birkhäuser, 1988, 295–326.

[Y] YI FAHUAI – An evolutionary continuous casting problem of two-phase and its periodic behaviour, *J. Part. Diff. Eq.*, 2(3) (1989), 7–22.

[YQ] YI, F. & QIN, Y. – *On the Stefan problem with prescribed convection* (to appear).

[Za] ZALTZMAN, B. – Multidimensional two-phase quasistationary Stefan problem, *Manuscripta Math.*, 78 (1993), 287–301.

[Z] ZEIDLER – *Nonlinear functional analysis and its applications*, Vol. II/B, Nonlinear monotone operators, Springer Verlag, New York, 1990.

Address: J.F. Rodrigues,
CMAF/Universidade de Lisboa, Av. Prof. Gama Pinto, 2
1699 LISBOA Codex, PORTUGAL

Numerical aspects of parabolic free boundary and hysteresis problems

Claudio Verdi

Dipartimento di Matematica, Università di Milano, Via Saldini 50, 20133 Milano, Italy

Introduction

In these notes we discuss various aspects of the numerical approximation of parabolic free boundary problems and parabolic equations with hysteresis by finite element methods. We consider the classical solid-liquid phase transition, described by the so-called two-phase Stefan problem, as a model example of parabolic free boundary problems, but the underlying ideas apply to a number of other free boundary problems as well. The motion by mean curvature of surfaces is another relevant free boundary problem addressed here. Finally we consider parabolic equations with hysteresis.

In Sect. 1 we recall the weak formulation of the two-phase Stefan problem in terms of energy density. Since the problem is degenerate parabolic, and so the solution exhibits a global lack of regularity, its numerical approximation is much harder than that of mildly nonlinear problems. It thus becomes imperative to know the regularity properties of the solution and exploit the proper functional setting. The analysis of the continuous dependence on the data serves to motivate the main techniques for error analysis of degenerate parabolic problems, namely, the inversion of the Laplacian, the integral-time method, and L^1 estimates. We also present, and discuss accuracy of, a couple of (continuous) approximation procedures based on either regularization of the energy-temperature constitutive relation or relaxation of the phase variable. The physical interpretation of the phase relaxation is discussed for a model of crystallization of polymers. The study of the travelling waves associated to the one-dimensional problems provides information about the smoothing effect of regularization and phase relaxation. Finally we recall the weak formulation of parabolic equation with a continuous hysteresis relation in the principal part of the operator.

In Sect. 2 we present a couple of time discretization techniques for the two-phase Stefan problem and parabolic equation with hysteresis. Based on backward-differences for the parabolic equation, enforcing the constitutive relations leads to the usual nonlinear methods, whereas linear methods may be obtained by a suitable violation of that constraint. Note that these linearization techniques are not based on Taylor's formula, because of the lack of regularity of the solution. The well-posedness of the discrete-time schemes relies either on the monotonicity of the energy-temperature constitutive relation or on the piecewise monotonicity of the hysteresis operators. We also study semidiscrete travelling waves for the two-phase Stefan problem. They provide valuable information about the regularizing effect of time discretization, which converts sharp interfaces into thin transition regions. Their width, as well as the profile and local regularity of the physical variables within them, turns out to be crucial in the design of adaptive finite element strategies.

In Sect. 3 we formulate the fully discrete methods and discuss their implementation on a computer. We derive several *a priori* estimates which, apart from their intrinsic interest, are extremely useful in the error analysis. We carry out the error analysis for the two-phase Stefan problem using the Laplacian inversion method for the nonlinear scheme and the integral method for the linear algorithm. It has to be emphasized that this analysis requires regularity properties compatible with the free boundary problem at hand, as expressed in the discrete stability estimates, whereas the standard approach to mildly nonlinear parabolic equations only assumes sufficient regularity on the continuous solution. The approximation of free boundaries is a challenging problem and we present some result based on nondegeneracy properties of the solution.

Even though the interface does not play any explicit role in the methods above, it is responsible for the global numerical pollution observed on the discrete solution. Accuracy and efficiency of fixed domain methods may be improved by mesh adaptation, as discussed in Sect. 4. We give a heuristic motivation for the local refinement and mesh selection strategies, which are based on equidistributing pointwise errors and following the interface motion; this, in turn, leads to highly graded meshes and to an optimal balance accuracy/degrees of freedom. We examine stability and accuracy and, finally, illustrate the performance of the adaptive strategy in light of various numerical experiments.

In Sect. 5 we study the motion of surfaces by mean curvature (with forcing) and its convergent approximation via singularly perturbed reaction-diffusion equations, which replaces interfaces by thin transition layers. Formal asymptotics motivates the use of double obstacles, which retains the local character of the original problem and is thus appealing from the numerical viewpoint. Convergence and interface error estimates, even past singularities, are derived using maximum principle and constructing suitable barriers. A space-time variable relaxation parameter gives rise to a transition layer of variable width which, in turn, enhances singularity resolution, as numerical experiments reveal. The fully discrete scheme can be implemented as a dynamic mesh algorithm, thus saving memory allocation and computing time. In addition this shows the potential to handle stiff systems, as the phase field model.

The following are some basic references for the course. Refs. [46,84,125] and [29,143] for Sect. 1; Refs. [101,113,114] for Sects. 2 and 3; Ref. [103] for Sect. 4; Refs. [56,57,102] for Sect. 5. We also refer to [38] for an introduction to the finite element method and functional space notation, to [132] for finite element approximations of parabolic problems, to [11,67] for adaptive finite element methods, and to [118] for the solution of nonlinear systems.

We indicate with $C > 0$ a universal constant that may change at the various occurrences; the notation $\circ = \mathcal{O}(*)$ stands for $|\circ| \leq C*$.

Table of contents

1. Continuous problems

In this section we recall the weak formulation of the two-phase Stefan problem together with existence, continuous dependence, and regularity properties of the solution (see, e.g., [46,84,93,125] and the references therein). We also present two approximation procedures produced by either regularization or phase relaxation. Finally we recall the weak formulation of parabolic equations with hysteresis (see [29,143]).

1.1. Two-phase Stefan problem

The physical motivation of the solid-liquid phase transition and the classical formulation of the two-phase Stefan problem can be found in [43,93,125].

Let $\Omega \subset \mathbf{R}^d$ ($d \leq 3$) be a $C^{1,1}$ bounded (or polyhedral and convex) domain occupied by water and ice and set $Q = \Omega \times (0,T)$, where $0 < T < \infty$ is fixed. We denote the energy density by u, the relative temperature by θ, and the phase variable or water concentration by χ. Hence, $\chi \in H(\theta)$, where H is the Heaviside graph: $H(s) = 0$ if $s < 0$, $H(s) = 1$ if $s > 0$, and $H(0) = [0,1]$. We also denote the specific heat per unit volume by c and the heat conductivity by k; both may depend on θ. Then the following constitutive relation holds: $u(\theta) = \int_0^\theta c(s)ds + \lambda\chi(\theta)$, where λ denotes the latent heat per unit volume. In the classical situation, that is neglecting supercooling/superheating and surface tension effects, heat transfer within the liquid $\{\theta > 0\}$ and solid $\{\theta < 0\}$ phases, separately, is governed by the heat equation

$$c(\theta)\partial_t\theta - \operatorname{div}\big(k(\theta)\nabla\theta\big) = 0, \tag{1.1}$$

whereas the interface I of separation between solid and liquid phases is given by

$$I(t) = \{\mathbf{x} \in \Omega : \theta(x,t) = 0\}. \tag{1.2}$$

Let $\nu(\mathbf{x})$ indicate the unit vector normal to $I(t)$ at $\mathbf{x}$ towards the solid phase and $V(\mathbf{x})$ the normal interface velocity. The energy balance about $\mathbf{x}$ leads to the Stefan condition

$$[\![k\nabla\theta(\mathbf{x},t)]\!]_{\text{water}}^{\text{ice}} \cdot \nu(\mathbf{x}) = \lambda V(\mathbf{x}). \tag{1.3}$$

Equations (1.1)–(1.3) constitute the classical formulation of the two-phase Stefan problem. The drawback of this setting is that I may exhibit singularities such as cusps and mushy regions, the latter being zones of positive measure in which $0 < \chi < 1$. The Stefan condition (1.3) has been used numerically to track the motion of $I(t)$ and then to solve decoupled heat equations (1.1) in each phase, giving rise to the so-called front-tracking methods (see, e.g., [37,49,58,145]). We will not discuss these methods here however.

The weak formulation in terms of energy density, instead, incorporates (1.1)–(1.3) into a single nonlinear equation, thus making I disappear as an explicit unknown; I is recovered *a posteriori* as zero level set of the temperature. The ensuing numerical methods are called fixed domain methods. Just for the sake of simplicity assume that all material properties are normalized to one. Let us introduce the maximal monotone graph γ and its inverse function $\beta = \gamma^{-1}$:

$$\gamma(s) = s + H(s), \quad \beta(s) = \min(s,0) + \max(s-1,0), \quad \forall\, s \in \mathbf{R}, \tag{1.4}$$

and observe that $\gamma'(s) = 1$ for all $s \neq 0$ and that $0 \leq \beta'(s) \leq 1$ for almost every $s \in \mathbf{R}$ and $\beta'(s) = 1$ for all $s \notin [0,1]$ (see Figure 1.1). The energy formulation of the two-phase Stefan problem reads:

$$\partial_t u - \Delta\theta = 0, \quad \theta = \beta(u), \quad \text{in } \mathcal{D}'(Q) \tag{1.5}$$

or, equivalently, since $u = \theta + \chi$:

$$\partial_t(\theta + \chi) - \Delta\theta = 0, \quad \chi \in H(\theta), \quad \text{in } \mathcal{D}'(Q). \tag{1.6}$$

Let the initial energy density $u_0 \in L^2(\Omega)$ and, just for simplicity, the boundary condition be $\theta = 0$ on $\partial\Omega \times (0,T)$. Let $\langle\cdot,\cdot\rangle$ denote both the inner product in $L^2(\Omega)$ and the duality pairing between $H^{-1}(\Omega)$ and $H_0^1(\Omega)$. The weak form of (1.5) reads:

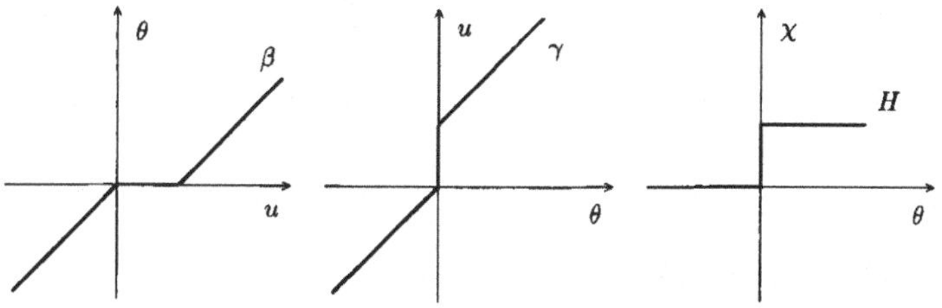

FIGURE 1.1. Constitutive relations $\theta = \beta(u)$, $u \in \gamma(\theta)$, and $\chi \in H(\theta)$ for the two-phase Stefan problem.

Seek $\{u, \theta\}$ *such that*

$$u \in L^\infty(0,T;L^2(\Omega)) \cap H^1(0,T;H^{-1}(\Omega)), \quad \theta \in L^2(0,T;H_0^1(\Omega)), \tag{1.7}$$

$$\theta(\mathbf{x},t) = \beta(u(\mathbf{x},t)) \quad \text{for a.e. } (\mathbf{x},t) \in Q, \tag{1.8}$$

$$u(\cdot,0) = u_0(\cdot) \quad (\text{in } H^{-1}(\Omega)), \tag{1.9}$$

and, for a.e. $t \in (0,T)$, *the following equation holds:*

$$\langle \partial_t u, \varphi \rangle + \langle \nabla\theta, \nabla\varphi \rangle = 0 \quad \forall\, \varphi \in H_0^1(\Omega). \tag{1.10}$$

REMARK 1.1. Formally, equation (1.5) can be rewritten as $\partial_t u - \operatorname{div}\big(\beta'(u)\nabla u\big) = 0$ which, in turn, reveals the degenerate character of the problem at hand because diffusion $\beta'(u)$ vanishes for $u \in (0,1)$. A vanishing diffusion actually characterizes the presence of a free boundary I between different phases; $I(t)$ has a finite speed of propagation and the solution exhibits a lack of regularity across I.

REMARK 1.2. We consider the simplest two-phase Stefan problem (1.7)–(1.10) just to simplify the presentation, but most of the results in Sects. 1–4 are valid for more general operators, including convection $\mathbf{v}$ and distributed sources or sinks f, and boundary conditions (see [113,125]):

$$\partial_t u - \operatorname{div}(\nabla\theta + \mathbf{v}) = f \quad \text{in } \mathcal{D}'(Q)$$
$$\theta = g_D \quad \text{on } \Gamma_D \times (0,T), \qquad \partial_\nu\theta + \mathbf{v}\cdot\nu + p\theta = g_N \quad \text{on } \Gamma_N \times (0,T),$$

where $\Gamma_D \cup \Gamma_N = \partial\Omega$ and ν stands for the exterior unit vector normal to $\partial\Omega$. Also phase-change problems in a concentrated capacity can be studied (see [7,59,86,88]).

REMARK 1.3. By changing the definition of the maximal monotone graph γ, problem (1.5) may model other physical processes. The constitutive relations corresponding to examples of relevant interest in applications are shown in Figure 1.2: one-phase Stefan problem (u energy density, θ temperature), porous medium equation (u gas density) [9], Hele-Shaw or electrochemical machining problem (θ electric potential) [44], nonstationary filtration (θ pressure, u saturation) [4]. The latter is an elliptic-parabolic problem, because equation (1.5) becomes elliptic in the saturation region $\{\theta > 0\}$.

REMARK 1.4. The Stefan problem with surface tension has been studied in [2,83]; see [3] for numerical simulation.

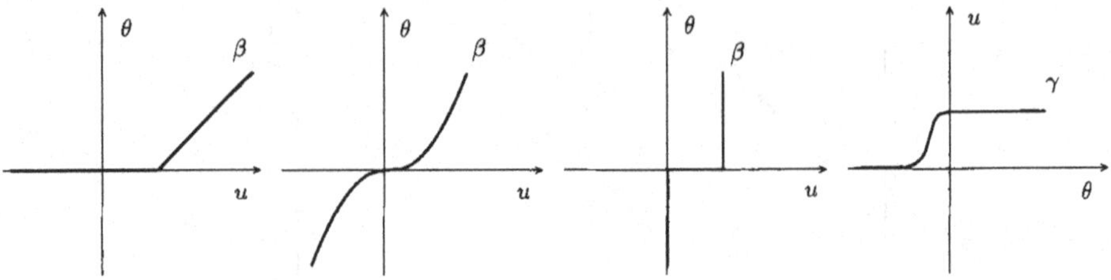

FIGURE 1.2. Constitutive relations for (a) one-phase Stefan problem (b) porous medium equation (c) Hele-Shaw problem (d) nonstationary filtration problem.

1.1.1. Existence, uniqueness, and regularity

Existence and uniqueness are well known for problem (1.7)–(1.10) (see [46,60,77,84,125] and the references therein). Existence can be proved, e.g., via regularization of both the initial datum u_0 and the function β with smooth strictly increasing functions β_ε, then proving *a priori* estimates for the corresponding classical solutions that are uniform in ε, and taking the limit by compactness and monotonicity procedures. Uniqueness is a by-product of the continuous dependence results of Sect. 1.1.2 (see also [77]).

Since (1.10) is not strictly parabolic, the solution $\{u, \theta\}$ exhibits a global lack of regularity; in fact, both u and $\nabla\theta$ are expected to have jump discontinuities across I. This behaviour makes the design and analysis of numerical methods for Stefan-like problems much harder than that for mildly nonlinear parabolic problems. In turn, the proper functional setting in which to measure global regularity plays a fundamental role. Since the *a priori* estimates for the (regularized) solutions have to be reproduced for the fully discrete problems, we sketch their proof.

THEOREM 1.1. *Depending on the regularity of the initial data u_0 and $\theta_0 = \beta(u_0)$, the following a priori estimates hold:*

$$\text{(1.11)} \quad \textit{if } u_0 \in L^2(\Omega): \quad \|u\|_{L^\infty(0,T;L^2(\Omega))} + \|\theta\|_{L^2(0,T;H^1(\Omega))} + \|u\|_{H^1(0,T;H^{-1}(\Omega))} \leq C,$$

$$\text{(1.12)} \quad \textit{if } \theta_0 \in H_0^1(\Omega): \quad \|\theta\|_{H^1(0,T;L^2(\Omega))} + \|\theta\|_{L^\infty(0,T;H^1(\Omega))} \leq C,$$
$$\|u\|_{H^{3/2-\delta}(0,T;H^{-1}(\Omega))} \leq C_\delta \quad \forall\, 0 < \delta \ll 1,$$

$$\text{(1.13)} \quad \textit{if } u_0 \in L^\infty(\Omega): \quad \|u\|_{L^\infty(Q)} + \|\theta\|_{L^\infty(Q)} \leq C,$$

$$\text{(1.14)} \quad \textit{if } \Delta\theta_0 \in L^1(\Omega) + M_{I_0}(\Omega): \quad \|\partial_t u\|_{L^\infty(0,T;M(\Omega))} \leq C.$$

$$\text{(1.15)} \quad \textit{if } u_0 \in BV(\Omega) \text{ and } \theta > 0 \text{ on } \partial\Omega \times (0,T): \quad \|u\|_{L^\infty(0,T;BV(\Omega))} \leq C.$$

Here $M_{I_0}(\Omega)$ denotes the space of finite regular Baire measures with support contained within the initial interface $I_0 = \{\mathbf{x} \in \Omega : \theta_0(\mathbf{x}) = 0\}$. Note that the assumptions on u_0 and θ_0 are compatible with the nature of the Stefan problem, because they allow jump discontinuities of u_0 and $\nabla\theta_0$ across I_0.

Proof. i) Proof of (1.11). Take $\varphi = \theta(t) \in H_0^1(\Omega)$ in (the regularized version of) (1.10) and integrate the resulting expression from 0 to $\bar{t} \leq T$. Set $\Phi_\beta(s) = \int_0^s \beta(r)\,dr$ and note that (1.4) yields $\beta^2(s) \leq 2\Phi_\beta(s) \leq s^2$ for all $s \in \mathbb{R}$. Estimate (1.11) then follows using

$$\langle \partial_t u, \theta \rangle = \langle \partial_t u, \beta(u) \rangle = \partial_t \int_\Omega \Phi_\beta(u).$$

ii) Proof of (1.12). To prove the bound for temperature, take $\varphi = \partial_t\theta(t) \in H_0^1(\Omega)$ in (the regularized version of) (1.10), integrate the resulting expression from 0 to $\bar{t} \le T$, and note that

$$\langle \partial_t u, \partial_t \theta \rangle \ge \|\partial_t \theta\|^2_{L^2(\Omega)}, \quad \langle \nabla\theta, \nabla\partial_t\theta \rangle = \tfrac{1}{2}\partial_t \|\nabla\theta\|^2_{L^2(\Omega)}.$$

The estimate for enthalpy has been recently obtained via semidiscretization in [129].

iii) Proof of (1.13). Take $\varphi = \max\big(0, u(t) - \|u_0\|_{L^\infty(\Omega)}\big) \in H_0^1(\Omega)$ in (the regularized version of) (1.10) and integrate the resulting expression from 0 to $\bar{t} \le T$ to get

$$u(\bar{t}) \le \|u_0\|_{L^\infty(\Omega)} \quad \text{a.e. in } \Omega.$$

iv) Proof of (1.14). See [72]. Differentiate in time (the regularized version of) (1.10). Then take $\varphi = \mathrm{sgn}_\sigma(\partial_t\theta(t))$ and integrate from 0 to $\bar{t} \le T$, where $\mathrm{sgn}_\sigma(s) = \frac{s}{\sqrt{\sigma + s^2}}$ is a regularization of the sign function. First, note that

$$\langle \nabla\partial_t\theta, \nabla \mathrm{sgn}_\sigma(\partial_t\theta) \rangle \ge 0,$$

then take the limit as $\sigma \downarrow 0$ using the Lebesgue theorem to get $\langle \partial_{tt}u, \mathrm{sgn}(\partial_t\theta)\rangle \le 0$. Since the regularized β is strictly increasing, we have $\mathrm{sgn}(\partial_t\theta) = \mathrm{sgn}(\partial_t u)$, whence

$$0 \ge \langle \partial_{tt}u, \mathrm{sgn}(\partial_t\theta)\rangle = \langle \partial_{tt}u, \mathrm{sgn}(\partial_t u)\rangle = \partial_t \|\partial_t u\|_{L^1(\Omega)}.$$

v) Proof of (1.15). See [65,70] and also [89] for stability in Nikolskiĭ spaces. For any $\mathbf{r} \in \mathbb{R}^d$ set $\theta_{\mathbf{r}}(\mathbf{x}, \cdot) = \theta(\mathbf{x} + \mathbf{r}, \cdot)$. Select $\varphi = \mathrm{sgn}_\sigma(\theta - \theta_{-\mathbf{r}})$ and $\varphi = \mathrm{sgn}_\sigma(\theta_{\mathbf{r}} - \theta)$ in (the regularized version of) (1.10), then take the difference of the resulting expressions and integrate from 0 to $\bar{t} \le T$. □

REMARK 1.5. In light of (1.11) and (1.12), that are sharp because $\partial_t u, \nabla u \notin L^2(Q)$ and thus $\theta \notin L^2(0,T; H^2(\Omega))$, we conclude that the natural energy spaces to examine the accuracy of any approximation to problem (1.7)–(1.10) are $L^\infty(0,T; H^{-1}(\Omega))$ for u and $L^2(Q)$ for θ. Morever, since integration in time of (1.5) yields, for all $t \in (0,T)$,

$$u(t) - \Delta \textstyle\int_0^t \theta(s)\,ds = u_0, \tag{1.16}$$

and so $\int_0^t \theta \in L^\infty(0,T; H^2(\Omega))$, we infer that the natural space for the variable $\int_0^t \nabla\theta(s)\,ds$, related to the heat flux, is $L^\infty(0,T; L^2(\Omega))$. On the other hand, $L^\infty(0,T; L^1(\Omega))$ error estimates for enthalpy may be derived from the BV stability (1.15) (see [89,137]).

We close this section with the concept of nondegeneracy, which is important in the approximation of interfaces as well as in the analysis of any regularization process. In fact, fixed domain methods give an approximation of both the physical unknowns u and θ, whereas the approximate free boundary is recovered *a posteriori* as a level set of the approximate solution which, in general, may not converge to the original interface. We say that problem (1.7)–(1.10) satisfies a nondegeneracy property if θ exhibits a linear growth rate away from the interface I, namely:

$$\mathrm{meas}\big(\{(\mathbf{x},t) \in Q : 0 \le \theta(\mathbf{x},t) \le \varepsilon\}\big) \le C\varepsilon. \tag{1.17}$$

This condition, which excludes the formation of mushy regions, is known to hold under certain qualitative properties of the data [99]. Nonappearance of mushy regions is discussed also in [21,65].

1.1.2. Continuous dependence on the data

Here we study the Lipschitz continuous dependence of the solution of problem (1.7)–(1.10) on the initial data. Despite its intrinsic interest, this study will motivate the two main techniques for the error analysis of the fully discrete schemes in a simpler setting.

THEOREM 1.2. *Let $\{u_1, \theta_1\}$ and $\{u_2, \theta_2\}$ be the solutions to (1.7)-(1.10) corresponding to initial data u_{01} and u_{02}, respectively. Then*

$$\|u_1 - u_2\|_{L^\infty(0,T;H^{-1}(\Omega))} + \|\theta_1 - \theta_2\|_{L^2(Q)} + \Big\|\int_0^t \nabla(\theta_1 - \theta_2)\Big\|_{L^\infty(0,T;L^2(\Omega))} \le C\|u_{01} - u_{02}\|_{H^{-1}(\Omega)}, \tag{1.18}$$

$$\|u_1 - u_2\|_{L^\infty(0,T;L^1(\Omega))} \le \|u_{01} - u_{02}\|_{L^1(\Omega)}. \tag{1.19}$$

Proof. Set $\tilde{u} = u_1 - u_2$, $\tilde{\theta} = \theta_1 - \theta_2$, and $\tilde{u}_0 = u_{01} - u_{02}$. From (1.10) we get the error equation, valid for a.e. $t \in (0,T)$,

$$\langle \partial_t \tilde{u}, \varphi \rangle + \langle \nabla \tilde{\theta}, \nabla \varphi \rangle = 0 \quad \forall\, \varphi \in H_0^1(\Omega) \tag{1.20}$$

and also, after integration in time, for all $t \in (0,T)$,

$$\langle \tilde{u}(t), \varphi \rangle + \langle \nabla \textstyle\int_0^t \tilde{\theta}(s)\,ds, \nabla \varphi \rangle = \langle \tilde{u}_0, \varphi \rangle \quad \forall\, \varphi \in H_0^1(\Omega). \tag{1.21}$$

i) Proof of (1.18). 1st procedure: inversion of the Laplacian. Let $G : H^{-1}(\Omega) \to H_0^1(\Omega)$ denote the Green's operator, defined for all $\psi \in H^{-1}(\Omega)$ by

$$\langle \nabla G\psi, \nabla \varphi \rangle = \langle \psi, \varphi \rangle \quad \forall\, \varphi \in H_0^1(\Omega). \tag{1.22}$$

The norm in $H^{-1}(\Omega)$ can thus be represented in terms of G as

$$\|\psi\|_{H^{-1}(\Omega)} = \|\nabla G\psi\|_{L^2(\Omega)} = \langle \psi, G\psi \rangle^{1/2}. \tag{1.23}$$

In addition, G is regular, that is $\|G\psi\|_{H^{2-s}(\Omega)} \le C\|\psi\|_{H^{-s}(\Omega)}$ $(s = 0,1)$, provided Ω is convex or smooth [38,p.138].

In view of (1.22), the error equation (1.20) equivalently reads, for a.e. $t \in (0,T)$,

$$\langle \partial_t \tilde{u}, G\psi \rangle + \langle \tilde{\theta}, \psi \rangle = 0 \quad \forall\, \psi \in H^{-1}(\Omega). \tag{1.24}$$

Take $\psi = \tilde{u}(t)$ in (1.24) and integrate from 0 to $\bar{t} \le T$. Exploit the symmetry of G, namely $\langle G\psi_1, \psi_2 \rangle = \langle \psi_1, G\psi_2 \rangle$ and (1.23), to note that

$$\langle \partial_t \tilde{u}, G\tilde{u} \rangle = \tfrac{1}{2}\partial_t \langle \tilde{u}, G\tilde{u} \rangle = \tfrac{1}{2}\partial_t \|\tilde{u}\|^2_{H^{-1}(\Omega)}.$$

This, in conjunction with the inequality

$$\langle \tilde{\theta}, \tilde{u} \rangle \ge \|\tilde{\theta}\|^2_{L^2(\Omega)},$$

which follows from (1.4), leads to the estimate

$$\tfrac{1}{2}\|\tilde{u}\|^2_{L^\infty(0,T;H^{-1}(\Omega))} + \|\tilde{\theta}\|^2_{L^2(Q)} \le \tfrac{1}{2}\|\tilde{u}_0\|^2_{H^{-1}(\Omega)}. \tag{1.25}$$

2nd *procedure: integral-time method.* Take $\varphi = \tilde{\theta}(t)$ in (1.21) and integrate from 0 to $\bar{t} \le T$. Then note that

$$\langle \tilde{u}, \tilde{\theta} \rangle \ge \|\tilde{\theta}\|^2_{L^2(\Omega)}, \quad \langle \textstyle\int_0^t \nabla\tilde{\theta}, \nabla\tilde{\theta} \rangle = \tfrac{1}{2}\partial_t \Big\|\textstyle\int_0^t \nabla\tilde{\theta}\Big\|^2_{L^2(\Omega)}.$$

Using the elementary inequality

$$(1.26) \qquad 2ab \le \tfrac{1}{\sigma}a^2 + \sigma b^2 \quad \forall\, a,b \in \mathbf{R},\ \sigma > 0,$$

in conjunction with the Poincaré inequality, we also have

$$\int_0^{\bar t}\langle \tilde u_0, \tilde\theta\rangle = \left\langle \tilde u_0, \int_0^{\bar t}\tilde\theta\right\rangle \le \|\tilde u_0\|_{H^{-1}(\Omega)}\left\|\int_0^t\tilde\theta\right\|_{H_0^1(\Omega)} \le C\|\tilde u_0\|^2_{H^{-1}(\Omega)} + \tfrac14\left\|\int_0^{\bar t}\nabla\tilde\theta\right\|^2_{L^2(\Omega)}.$$

The previous estimates leads to

$$(1.27) \qquad \|\tilde\theta\|^2_{L^2(Q)} + \tfrac14\left\|\int_0^t\nabla\tilde\theta\right\|^2_{L^\infty(0,T;L^2(\Omega))} \le C\|\tilde u_0\|^2_{H^{-1}(\Omega)}.$$

To conclude the proof of (1.18), note that the bounds for $\tilde u$ in $L^\infty(0,T;H^{-1}(\Omega))$ and $\int_0^t\nabla\tilde\theta$ in $L^\infty(0,T;L^2(\Omega))$ are equivalent. In fact, take $\varphi = \int_0^t\tilde\theta$ in (1.21) and use (1.25), in conjunction with (1.26) and the Poincaré inequality, to get, for all $t \in (0,T)$,

$$\left\|\int_0^t\nabla\tilde\theta\right\|^2_{L^2(\Omega)} \le \|\tilde u(t) - \tilde u_0\|_{H^{-1}(\Omega)}\left\|\int_0^t\tilde\theta\right\|_{H_0^1(\Omega)} \le C\|\tilde u_0\|^2_{H^{-1}(\Omega)} + \tfrac12\left\|\int_0^t\nabla\tilde\theta\right\|^2_{L^2(\Omega)}.$$

Viceversa, take $\varphi = G\tilde u$ in (1.21) and use (1.27) and (1.23) to get, for all $t \in (0,T)$, $\|\tilde u(t)\|_{H^{-1}(\Omega)} \le C\|\tilde u_0\|_{H^{-1}(\Omega)}$.

ii) Proof of (1.19). See [45]. Since the estimate is independent of the minimum slope of β, we deal with a regularized problem with $\beta'(s) > 0$ for a.e. $s \in \mathbf{R}$. Let $\mathrm{sgn}_\sigma(s) = \frac{s}{\sqrt{\sigma+s^2}}$ for all $s \in \mathbf{R}$ be a regularization of the sign function. Select $\varphi = \mathrm{sgn}_\sigma(\tilde\theta(t))$ in (1.20) and note that $\langle\nabla\tilde\theta, \nabla\mathrm{sgn}_\sigma(\tilde\theta)\rangle \ge 0$; then take the limit as $\sigma \downarrow 0$ using the Lebesgue dominated convergence theorem to get

$$0 \ge \langle\partial_t\tilde u, \mathrm{sgn}(\tilde\theta)\rangle = \langle\partial_t\tilde u, \mathrm{sgn}(\tilde u)\rangle = \partial_t\|\tilde u\|_{L^1(\Omega)}.$$

After integration in time from 0 to $\bar t \le T$, taking the lim inf as the regularization parameter vanishes, we get (1.19). □

1.1.3. Regularization

Since the solution $\{u,\theta\}$ has low regularity properties, sometimes a smoothing procedure is used prior to space and time discretization. The usual one consists in replacing the constitutive function β by strictly increasing functions β_ε, where $\varepsilon > 0$ indicates the regularization parameter. The regularized problem now behaves as a mildly nonlinear parabolic equation, but the global regularity of its solution may depend on ε. In this section we precisely study the effects of such a smoothing procedure.

Let $\varepsilon > 0$ indicate the regularization parameter. Let us introduce a regularization H_ε to the Heaviside graph:

$$(1.28) \qquad H_\varepsilon(s) = H(s) \quad \text{if } s < 0 \text{ or } s > \varepsilon, \qquad H_\varepsilon(s) = \tfrac{1}{\varepsilon}s \quad \text{if } 0 \le s \le \varepsilon,$$

and set

$$(1.29) \qquad \gamma_\varepsilon(s) = s + H_\varepsilon(s), \qquad \beta_\varepsilon(s) = \gamma_\varepsilon^{-1}(s), \qquad \forall\, s \in \mathbf{R}.$$

Set $\hat\varepsilon = \frac{\varepsilon}{1+\varepsilon}$ and note that $\beta'_\varepsilon(s) \ge \hat\varepsilon$ for a.e. $s \in \mathbf{R}$ (see Figure 1.3).

Let $\{u_\varepsilon, \theta_\varepsilon\}$ denote the solution to the problem obtained from (1.5) by replacing the constitutive relation $\theta = \beta(u)$ by $\theta_\varepsilon = \beta_\varepsilon(u_\varepsilon)$. The regularized problem thus reads:

$$(1.30) \qquad \partial_t u_\varepsilon - \Delta\theta_\varepsilon = 0, \quad \theta_\varepsilon = \beta_\varepsilon(u_\varepsilon), \quad \text{in } Q$$

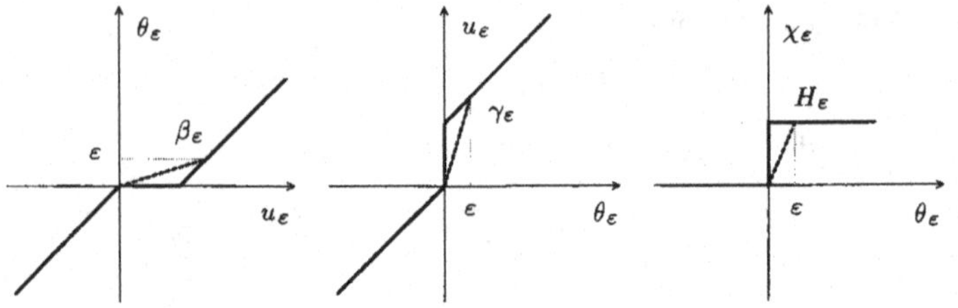

FIGURE 1.3. Regularized constitutive relations $\theta_\varepsilon = \beta_\varepsilon(u_\varepsilon)$, $u_\varepsilon = \gamma(\theta_\varepsilon)$, and $\chi_\varepsilon = H_\varepsilon(\theta_\varepsilon)$.

or, equivalently in terms of θ_ε and $\chi_\varepsilon = u_\varepsilon - \theta_\varepsilon$:

$$\partial_t(\theta_\varepsilon + \chi_\varepsilon) - \Delta\theta_\varepsilon = 0, \quad \chi_\varepsilon = H_\varepsilon(\theta_\varepsilon), \quad \text{in } Q, \tag{1.31}$$

which are to be interpreted in the weak sense.

REMARK 1.6. The a priori estimates of Theorem 1.1 hold uniformly in the regularization parameter ε provided the initial datum $u_{0\varepsilon}$ is properly chosen: $u_{0\varepsilon} = u_0$ for (1.11) and (1.13); $u_{0\varepsilon} = \gamma_\varepsilon(\theta_0)$ for (1.12) and (1.14), which are not valid if $u_{0\varepsilon} = u_0$. Now recall that (1.29) yields $\beta_\varepsilon'(s) \geq \frac{\varepsilon}{1+\varepsilon} = \hat{\varepsilon}$ for a.e. $s \in \mathbf{R}$ and, since no confusion is possible, remove the *hat* and set $\varepsilon = \hat{\varepsilon}$. The following *a priori* estimates hold:

$$\textit{if } u_0 \in L^2(\Omega): \quad \|\nabla u_\varepsilon\|_{L^2(Q)} \leq C\varepsilon^{-1/2}, \tag{1.32}$$

$$\textit{if } \theta_0 \in H_0^1(\Omega): \quad \|\partial_t u_\varepsilon\|_{L^2(Q)} + \|\theta_\varepsilon\|_{L^2(0,T;H^2(\Omega))} \leq C\varepsilon^{-1/2}. \tag{1.33}$$

In light of the study of travelling wave solutions in Sect. 1.1.3.1, we see that both (1.32) and (1.33) are sharp. The proof proceeds as for estimates (1.11) and (1.12), respectively. To prove (1.32), take $\varphi = u_\varepsilon = \gamma_\varepsilon(\theta_\varepsilon) \in H_0^1(\Omega)$ in the weak formulation of (1.30) and use that $\langle \nabla\theta_\varepsilon, \nabla u_\varepsilon \rangle \geq \varepsilon \|\nabla u_\varepsilon\|_{L^2(Q)}^2$. To prove (1.33), take $\varphi = \partial_t\theta_\varepsilon$ in (1.30) and use that $\langle \partial_t u_\varepsilon, \partial_t\theta_\varepsilon \rangle \geq \varepsilon \|\partial_t u_\varepsilon\|_{L^2(Q)}^2$. In view of (1.32) and (1.33) we readily deduce that

$$\operatorname{meas}\left(\{(\mathbf{x},t) \in Q : |\partial_t u_\varepsilon|, |\nabla u_\varepsilon| \geq \varepsilon^{-1}\}\right) \leq C\varepsilon.$$

1.1.3.1. Travelling waves for the regularized problem. We study the travelling waves for the regularized problem in one space dimension in order to understand the effects produced by this smoothing procedure on both free boundary and physical unknowns. Since the effects of time discretization are similar, it turns out that this study is crucial in the design of adaptive finite element methods, where the relation between local meshsize and time step as well as the local velocity indicators are dictated by local regularity and profile of the discrete variables.

Let us first introduce the travelling waves for the Stefan problem (1.6) in one space dimension. Let $\bar{\theta} \in C^{0,1}(\mathbf{R})$ and $\bar{\chi} \in H(\bar{\theta})$ a.e. in $\mathbf{R}$. The travelling waves moving towards the left with velocity $V > 0$ are solutions of (1.6) of the form

$$\theta(x,t) = \bar{\theta}(x + Vt), \quad \chi(x,t) = \bar{\chi}(x + Vt), \quad \text{in } \mathbf{R} \times (0,T).$$

Since no confusion is possible, we drop the *overbar* from now on. We thus have

$$V\theta' + V\chi' - \theta'' = 0 \quad \text{in } \mathcal{D}'(\mathbf{R}). \tag{1.34}$$

Let us define the solid (or crystal), interface (or mushy), and liquid regions as

$$\mathcal{C} = \{x \in \mathbf{R} : \theta(x) < 0\}, \quad \mathcal{T} = \{x \in \mathbf{R} : \theta(x) = 0\}, \quad \mathcal{L} = \{x \in \mathbf{R} : \theta(x) > 0\},$$

and note that $\chi = 0$ in $\mathcal{C}$, $\chi = 1$ in $\mathcal{L}$, and $0 \le \chi \le 1$ in $\mathcal{T}$. Then (1.34) becomes $V\theta'(x) - \theta''(x) = 0$ within the open set $\mathcal{C} \cup \mathcal{L}$ and $\theta \in \operatorname{span}\{e^{Vx}, 1\}$ in $\mathcal{C}$ and $\mathcal{L}$. It is easy to see that the travelling waves for the melting process correspond to $\mathcal{L} = (0, +\infty)$ and either $\mathcal{T} = \{0\}$ (two-phase process) or $\mathcal{T} = (-\infty, 0]$ (one-phase process). Consider only the two-phase problem. Since $\theta \in C^{0,1}(\mathbf{R})$, on imposing $\theta(0) = 0$ and the Stefan condition $V = [\![\theta']\!]$ at $x = 0$, for any $\sigma = \theta'(0^-) > 0$ we obtain the explicit representation:

$$\theta(x) = \begin{cases} \frac{\sigma}{V}(e^{Vx} - 1) \\ \frac{\sigma + V}{V}(e^{Vx} - 1) \end{cases} \qquad \chi(x) = \begin{cases} 0 & \text{if } x \in \mathcal{C} \cup \mathcal{T} = (-\infty, 0] \\ 1 & \text{if } x \in \mathcal{L} = (0, +\infty) \end{cases}$$

(see Figure 1.4). Also, θ' is discontinuous at $x = 0$, and thus $\theta \in C^{0,1}(\mathbf{R}) \backslash C^1(\mathbf{R})$.

Let now $\bar{\theta}_\varepsilon \in C^{1,1}(\mathbf{R})$ and $\bar{\chi}_\varepsilon = H_\varepsilon(\bar{\theta}_\varepsilon) \in C^{0,1}(\mathbf{R})$. Let

$$\theta_\varepsilon(x,t) = \bar{\theta}_\varepsilon(x + Vt), \quad \chi_\varepsilon(x,t) = \bar{\chi}_\varepsilon(x + Vt), \quad \text{in } \mathbf{R} \times (0,T),$$

be travelling waves for the regularized problem (1.31) in one space dimension moving towards the left with velocity $V > 0$. For simplicity of notation, we drop the *overbar* in the sequel. Then θ_ε satisfies the following ordinary differential equation:

$$V(1 + H'_\varepsilon(\theta_\varepsilon))\theta'_\varepsilon - \theta''_\varepsilon = 0 \quad \text{in } \mathcal{D}'(\mathbf{R}). \tag{1.35}$$

We define the solid, transition, and liquid regions as

$$\mathcal{C}_\varepsilon = \{x \in \mathbf{R} : \chi_\varepsilon(x) = 0\}, \quad \mathcal{T}_\varepsilon = \{x \in \mathbf{R} : 0 < \chi_\varepsilon(x) < 1\}, \quad \mathcal{L}_\varepsilon = \{x \in \mathbf{R} : \chi_\varepsilon(x) = 1\},$$

and note that $\theta_\varepsilon \le 0$ in $\mathcal{C}_\varepsilon$, $\theta_\varepsilon \ge \varepsilon$ in $\mathcal{L}_\varepsilon$, and $0 < \theta_\varepsilon < \varepsilon$ in $\mathcal{T}_\varepsilon$. In view of the maximum principle, θ_ε is exponentially monotone in $\mathbf{R}$; thus $\mathcal{T}_\varepsilon$ is a single (possibly unbounded) open interval and the melting process correspond to either $\mathcal{T}_\varepsilon = (0, x_\varepsilon)$ and $\mathcal{L}_\varepsilon = [x_\varepsilon, +\infty)$ (two-phase process) or $\mathcal{T}_\varepsilon = (-\infty, 0)$ and $\mathcal{L}_\varepsilon = [0, +\infty)$ (one-phase process). Owing to (1.28), equation (1.35) becomes $V\theta'_\varepsilon(x) - \theta''_\varepsilon(x) = 0$ within $\mathcal{C}_\varepsilon \cup \mathcal{L}_\varepsilon$ and $\frac{V}{\hat{\varepsilon}}\theta'_\varepsilon(x) - \theta''_\varepsilon(x) = 0$ in $\mathcal{T}_\varepsilon$; then $\theta_\varepsilon \in \operatorname{span}\{e^{Vx}, 1\}$ in $\mathcal{C}_\varepsilon$ and $\mathcal{L}_\varepsilon$ and $\theta_\varepsilon \in \operatorname{span}\{e^{Vx/\hat{\varepsilon}}, 1\}$ in $\mathcal{T}_\varepsilon$. Since $\theta_\varepsilon \in C^{1,1}(\mathbf{R})$, on imposing $\theta_\varepsilon(0) = \chi_\varepsilon(0) = 0$ and $\theta_\varepsilon(x_\varepsilon) = \varepsilon$ (i.e., $\chi_\varepsilon(x_\varepsilon) = 1$), we obtain the explicit representation of the travelling waves for the two-phase problem:

$$\theta_\varepsilon(x) = \begin{cases} \theta(x) \\ \frac{\sigma}{V}\hat{\varepsilon}(e^{Vx/\hat{\varepsilon}} - 1) \\ \frac{\sigma + (1+\varepsilon)V}{V}(e^{V(x - x_\varepsilon)} - 1) + \varepsilon \end{cases} \qquad \chi_\varepsilon(x) = \begin{cases} \chi(x) & \text{if } x \in \mathcal{C}_\varepsilon = (-\infty, 0] \\ \frac{\sigma}{V}\frac{\hat{\varepsilon}}{\varepsilon}(e^{Vx/\hat{\varepsilon}} - 1) & \text{if } x \in \mathcal{T}_\varepsilon = (0, x_\varepsilon) \\ \chi(x) & \text{if } x \in \mathcal{L}_\varepsilon = [x_\varepsilon, +\infty) \end{cases}$$

(see Figure 1.4). Both χ'_ε and θ''_ε exhibit jump discontinuities at $x = 0, x_\varepsilon$, and thus $\theta_\varepsilon \in C^{1,1}(\mathbf{R}) \backslash C^2(\mathbf{R})$ and $\chi_\varepsilon \in C^{0,1}(\mathbf{R}) \backslash C^1(\mathbf{R})$.

We realize that the free boundary is spread out into a thin transition layer of width $x_\varepsilon = \frac{\hat{\varepsilon}}{V} \log\left(1 + \frac{V}{\sigma}(1+\varepsilon)\right) = \mathcal{O}(\varepsilon)$. A smearing effect is observed, which does not prevent the Stefan condition $[\![\theta']\!] = V$ at $x = 0$ to be recovered at the ends of the transition region, because $\theta'_\varepsilon(x_\varepsilon) - \theta'_\varepsilon(0) = V(1 + \varepsilon)$. Note that $\theta_\varepsilon(x) = \theta(x) + e^{Vx}\mathcal{O}(\varepsilon)$ in $\mathcal{L}_\varepsilon$.

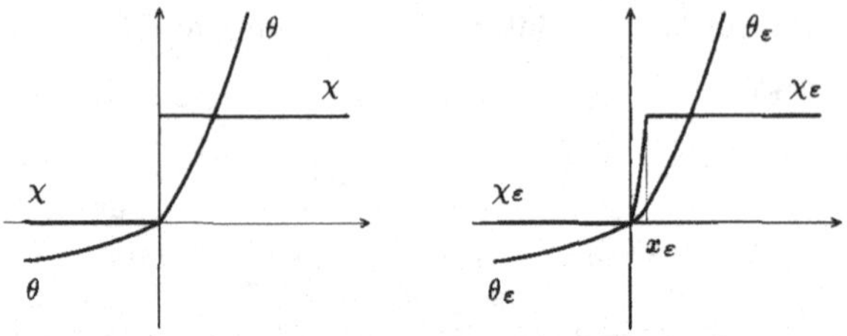

FIGURE 1.4. Travelling waves for (a) the two-phase Stefan problem ($\sigma = 0.5$) and (b) the regularized problem ($\varepsilon = 0.15$) melting towards the left with velocity $V = 1$.

Moreover, $\|\theta_\varepsilon''\|_{L^1(-\infty,x)} \leq Ce^{Vx}$ is valid uniformly in ε for $x > 0$, and resembles the estimate (1.14) for the Stefan problem.

1.1.3.2. Error analysis for the regularized problem. We study the error between the solution $\{u, \theta\}$ of problem (1.7)–(1.10) and the regularized solutions $\{u_\varepsilon, \theta_\varepsilon\}$.

THEOREM 1.3. *Let $u_0 \in L^2(\Omega)$ and $u_{0\varepsilon} = u_0$. Then the following error estimate holds:*

$$
\begin{aligned}
E_\varepsilon := \varepsilon^{1/2}\|u - u_\varepsilon\|_{L^2(Q)} + \|\theta - \theta_\varepsilon\|_{L^2(Q)} + \|u - u_\varepsilon\|_{L^\infty(0,T;H^{-1}(\Omega))} \\
+ \left\|\int_0^t \nabla(\theta - \theta_\varepsilon)\right\|_{L^\infty(0,T;L^2(\Omega))} \leq C\big(\varepsilon \operatorname{meas}(A_\varepsilon)\big)^{1/2},
\end{aligned}
\tag{1.36}
$$

where $A_\varepsilon = \{(\mathbf{x}, t) \in Q : 0 \leq \theta(\mathbf{x}, t) \leq \varepsilon\} = \{(\mathbf{x}, t) \in Q : 0 \leq u(\mathbf{x}, t) \leq 1 + \varepsilon\}$.

The general rate of convergence is $\mathcal{O}(\varepsilon^{1/2})$, which includes the formation of mushy regions [54,73,96]. If the nondegeneracy property (1.17) is valid, then the rate becomes $\mathcal{O}(\varepsilon)$, which is optimal according to the regularity being dealt with (see [96,98]); in addition, the error estimate $\|u - u_\varepsilon\|_{L^2(Q)} \leq C\varepsilon^{1/2}$ holds.

Proof. Set $\tilde{u} = u - u_\varepsilon$ and $\tilde{\theta} = \theta - \theta_\varepsilon$. Take the difference between (1.10) and the weak form of (1.30) to get the error equation, for a.e. $t \in (0, T)$,

$$
\langle \partial_t \tilde{u}, \varphi \rangle + \langle \nabla\tilde{\theta}, \nabla\varphi \rangle = 0 \quad \forall\, \varphi \in H_0^1(\Omega). \tag{1.37}
$$

Take $\varphi = G\tilde{u}(t)$ in (1.37) and integrate from 0 to $\bar{t} \leq T$. Using (1.22) and the relation $\partial_t\|\tilde{u}\|^2_{H^{-1}(\Omega)} = 2\langle \partial_t\tilde{u}, G\tilde{u}\rangle$, we arrive at

$$
\|\tilde{u}(\bar{t})\|^2_{H^{-1}(\Omega)} + \int_0^{\bar{t}} \langle \tilde{\theta}(t), \tilde{u}(t)\rangle\, dt = \|\tilde{u}(0)\|^2_{H^{-1}(\Omega)} = 0.
$$

Now we write, for a.e. $(\mathbf{x}, t) \in Q$,

$$
\langle \tilde{\theta}, \tilde{u} \rangle = \langle \beta(u) - \beta_\varepsilon(u_\varepsilon), u - u_\varepsilon \rangle = \langle \beta_\varepsilon(u) - \beta_\varepsilon(u_\varepsilon), u - u_\varepsilon\rangle + \langle \beta(u) - \beta_\varepsilon(u), u - u_\varepsilon \rangle.
$$

Note that the property $\varepsilon \leq \beta_\varepsilon'(s) \leq 1$ for a.e. $s \in \mathbf{R}$, in conjunction with (1.26), yields

$$
\begin{aligned}
2\langle \beta_\varepsilon(u) - \beta_\varepsilon(u_\varepsilon), u - u_\varepsilon \rangle &\geq \varepsilon\|\tilde{u}\|^2_{L^2(\Omega)} + \|\beta_\varepsilon(u) - \beta_\varepsilon(u_\varepsilon)\|^2_{L^2(\Omega)} \\
&\geq \varepsilon\|\tilde{u}\|^2_{L^2(\Omega)} + \tfrac{1}{2}\|\tilde{\theta}\|^2_{L^2(\Omega)} - \|\beta(u) - \beta_\varepsilon(u)\|^2_{L^2(\Omega)},
\end{aligned}
$$

whereas

$$
\langle \beta(u) - \beta_\varepsilon(u), u - u_\varepsilon \rangle \geq -\tfrac{\varepsilon}{4}\|\tilde{u}\|^2_{L^2(\Omega)} - \tfrac{1}{\varepsilon}\|\beta(u) - \beta_\varepsilon(u)\|^2_{L^2(\Omega)}.
$$

Finally exploit the property $\beta(u) = \beta_\varepsilon(u)$ if $(\mathbf{x}, t) \notin A_\varepsilon$ to get

$$
\|\beta(u) - \beta_\varepsilon(u)\|^2_{L^2(\Omega)} \leq \varepsilon^2 \operatorname{meas}(A_\varepsilon).
$$

Collecting the previous estimates leads to the desired bounds for $\tilde{u}$ and $\tilde{\theta}$. Now integrate (1.37) in time,

$$\langle \tilde{u}(t), \varphi \rangle + \left\langle \int_0^t \nabla \tilde{\theta}, \nabla \varphi \right\rangle = 0 \quad \forall\, \varphi \in H_0^1(\Omega),$$

and select $\varphi = \int_0^t \tilde{\theta}$. Invoking the Poincaré inequality, in conjunction with the above estimate for $\tilde{u}$, we get the asserted bound for $\int_0^t \nabla \tilde{\theta}$. □

REMARK 1.7. Let $u_{0\varepsilon} = \gamma_\varepsilon(\theta_0)$ and $\text{meas}(\{\mathbf{x} \in \Omega : 0 \le \theta_0(\mathbf{x}) \le \varepsilon\} \le C\varepsilon$. Then the following error estimate for the initial energy densities holds [115]:

$$\|u_0 - u_{0\varepsilon}\|_{H^{-1}(\Omega)} \le C\varepsilon |\log \varepsilon|^{1/2}.$$

To prove it, use the initial nondegeneracy property to derive

$$\|u_0 - u_{0\varepsilon}\|_{L^p(\Omega)} \le C\varepsilon^{1/p} \quad \forall\, 1 \le p < \infty$$

and then apply the well-known two dimensional Poincaré-Sobolev inequality

$$\|\varphi\|_{L^p(\Omega)} \le Cp^{1/2} \|\varphi\|_{H_0^1(\Omega)} \quad \forall\, \varphi \in H_0^1(\Omega).$$

Hence, the error estimate (1.36) becomes $E_\varepsilon \le C\big(\varepsilon \text{meas}(A_\varepsilon) + \varepsilon^2 |\log \varepsilon|\big)^{1/2}$.

1.1.4. Phase relaxation

We present a different approximation procedure based on a suitable violation of the constitutive relation $\chi \in H(\theta)$ or, equivalently, $\theta \in H^{-1}(\chi)$. Let $\varepsilon > 0$ indicate a relaxation parameter. The following dynamic phase condition was proposed by Visintin [142] to model supercooling/superheating effects in phase transitions:

$$\varepsilon \partial_t \chi_\varepsilon + H^{-1}(\chi_\varepsilon) \ni \theta_\varepsilon \tag{1.38}$$

or, equivalently:

$$\varepsilon \partial_t \chi_\varepsilon = \begin{cases} \max(\theta_\varepsilon, 0) & \text{if } \chi_\varepsilon = 0 \\ \theta_\varepsilon & \text{if } 0 < \chi_\varepsilon < 1 \\ \min(\theta_\varepsilon, 0) & \text{if } \chi_\varepsilon = 1. \end{cases}$$

Problem (1.6) is thus replaced by the system

$$\partial_t \theta_\varepsilon + \partial_t \chi_\varepsilon - \Delta \theta_\varepsilon = 0, \quad \varepsilon \partial_t \chi_\varepsilon + H^{-1}(\chi_\varepsilon) \ni \theta_\varepsilon, \quad \text{in } Q, \tag{1.39}$$

which is to be interpreted in the weak sense. Existence and uniqueness for (1.39) are well known [142].

Apart from its own physical relevance, (1.39) can be viewed as an approximation to the Stefan problem; in this case the initial conditions for θ_ε and χ_ε are inherited from that of the original problem (1.7)–(1.10), namely $\theta_0 = \beta(u_0)$ and $u_0 - \theta_0$, respectively. The rate of convergence of such an approximation as $\varepsilon \downarrow 0$ is studied in Sect. 1.1.4.2, whereas the smoothing effects produced on the free boundary and the physical variables are discussed in Sect. 1.1.4.1 for travelling wave solutions in one dimension.

REMARK 1.8. *Crystallization of polymers.* The dynamic constitutive relation (1.38) may be generalized to model various phenomena, such as polymer solidification (see, e.g., [8,15,22,91,92,133,134] and the references therein). Denote by $w = 1 - \chi$ the crystalline volume fraction of the polymer and consider the global crystallization kinetics

$$\partial_t w = K(\theta) R(w), \tag{1.40}$$

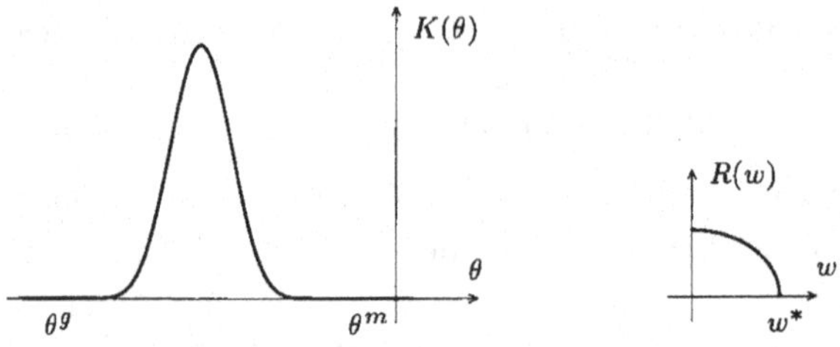

FIGURE 1.5. Kinetic constant $K(\theta)$ and reduction factor $R(w)$ in the global crystallization kinetics.

coupled with the heat equation

$$\partial_t\theta - \Delta\theta = \partial_t w, \tag{1.41}$$

and with suitable initial and boundary conditions. The typical profile of the kinetic constant $K(\theta)$ and reduction factor $R(w)$ are depicted in Figure 1.5. At temperatures bigger than θ^m, the equilibrium melting temperature, the polymer is liquid and $w = 0$. As soon as the temperature θ enters the range $[\theta^g, \theta^m]$, where the kinetic constant $K(\theta)$ is positive, the crystallization starts because $\partial_t w > 0$. The process concludes either when θ reaches the glass transition temperature θ^g or the crystallinity w reaches the equilibrium value $0 < w^* \leq 1$ because $R(w^*) = 0$. Note that a non-Lipschitz continuous reduction factor $R(w)$ allows the crystallinity w to attain w^*. In that case the continuous dependence on the data can be proved using L^1-techniques (see [8]; see also [76] for the error analysis of a fully discrete approximation to a model of polymer crystallization).

THEOREM 1.4. *Let $\{\theta_1, w_1\}$ and $\{\theta_2, w_2\}$ be the solutions to (1.40),(1.41) corresponding to initial data $\{\theta_{01}, w_{01}\}$ and $\{\theta_{02}, w_{02}\}$, respectively. Then*

$$\|\theta_1-\theta_2\|_{L^\infty(0,T;L^1(\Omega))}+\|w_1-w_2\|_{L^\infty(0,T;L^1(\Omega))}\leq C\big(\|\theta_{01}-\theta_{02}\|_{L^1(\Omega)}+\|w_{01}-w_{02}\|_{L^1(\Omega)}\big).$$

Proof. Set $\tilde\theta = \theta_1 - \theta_2$, $\tilde w = w_1 - w_2$, $\tilde\theta_0 = \theta_{01} - \theta_{02}$, and $\tilde w_0 = w_{01} - w_{02}$. From equations (1.40) and (1.41) we get the error equations

$$\partial_t\tilde\theta - \Delta\tilde\theta = \partial_t\tilde w = K(\theta_1)R(w_1) - K(\theta_2)R(w_2), \tag{1.42}$$

$$\partial_t\tilde w = K(\theta_1)R(w_1) - K(\theta_2)R(w_2). \tag{1.43}$$

Now we multiply (1.42) by the usual regularization of the sign function $\mathrm{sgn}_\sigma(\tilde\theta)$ and (1.43) by $\mathrm{sgn}(\tilde w)$. First note that $\langle\nabla\tilde\theta, \nabla\mathrm{sgn}_\sigma(\tilde\theta)\rangle \geq 0$ and take the limit as $\sigma \downarrow 0$ using the Lebesgue dominated convergence theorem. Then use that $\langle\partial_t\tilde\theta, \mathrm{sgn}(\tilde\theta)\rangle = \partial_t\|\tilde\theta\|_{L^1(\Omega)}$ and $\langle\partial_t\tilde w, \mathrm{sgn}(\tilde w)\rangle = \partial_t\|\tilde w\|_{L^1(\Omega)}$. Upon splitting the right hand side of both (1.42) and (1.43) as $K(\theta_1)R(w_1) - K(\theta_2)R(w_2) = K(\theta_1)\big(R(w_1) - R(w_2)\big) + \big(K(\theta_1) - K(\theta_2)\big)R(w_2)$, we get

$$\partial_t\|\tilde\theta\|_{L^1(\Omega)} \leq \int_\Omega K(\theta_1)|R(w_1) - R(w_2)| + L_K\|\tilde\theta\|_{L^1(\Omega)},$$

$$\partial_t\|\tilde w\|_{L^1(\Omega)} + \int_\Omega K(\theta_1)|R(w_1) - R(w_2)| \leq L_K\|\tilde\theta\|_{L^1(\Omega)},$$

because $\big(R(w_1) - R(w_2)\big)\mathrm{sgn}(w_1 - w_2) = -|R(w_1) - R(w_2)|$. Here L_K is the Lipschitz constant of K. After adding the two inequality above, we integrate the resulting expression from 0 to $\bar t \leq T$ and use Gronwall's lemma to get the asserted estimate. □

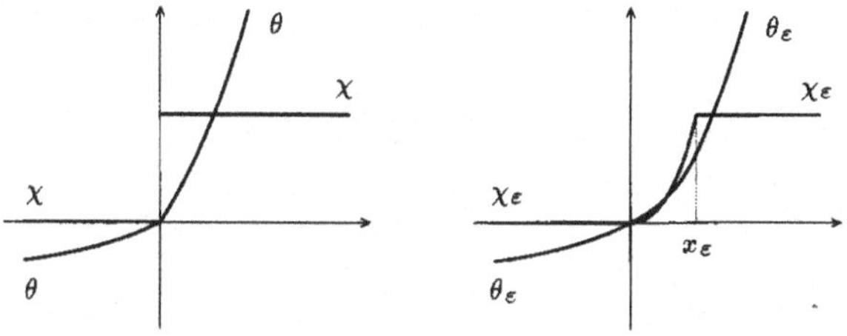

FIGURE 1.6. Travelling waves for (a) the two-phase Stefan problem ($\sigma = 0.5$) and (b) the relaxed problem ($\varepsilon = 0.15$) melting towards the left with velocity $V = 1$.

1.1.4.1. Travelling waves for the relaxed problem. We derive explicit representations of the travelling waves for the relaxed problem (1.39) in one space dimension (see [108]). Let $\bar{\theta}_\varepsilon \in C^{1,1}(\mathbf{R})$ and $\bar{\chi}_\varepsilon \in C^{0,1}(\mathbf{R})$, $0 \le \bar{\chi}_\varepsilon \le 1$ a.e. in $\mathbf{R}$. Let

$$\theta_\varepsilon(x,t) = \bar{\theta}_\varepsilon(x+Vt), \quad \chi_\varepsilon(x,t) = \bar{\chi}_\varepsilon(x+Vt), \quad \text{in } \mathbf{R}\times(0,T),$$

be the desired travelling waves moving towards the left with velocity $V > 0$. For simplicity of notation, we drop the *overbar* in the sequel. Then θ_ε and χ_ε satisfy the following ordinary differential equations in $\mathcal{D}'(\mathbf{R})$:

$$V\chi_\varepsilon' + V\theta_\varepsilon' - \theta_\varepsilon'' = 0, \quad \varepsilon V\chi_\varepsilon' = \begin{cases} \max(\theta_\varepsilon, 0) & \text{if } \chi_\varepsilon = 0 \\ \theta_\varepsilon & \text{if } 0 < \chi_\varepsilon < 1 \\ \min(\theta_\varepsilon, 0) & \text{if } \chi_\varepsilon = 1. \end{cases} \tag{1.44}$$

We define the solid, transition, and liquid regions as follows:

$$\mathcal{C}_\varepsilon = \{x \in \mathbf{R} : \chi_\varepsilon(x) = 0\}, \quad \mathcal{T}_\varepsilon = \{x \in \mathbf{R} : 0 < \chi_\varepsilon(x) < 1\}, \quad \mathcal{L}_\varepsilon = \{x \in \mathbf{R} : \chi_\varepsilon(x) = 1\}.$$

It is possible to prove that $\theta_\varepsilon \le 0$ in $\mathcal{C}_\varepsilon$ and $\theta_\varepsilon \ge 0$ in $\mathcal{L}_\varepsilon$ and that the transition region $\mathcal{T}_\varepsilon$ is a single (possibly unbounded) open interval. The melting process thus correspond to either $\mathcal{T}_\varepsilon = (0, x_\varepsilon)$ and $\mathcal{L}_\varepsilon = [x_\varepsilon, +\infty)$ (two-phase process) or $\mathcal{T}_\varepsilon = (-\infty, 0)$ and $\mathcal{L}_\varepsilon = [0, +\infty)$ (one-phase process). The first equation in (1.44) becomes $V\theta_\varepsilon'(x) - \theta_\varepsilon''(x) = 0$ within $\mathcal{C}_\varepsilon \cup \mathcal{L}_\varepsilon$ and $\theta_\varepsilon \in \operatorname{span}\{e^{Vx}, 1\}$ in $\mathcal{C}_\varepsilon$ and $\mathcal{L}_\varepsilon$. On the other hand, (1.44) becomes

$$\tfrac{1}{\varepsilon}\theta_\varepsilon(x) + V\theta_\varepsilon'(x) - \theta_\varepsilon''(x) = 0, \quad \varepsilon V\chi_\varepsilon'(x) = \theta_\varepsilon(x), \quad \text{in } \mathcal{T}_\varepsilon,$$

so that $\theta_\varepsilon \in \operatorname{span}\{e^{\delta_1 x}, e^{\delta_2 x}\}$ and $\chi_\varepsilon \in \operatorname{span}\{e^{\delta_1 x}, e^{\delta_2 x}\} + c$. Here $0 < -\delta_2 < \delta_1$ are the roots of the algebraic equation $\delta^2 - V\delta - \frac{1}{\varepsilon} = 0$.

Consider only the two-phase process. Since $\chi_\varepsilon(0) = \theta_\varepsilon(0) = \chi_\varepsilon'(0^+) = 0$ (see [108]) and $\theta_\varepsilon \in C^{1,1}(\mathbf{R})$, we get

$$\theta_\varepsilon(x) = \begin{cases} \theta(x) \\ \frac{2\sigma}{\delta_1-\delta_2} e^{(\delta_1+\delta_2)x/2} \sinh\left(\frac{\delta_1-\delta_2}{2}x\right) \\ \frac{\theta_\varepsilon'(x_\varepsilon)}{V}\left(e^{V(x-x_\varepsilon)} - 1\right) + \theta_\varepsilon(x_\varepsilon) \end{cases} \quad \chi_\varepsilon(x) = \begin{cases} \chi(x) & \text{if } x \in \mathcal{C}_\varepsilon = (-\infty, 0] \\ \frac{1}{V\varepsilon}\int_0^x \theta_\varepsilon(s)\,ds & \text{if } x \in \mathcal{T}_\varepsilon = (0, x_\varepsilon) \\ \chi(x) & \text{if } x \in \mathcal{L}_\varepsilon = [x_\varepsilon, +\infty) \end{cases}$$

where x_ε is defined by $\chi_\varepsilon(x_\varepsilon) = 1$ (see Figure 1.6). Both χ_ε' and θ_ε'' exhibit jump discontinuities at $x = x_\varepsilon$, and thus $\theta_\varepsilon \in C^{1,1}(\mathbf{R}) \backslash C^2(\mathbf{R})$ and $\chi_\varepsilon \in C^{0,1}(\mathbf{R}) \backslash C^1(\mathbf{R})$.

We observe that the free boundary is spread out into a transition layer of width $x_\varepsilon = \varepsilon^{1/2}\operatorname{arccosh}(1 + \frac{V}{\sigma}) + \mathcal{O}(\varepsilon)$. The Stefan condition $[\![\theta']\!] = V$ at $x = 0$ is replaced by $\theta_\varepsilon'(x_\varepsilon) - \theta_\varepsilon'(0) = V + \mathcal{O}(\varepsilon^{1/2})$. Moreover, the constitutive relation $\theta_\varepsilon = \beta(\theta_\varepsilon + \chi_\varepsilon)$ only holds in $\mathbb{R}\backslash\mathcal{T}_\varepsilon$. Note that $\theta_\varepsilon(x) = \theta(x) + e^{Vx}\mathcal{O}(\varepsilon^{1/2})$ in $\mathcal{L}_\varepsilon$ and that $\|\theta''_\varepsilon\|_{L^1(-\infty,x)} \leq Ce^{Vx}$ is valid uniformly in ε for $x > 0$.

1.1.4.2. Error analysis for the relaxed problem. We study the rate of convergence of the phase relaxation approximation (1.39) to the two-phase Stefan problem (1.6) as $\varepsilon \downarrow 0$. To this end, we first need a priori estimates for $\{\theta_\varepsilon, \chi_\varepsilon\}$. Since $\chi_\varepsilon \to \chi$, we cannot expect $\chi_\varepsilon \in H^1(0,T;L^2(\Omega))$. Here is the precise way χ_ε degenerates as $\varepsilon \downarrow 0$. Note that (1.45) is sharp.

THEOREM 1.5. *Let $u_0 \in L^2(\Omega)$. Then the following a priori estimate holds:*

$$\|\theta_\varepsilon\|_{L^\infty(0,T;L^2(\Omega))} + \|\theta_\varepsilon\|_{L^2(0,T;H^1(\Omega))} + \varepsilon^{1/2}\|\partial_t\chi_\varepsilon\|_{L^2(Q)} \leq C. \tag{1.45}$$

Proof. Multiply the first equation in (1.39) by θ_ε and the second one by $\partial_t\chi_\varepsilon$. Then add the resulting expressions and note that terms $\langle\partial_t\chi_\varepsilon, \theta_\varepsilon\rangle$ cancel. The indicatrix function $\mathcal{I}_{[0,1]}$ of the interval $[0,1]$ is a lower semicontinuous convex function whose subdifferential coincides with H^{-1} We thus have, for all $h_\varepsilon \in H^{-1}(\chi_\varepsilon)$,

$$\langle h_\varepsilon, \partial_t\chi_\varepsilon\rangle = \partial_t \textstyle\int_\Omega \mathcal{I}_{[0,1]}(\chi_\varepsilon) = 0,$$

because $0 \leq \chi_\varepsilon \leq 1$ for a.e. $(\mathbf{x}, t) \in Q$. It follows that

$$\tfrac{1}{2}\partial_t\|\theta_\varepsilon\|^2_{L^2(\Omega)} + \|\nabla\theta_\varepsilon\|^2_{L^2(\Omega)} + \varepsilon\|\partial_t\chi_\varepsilon\|^2_{L^2(\Omega)} \leq 0.$$

After integration from 0 to $\bar{t} \leq T$ we get the asserted a priori estimate. □

We are now in a position to carry out the desired error analysis as $\varepsilon \downarrow 0$. The general rate of convergence, valid for $u_0 \in L^2(\Omega)$, can be improved by strengthening the assumptions on the initial data.

THEOREM 1.6. *Let $u_0 \in L^2(\Omega)$ and $\theta_\varepsilon(0) = \theta_0 = \beta(u_0)$, $\chi_\varepsilon(0) = u_0 - \theta_0$. Then*

$$\begin{aligned} E_\varepsilon = \|\theta - \theta_\varepsilon\|_{L^2(Q)} &+ \left\|\textstyle\int_0^t \nabla(\theta - \theta_\varepsilon)\right\|_{L^\infty(0,T;L^2(\Omega))} \\ &+ \|u - (\theta_\varepsilon + \chi_\varepsilon)\|_{L^\infty(0,T;H^{-1}(\Omega))} \leq C\varepsilon^{1/4}. \end{aligned} \tag{1.46}$$

If $\theta_0 \in H_0^1(\Omega)$ and $\Delta\theta_0 \in L^1(\Omega) + M_{I_0}(\Omega)$, then

$$E_\varepsilon \leq C\varepsilon^{1/2}. \tag{1.47}$$

Proof. Set $\tilde{\theta} = \theta - \theta_\varepsilon$ and $\tilde{\chi} = \chi - \chi_\varepsilon$. The weak formulation of the error equations reads, for a.e. $t \in (0,T)$,

$$\langle\partial_t\tilde{\theta}, \varphi\rangle + \langle\partial_t\tilde{\chi}, \varphi\rangle + \langle\nabla\tilde{\theta}, \nabla\varphi\rangle = 0 \quad \forall\, \varphi \in H_0^1(\Omega), \tag{1.48}$$

$$\langle\tilde{\theta}, \zeta\rangle = \langle\tilde{h}, \zeta\rangle - \varepsilon\langle\partial_t\chi_\varepsilon, \zeta\rangle \quad \forall\, \zeta \in L^2(\Omega), \tag{1.49}$$

where $\tilde{h} = h - h_\varepsilon$ and $h \in H^{-1}(\chi)$, $h_\varepsilon \in H^{-1}(\chi_\varepsilon)$. Integrate (1.48) from 0 to $t \leq T$ and note that $\tilde{\theta}(0) = \tilde{\chi}(0) = 0$ to get, for all $t \in (0,T)$,

$$\langle\tilde{\theta}(t), \varphi\rangle + \langle\tilde{\chi}(t), \varphi\rangle + \left\langle\textstyle\int_0^t \nabla\tilde{\theta}, \nabla\varphi\right\rangle = 0 \quad \forall\, \varphi \in H_0^1(\Omega). \tag{1.50}$$

Next choose $\varphi = \tilde{\theta}(t)$ in (1.50) and use (1.49) with $\zeta = \tilde{\chi}(t)$ for the middle term. Note that $\langle\tilde{h}, \tilde{\chi}\rangle \geq 0$ because of the monotonicity of H^{-1}. After integration from 0 to $\bar{t} \leq T$, since $\left\langle \int_0^t \nabla\tilde\theta, \nabla\tilde\theta(t)\right\rangle = \frac{1}{2}\partial_t \left\| \int_0^t \nabla\tilde\theta \right\|^2_{L^2(\Omega)}$, we have

$$\|\tilde\theta\|^2_{L^2(0,\bar t;L^2(\Omega))} + \tfrac{1}{2}\left\| \int_0^{\bar t} \nabla\tilde\theta \right\|^2_{L^2(\Omega)} \le \varepsilon \int_0^{\bar t} \langle \partial_t \chi_\varepsilon, \tilde\chi\rangle. \tag{1.51}$$

Owing to the *a priori* estimate (1.45), we readily obtain

$$\varepsilon \int_0^{\bar t} \langle \partial_t \chi_\varepsilon, \tilde\chi\rangle \le \varepsilon \|\partial_t \chi_\varepsilon\|_{L^2(Q)} \|\tilde\chi\|_{L^2(Q)} \le C\varepsilon^{1/2},$$

which yields the asserted estimates for $\tilde\theta$ and $\int_0^t \nabla\tilde\theta$ in (1.46).

Under the additional assumptions on the initial data stated before, we can modify the treatment of the right hand side of (1.51) and improve the final estimate. In fact, integration by parts yields

$$\begin{aligned}\varepsilon \int_0^{\bar t} \langle \partial_t \chi_\varepsilon, \tilde\chi\rangle &= \varepsilon \int_0^{\bar t} \langle \partial_t \chi_\varepsilon, \chi\rangle - \varepsilon \int_0^{\bar t} \langle \partial_t \chi_\varepsilon, \chi_\varepsilon\rangle \\ &= \varepsilon \langle \chi_\varepsilon, \chi\rangle\big|_0^{\bar t} - \varepsilon \int_0^{\bar t} \langle \chi_\varepsilon, \partial_t \chi\rangle - \tfrac{\varepsilon}{2}\left(\|\chi_\varepsilon(\bar t)\|^2_{L^2(\Omega)} - \|u_0 - \theta_0\|^2_{L^2(\Omega)}\right).\end{aligned}$$

First and third term on the right hand side are clearly $\mathcal{O}(\varepsilon)$ whereas, owing to (1.12) and (1.14), the middle term can be bounded by

$$-\varepsilon \int_0^{\bar t} \langle \chi_\varepsilon, \partial_t (u-\theta)\rangle \le \varepsilon \int_0^{\bar t} \|\chi_\varepsilon\|_{C^0(\bar\Omega)} \|\partial_t u\|_{M(\Omega)} + \varepsilon \int_0^{\bar t} \|\chi_\varepsilon\|_{L^2(\Omega)} \|\partial_t \theta\|_{L^2(\Omega)} \le C\varepsilon,$$

which clearly implies the asserted estimates for $\tilde\theta$ and $\int_0^t \nabla\tilde\theta$ in (1.47). Note that $\chi_\varepsilon \in C^0(\bar\Omega)$ may fail to hold; in order to make the duality between $C^0(\bar\Omega)$ and $M(\Omega)$, we can simply regularize χ_ε.

The error bound for $\tilde u = \tilde\theta + \tilde\chi$ can be finally derived from (1.50), selecting $\varphi = G\tilde u(t)$, by virtue of the error estimate for $\int_0^t \nabla\tilde\theta$. □

1.2. Parabolic equations with hysteresis

We refer to [29,80,143] for the precise definition and examples of hysteresis operators and for the analysis of differential models of hysteresis. Even though the analytical formulations of continuous and discontinuous hysteresis functionals are different, from the numerical viewpoint a unified approach can be followed. Therefore, here we just recall the weak formulation of parabolic equation with a continuous hysteresis relation in the principal part of the differential operator.

Hysteresis between two time-dependent variables χ and θ arises when the output $\chi(t)$ is not uniquely determined by the input $\theta(t)$ at the same instant $t \in [0,T]$, but instead $\chi(t)$ depends on the evolution of θ in the whole time interval $[0,t]$ and on $\chi(0) = \chi_0$: $\chi(t) = \mathcal{F}(\theta(\cdot), \chi_0)(t)$. Here $\mathcal{F}$ is a causal or Volterra functional, namely the output $\chi(t)$ does not depend on $\theta|_{[t,T]}$. Thus hysteresis is a memory effect. In order to exclude different memory effects like viscosity, we require that $\mathcal{F}$ is rate-independent, i.e., $\chi(t)$ depends just on the range of θ in $[0,t]$ and on the order in which these values are assumed, not on its velocity. Finally the piecewise monotonicity property turns out to be crucial in the analysis of differential models with hysteresis: if θ is monotone in $[t_1, t_2]$ then so is also χ in the same time interval. Usually the couple $\big(\theta(t), \chi(t)\big)$ is confined to a set $\mathcal{S} \subset \mathsf{R}^2$, corresponding to the hysteresis loop which includes also its internal points (see Figure 1.7).

Now we consider distributed systems; in the hysteresis relation the space variable $\mathbf{x} \in \Omega$ appears just as a parameter. Let $\mathcal{S} \subset \mathsf{R}^2$ and $(\mathcal{S}, \mathcal{F})$ be a continuous and piecewise

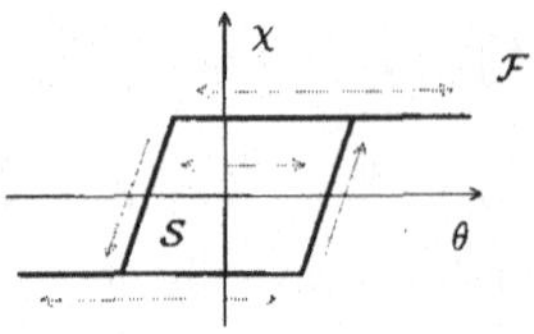

FIGURE 1.7. Hysteresis operator $(\mathcal{S}, \mathcal{F})$.

monotone hysteresis operator. Let $\theta_0 \in H_0^1(\Omega)$, $\chi_0 \in L^2(\Omega)$ be such that $(\theta_0, \chi_0) \in \mathcal{S}$. We consider the problem:

Seek $\{\theta, \chi\}$ *such that*

$$\theta \in L^\infty(0,T; H_0^1(\Omega)) \cap H^1(0,T; L^2(\Omega)), \quad \chi \in L^\infty(0,T; L^2(\Omega)) \cap L^2(\Omega; C^0[0,T]), \tag{1.52}$$

$$\chi(\mathbf{x}, t) = \mathcal{F}\big(\theta(\mathbf{x}, \cdot), \chi_0(\mathbf{x})\big)(t) \quad \forall\, t \in (0,T), \ \text{ for a.e. } \mathbf{x} \in \Omega, \tag{1.53}$$

$$\theta(\cdot, 0) = \theta_0(\cdot), \quad \chi(\cdot, 0) = \chi_0(\cdot), \quad \text{a.e. in } \Omega, \tag{1.54}$$

and, for a.e. $t \in (0,T)$, *the following equation holds:*

$$\langle \partial_t \theta + \partial_t \chi, \varphi \rangle + \langle \nabla\theta, \nabla\varphi \rangle = 0 \quad \forall\, \varphi \in H_0^1(\Omega). \tag{1.55}$$

Note that condition (1.53) is meaningful because $L^\infty(0,T; H_0^1(\Omega)) \cap H^1(0,T; L^2(\Omega)) \subset L^2(\Omega; C^0[0,T])$ with compact injection. Existence is well known for problem (1.52)–(1.55) (see [29,143]). It can be proved via time discretization, then proving *a priori* estimates for the semidiscrete solutions that are independent of the time step, and finally using suitable limit procedures. Uniqueness holds for some general classes of hysteresis operators (see [29,68,143]).

REMARK 1.9. *Preisach operator.* The Preisach operator is probably the most useful model of hysteresis functionals. It can be defined in terms of the relay operator $\mathcal{F}_\rho$ by a finite measure μ over the half-plane of the admissible thresholds, i.e., $R = \{\rho = (\rho_1, \rho_2) \in \mathbb{R}^2 : \rho_1 < \rho_2\}$. The Preisach operator associated to μ is given by $\mathcal{F}_\mu\big(\theta(\cdot), \chi_0\big)(t) = \int_R \mathcal{F}_\rho\big(\theta(\cdot), \chi_0\big)\, d\mu$. Taking advantage of the geometrical properties of the Preisach model a simple approximation procedure for $\mathcal{F}_\mu$ can be designed. More precisely, assuming that μ is absolutely continuous and with compact support in R, the underlying density function r can be approximated by piecewise constant functions r^δ and the functional $\mathcal{F}_\mu$ is thus replaced by F_μ^δ (see [69]). In view of the piecewise monotonicity, as the control θ increases or decreases monotonically in time from $\theta(t)$, the output χ moves from $\chi(t)$ along a monotone curve in the (θ, χ)-plane. In turn, this curve is piecewise linear and can be constructed easily (see [141]).

2. Discrete-time schemes

In order to motivate the fully discrete schemes below we first discuss discrete-time approximations. They are simpler but at the same time possess the major characteristic features of interest. We also derive semi-explicit formulae for the travelling waves supported by the discrete-time problems in one space dimension. Layer width estimates, pointwise error estimates, and asymptotic expressions for the profile of the physical variables play a major role in designing adaptive numerical methods.

Let $\tau = \frac{T}{N}$ denote the time step and set $t^n = n\tau$, $I^n = (t^{n-1}, t^n]$ for $1 \le n \le N$. Also set $\partial z^n = \frac{z^n - z^{n-1}}{\tau}$ for any sequence $\{z^n\}_{n=0}^N$.

If we simply use backward differences to discretize (1.5) in time and enforce the constitutive relation β between semidiscrete energy U^n and temperature Θ^n, we obtain a set of nonlinear elliptic partial differential equation: $\partial U^n - \Delta\beta(U^n) = 0$. This scheme is stable and corresponds to the generation of a nonlinear semigroup of contractions in $L^1(\Omega)$ [42], and in $H^{-1}(\Omega)$ [25]. It was introduced and succesfully used for theoretical purposes in many pioneering papers about the two-phase Stefan problem. Upon finite element discretization in space, it becomes a standard nonlinear algorithm. Its main drawback is the strong nonlinearity present in $\Theta^n = \beta(U^n)$, which makes powerful iterative methods such Newton's method fail, at least in a general setting. On the other hand, using forward differences to discretize (1.5) in time leads to an unstable method; after space discretization, the severe parabolic stability constraint $\tau \le Ch^2$, where h stand for the meshsize, must be enforced (see [40]). It was long believed that the rate of convergence of the backward difference method for degenerate parabolic problems could be at most $\mathcal{O}(\tau^{1/2})$ [42]. Recently, Rulla [126] has proved that such a rate is in fact $\mathcal{O}(\tau)$ combining that the operator $-\Delta\beta$ is a subgradient in the Hilbert space $H^{-1}(\Omega)$ [25] and the initial datum u_0 belongs to its domain $\{\zeta \in L^2(\Omega) : \beta(\zeta) \in H_0^1(\Omega)\}$. Rulla's results extend those obtained in [12,128] for parabolic variational inequalities.

Linearization of nonlinear parabolic problems is an extremely useful numerical tool in that one can use efficient linear solvers to compute the solution of the ensuing linear systems. The success of the standard linearization approaches based on Taylor's expansions essentially requires strong regularity properties of the underlying solution, which are often quite unrealistic for free boundary problems even with a preliminary regularization. For instance, assuming nondegeneracy properties of the discrete solutions, a difficult task to prove, one can consider the following linearization schemes, which are based on (formally) equivalent formulations of (1.5). More precisely, the equation $\partial_t u - \operatorname{div}(\beta'(u)\nabla u) = 0$ suggests the scheme (see [6]): $\partial U^n - \operatorname{div}(\beta'(U^{n-1})\nabla U^n) = 0$, whereas, after regularization, the formulation $\gamma_\varepsilon'(\theta_\varepsilon)\partial_t\theta_\varepsilon - \Delta\theta_\varepsilon = 0$ suggests the scheme: $\gamma_\varepsilon'(\Theta_\varepsilon^{n-1})\partial\Theta_\varepsilon^n - \Delta\Theta_\varepsilon^n = 0$.

An appealing linearization algorithm is suggested by the so-called nonlinear Chernoff formula, which arises in the theory of nonlinear semigroups of contractions (see [27]) and was first used numerically in [20] (see also [71,85,87]). In Sect. 2.2.1 we identify the variational properties of the nonlinear Chernoff formula in relation with the discrete-time approximation of the phase relaxation model introduced in Sect. 1.1.4. This show that a suitable violation of the strongly nonlinear constitutive relation β, which preserves the natural stability properties, turns out to be a potential linearization technique. This crucial idea was exploited also for introducing an extrapolation method in [100]. Recently, an optimal linear rate of convergence for an implicit time discretization of the phase relaxation model has been proved in [75].

2.1. Nonlinear methods

In this section we introduce the backward difference method for both Stefan-like problems and parabolic equations with hysteresis.

2.1.1. Nonlinear method for the two-phase Stefan problem

The nonlinear discrete-time scheme for problem (1.7)–(1.10) reads:

Set $U^0 = u_0$ *and, for* $1 \le n \le N$, *seek* $\{U^n, \Theta^n\}$ *such that* $U^n \in L^2(\Omega)$, $\Theta^n \in H_0^1(\Omega)$, *and*

$$\partial U^n - \Delta\Theta^n = 0, \quad \Theta^n = \beta(U^n), \quad \text{in } \mathcal{D}'(\Omega). \tag{2.1}$$

Existence and uniqueness of the solution $\{U^n, \Theta^n\}$ to (2.1) is well known [26] in view of the monotonicity of β.

2.1.1.1. Semidiscrete travelling waves for the nonlinear method. See [108]. Let $\theta_\tau \in C^{1,1}(\mathbb{R})$, $\chi_\tau \in H(\theta_\tau)$ a.e. in $\mathbb{R}$, and $u_\tau = \theta_\tau + \chi_\tau$. The semidiscrete travelling waves moving towards the left with velocity $V > 0$ are given, for all $0 \le n \le N$, by

$$\Theta^n(x) = \theta_\tau(x + Vt^n), \quad X^n(x) = \chi_\tau(x + Vt^n), \quad U^n(x) = u_\tau(x + Vt^n), \quad \text{in } \mathbb{R}.$$

Inserting the above expressions into (2.1) leads to the following delay ordinary differential equation in $\mathcal{D}'(\mathbb{R})$: $u_\tau(x) - \tau\theta_\tau''(x) = u_\tau(x - V\tau)$, $\theta_\tau(x) = \beta(u_\tau(x))$ or, equivalently:

$$\theta_\tau(x) - \tau\theta_\tau''(x) = \theta_\tau(x - V\tau) + \chi_\tau(x - V\tau) - \chi_\tau(x), \quad \chi_\tau(x) \in H(\theta_\tau(x)). \tag{2.2}$$

The structure of the semidiscrete travelling waves is reacher than that of the continuous counterpart in that oscillating solution may arise [108]. Since we are interested in non-oscillating travelling waves, some natural assumptions at $-\infty$, say in $(-\infty, 0)$ up to translations, are required to mimic the behaviour described in Sect. 1.1.3.1. The solution $\{\theta_\tau, \chi_\tau\}$ can then be constructed iteratively in $[Vt^n, Vt^{n+1}]$ for $n \ge 0$ from (2.2) by imposing a C^1 initial condition for θ_τ at $x = Vt^n$.

We define the discrete solid, transition, and liquid regions as

$$\mathcal{C}_\tau = \{x \in \mathbb{R} : \theta_\tau(x) < 0\}, \quad \mathcal{T}_\tau = \{x \in \mathbb{R} : \theta_\tau(x) = 0\}, \quad \mathcal{L}_\tau = \{x \in \mathbb{R} : \theta_\tau(x) > 0\},$$

and note that $\chi_\tau(x) = 0, u_\tau(x) = \theta_\tau(x) < 0$ for $x \in \mathcal{C}_\tau$, $0 \le \chi_\tau(x) = u_\tau(x) \le 1$ for $x \in \mathcal{T}_\tau$, and $\chi_\tau(x) = 1, u_\tau(x) = \theta_\tau(x) + 1 > 1$ for $x \in \mathcal{L}_\tau$, because of the constitutive relation in (2.2). The melting process is specified through the requirement that either $\mathcal{C}_\tau \supset (-\infty, 0)$ (two-phase process) or $\mathcal{T}_\tau \supset (-\infty, 0]$ (one-phase process). In the sequel we examine only the two-phase problem. Equation (2.2) becomes $\theta_\tau(x) - \tau\theta_\tau''(x) = \theta_\tau(x - V\tau)$ in $\mathcal{C}_\tau$ so that $\theta_\tau \in \operatorname{span}\{e^{V_\tau x}, 1\}$, where $V_\tau > 0$ is the positive root to $1 - \tau s^2 = e^{-sV\tau}$. Note that $V_\tau = V - V^3\frac{\tau}{2} + \mathcal{O}(\tau^2)$. Since $\theta_\tau \in C^{1,1}(\mathbb{R})$, on imposing $\theta_\tau(0) = 0$, we thus have $\sigma = \theta_\tau'(0) > 0$ and

$$\theta_\tau(x) = \tfrac{\sigma}{V_\tau}(e^{V_\tau x} - 1), \quad \chi_\tau(x) = 0, \quad \text{if } x \in (-\infty, 0).$$

An elementary comparison result for delay ordinary differential equations (see [108]) is used to prove that $\theta_\tau(x) > 0$ and $\chi_\tau(x) = 1$ for $x \in (0, +\infty)$ and so $\mathcal{L}_\tau = (0, +\infty)$ and $\mathcal{T}_\tau = \{0\}$. Equation (2.2) then becomes $\theta_\tau(x) - \tau\theta_\tau''(x) = \theta_\tau(x - V\tau) + \chi_\tau(x - V\tau) - 1$ in $\mathcal{L}_\tau$, but its solution can be computed explicitly only in the first interval $[0, V\tau]$, where $\theta_\tau(x) = \frac{\sigma}{V_\tau}(e^{V_\tau x} - 1) + \cosh\frac{x}{\sqrt{\tau}} - 1$. Note that $\theta_\tau \in C^{1,1}(\mathbb{R})\backslash C^2(\mathbb{R})$ because $\tau\theta_\tau''$ exhibits unit jump discontinuities at $x = 0, h$.

The semidiscrete travelling wave can be compared with the continuous counterpart in order to quantify the distorsion produced by time discretization (see Figure 2.1). The asymptotic estimate $\theta_\tau(x) = \theta(x) + (1 + x)e^{Vx}\mathcal{O}(\tau)$ for $x > 0$ makes the notion of truncation error meaningful without the requirement of pointwise regularity of θ, typical of Taylor's expansions. We also have a lower bound for the difference $\theta - \theta_\tau$, because

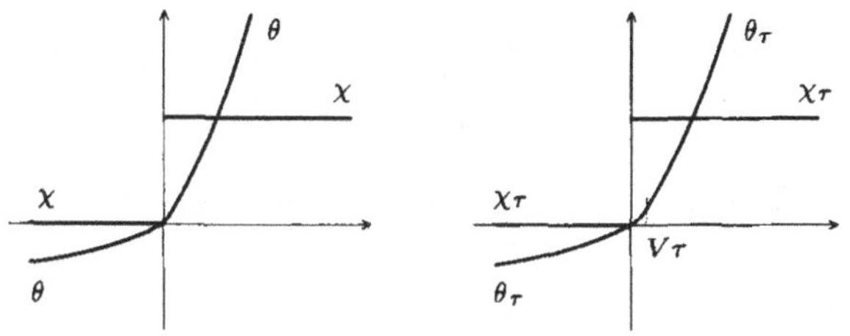

FIGURE 2.1. Travelling waves for (a) the two-phase Stefan problem ($\sigma = 0.5$) and (b) the discrete-time nonlinear problem ($\tau = 0.15$) melting towards the left with velocity $V = 1$.

$\theta(V\tau) - \theta_\tau(V\tau) = V^2\frac{\tau}{2} + \mathcal{O}(\tau^2)$. We see that the first derivative of θ_τ does not exhibit a jump discontinuity at $x = 0$ but rather a rapid variation from σ to $\sigma + V + \mathcal{O}(\tau)$ within the interval $(0, V\tau)$, where $\theta_\tau''(x) = \frac{1}{\tau} + \mathcal{O}(1)$. This smearing effect was first observed numerically [103,104,106]. It is then not surprising that such an effect, together with the estimate $\|\theta_\tau''\|_{L^1(-\infty,x)} \le Cxe^{Vx}$ for $x > 0$, plays an important role in the design and analysis of adaptive finite element methods.

2.1.2. Nonlinear method for parabolic equations with hysteresis

Let $\hat{\Theta}$ indicate the piecewise linear in time function such that $\hat{\Theta}(\cdot, t^n) = \Theta^n(\cdot)$ in Ω, for all $0 \le n \le N$. The nonlinear scheme for problem (1.52)–(1.55) then reads:

Set $\Theta^0 = \theta_0$, $X^0 = \chi_0$ *and, for* $1 \le n \le N$, *seek* $\{\Theta^n, X^n\}$ *such that* $\Theta^n \in H_0^1(\Omega)$, $X^n \in L^2(\Omega)$, *and*

$$\partial\Theta^n + \partial X^n - \Delta\Theta^n = 0, \quad X^n = \mathcal{F}(\hat{\Theta}(\cdot), X^0)(t^n), \quad \text{in } \mathcal{D}'(\Omega). \tag{2.3}$$

Note that at the time step n, the functions X^0 and $\Theta^0, ..., \Theta^{n-1}$ are known. Therefore, by the causality of $\mathcal{F}$, for a.e. $\mathbf{x} \in \Omega$, $X^n(\mathbf{x})$ depends only on $\Theta^n(\mathbf{x})$ and $\mathbf{x}$, i.e., there exists a function $F^n : \mathbf{R} \times \Omega \to \mathbf{R}$ such that

$$X^n(\mathbf{x}) = F^n(\Theta^n(\mathbf{x}); \mathbf{x}) \quad \text{for a.e. } \mathbf{x} \in \Omega.$$

Moreover, a.e. $\mathbf{x} \in \Omega$, the unknown function $\hat{\Theta}$ is linear, then either nonincreasing or nondecreasing, in $[t^{n-1}, t^n]$; in view of the piecewise monotonicity of $\mathcal{F}$, it turns out that F^n is a monotone nondecreasing function. This allows us to solve (2.3) by means of a standard procedure. Note that $F^n(\Theta^{n-1}(\mathbf{x}); \mathbf{x}) = X^{n-1}(\mathbf{x})$ a.e. $\mathbf{x} \in \Omega$.

REMARK 2.1. Let us introduce, for any $1 \le n \le N$ and for a.e. $\mathbf{x} \in \Omega$, the Lipschitz continuous function

$$\beta^n(\cdot; \mathbf{x}) = \left(I + F^n(\cdot; \mathbf{x})\right)^{-1},$$

and set $U^n = \Theta^n + X^n$. Then (2.3) can be written equivalently

$$\partial U^n - \Delta\Theta^n = 0, \quad \Theta^n(\cdot) = \beta^n(U^n(\cdot); \cdot), \quad \text{in } \mathcal{D}'(\Omega).$$

This simple transformation allows the numerical treatment of parabolic equations with discontinuous hysteresis functionals in the same framework as for continuous hysteresis (see [138,141]).

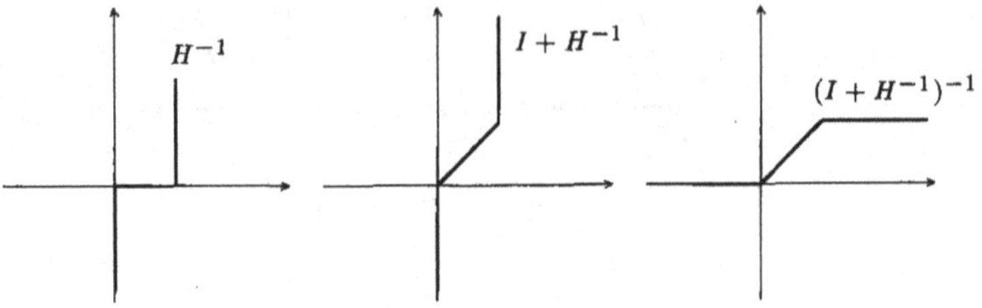

FIGURE 2.2. Graphs H^{-1}, $I + H^{-1}$, and function $(I + H^{-1})^{-1} = I - \beta$.

2.2. Linear methods

In this section we introduce the nonlinear Chernoff formula, which arises in the theory of nonlinear semigroups of contractions [27], and study its variational structure.

2.2.1. Nonlinear Chernoff formula for the two-phase Stefan problem

We consider the following nonlinear Chernoff formula (see [20,87]):

Set $U^0 = u_0$ *and, for* $1 \le n \le N$, *seek* $\{U^n, \Theta^n\}$ *such that* $U^n \in L^2(\Omega)$, $\Theta^n \in H_0^1(\Omega)$, *and*

$$\Theta^n - \tau\Delta\Theta^n = \beta(U^{n-1}), \quad U^n = U^{n-1} + \Theta^n - \beta(U^{n-1}), \quad \text{in } \mathcal{D}'(\Omega). \tag{2.4}$$

This scheme consists of solving a linear elliptic equation in Θ^n and performing next a pointwise explicit correction to account for the nonlinearity and compute U^n; it turns out that the algorithm is well posed and computationally appealing. Note that replacing $-\Delta\Theta^n$ by $-\Delta\big(U^n - U^{n-1} + \beta(U^{n-1})\big)$ suggests the nonlinear Chernoff formula as a stable correction of the forward difference time discretization (see also [50]).

To understand its variational properties, we relate the Chernoff formula to the following time discretization of the phase relaxation approximation (1.39) (see [139,140]):

$$\partial\Theta^n + \partial X^n - \Delta\Theta^n = 0, \quad \varepsilon\partial X^n + H^{-1}(X^n) \ni \Theta^{n-1}. \tag{2.5}$$

Note that the presence of Θ^{n-1} rather than Θ^n in the second equation makes it explicit in X^n:

$$X^n = (I + H^{-1})^{-1}\big(\tfrac{\tau}{\varepsilon}\Theta^{n-1} + X^{n-1}\big),$$

so that the first equation becomes linear in Θ^n. The scheme (2.5) is stable under the mild constraint $\tau \le \varepsilon$ (see [140]). If we choose $\varepsilon = \tau$ and set $U^n = \Theta^n + X^n$, the second equation in (2.5) can be written equivalently $X^n = U^{n-1} - \beta(U^{n-1})$, because $(I + H^{-1})^{-1} = I - \beta$ (see Figure 2.2). Consequently we obtain $U^n = \Theta^n + X^n = U^{n-1} + \Theta^n - \beta(U^{n-1})$, which is the nonlinear correction in (2.4). Since the partial differential equation in both (2.5) and (2.4) clearly coincide because $U^n - U^{n-1} = \Theta^n - \beta(U^{n-1})$, we realize that the nonlinear Chernoff formula is a proper combination of phase relaxation and time discretization.

2.2.1.1. Semidiscrete travelling waves for the Chernoff formula. See [108]. Let $\theta_\tau \in C^{2,1}(\mathbf{R})$, $\chi_\tau \in C^{0,1}(\mathbf{R})$, $0 \le \chi_\tau \le 1$ a.e. in $\mathbf{R}$, and $u_\tau = \theta_\tau + \chi_\tau$. The semidiscrete travelling waves moving to the left with velocity $V > 0$ are given, for all $0 \le n \le N$, by

$$\Theta^n(x) = \theta_\tau(x + Vt^n), \quad X^n(x) = \chi_\tau(x + Vt^n), \quad U^n(x) = u_\tau(x + Vt^n), \quad \text{in } \mathbf{R}.$$

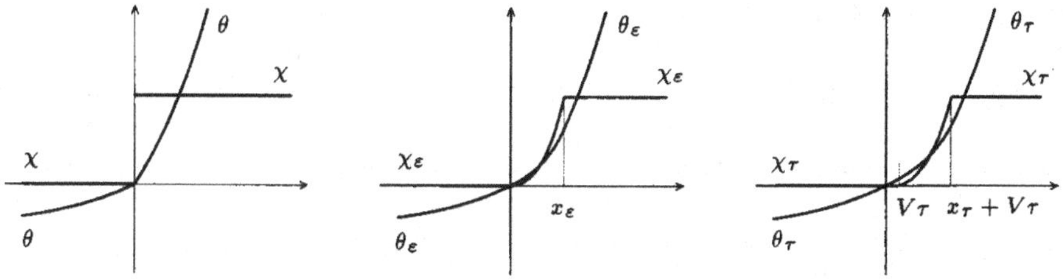

FIGURE 2.3. Travelling waves for (a) the two-phase Stefan problem ($\sigma = 0.5$), (b) the relaxed problem ($\varepsilon = 0.15$), and (c) the nonlinear Chernoff formula ($\tau = 0.15$) melting towards the left with velocity $V = 1$.

Inserting the above expressions into (2.4) leads to the following delay ordinary differential equation in $\mathcal{D}'(\mathbf{R})$:

$$(2.6) \qquad \theta_\tau(x) - \tau\theta''_\tau(x) = \beta(u_\tau(x - V\tau)), \quad u_\tau(x) = u_\tau(x - V\tau) + \theta_\tau(x) - \beta(u_\tau(x - V\tau))$$

or, equivalently:

$$\theta_\tau(x) - \tau\theta''_\tau(x) = \theta_\tau(x - V\tau) + \chi_\tau(x - V\tau) - \chi_\tau(x), \quad \chi_\tau(x) = (I - \beta)(u_\tau(x - V\tau)).$$

We define the discrete solid, transition, and liquid regions as

$$\mathcal{C}_\tau = \{x \in \mathbf{R} : \chi_\tau(x) = 0\}, \quad \mathcal{T}_\tau = \{x \in \mathbf{R} : 0 < \chi_\tau(x) < 1\}, \quad \mathcal{L}_\tau = \{x \in \mathbf{R} : \chi_\tau(x) = 1\},$$

and note that $u_\tau(x - V\tau) = \theta_\tau(x - V\tau) \le 0$ for $x \in \mathcal{C}_\tau$, $0 < u_\tau(x - V\tau) < 1$ for $x \in \mathcal{T}_\tau$, and $u_\tau(x - V\tau) = \theta_\tau(x - V\tau) + 1 > 1$ for $x \in \mathcal{L}_\tau$. The melting process is specified by requiring that either $\mathcal{C}_\tau \supset (-\infty, V\tau]$ (two-phase process) or $\mathcal{T}_\tau \supset (-\infty, 0)$ (one-phase process). In the sequel we examine only the two-phase problem. Since $\chi_\tau \in C^{0,1}(\mathbf{R})$, then $0 < \chi_\tau(x) < 1$ for $x \in (V\tau, x_\tau + V\tau)$, where $x_\tau > 0$ is to be determined. Equation (2.6) becomes $\theta_\tau(x) - \tau\theta''_\tau(x) = \theta_\tau(x - V\tau)$ in $\mathcal{C}_\tau$ and $\theta_\tau \in \operatorname{span}\{e^{V_\tau x}, 1\}$, where $V_\tau > 0$ is the positive root to $1 - \tau s^2 = e^{-sV\tau}$. Since $\theta_\tau(x) = u_\tau(x) \le 0$ for $x \in (-\infty, 0]$ and $\theta_\tau(x) = u_\tau(x) > 0$ for $x \in (0, \min(V\tau, x_\tau))$, then $\theta_\tau(0) = 0$, $\sigma = \theta'_\tau(0) > 0$ and so

$$\theta_\tau(x) = \tfrac{\sigma}{V_\tau}(e^{V_\tau x} - 1), \quad \chi_\tau(x) = 0, \quad \text{if } x \in (-\infty, V\tau].$$

Since $0 < u_\tau(x - V\tau) < 1$ for $x \in \mathcal{T}_\tau$, equation (2.6) becomes $\theta_\tau(x) - \tau\theta''_\tau(x) = 0$, $\chi_\tau(x) = \chi_\tau(x - V\tau) + \theta_\tau(x - V\tau)$ in $\mathcal{T}_\tau$. On imposing C^1 initial conditions for θ_τ at $x = h$ and noting that $\chi_\tau(x - i_x V\tau) = 0$ (where i_x denotes the integer part of $\frac{x}{V\tau}$) we get, for $x \in (V\tau, x_\tau + V\tau) \subset \mathcal{T}_\tau$,

$$\theta_\tau(V\tau) = \theta'_\tau(V\tau)\sqrt{\tau}\sinh\tfrac{x - V\tau}{\sqrt{\tau}} + \theta_\tau(V\tau)\cosh\tfrac{x - V\tau}{\sqrt{\tau}}, \quad \chi_\tau(x) = \textstyle\sum_{l=1}^{i_x}\theta_\tau(x - lV\tau).$$

In turn, there exists a unique point x_τ such that $\chi_\tau(x_\tau + V\tau) = 1$, because χ_τ is strictly increasing in $(V\tau, x_\tau + V\tau)$. An elementary comparison result for delay ordinary differential equations (see [108]) is used to prove that $\chi_\tau(x) = 1$ for $x \in [x_\tau + V\tau, +\infty)$ and so $\mathcal{T}_\tau = (V\tau, x_\tau + V\tau)$ and $\mathcal{L}_\tau = [x_\tau + V\tau, +\infty)$. Equation (2.2) then becomes $\theta_\tau(x) - \tau\theta''_\tau(x) = \theta_\tau(x - V\tau) + \chi_\tau(x - V\tau) - 1$ in $\mathcal{L}_\tau$; its solution cannot be computed explicitly but only compared with θ (see Figure 2.3): $\theta_\tau(x) = \theta(x) + (1 + x)e^{Vx}\mathcal{O}(\tau^{1/2})$. Note that χ'_τ exhibits jump discontinuities at each point $x = i_x V\tau \in [V\tau, x_\tau + V\tau), x_\tau + V\tau$, whereas θ'''_τ does so at $x = V\tau, x_\tau + V\tau, i_{x_\tau + 2V\tau}V\tau, x_\tau + 2V\tau$. Therefore, $\theta_\tau \in C^{2,1}(\mathbf{R})\backslash C^3(\mathbf{R})$ and $\chi_\tau \in C^{0,1}(\mathbf{R})\backslash C^1(\mathbf{R})$.

We see that the Stefan condition $[\![\theta']\!] = V$ at $x = 0$ is substituted by $\theta'_\tau(x_\tau + V\tau) - \theta'_\tau(V\tau) = V + \mathcal{O}(\tau^{1/2})$. Such a smearing effect is also quantified by the transition region width $x_\tau = \sqrt{\tau}\operatorname{arccosh}(1+\frac{V}{\sigma}) + \mathcal{O}(\tau) = \mathcal{O}(\tau^{1/2})$, which happens to be much bigger than that for the nonlinear scheme, namely $V\tau$. This, in turn, means that much more artificial diffusion is incorporated by the linear technique, which is consistent with numerical evidence [106,107]. It is worth noting that the constitutive relation $\theta_\tau(x) = \beta(u_\tau(x))$ is not valid in a region even bigger than $\mathcal{T}_\tau$, a sort of enlarged transition region $(0, x_\tau + V\tau)$. This property, in conjunction with the a priori estimate $\|\theta''_\tau\|_{L^1(-\infty,x)} \le Cxe^{Vx}$ for $x > 0$ turns out to be crucial in designing adaptive numerical methods (see [106,107]).

2.2.2. Linear method for parabolic equations with hysteresis

In view of Remark 2.1, a sort of nonlinear Chernoff formula can be proposed for the time discretization of problem (1.52)–(1.55) (see [141]):

Set $\Theta^0 = \theta_0$, $X^0 = \chi_0$, $U^0 = \Theta^0 + X^0$ and, for $1 \le n \le N$, seek $\{U^n, \Theta^n\}$ such that $U^n \in L^2(\Omega)$, $\Theta^n \in H^1_0(\Omega)$, and

$$\Theta^n - \tau\Delta\Theta^n = \beta^n(U^{n-1};\cdot), \quad U^n = U^{n-1} + \Theta^n - \beta^n(U^{n-1};\cdot), \quad \text{in } \mathcal{D}'(\Omega). \tag{2.7}$$

This scheme is well posed because (2.7) consists of solving a linear elliptic equation in Θ^n and performing next a pointwise explicit correction to account for the nonlinearity and compute U^n and $X^n = U^n - \Theta^n$.

3. Fully discrete schemes for the two-phase Stefan problem

In this section we introduce a proper space discretization by finite elements of the previous discrete-time schemes for Stefan-like problems. Its extension to the schemes for parabolic equations with hysteresis is straightforward. We approximate U^n and Θ^n with at most continuous piecewise linear functions, because the regularity of the underlying continuous variables does not justify the use of higher order elements. The stability analysis of the fully discrete schemes proceed along the same lines as that for the continuous problem. The error analysis is performed using either the inversion of the Laplacian, for the nonlinear scheme, or the integral-time method, for the linear scheme, both illustrated in Sect.1.1.2 in a simpler setting. Let us start by introducing the finite element spaces and some basic properties.

3.1. Finite element spaces

Let $\{\mathcal{S}_h\}_h$ be a family of regular partitions of Ω into simplicial finite elements [38,p.132]; as usual $h > 0$ stands for the meshsize. We assume, for simplicity, that $\bar{\Omega} = \Omega_h = \cup_{S\in\mathcal{S}_h} S$ and refer to [113] for a precise analysis of the discrepance between Ω and Ω_h. Since quasi-uniformity [38,p.140] is not required, local refinement are allowed. The finite element spaces we work with are $V_h^1 \subset H_0^1(\Omega)$ and $V_h^0 \subset L^2(\Omega)$ which satisfy

$$V_h^1|_S = P^1(S), \quad V_h^0|_S = P^0(S), \quad \forall\, S \in \mathcal{S}_h,$$

where $P^i(S)$ indicates the space of polynomials of degree at most i.

Let $\{\mathbf{x}_j\}_{j=1}^J$ indicate the set of internal nodes and $\{\mathbf{b}_k\}_{k=1}^K$ the barycentres of $\mathcal{S}_h$. Let $\{\phi_j\}_{j=1}^J$ denote the canonical basis of V_h^1 and $\{\sigma_k\}_{k=1}^K$ the canonical basis of V_h^0. Let Π_h

be the local Lagrange interpolation operator $\Pi_h|_S : C^0(S) \to P^1(S)$ for all $S \in \mathcal{S}_h$. We associate with $\langle\cdot,\cdot\rangle$ a discrete inner product in V_h^1 defined, for any piecewise uniformly continuous functions ζ and ϕ, by

$$(3.1)\qquad \langle\zeta,\phi\rangle_h = \sum_{S\in\mathcal{S}_h} \int_S \Pi_h|_S(\zeta\phi)\,d\mathbf{x};$$

note that $\langle\cdot,\cdot\rangle_h$ satisfies [124,p.260]:

$$(3.2)\qquad \|\phi\|^2_{L^2(\Omega)} \le \langle\phi,\phi\rangle_h \le C\|\phi\|^2_{L^2(\Omega)} \quad \forall\,\phi\in V_h^1.$$

Note that $\langle\zeta,\phi\rangle_h$ can be evaluated elementwise by the vertex quadrature rule, which is exact for piecewise linear functions. The discrepance between $\langle\cdot,\cdot\rangle$ and $\langle\cdot,\cdot\rangle_h$ can be bounded as follows:

$$(3.3)\quad |\langle\zeta,\phi\rangle-\langle\zeta,\phi\rangle_h|\le Ch^2\|\nabla\zeta\|_{L^2(\Omega)}\|\nabla\phi\|_{L^2(\Omega)}\le Ch\|\zeta\|_{L^2(\Omega)}\|\nabla\phi\|_{L^2(\Omega)} \quad \forall\,\zeta,\phi\in V_h^1.$$

We can define the following stiffness and mass matrices:

$$\mathbf{K} = \{\langle\nabla\phi_j,\nabla\phi_l\rangle\}_{j,l=1}^{J},\quad \mathbf{M} = \{\langle\phi_j,\phi_l\rangle_h\}_{j,l=1}^{J},\quad \mathbf{N} = \{\langle\phi_j,\sigma_k\rangle\}_{j=1,k=1}^{J\quad K}.$$

Note that $\mathbf{K}$ is symmetric and positive definite, whereas $\mathbf{M}$ is diagonal.

We also need some projection operators associated with the above discrete spaces. The L^2-projections operators P_h^i onto V_h^i $(i=0,1)$ are defined, for all $\zeta\in L^2(\Omega)$, by

$$(3.4)\qquad \langle P_h^0\zeta,\sigma\rangle = \langle\zeta,\sigma\rangle \quad \forall\,\sigma\in V_h^0,$$
$$(3.5)\qquad \langle P_h^1\zeta,\phi\rangle_h = \langle\zeta,\phi\rangle \quad \forall\,\phi\in V_h^1,$$

and satisfy the following approximation properties [38]:

$$(3.6)\qquad \|P_h^0\zeta-\zeta\|_{H^{-s}(\Omega)} \le Ch^{r+s}\|\zeta\|_{H^r(\Omega)} \quad \forall\,\zeta\in H^r(\Omega)\ \ 0\le r,s\le 1,$$
$$(3.7)\qquad \|P_h^1\zeta-\zeta\|_{H^{-1}(\Omega)} \le Ch\|\zeta\|_{L^2(\Omega)} \quad \forall\,\zeta\in L^2(\Omega).$$

Note the use of quadratures in (3.5). The H^1-projection operator R_h onto V_h^1 is defined, for all $\varphi\in H_0^1(\Omega)$, by

$$(3.8)\qquad \langle\nabla R_h\varphi,\nabla\phi\rangle = \langle\nabla\varphi,\nabla\phi\rangle \quad \forall\,\phi\in V_h^1.$$

Note that $\|\nabla R_h\varphi\|_{L^2(\Omega)} \le \|\nabla\varphi\|_{L^2(\Omega)}$ for all $\varphi\in H_0^1(\Omega)$. Under the assumption of regularity of both $\{\mathcal{S}_h\}_h$ and G, R_h satisfies the following error estimates [38,pp.138,139]:

$$(3.9)\quad \|R_h\varphi-\varphi\|_{H^{1-s}(\Omega)}\le Ch^{r+s}\|\varphi\|_{H^{1+r}(\Omega)} \quad \forall\,\varphi\in H_0^1(\Omega)\cap H^{1+r}(\Omega),\ \ 0\le r,s\le 1.$$

Note that (3.9) and (3.6) are superconvergence estimates for $s>0$. The discrete Green's operator $G_h : H^{-1}(\Omega)\to V_h^1$ is then defined by $G_h = R_h \circ G$, namely, for all $\psi\in H^{-1}(\Omega)$,

$$(3.10)\qquad \langle\nabla G_h\psi,\nabla\phi\rangle = \langle\psi,\phi\rangle, \quad \forall\,\phi\in V_h^1.$$

In view of (3.9) and the regularity of G, G_h satisfies the estimates

$$(3.11)\qquad \|(G-G_h)\psi\|_{H^{1-s}(\Omega)} \le Ch^{r+s}\|\psi\|_{H^{r-1}(\Omega)} \quad \forall\,\psi\in H^{-s}(\Omega),\ \ 0\le r,s\le 1.$$

Note that $\|\nabla G_h\psi\|_{L^2(\Omega)} \le \|\psi\|_{H^{-1}(\Omega)}$ for all $\psi\in H^{-1}(\Omega)$ and also that $\|\nabla G_h\zeta\|_{L^2(\Omega)} \ge \|\zeta\|_{H^{-1}(\Omega)} - Ch\|\zeta\|_{L^2(\Omega)}$ for all $\zeta\in L^2(\Omega)$.

On several occasions, to compensate for the lack of regularity, we exploit monotonicity properties via the discrete maximum principle [39], which holds if $\mathcal{S}_h$ is acute (or weakly acute in two space dimensions). We say that $\mathcal{S}_h$ is weakly acute if for any pair of adjacent triangles the sum of the opposite angles relative to the common side does not exceed π. In that case $\mathbf{K}$ is an M-matrix [23], and so

$$\langle \nabla \phi, \nabla \phi_j \rangle \geq 0, \tag{3.12}$$

if $\phi \in V_h^1$ attains its maximum at the internal node $\mathbf{x}_j$. Another consequence of the discrete maximum principle is the following statement. Let $\xi : \mathbf{R} \to \mathbf{R}$ be a Lipschitz continuous function which satisfies $\xi(0) = 0$ and $0 \leq \xi'(s) \leq L_\xi$ for a.e. $s \in \mathbf{R}$. Then

$$\|\nabla \Pi_h \xi(\phi)\|^2_{L^2(\Omega)} \leq L_\xi \langle \nabla \phi, \nabla \Pi_h \xi(\phi) \rangle \quad \forall\, \phi \in V_h^1. \tag{3.13}$$

3.2. Fully discrete nonlinear scheme

The finite element method associated with (2.1) reads as follows:

Set $U^0 = P_h^1 u_0$ and, for all $1 \leq n \leq N$, seek $U^n, \Theta^n \in V_h^1$ such that

$$\Theta^n = \Pi_h \beta(U^n), \tag{3.14}$$

$$\langle U^n, \phi \rangle_h + \tau \langle \nabla \Theta^n, \nabla \phi \rangle = \langle U^{n-1}, \phi \rangle_h \quad \forall\, \phi \in V_h^1. \tag{3.15}$$

Note that it is useful to assume $U^n = 0$ on $\partial\Omega$ even if boundary conditions for U^n actually do not make sense because $\gamma(\Theta^n) = \gamma(0) = [0,1]$ on $\partial\Omega$. Equation (3.15) is a strongly nonlinear system in $\mathbf{R}^J$ which, by virtue of the numerical quadrature that enforces the constitutive relation (3.14) just at the nodes, can be easily reduced in matrix form (see [53,113,120]). We identify all piecewise linear functions of V_h^1 with the vectors of their nodal values and thus (3.15) reads:

$$\mathbf{M} U^n + \tau \mathbf{K} \Theta^n = \mathbf{M} U^{n-1}. \tag{3.16}$$

Set $\mathcal{G}(\mathbf{s}) = \mathbf{M}\mathbf{s} + \tau \mathbf{K} \beta(\mathbf{s})$ for all $\mathbf{s} \in \mathbf{R}^J$, where $\beta(\mathbf{s}) = \{\beta(s_j)\}_{j=1}^J$, and note that the nonlinear system (3.16) also reads $\mathcal{G}(U^n) = \mathbf{M} U^{n-1}$. Since $\mathcal{G}$ is coercive, that is $\mathbf{s}^T \mathcal{G}(\mathbf{s}) \geq C(|\mathbf{s}|^2 - 1)$, then $\mathcal{G}(\mathbf{R}^J) = \mathbf{R}^J$ (see [118,p.166]); uniqueness of the solution $\{U^n, \Theta^n\}$ follows using both that $\mathcal{G}$ is strictly monotone in $\beta(\mathbf{s})$, that is

$$\big(\mathcal{G}(\mathbf{s}_1) - \mathcal{G}(\mathbf{s}_2)\big)\big(\beta(\mathbf{s}_1) - \beta(\mathbf{s}_2)\big) \geq \min_{1 \leq j \leq J} m_{jj} |\beta(\mathbf{s}_1) - \beta(\mathbf{s}_2)|^2,$$

and equation (3.16). Since $\mathbf{M}$ is diagonal, if we assume that $\mathcal{S}_h$ is (weakly) acute, then it is easy to prove that $\mathcal{G}$ is an M-function, that is $\mathcal{G}$ is inverse isotone and off-diagonal antitone (see [118,p.468]). It turns out that the system (3.16) can be easily and efficiently solved via the following nonlinear overrelaxation method, which globally converges for $0 < \omega < 2$ (see [118,p.467] and [104,113,120]):

Set $U^n_{\cdot,0} = U^{n-1}$ and for $m \geq 1$ compute $U^n_{j,m}$, $1 \leq j \leq J$, as

$$U^n_{j,m} = (1-\omega) U^n_{j,m-1} + \omega \beta_j^{-1} \Big(m_{jj} U_j^{n-1} - \tau \sum_{l=1}^{j-1} k_{jl} \beta(U^n_{l,m}) - \tau \sum_{l=j+1}^{J} k_{jl} \beta(U^n_{l,m-1}) \Big).$$

Here $\beta_j = m_{jj} I + \tau k_{jj} \beta$ is a strictly increasing function. Since the system (3.16) can be equivalently written in terms of the maximal monotone graph γ as follows:

$$\mathbf{M} \Pi_h \gamma(\Theta^n) + \tau \mathbf{K} \Theta^n \ni \mathbf{M} U^{n-1},$$

we can also use the following more convenient SOR algorithm:

Set $\Theta^n_{\cdot,0} = \Pi_h\beta(U^{n-1})$ *and for* $m \geq 1$ *compute* $\Theta^n_{j,m}$, $1 \leq j \leq J$, *as*

(3.17) $\Theta^n_{j,m} = (1-\omega)\Theta^n_{j,m-1} + \omega\gamma_j^{-1}\left(m_{jj}U_j^{n-1} - \tau\sum_{l=1}^{j-1} k_{jl}\Theta^n_{l,m} - \tau\sum_{l=j+1}^{J} k_{jl}\Theta^n_{l,m-1}\right)$

and, after stopping the iterations, set

$$U^n = U^{n-1} - \tau\mathbf{M}^{-1}\mathbf{K}\Theta^n.$$

Here $\gamma_j = m_{jj}\gamma + \tau k_{jj}I$. Note that γ_j^{-1} is a nondecreasing Lipschitz continuous function because $\gamma_j'(s) \geq m_{jj} + \tau k_{jj}$ for a.e. $s \in \mathbf{R}$. The velocity of convergence of this scheme can be estimated from below by the velocity of the linear overrelaxation method for the M-matrix $\mathbf{A} = \mathbf{M} + \tau\mathbf{K}$ (see [135,147]). In fact, since Θ^n satisfies

$$\Theta^n_j = (1-\omega)\Theta^n_j + \omega\gamma_j^{-1}\left(m_{jj}U_j^{n-1} - \tau\sum_{l=1}^{j-1} k_{jl}\Theta^n_l - \tau\sum_{l=j+1}^{J} k_{jl}\Theta^n_l\right),$$

taking the difference between the latter and (3.17), and using the Lipschitz continuity of γ_j^{-1} in conjunction with the property $k_{jl} \leq 0$ for $j \neq l$, we get

$$|\tilde\Theta^n_{j,m}| \leq (1-\omega)|\tilde\Theta^n_{j,m-1}| - \omega\frac{\tau}{m_{jj}+\tau k_{jj}}\left(\sum_{l=1}^{j-1} k_{jl}|\tilde\Theta^n_{l,m}| + \sum_{l=j+1}^{J} k_{jl}|\tilde\Theta^n_{l,m-1}|\right),$$

where $\tilde\Theta^n_{\cdot,m} = \Theta^n - \Theta^n_{\cdot,m}$. Hence, letting $\mathbf{A} = \mathbf{D_A} - \mathbf{L_A} - \mathbf{U_A}$ the usual splitting of $\mathbf{A}$, it follows that

$$(\mathbf{D_A} - \omega\mathbf{L_A})\tilde\Theta^n_{\cdot,m} \leq ((1-\omega)\mathbf{D_A} + \omega\mathbf{U_A})\tilde\Theta^n_{\cdot,m-1},$$

whence, since $(\mathbf{D_A} - \omega\mathbf{L_A})^{-1} \geq 0$,

$$\tilde\Theta^n_{\cdot,m} \leq \mathbf{A}_{GS}\tilde\Theta^n_{\cdot,m-1}.$$

Here $\mathbf{A}_{GS} = (\mathbf{D_A} - \omega\mathbf{L_A})^{-1}((1-\omega)\mathbf{D_A} + \omega\mathbf{U_A})$ is the iteration matrix of the linear SOR method for $\mathbf{A}$. Therefore, the nonlinear SOR iterations (3.17) converge as fast as the corresponding SOR iterations for the M-matrix $\mathbf{A}$. It turns out that, as usual for the linear SOR, an optimal value ω_* of the relaxation parameter ω can be (empirically) evaluated so as to accelerate the convergence of the proposed method (see [104]).

3.2.1. Stability for the nonlinear scheme

Our aim now is to derive several *a priori* estimates for the above fully discrete nonlinear scheme. The techniques employed are just discrete analogues to those introduced in Sect.1.1.1. In stricking contrast to mildly nonlinear parabolic equations, for which the usual approach is to assume regularity on the continuous solution, the discrete stability estimates play a major role in the error analysis.

Given a Lipschitz continuous function $\xi : \mathbf{R} \to \mathbf{R}$ so that $\xi(0) = 0$ and $0 \leq \xi'(s) \leq L_\xi$ for a.e. $s \in \mathbf{R}$, Φ_ξ stands for the convex function defined by

$$\Phi_\xi(s) = \int_0^s \xi(r)\,dr \quad \forall\, s \in \mathbf{R},$$

and clearly satisfies

(3.18) $$\frac{1}{2L_\xi}\xi^2(s) \leq \Phi_\xi(s) \leq \frac{L_\xi}{2}s^2 \quad \forall\, s \in \mathbf{R}.$$

THEOREM 3.1. *Let $\{U^n,\Theta^n\}_{n=0}^N$ be the solution to problem (3.14),(3.15). Then*

$$\max_{1\le n\le N}\|U^n\|_{L^2(\Omega)}+\sum_{n=1}^N\tau\|\nabla\Theta^n\|^2_{L^2(\Omega)}\le C\|U^0\|_{L^2(\Omega)}, \tag{3.19}$$

$$\sum_{n=1}^N\tau\|\partial\Theta^n\|^2_{L^2(\Omega)}+\max_{1\le n\le N}\|\nabla\Theta^n\|_{L^2(\Omega)}\le C\|\nabla\Theta^0\|_{L^2(\Omega)}, \tag{3.20}$$

If S_h is (weakly) acute, then the additional estimates hold:

$$\sum_{n=1}^N\|U^n-U^{n-1}\|^2_{L^2(\Omega)}\le C\|U^0\|_{L^2(\Omega)}, \tag{3.21}$$

$$\max_{1\le n\le N}\|U^n\|_{L^\infty(\Omega)}\le\|U^0\|_{L^\infty(\Omega)}. \tag{3.22}$$

These *a priori* estimates make sense according to whether or not the underlying norms of the discrete initial data are bounded independently of the meshsize h. The weak estimates (3.19) and (3.21) just require that $u_0\in L^2(\Omega)$, because $\|U^0\|_{L^2(\Omega)}=\|P_h^1u_0\|_{L^2(\Omega)}\le\|u_0\|_{L^2(\Omega)}$, whereas the strong estimate (3.20) is meaningful only if $\|\nabla\Theta^0\|_{L^2(\Omega)}\le C$. Since $\Theta^0=\Pi_h\beta(U^0)$, such a constraint may not be satisfied, even with $\theta_0\in H_0^1(\Omega)$. The following is just a sufficient, and still reasonable, condition for such a bound to hold: let $d=2$, $\theta_0\in W_0^{1,\infty}(\Omega)$, and I_0 be a Lipschitz curve, and set $U^0(\mathbf{x}_j)=\limsup_{\mathbf{x}\to\mathbf{x}_j}u_0(\mathbf{x})$; then $\|\nabla\Theta^0\|_{L^\infty(\Omega)}\le\|\nabla\theta_0\|_{L^\infty(\Omega)}\le C$ and $\|U^0-u_0\|_{H^{-1}(\Omega)}\le Ch|\log h|^{1/2}$. Finally the maximum norm estimate (3.22) is meaningful if $u_0\in L^\infty(\Omega)$, because P_h^1 is bounded in $L^\infty(\Omega)$. Note that (3.21) is a sort of discrete $H^{1/2}(0,T;L^2(\Omega))$ estimate.

Proof. i) Proof of (3.19). Take $\phi=\Theta^n\in V_h^1$ in (3.15) and add the resulting expressions on n from 1 to $m\le N$. We have

$$\sum_{n=1}^m\langle U^n-U^{n-1},\Theta^n\rangle_h+\tau\sum_{n=1}^m\|\nabla\Theta^n\|^2_{L^2(\Omega)}=0.$$

Since $\Theta_j^n=\beta(U_j^n)$ nodewise, the convexity of Φ_β together with (3.18) leads to

$$\begin{aligned}\sum_{n=1}^m(U_j^n-U_j^{n-1})\Theta_j^n&\ge\sum_{n=1}^m\left(\Phi_\beta(U_j^n)-\Phi_\beta(U_j^{n-1})\right)\\&=\Phi_\beta(U_j^m)-\Phi_\beta(U_j^0)\ge\tfrac12(\Theta_j^m)^2-\tfrac12(U_j^0)^2.\end{aligned}$$

Then, on using (3.2), we arrive at

$$\sum_{n=1}^m\langle U^n-U^{n-1},\Theta^n\rangle_h=\sum_{n=1}^m\sum_{j=1}^J m_{jj}(U_j^n-U_j^{n-1})\Theta_j^n\ge\tfrac12\|\Theta^m\|^2_{L^2(\Omega)}-C\|U^0\|^2_{L^2(\Omega)}.$$

Since a bound on $\|\Theta^n\|_{L^2(\Omega)}$ readily extends to a bound on $\|U^n\|_{L^2(\Omega)}$ because $|\beta(s)|\ge|s|-1$, we get the weak estimate (3.19).

ii) Proof of (3.21). Take $\phi=U^n\in V_h^1$ in (3.15) and add on n from 1 to $m\le N$. We get

$$\sum_{n=1}^m\langle U^n-U^{n-1},U^n\rangle_h+\tau\sum_{n=1}^m\langle\nabla\Theta^n,\nabla U^n\rangle=0.$$

By virtue of the elementary relation

$$2a(a-b)=a^2-b^2+(a-b)^2\quad\forall\,a,b\in\mathbf{R}, \tag{3.23}$$

together with (3.2), we can handle the first term above as follows:

$$2\sum_{n=1}^m\langle U^n-U^{n-1},U^n\rangle_h\ge\|U^m\|^2_{L^2(\Omega)}-C\|U^0\|^2_{L^2(\Omega)}+\sum_{n=1}^m\|U^n-U^{n-1}\|^2_{L^2(\Omega)}.$$

If the mesh S_h is (weakly) acute, then we can use (3.13) with $\xi=\beta$ to deduce

$$\tau\sum_{n=1}^m\langle\nabla\Theta^n,\nabla U^n\rangle=\tau\sum_{n=1}^m\langle\nabla\Pi_h\beta(U^n),\nabla U^n\rangle\ge\tau\sum_{n=1}^m\|\nabla\Theta^n\|^2_{L^2(\Omega)}.$$

iii) Proof of (3.20). Take $\phi = \tau\partial\Theta^n \in V_h^1$ in (3.15) and add on n from 1 to $m \leq N$. Use (3.2) and (3.23), in conjunction with the property $\beta'(s) \leq 1$ for a.e. $s \in \mathbf{R}$, to get

$$\begin{aligned} 0 &= \tau \textstyle\sum_{n=1}^m \langle \partial U^n, \partial\Theta^n\rangle_h + \sum_{n=1}^m \langle \nabla\Theta^n, \nabla(\Theta^n - \Theta^{n-1})\rangle \\ &\geq \tau \textstyle\sum_{n=1}^m \|\partial\Theta^n\|^2_{L^2(\Omega)} + \frac{1}{2}\|\nabla\Theta^m\|^2_{L^2(\Omega)} - \frac{1}{2}\|\nabla\Theta^0\|^2_{L^2(\Omega)}, \end{aligned}$$

which leads to the bound (3.20) in terms of $\|\nabla\Theta^0\|_{L^2(\Omega)}$.

iv) Proof of (3.22). We argue by induction and suppose $\|U^{n-1}\|_{L^\infty(\Omega)} \leq \|U^0\|_{L^\infty(\Omega)}$ for $1 \leq n \leq N$. Since both U^n and Θ^n are piecewise linear and β is nondecreasing, U^n and Θ^n attain their maximum at a node of $\mathcal{S}_h$, say $\mathbf{x}_j$. We can assume that $\mathbf{x}_j$ is interior, for otherwise $U^n(\mathbf{x}_j) = \Theta^n(\mathbf{x}_j) = 0$. Since $\mathcal{S}_h$ is (weakly) acute, the discrete maximum principle (3.12) applies and (3.15) yields

$$m_{jj}U_j^n = \langle U^n, \phi_j\rangle_h \leq \langle U^n, \phi_j\rangle_h + \langle\nabla\Theta^n, \nabla\phi_j\rangle = \langle U^{n-1}, \phi_j\rangle_h \leq m_{jj}\|U^{n-1}\|_{L^\infty(\Omega)}.$$

A similar argument produces the bound from below for the minimum of U^n. □

REMARK 3.1. The *a priori* estimate (3.20) also holds for the fully discrete nonlinear approximation of parabolic equations with hysteresis, in view of the crucial property

$$X^n(\mathbf{x}) = \dot{F}^n(\Theta^n(\mathbf{x}); \mathbf{x}), \quad X^{n-1}(\mathbf{x}) = F^n(\Theta^{n-1}(\mathbf{x}); \mathbf{x}), \quad \text{for a.e. } \mathbf{x} \in \Omega.$$

3.2.2. Error analysis for the nonlinear scheme

In this section we prove error estimates for the physical variables u, θ, and $\nabla\theta$ in the natural energy spaces, using the Laplacian inversion technique (see [113,136]).

Let us define the error functions $\tilde{u}$ and $\tilde{\theta}$ as

$$\tilde{u}(\cdot,t) = u(\cdot,t) - U^n(\cdot), \quad \tilde{\theta}(\cdot,t) = \theta(\cdot,t) - \Theta^n(\cdot), \quad \forall\, t \in I^n = (t^{n-1}, t^n], \ 1 \leq n \leq N,$$

and set $\tilde{u}^n = \tilde{u}(t^n)$.

THEOREM 3.2. *Let $u_0 \in L^2(\Omega)$ and $\mathcal{S}_h$ be (weakly) acute. Let $\tau = C^*h$, with $C^* > 0$ arbitrary. Then*

$$E_h = \|\tilde{u}\|_{L^\infty(0,T;H^{-1}(\Omega))} + \|\tilde{\theta}\|_{L^2(Q)} + \left\|\textstyle\int_0^t \nabla\tilde{\theta}\right\|_{L^\infty(0,T;L^2(\Omega))} \leq Ch^{1/2}. \tag{3.24}$$

Note that we deal with the minimal regularity on the initial datum, namely $u_0 \in L^2(\Omega)$.

Proof. Upon integrating (1.10) on the interval I^n, we introduce the discrete-time equations satisfied by $\{u, \theta\}$:

$$\langle u(t^n) - u(t^{n-1}), \varphi\rangle + \textstyle\int_{I^n} \langle\nabla\theta(t), \nabla\varphi\rangle dt = 0 \quad \forall\, \varphi \in H_0^1(\Omega), \ 1 \leq n \leq N. \tag{3.25}$$

On the other hand, the discrete equation (3.15) can be written equivalently, for all $\varphi \in H_0^1(\Omega)$, $\phi \in V_h^1$, and $1 \leq n \leq N$, as

$$\begin{aligned} \langle U^n - U^{n-1}, \varphi\rangle + \tau\langle\nabla\Theta^n, \nabla\varphi\rangle &= \tau\langle\nabla\Theta^n, \nabla(\varphi - \phi)\rangle \\ &+ \langle U^n - U^{n-1}, \varphi - \phi\rangle + \langle U^n - U^{n-1}, \phi\rangle - \langle U^n - U^{n-1}, \phi\rangle_h. \end{aligned} \tag{3.26}$$

Subtraction of (3.26) from (3.25) yields the error equation, for all $\varphi \in H_0^1(\Omega)$ and $\phi \in V_h^1$,

$$\begin{aligned} \langle \tilde{u}^n - \tilde{u}^{n-1}, \varphi\rangle + \textstyle\int_{I^n} \langle\nabla\tilde{\theta}(t), \nabla\varphi\rangle dt &= \tau\langle\nabla\Theta^n, \nabla(\phi - \varphi)\rangle \\ &+ \langle U^n - U^{n-1}, \phi - \varphi\rangle + \langle U^n - U^{n-1}, \phi\rangle_h - \langle U^n - U^{n-1}, \phi\rangle \end{aligned} \tag{3.27}$$

Take now $\varphi = G\tilde{u}^n \in H_0^1(\Omega)$ and $\phi = G_h\tilde{u}^n \in V_h^1$ in (3.27) and then add on n from 1 to $m \leq N$. In view of (1.22) and (3.10), we get

$$
\begin{aligned}
(3.28)\quad \textstyle\sum_{n=1}^m \langle \tilde{u}^n - \tilde{u}^{n-1}, G\tilde{u}^n\rangle + \sum_{n=1}^m \int_{I^n} \langle \tilde{\theta}(t), \tilde{u}^n\rangle dt = \sum_{n=1}^m \langle U^n - U^{n-1}, (G_h - G)\tilde{u}^n\rangle \\
+ \textstyle\sum_{n=1}^m \left(\langle U^n - U^{n-1}, G_h\tilde{u}^n\rangle_h - \langle U^n - U^{n-1}, G_h\tilde{u}^n\rangle\right),
\end{aligned}
$$

that is I + II = III + IV. We now evaluate each term separately. By virtue of (3.7), we have $\|U^0 - u_0\|_{H^{-1}(\Omega)} \leq Ch\|u_0\|_{L^2(\Omega)} \leq Ch$. The elementary identity (3.23), in conjunction with (1.22), then yields

$$
2\mathrm{I} = \|\tilde{u}^m\|^2_{H^{-1}(\Omega)} - \|u_0 - U^0\|^2_{H^{-1}(\Omega)} + \textstyle\sum_{n=1}^m \|\tilde{u}^n - \tilde{u}^{n-1}\|^2_{H^{-1}(\Omega)} \geq \|\tilde{u}^m\|^2_{H^{-1}(\Omega)} - Ch^2.
$$

In view of the constitutive relations (1.8) and (3.14), term II can be further split as

$$
\begin{aligned}
\mathrm{II}_1 + \mathrm{II}_2 + \mathrm{II}_3 = \textstyle\sum_{n=1}^m \int_{I^n} \langle \tilde{\theta}(t), u(t^n) - u(t)\rangle dt + \sum_{n=1}^m \int_{I^n} \langle \beta(U^n) - \Pi_h\beta(U^n), \tilde{u}(t)\rangle dt \\
+ \textstyle\sum_{n=1}^m \int_{I^n} \langle \beta(u(t)) - \beta(U^n), u(t) - U^n\rangle dt.
\end{aligned}
$$

We first make use of the *a priori* estimates (1.11) and (3.19), in conjunction with the Cauchy-Schwarz inequality, to estimate term II_1 as follows:

$$
\begin{aligned}
|\mathrm{II}_1| &\leq \textstyle\sum_{n=1}^m \int_{I^n} \|\nabla\tilde{\theta}(t)\|_{L^2(\Omega)} \left\|\int_t^{t^n} \partial_t u(s) ds\right\|_{H^{-1}(\Omega)} dt \\
&\leq \tau \|\nabla\tilde{\theta}\|_{L^2(Q)} \|\partial_t u\|_{L^2(0,T;H^{-1}(\Omega))} \leq C\tau.
\end{aligned}
$$

We next use the following elementary interpolation estimate (see [53]):

$$
(3.29)\quad \|\xi(\phi) - \Pi_h\xi(\phi)\|_{L^p(S)} \leq Ch_S\|\nabla\Pi_h\xi(\phi)\|_{L^p(S)} \quad \forall\, S \in \mathcal{S}_h, \phi \in V_h^1, 1 \leq p \leq \infty,
$$

where $\xi : \mathbf{R} \to \mathbf{R}$ is a continuous and nondecreasing function. Since $\Theta^n = \Pi_h\beta(U^n)$, applying (3.29) with $\xi = \beta$, $\phi = U^n$, and $p = 2$, owing to (1.11) and (3.19), we get

$$
|\mathrm{II}_2| \leq Ch \textstyle\sum_{n=1}^m \int_{I^n} \|\nabla\Theta^n\|_{L^2(\Omega)} \|\tilde{u}(t)\|_{L^2(\Omega)} dt \leq Ch.
$$

Using further that $\beta'(s) \leq 1$ for a.e. $s \in \mathbf{R}$ and the inequality $a^2 \leq 2(a-b)^2 + 2b^2$ for all $a, b \in \mathbf{R}$, we bound term II_3 from below as follows:

$$
\begin{aligned}
\mathrm{II}_2 &\geq \textstyle\sum_{n=1}^m \int_{I^n} \|\beta(u(t)) - \beta(U^n)\|^2_{L^2(\Omega)} dt \\
&\geq \tfrac{1}{2} \textstyle\sum_{n=1}^m \int_{I^n} \|\tilde{\theta}(t)\|^2_{L^2(\Omega)} dt - Ch^2 \sum_{n=1}^m \tau \|\nabla\Theta^n\|^2_{L^2(\Omega)} \geq \tfrac{1}{2}\|\tilde{\theta}\|^2_{L^2(0,t^m;L^2(\Omega))} - Ch^2.
\end{aligned}
$$

To examine term III we make use of (3.11) with $s = 1$, $r = 0$, in conjunction with the discrete $H^{1/2}$ *a priori* estimate (3.21) and the elementary inequality (1.26) to get

$$
\begin{aligned}
|\mathrm{III}| &\leq Ch \textstyle\sum_{n=1}^m \|U^n - U^{n-1}\|_{L^2(\Omega)} \|\tilde{u}^n\|_{H^{-1}(\Omega)} \\
&\leq \tfrac{1}{8}\max_{1\leq n\leq m} \|\tilde{u}^n\|^2_{H^{-1}(\Omega)} + C\tfrac{h^2}{\tau} \textstyle\sum_{n=1}^m \|U^n - U^{n-1}\|^2_{L^2(\Omega)} \\
&\leq \tfrac{1}{8}\max_{1\leq n\leq m} \|\tilde{u}^n\|^2_{H^{-1}(\Omega)} + C\tfrac{h^2}{\tau}.
\end{aligned}
$$

Term IV accounts for the effects of numerical integration. Hence we use (3.3) and then proceed as for term III to find

$$
|\mathrm{IV}| \leq Ch \textstyle\sum_{n=1}^m \|U^n - U^{n-1}\|_{L^2(\Omega)} \|\nabla G_h\tilde{u}^n\|_{L^2(\Omega)} \leq \tfrac{1}{8}\max_{1\leq n\leq m} \|\tilde{u}^n\|^2_{H^{-1}(\Omega)} + C\tfrac{Ch^2}{\tau},
$$

because $\|\nabla G_h\psi\|_{L^2(\Omega)} \leq \|\psi\|_{H^{-1}(\Omega)}$ for all $\psi \in H^{-1}(\Omega)$. Note that given a sequence $\{a_n\}_{n=1}^N$ such that $a_m \leq C + \frac{1}{2}\max_{1\leq n\leq m} a_n$ for all $1 \leq m \leq N$, then $\max_{1\leq n\leq N} a_n \leq$

$2C$. Hence, inserting the above estimates back into (3.28), on imposing the relation $\tau = C^*h$ leads to

$$\max_{1\le n\le N}\|\tilde u^n\|^2_{H^{-1}(\Omega)} + \|\tilde\theta\|^2_{L^2(Q)} \le C\big(\tau + h + \tfrac{h^2}{\tau}\big) = Ch. \tag{3.30}$$

Since $u \in H^1(0,T;H^{-1}(\Omega)) \subset C^{0,1/2}([0,T];H^{-1}(\Omega))$, we have $\max_{1\le n\le N}\|\tilde u^n\|^2_{H^{-1}(\Omega)} \ge \|\tilde u\|^2_{L^\infty(0,T;H^{-1}(\Omega))} - C\tau$, and this yields the asserted error estimates for $\tilde\theta$ and $\tilde u$.

To deduce the remaining estimate for $\int_0^t\nabla\tilde\theta$ we integrate the error equation (3.27) in time, that is we sum (3.27) on n from 1 to $m \le N$, and next choose the test functions $\varphi = \int_0^{t^m}\tilde\theta(t)dt \in H^1_0(\Omega)$ and $\phi = R_h\int_0^{t^m}\tilde\theta(t)dt \in V^1_h$. We have

$$\begin{aligned}\Big\|\int_0^{t^m}\nabla\tilde\theta(t)dt\Big\|^2_{L^2(\Omega)} &= \Big\langle \tilde u^0 - \tilde u^m, \int_0^{t^m}\tilde\theta(t)dt\Big\rangle + \Big\langle U^0 - U^m, (I-R_h)\int_0^{t^m}\tilde\theta(t)dt\Big\rangle \\ &\quad + \Big\langle U^m - U^0, R_h\int_0^{t^m}\tilde\theta(t)dt\Big\rangle_h - \Big\langle U^m - U^0, R_h\int_0^{t^m}\tilde\theta(t)dt\Big\rangle.\end{aligned}$$

Using (3.9) for $s=1$, $r=0$, (3.3), the fact that $\|\nabla R_h\varphi\|_{L^2(\Omega)} \le \|\nabla\varphi\|_{L^2(\Omega)}$ for all $\varphi \in H^1_0(\Omega)$, and also the error estimate (3.30), we obtain

$$\Big\|\int_0^{t^m}\nabla\tilde\theta(t)dt\Big\|_{L^2(\Omega)} \le \|\tilde u^0\|_{H^{-1}(\Omega)} + \|\tilde u^m\|_{H^{-1}(\Omega)} + Ch\big(\|U^m\|_{L^2(\Omega)} + \|U^0\|_{L^2(\Omega)}\big) \le Ch.$$

The regularity $\int_0^t\nabla\theta \in H^1(0,T;H^1(\Omega))$ allows us to replace $\max_{1\le n\le N}\Big\|\int_0^{t^n}\nabla\tilde\theta\Big\|_{L^2(\Omega)}$ with $\Big\|\int_0^t\nabla\tilde\theta\Big\|_{L^\infty(0,T;L^2(\Omega))}$ and thus yields the desired error estimate. □

REMARK 3.2. The preliminary regularization (1.29) on β can be used, in the case the nondegeneracy property (1.17) holds, to improve the error estimate (3.24) (see [113]). In fact, exploiting the property $\beta'_\varepsilon(s) \ge \varepsilon$ for a.e. $s \in \mathbb{R}$, the analysis of terms II_2, III, and IV can be modified to allow the control of $\tilde u$ in $L^2(\Omega)$ instead of $H^{-1}(\Omega)$, so as to prove the following error estimate between the solutions of the regularized problem and its fully discrete version:

$$\begin{aligned}&\|\tilde u_\varepsilon\|_{L^\infty(0,T;H^{-1}(\Omega))} + \|\tilde\theta_\varepsilon\|_{L^2(Q)} + \Big\|\int_0^t\nabla\tilde\theta_\varepsilon\Big\|_{L^\infty(0,T;L^2(\Omega))} \\ &\qquad + \varepsilon^{1/2}\|\tilde u_\varepsilon\|_{L^2(Q)} \le C\big(\tfrac{h}{\varepsilon^{1/2}} + \tfrac{h^2}{\varepsilon\tau^{1/2}} + \tau^{1/2}\big).\end{aligned} \tag{3.31}$$

Combining estimate (3.31) with (1.36), under the relation $\varepsilon, \tau = C^*h$ we obtain the same $\mathcal{O}(h^{1/2})$ global rate of convergence in the general situation. But, under the nondegeneracy assumption (1.17), by relating $\tau = C^*h^{4/3}$ and $\varepsilon = C^*h^{2/3}$, the estimate is improved up to $\mathcal{O}(h^{2/3})$. In addition, the following estimate for $\tilde u$ holds: $\|\tilde u\|_{L^2(Q)} \le Ch^{1/3}$. Also, the discrete maximum principle assumption, $\mathcal{S}_h$ (weakly) acute, can be removed.

REMARK 3.3. A quasi-optimal rate of convergence $E_h \le Ch|\log h|^{3/2}$ under the relation $\tau = C^*h$ has been proved just recently in [127] for fully discrete backward differences, in two space dimensions, but without numerical integration.

3.3. Fully discrete nonlinear Chernoff formula

The finite element discretization of the nonlinear Chernoff formula (2.4) reads:

Set $U^0 = P^0_h u_0$ and, for any $1 \le n \le N$, seek $U^n \in V^0_h$ and $\Theta^n \in V^1_h$ such that

$$\langle\Theta^n, \phi\rangle_h + \tau\langle\nabla\Theta^n, \nabla\phi\rangle = \langle\beta(U^{n-1}), \phi\rangle \quad \forall\, \phi \in V^1_h, \tag{3.32}$$

$$U^n = U^{n-1} + P^0_h\Theta^n - \beta(U^{n-1}). \tag{3.33}$$

Equation (3.32) is equivalent to the following symmetric positive definite linear system:

$$(\mathbf{M}+\tau\mathbf{K})\Theta^n = \mathbf{N}\beta(U^{n-1}), \tag{3.34}$$

whereas, since both U^n and $P_h^0\Theta^n$ are piecewise constant ($P_h^0\phi = \sum_{k=1}^K \phi(\mathbf{b}_k)\sigma_k$ for any $\phi \in V_h^1$), equation (3.33) is just a set of elementwise scalar corrections:

$$U_k^n = U_k^{n-1} + \Theta^n(\mathbf{b}_k) - \beta(U_k^{n-1}).$$

This yields unique solvability for the fully discrete nonlinear Chernoff formula. The efficiency of the method is clearly dictated by the resolution of the linear system (3.34). This can be done using the preconditioned, via incomplete Cholesky factorization (see [10,90]), conjugate gradient method. Note that, since

$$Ch^d \le \tfrac{\mathbf{s}^T\mathbf{A}\mathbf{s}}{|\mathbf{s}|^2} \le C^{-1}h^d(1+\tau h^{-2}) \quad \forall\, \mathbf{s}\in\mathbf{R}^J,$$

and $\tau = C^*h$ will be enforced, the condition number $k(\mathbf{A})$ of $\mathbf{A} = \mathbf{M}+\tau\mathbf{K}$ verifies $k(\mathbf{A}) = \mathcal{O}(h^{-1})$. The conjugate gradient iterations can be stopped safely, that is still preserving stability and accuracy of the method, if a suitable error reduction is attained. This reduces considerably the computational labor (see [51,100]).

REMARK 3.4. We can also use piecewise linear finite elements for both variables U^n and Θ^n, that is:

Set $U^0 = P_h^1 u_0$ and, for any $1 \le n \le N$, seek $U^n, \Theta^n \in V_h^1$ such that

$$\langle\Theta^n,\phi\rangle_h + \tau\langle\nabla\Theta^n,\nabla\phi\rangle = \langle\beta(U^{n-1}),\phi\rangle_h \quad \forall\,\phi\in V_h^1, \tag{3.35}$$

$$U^n = U^{n-1} + \Theta^n - \Pi_h\beta(U^{n-1}). \tag{3.36}$$

The scheme (3.35), (3.36) is still easy to implement and experimental accuracy is improved, but the corresponding stability and error analysis is much more intricate than for the previous scheme (3.32), (3.33) (see [74,106,107]).

3.3.1. Stability for the nonlinear Chernoff formula

We examine the stability of the fully discrete nonlinear Chernoff formula. The fact that the constitutive relation $\Theta^n = \Pi_h\beta(U^n)$ is no longer enforced makes the analysis much more intricate than for the nonlinear method (3.14), (3.15). Note that (3.32) and (3.33) can be combined to give

$$\langle U^n - U^{n-1},\phi\rangle + \tau\langle\nabla\Theta^n,\nabla\phi\rangle = \langle(P_h^0 - I)\Theta^n,\phi\rangle_h \quad \forall\,\phi\in V_h^1. \tag{3.37}$$

Let $\alpha:\mathsf{R}\to\mathsf{R}$ be defined by $\alpha = I-\beta$; thus $0\le\alpha'(s)\le 1$ for a.e. $s\in\mathsf{R}$.

THEOREM 3.3. *Let $\{U^n,\Theta^n\}_{n=0}^N$ be the solution to problem (3.32),(3.33). Then*

$$\begin{aligned}\max_{1\le n\le N}\|U^n\|_{L^2(\Omega)} &+ \textstyle\sum_{n=1}^N \|U^n - U^{n-1}\|_{L^2(\Omega)}^2 \\ &+ \textstyle\sum_{n=1}^N \tau\|\nabla\Theta^n\|_{L^2(\Omega)}^2 \le C\|U^0\|_{L^2(\Omega)}.\end{aligned} \tag{3.38}$$

If S_h is (weakly) acute, then the additional estimate holds:

$$\max_{1\le n\le N}\|U^n\|_{L^\infty(\Omega)} \le \|U^0\|_{L^\infty(\Omega)}. \tag{3.39}$$

Note that, since P_h^0 is bounded in $L^\infty(\Omega)$ and $U^0 = P_h^0 u_0$, if $u_0\in L^\infty(\Omega)$ then $\|U^0\|_{L^\infty(\Omega)} \le \|u_0\|_{L^\infty(\Omega)}$. It is also possible to prove a strong stability estimate that resembles (3.20) and is useful in analyzing the incomplete iteration process mentioned above. The choice of the initial data is again a crucial issue (see [101]).

Proof. i) Proof of (3.38). Take $\phi = \Theta^n \in V_h^1$ in (3.37) and add the resulting expressions on n from 1 to $m \le N$. In light of the definitions (3.4) of P_h^0 and (3.1) of $\langle \cdot, \cdot \rangle_h$, we get

$$\sum_{n=1}^m \langle U^n - U^{n-1}, P_h^0 \Theta^n \rangle + \tau \sum_{n=1}^m \|\nabla \Theta^n\|_{L^2(\Omega)}^2 \le 0, \tag{3.40}$$

because $\langle (P_h^0 - I)\Theta^n, \Theta^n \rangle_h = \langle (P_h^0 - I)\Theta^n, (I - P_h^0)\Theta^n \rangle_h \le 0$. Let us split $P_h^0 \Theta^n$ as

$$\begin{aligned} P_h^0 \Theta^n &= U^n - U^{n-1} + \beta(U^{n-1}) \\ &= \tfrac{1}{2}\beta(U^n) - \tfrac{1}{2}\alpha(U^{n-1}) + \tfrac{1}{2}U^n + \tfrac{1}{2}\big(\alpha(U^n) - \alpha(U^{n-1})\big). \end{aligned} \tag{3.41}$$

On using the convexity of both Φ_β and Φ_α as well as the identity (3.23), the first term in (3.40) can be bounded from below as follows:

$$\begin{aligned} 2\sum_{n=1}^m \langle U^n - U^{n-1}, P_h^0 \Theta^n \rangle \ge \sum_{n=1}^m \Big(&\int_\Omega \big(\Phi_\beta(U^n) - \Phi_\beta(U^{n-1})\big) \\ &+ \int_\Omega \big(\Phi_\alpha(U^{n-1}) - \Phi_\alpha(U^n)\big) + \tfrac{1}{2}\big(\|U^n\|_{L^2(\Omega)}^2 - \|U^{n-1}\|_{L^2(\Omega)}^2 + \|U^n - U^{n-1}\|_{L^2(\Omega)}^2\big)\Big). \end{aligned}$$

Here we have also used that α is nondecreasing in order to eliminate the contribution of the last term in (3.41). The terms involving Φ_β and Φ_α can be further bounded by means of (3.18), which leads to

$$\begin{aligned} 4\sum_{n=1}^m \langle U^n - U^{n-1}, P_h^0 \Theta^n \rangle &\ge \|\beta(U^m)\|_{L^2(\Omega)}^2 - \|U^0\|_{L^2(\Omega)}^2 + \|\alpha(U^0)\|_{L^2(\Omega)}^2 - \|\alpha(U^m)\|_{L^2(\Omega)}^2 \\ &\quad + \|U^m\|_{L^2(\Omega)}^2 - \|U^0\|_{L^2(\Omega)}^2 + \sum_{n=1}^m \|U^n - U^{n-1}\|_{L^2(\Omega)}^2 \\ &\ge \|\beta(U^m)\|_{L^2(\Omega)}^2 + \sum_{n=1}^m \|U^n - U^{n-1}\|_{L^2(\Omega)}^2 - 2\|U^0\|_{L^2(\Omega)}^2. \end{aligned}$$

Finally the estimate for $\|\beta(U^n)\|_{L^2(\Omega)}$ yields a bound for $\|U^n\|_{L^2(\Omega)}$ because $\beta(s) \ge |s| - 1$.

ii) Proof of (3.39). We argue by induction and suppose $\|U^{n-1}\|_{L^\infty(\Omega)} \le \|U^0\|_{L^\infty(\Omega)} = C_0$ for $1 \le n \le N$. Let $\mathbf{x}_j$ be an internal node where Θ^n attains its maximum. Since $\mathcal{S}_h$ is (weakly) acute, the discrete maximum principle (3.12) applies and (3.32) yields

$$m_{jj}\Theta_j^n = \langle \Theta^n, \phi_j \rangle_h \le \langle \Theta^n, \phi_j \rangle_h + \tau \langle \nabla \Theta^n, \nabla \phi_j \rangle = \langle \beta(U^{n-1}, \phi_j \rangle \le m_{jj}\beta(C_0),$$

because β is nondecreasing. A similar argument produces a bound from below for the minimum of Θ^n. Therefore, we get the estimate $\beta(-C_0) \le P_h^0 \Theta^n \le \beta(C_0)$. Now, since $U^n = U^{n-1} + P_h^0\Theta^n - \beta(U^{n-1}) = \alpha(U^{n-1}) + P_h^0 \Theta^n$, and α is nondecreasing and verifies $\alpha + \beta = I$, we see that

$$-C_0 = \alpha(-C_0) + \beta(-C_0) \le U^n \le \alpha(C_0) + \beta(C_0) = C_0,$$

which, in turn, completes the induction argument. □

3.3.2. Error analysis for the nonlinear Chernoff formula

In this section we analyse the accuracy of the fully discrete nonlinear Chernoff formula, using the integral-time method. We deal with the weaker assumption $u_0 \in L^2(\Omega)$.

Let us define the discrete, piecewise constant in time, solutions U and Θ by

$$U(\cdot, t) = U^n(\cdot), \quad \Theta(\cdot, t) = \Theta^n(\cdot), \quad \forall\, t \in I^n,\ 1 \le n \le N,$$

and let the error functions $\tilde{u} = u - U$ and $\tilde{\theta} = \theta - \Theta$.

THEOREM 3.4. *Let $u_0 \in L^2(\Omega)$. Let $\tau = C^* h^{4/3}$, with $C^* > 0$ arbitrary. Then*

$$E_h = \|\tilde{u}\|_{L^\infty(0,T;H^{-1}(\Omega))} + \|\tilde{\theta}\|_{L^2(Q)} + \left\| \int_0^t \nabla \tilde{\theta} \right\|_{L^\infty(0,T;L^2(\Omega))} \leq C h^{1/3}. \tag{3.42}$$

REMARK 3.5. *i)* Assuming stronger regularity on the initial data, namely $u_0 \in L^\infty(\Omega)$ and $\Delta\theta_0 \in L^1(\Omega) + M_{I_0}(\Omega)$, and also that S_h is (weakly) acute, on imposing the relation $\tau = C^* h$, estimate (3.42) becomes $E_h \leq C h^{1/2}$ (see [114]).

ii) Consider the following modification of the fully discrete nonlinear Chernoff formula (3.32), (3.33):

Set $U^0 = P_h^0 u_0$, $\Theta^0 = P_h^1 \theta_0$, and, for any $1 \leq n \leq N$, seek $U^n \in V_h^0$ and $\Theta^n \in V_h^1$ such that

$$\langle \Theta^n, \phi \rangle_h + \tau \langle \nabla \Theta^n, \nabla \phi \rangle = \langle \beta(U^{n-1}), \phi \rangle + \langle (I - P_h^0)\Theta^{n-1}, \phi \rangle_h \quad \forall \phi \in V_h^1, \tag{3.43}$$

$$U^n = U^{n-1} + P_h^0 \Theta^n - \beta(U^{n-1}). \tag{3.44}$$

Under the sole assumption $u_0 \in L^2(\Omega)$ the scheme (3.43), (3.44) is stable and, provided $\tau = C^* h^2$, the error estimate (3.42) becomes

$$E_h \leq C h^{1/2}. \tag{3.45}$$

iii) The combination of regularization of β with the nonlinear Chernoff formula has been investigated in [115].

Proof. For any $1 \leq n \leq N$ and $t \in I^n$, we integrate (1.10) from 0 to t; then we add (3.37) over i from 1 to n. Taking the difference of the resulting expressions we get the error equation:

$$\begin{aligned} \langle \tilde{u}(t) - \tilde{u}(0), \phi \rangle + \left\langle \int_0^t \nabla \tilde{\theta}(s) ds, \nabla \phi \right\rangle &= \frac{1}{\tau} \left\langle (I - P_h^0) \int_0^{t^n} \Theta(s) ds, \phi \right\rangle_h \\ &\quad + (t^n - t) \langle \nabla \Theta^n, \nabla \phi \rangle \quad \forall \phi \in V_h^1. \end{aligned} \tag{3.46}$$

Take $\phi = R_h \tilde{\theta}(t) = R_h \theta(t) - \Theta(t) \in V_h^1$ and integrate from 0 to $\bar{t} \leq T$. After reordering we get

$$\begin{aligned} &\int_0^{\bar{t}} \langle \tilde{u}(t), \tilde{\theta}(t) \rangle dt + \int_0^{\bar{t}} \left\langle \int_0^t \nabla \tilde{\theta}(s) ds, \nabla R_h \tilde{\theta}(t) \right\rangle dt \\ &\quad \leq \int_0^{\bar{t}} \langle \tilde{u}(t), (I - R_h)\theta(t) \rangle dt + \left\langle u_0 - U^0, \int_0^{\bar{t}} R_h \tilde{\theta}(t) dt \right\rangle \\ &\quad + \frac{1}{\tau} \sum_{n=1}^N \int_{I^n} \left| \left\langle (I - P_h^0) \int_0^{t^n} \Theta(s) ds, R_h \tilde{\theta}(t) \right\rangle_h \right| dt \\ &\quad + \sum_{n=1}^N \int_{I^n} \left| (t^n - t) \langle \nabla \Theta^n, \nabla R_h \tilde{\theta}(t) \rangle \right| dt, \end{aligned} \tag{3.47}$$

that is I + II $\leq$ III + IV + V + VI. We now proceed to estimate each term separately. To begin with, note that (1.8) and (3.33) yield

$$u = \theta + \alpha(u), \quad U^n = P_h^0 \Theta^n + \alpha(U^{n-1}),$$

whence, for all $t \in I^n$, $1 \leq n \leq N$,

$$\tilde{u}(t) = \tilde{\theta}(t) + \big(\alpha(u(t)) - \alpha(U^{n-1})\big) + (I - P_h^0)\Theta^n, \tag{3.48}$$

$$\tilde{\theta}(t) = \big(\beta(u(t)) - \beta(U^{n-1})\big) - (U^n - U^{n-1}) - (I - P_h^0)\Theta^n. \tag{3.49}$$

Then, in view of (3.48) and (3.49), term I in (3.47) can be further split as follows:

$$\begin{aligned}\mathrm{I} = \|\tilde\theta\|^2_{L^2(0,\bar t;L^2(\Omega))} + \textstyle\int_0^{\bar t}\langle\alpha(u(t)) - \alpha(U^{n-1}), \beta(u(t)) - \beta(U^{n-1})\rangle dt\\ - \textstyle\int_0^{\bar t}\langle\alpha(u(t)) - \alpha(U^{n-1}), U^n - U^{n-1}\rangle dt - \int_0^{\bar t}\langle\alpha(u(t)) - \alpha(U^{n-1}), (I-P_h^0)\Theta^n\rangle dt\\ + \textstyle\int_0^{\bar t}\langle\tilde\theta(t), (I-P_h^0)\Theta^n\rangle dt = \mathrm{I}_1 + \cdots + \mathrm{I}_5.\end{aligned}$$

The fact that both α and β are nondecreasing functions implies that $\mathrm{I}_2 \geq 0$. Since $\alpha'(s) \leq 1$ for a.e. $s \in \mathbf{R}$, owing to (1.11) and (3.38) we get

$$|\mathrm{I}_3| \leq \left(\|u\|_{L^2(Q)} + \|U\|_{L^2(Q)}\right)\left(\textstyle\sum_{n=1}^N \tau\|U^n - U^{n-1}\|^2_{L^2(\Omega)}\right)^{1/2} \leq C\tau^{1/2}.$$

Next we can use (3.6) for $s = 0$, $r = 1$ and again (1.11) and (3.38) to obtain

$$\begin{aligned}|\mathrm{I}_4| + |\mathrm{I}_5| &\leq \left(\|u\|_{L^2(Q)} + \|U\|_{L^2(Q)} + \|\tilde\theta\|_{L^2(Q)}\right)\left(\textstyle\sum_{n=1}^N \tau\|(I-P_h^0)\Theta^n\|^2_{L^2(\Omega)}\right)^{1/2}\\ &\leq Ch\left(\textstyle\sum_{n=1}^N \tau\|\nabla\Theta^n\|^2_{L^2(\Omega)}\right)^{1/2} \leq Ch.\end{aligned}$$

Combining (3.8) and (3.9), for $s = 0$, $r = 1$, with $\int_0^t\theta \in L^\infty(0,T;H^2(\Omega))$, which comes from (1.16), we can bound term II from below as follows:

$$\begin{aligned}2\mathrm{II} = \left\|\nabla R_h \textstyle\int_0^{\bar t}\tilde\theta(t)dt\right\|^2_{L^2(\Omega)} &\geq \tfrac{1}{2}\left\|\textstyle\int_0^{\bar t}\nabla\tilde\theta(t)dt\right\|^2_{L^2(\Omega)} - \left\|\nabla(I-R_h)\textstyle\int_0^{\bar t}\theta(t)dt\right\|^2_{L^2(\Omega)}\\ &\geq \tfrac{1}{2}\left\|\textstyle\int_0^{\bar t}\nabla\tilde\theta(t)dt\right\|^2_{L^2(\Omega)} - Ch^2.\end{aligned}$$

Using (3.9) for $s = 1$, $r = 0$, in conjunction with (1.11) and (3.38), we get

$$|\mathrm{III}| \leq Ch\|\tilde u\|_{L^2(Q)}\|\theta\|_{L^2(0,T;H^1(\Omega))} \leq Ch.$$

Since $U^0 = P_h^0 u_0$, owing to (3.6), for $s = 1$, $r = 0$, and using again (1.11) and (3.38), a duality argument between $H^{-1}(\Omega)$ and $H_0^1(\Omega)$ leads to

$$|\mathrm{IV}| \leq Ch\|u_0\|_{L^2(\Omega)}\|\nabla R_h \textstyle\int_0^{\bar t}\tilde\theta(t)dt\|_{L^2(\Omega)} \leq Ch.$$

In order to get a bound for V, we first split this term in the more suitable form:

$$\begin{aligned}|\mathrm{V}| &\leq \tfrac{1}{\tau}\textstyle\sum_{n=1}^N \int_{I^n}\left|\left\langle (I-P_h^0)\int_0^{t^n}\Theta(s)ds, R_h\tilde\theta(t)\right\rangle\right|dt\\ &+ \tfrac{1}{\tau}\textstyle\sum_{n=1}^N \int_{I^n}\left|\left\langle \int_0^{t^n}\Theta(s)ds, R_h\tilde\theta(t)\right\rangle_h - \left\langle \int_0^{t^n}\Theta(s)ds, R_h\tilde\theta(t)\right\rangle\right|dt.\end{aligned}$$

Using a duality argument between $H^{-1}(\Omega)$ and $H_0^1(\Omega)$, we exploit (3.6) for $s = 1$, $r = 1$, to estimate the first term, whereas the second one is an error due to the quadrature rule and can be controlled by (3.3). Thus

$$\begin{aligned}|\mathrm{V}| &\leq C\tfrac{h^2}{\tau}\textstyle\sum_{n=1}^N \int_{I^n}\left\|\int_0^{t^n}\Theta\right\|_{H^1(\Omega)}\|\nabla R_h\tilde\theta(t)\|_{L^2(\Omega)}dt\\ &\leq C\tfrac{h^2}{\tau}\|\Theta\|_{L^2(0,T;H^1(\Omega))}\|\tilde\theta\|_{L^2(0,T;H^1(\Omega))} \leq C\tfrac{h^2}{\tau},\end{aligned}$$

where (1.11) and (3.38) have been used again. Finally, for term VI we readily get

$$|\mathrm{VI}| \leq \tau\|\Theta\|_{L^2(0,T;H^1(\Omega))}\|\tilde\theta\|_{L^2(0,T;H^1(\Omega))} \leq C\tau.$$

Collecting all the previous estimates, and inserting them back into (3.47), upon imposing the relation $\tau = C^* h^{4/3}$, gives the asserted estimates for $\tilde{\theta}$ and $\int_0^t \nabla\tilde{\theta}$, because

$$\|\tilde{\theta}\|^2_{L^2(0,\bar{t};L^2(\Omega))} + \tfrac{1}{4}\left\|\int_0^{\bar{t}} \nabla\tilde{\theta}(t)dt\right\|^2_{L^2(\Omega)} \le C\left(\tau^{1/2} + h + \tfrac{h^2}{\tau}\right).$$

In order to derive the error estimate for $\tilde{u}$ in $L^\infty(0,T;H^{-1}(\Omega))$, we select $\phi = G_h\tilde{u}(t)$ in (3.46) to realize that, for any $t \in I^n$, $1 \le n \le N$,

$$\|\nabla G_h\tilde{u}(t)\|^2_{L^2(\Omega)} \le \Big(\|\tilde{u}(0)\|_{H^{-1}(\Omega)} + \left\|\int_0^t \nabla\tilde{\theta}(s)ds\right\|_{L^2(\Omega)} + C\tfrac{h^2}{\tau}\left\|\int_0^{t^n} \nabla\Theta(s)ds\right\|_{L^2(\Omega)}$$
$$+ \tau\|\nabla\Theta^n\|_{L^2(\Omega)}\Big)\|\nabla G_h\tilde{u}(t)\|_{L^2(\Omega)};$$

this concludes the argument because $\|\tilde{u}(t)\|_{H^{-1}(\Omega)} - Ch\|\tilde{u}(t)\|_{L^2(\Omega)} \le \|\nabla G_h\tilde{u}(t)\|_{L^2(\Omega)}$.□

3.4. Approximation of free boundaries

Fixed domain methods provide an approximation of temperature θ and energy u but, convergence of the discrete solutions Θ and U may provide no information about convergence of discrete free boundaries, which are level sets of either Θ or U. This fact can be easily seen for flat solutions (e.g., for the one-phase Stefan problem — see Figure 3.1). Approximation of interfaces is in fact related to nondegeneracy. Roughly, a free boundary problem is nondegenerate if the solution leaves the free boundary with a proper speed. The crucial idea of nondegeneracy and interface approximability was introduced by Caffarelli [31] and first used in a numerical framework by Brezzi and Caffarelli [28].

Consider the following stationary obstacle problem:

Seek $u \in K = \{\varphi \in H^1(\Omega) : \varphi \ge 0 \text{ in } \Omega, v = g \ge 0 \text{ on } \partial\Omega\}$ *such that*

$$\langle \nabla u, \nabla(\varphi - u)\rangle \ge \langle f, \varphi - u\rangle \quad \forall\, \varphi \in K. \tag{3.50}$$

Define the positivity or noncontact set $P = \{\mathbf{x} \in \Omega : u(\mathbf{x}) > 0\}$ and the interface $I = \partial P \cap \Omega$. Introduce also the strip $S_\varepsilon(I) = \{\mathbf{x} \in \Omega : \operatorname{dist}(\mathbf{x}, I) \le \varepsilon\}$ and the set $A_\varepsilon = \{\mathbf{x} \in \Omega : 0 < u(\mathbf{x}) \le \varepsilon^2\}$. Finally let $B_\varepsilon(\mathbf{x})$ indicate the ball centered at $\mathbf{x} \in \Omega$ and radius ε.

If $f \le -\sigma < 0$ in Ω, then u satisfies the following nondegeneracy property (see [31,61]):

$$\sup_{\mathbf{x}\in B_\varepsilon(\bar{\mathbf{x}})} u(\mathbf{x}) \ge \tfrac{\sigma}{4d}\varepsilon^2 + u(\bar{\mathbf{x}}) \quad \forall\, \bar{\mathbf{x}} \in P \cup I. \tag{3.51}$$

The proof of (3.51) is based on the maximum principle. In fact, define the auxiliary function $w(\mathbf{x}) = u(\mathbf{x}) - u(\bar{\mathbf{x}}) - \frac{\sigma}{4d}|\mathbf{x} - \bar{\mathbf{x}}|^2$. Then, for all $\mathbf{x} \in B_\varepsilon(\bar{\mathbf{x}}) \cap P$, (3.50) yields $-\Delta w(\mathbf{x}) = -\Delta u(\mathbf{x}) + \frac{\sigma}{2} = f + \frac{\sigma}{2} < 0$, hence the sup of w is attained at the boundary of $B_\varepsilon(\bar{\mathbf{x}}) \cap P$. Since $w(\bar{\mathbf{x}}) = 0$ and $w \le 0$ on I, then the sup is attained at $\partial B_\varepsilon(\bar{\mathbf{x}}) \cap P$, and this concludes the argument.

Using (3.51) it is possible to prove that

$$\operatorname{meas}\big(S_\varepsilon(I) \cup A_\varepsilon\big) \le C\varepsilon, \tag{3.52}$$

whereas if I is (locally) Lipschitz continuous, the stronger property $A_\varepsilon \subset S_{C\varepsilon}(I)$ holds.

Let us introduce a finite element approximation of problem (3.50):

Seek $u_h \in K_h = \{\phi \in V_h^1 : \phi \ge 0 \text{ in } \Omega, \phi = \Pi_h g \text{ on } \partial\Omega\}$ *such that*

$$\langle \nabla u_h, \nabla(\phi - u_h)\rangle \ge \langle f, \phi - u_h\rangle \quad \forall\, \phi \in K_h. \tag{3.53}$$

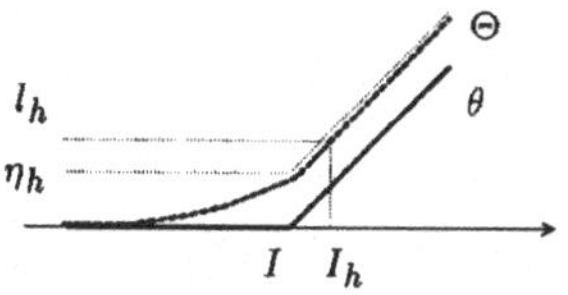

FIGURE 3.1. One-phase Stefan problem: temperature θ and discrete temperature Θ (dashed line).

The following pointwise error estimate is well known (see, e.g., [95]):

$$\|u - u_h\|_{L^\infty(\Omega)} \leq Ch^2|\log h| = \eta_h^2. \tag{3.54}$$

The discrete solution u_h to (3.53)satisfies the discrete analogue of the nondegeneracy property (3.51) provided $\mathcal{S}_h$ is acute (see [28]), that is

$$\sup_{\mathbf{x}\in B_\varepsilon(\bar{\mathbf{x}})} u_h(\mathbf{x}) \geq \sigma^*\varepsilon^2. \tag{3.55}$$

This, in turn, allows us to define naturally the discrete free boundary $I_h = \partial P_h \cap \Omega$, where $P_h = \{\mathbf{x} \in \Omega : u_h(\mathbf{x}) > 0\}$, and prove a rate of convergence in measure for interfaces:

$$\text{meas}(P \Delta P_h) \leq C\eta_h. \tag{3.56}$$

Let us sketch the proof of (3.56). *i)* Let $\bar{\mathbf{x}} \in P \backslash P_h$. Then $u_h(\bar{\mathbf{x}}) = 0$, whence, owing to (3.54), we have $0 < u(\bar{\mathbf{x}}) \leq \eta_h^2$, that is $P\backslash P_h \subset A_{\eta_h}$. *ii)* Let $\bar{\mathbf{x}} \in P_h \backslash P$. Then $u_h(\bar{\mathbf{x}}) > 0$ and, by virtue of (3.55), $\sup_{B_{\delta\eta_h}(\bar{\mathbf{x}})} u_h \geq \sigma^*\delta^2\eta_h^2$. Using again (3.54) it follows that $\sup_{B_{\delta\eta_h}(\bar{\mathbf{x}})} u \geq (\sigma^*\delta^2 - 1)\eta_h^2 > 0$ provided δ is sufficiently large. This entails $I \cap B_{\delta\eta_h}(\bar{\mathbf{x}}) \neq \emptyset$, that is $P_h \backslash P \subset \mathcal{S}_{\delta\eta_h}(I)$. Therefore we conclude that $P \Delta P_h \subset A_{\eta_h} \cup \mathcal{S}_{\delta\eta_h}(I)$, which yields the assertion by virtue of (3.52).

As a by-product of this argument we have that, if I is (locally) Lipschitz continuous, then $P \Delta P_h \subset \mathcal{S}_{C\eta_h}(I)$, that is the following error estimate in distance holds: $\text{dist}(I, I_h) \leq C\eta_h$, where dist stands for the Hausdorff distance.

A similar interface convergence result was proved in [123] for the one-phase Stefan problem set in terms of a parabolic variational inequality.

The main drawback of Brezzi and Caffarelli idea is that the nondegeneracy of discrete solutions and error estimates in maximum norm for solutions are both essential in proving the interface error estimates. Unfortunately these properties are not yet proved, as far as the author knows, for the energy formulation of the two-phase Stefan problem. In order to overcome this difficulty, we present a different idea (see [97] and [13]) which simply consists in changing the definition of discrete free boundary by shifting the normal level set of the discrete solution:

$$P_h = \{\mathbf{x} \in \Omega : u_h(\mathbf{x}) > l_h\}, \quad I_h = \partial P_h \cap \Omega. \tag{3.57}$$

We just assume a nondegeneracy property for the continuous solution and an L^p-error estimate for solutions to hold, namely,

$$\text{meas}\left(\{\mathbf{x} \in \Omega : 0 < u(\mathbf{x}) \leq \varepsilon^s\} \cup I\right) \leq C\varepsilon, \tag{3.58}$$

$$\|u - u_h\|_{L^p(\Omega)} \leq Ch^r = \eta_h^r. \tag{3.59}$$

The level l_h in the definition (3.57) of discrete free boundary is $l_h = \delta\eta_h^{rsp/(1+sp)}$ if $1 \le p < \infty$ and $l_h = \delta\eta_h^r$ if $p = \infty$ ($\delta \ge 1$). The following error estimates in measure for interfaces hold (see [97]):

$$\text{(3.60)} \quad \operatorname{meas}(P \triangle P_h) \le C\eta_h^{rp/(1+sp)} \quad \text{if } 1 \le p < \infty, \quad \operatorname{meas}(P \triangle P_h) \le C\eta_h^{r/s} \quad \text{if } p = \infty.$$

We just sketch the proof of (3.60) in the case $p = \infty$. *i)* Let $\bar{\mathbf{x}} \in P \backslash P_h$. Then $u_h(\bar{\mathbf{x}}) \le \delta\eta_h^r$, whence, owing to (3.59), we have $0 < u(\bar{\mathbf{x}}) \le (\delta + 1)\eta_h^r$, that is $P \backslash P_h \subset A_{C\eta_h^{r/s}}$. *ii)* Let $\bar{\mathbf{x}} \in P_h \backslash P$. Then $u_h(\bar{\mathbf{x}}) > \delta\eta_h^r \ge \eta_h^r$, which contradicts (3.59). Thus $P_h \backslash P = \emptyset$. In view of (3.58) this concludes the proof of the asserted estimate (3.60).

If the stronger nondegeneracy property $\{\mathbf{x} \in \Omega : 0 < u(\mathbf{x}) \le \varepsilon^s\} \subset S_{C\varepsilon}(I)$ holds and $p = \infty$ in (3.59), then the following interface error estimate in distance is valid: $I_h \subset S_{C\eta_h^{r/s}}(I)$.

We now apply these general results to the two-phase Stefan problem, for which the nondegeneracy property (1.17) of order $s = 1$ was proved by Nochetto [99] under proper qualitative assumptions on the data. In view of the error estimates (3.24) and (3.45), for which $p = 2$ and $r = \frac{1}{2}$, we deduce the following rate of convergence in measure for interfaces (see [113,114]):

$$\operatorname{meas}\big(\{(\mathbf{x}, t) \in Q : \theta(\mathbf{x}, t) > 0\} \triangle \{(\mathbf{x}, t) \in Q : \Theta(\mathbf{x}, t) > \eta_h^{1/3}\}\big) \le Ch^{1/3}.$$

This result can be improved up to $\mathcal{O}(h^{4/9})$ by combining regularization of β and discretization, as observed in Remark 3.5, iii). It has to be stressed that the error estimates proved for both solutions and interfaces are pessimistic in most cases. This has been observed also experimentally (see Sect. 4.6).

4. Adaptive finite element methods for the two-phase Stefan problem

In dealing with the fixed domain formulation to the two-phase Stefan problem, the interface does not play an explicit role but, whenever the singularities located there are not properly resolved, it is responsible for the global numerical pollution of finite element approximations, which perform worse than expected according to the interpolation theory [53,113,114,120,136]. Methods studied so far are not completely satisfactory in that they do no take advantage of the fact that singularities are located in a small region compared with the entire domain. Consequently, the use of adaptive highly graded meshes is extremely important in parabolic free boundary problems and seems a proper remedy to the lack of effectiveness of the standard finite element methods [103,104,105,106,107]. The local refinement strategy, being based based on equidistributing pointwise interpolation errors in space and balancing with the truncation errors in time, reflects the regularity of the underlying solutions. This is accomplished by performing various tests on the discrete temperature to extract information about the discrete derivatives as well as to predict the free boundary evolution. A typical triangulation is coarse away from the discrete interface, where meshsize and time step satisfy a parabolic relation, whereas it is locally refined in the vicinity of the discrete interface for the relation to become hyperbolic; in turn, a drastic reduction of spatial degrees of freedom is obtained with highly graded meshes that accompany the layer motion. Mesh changes incorporate an interpolation error which eventually accumulates in time. Its control imposes several constraints on admissible meshes and allowable mesh changes, and leads to the mesh selection algorithm. A new mesh is not produced by enrichment/coarsening strategies, that are in general strikingly complex, but is simply completely regenerated; thus consecutive meshes are not compatible. In turn, an interpolation theory for noncompatible meshes is developed. The proposed adaptive scheme is stable in various Sobolev norms. A rate of convergence of essentially $\mathcal{O}(\tau^{1/2})$ is derived in the natural energy spaces for both temperature and energy density. Here we deal with the adaptive nonlinear algorithm and refer to [107] for an adaptive linear scheme. Mesh adaptation is extremely important also in phase-change problems that exhibit thin transition regions. Prototype problems are the phase relaxation model of Sect. 1.1.4 and the phase field model (see [32,79]). The latter reads, in terms of temperature θ and phase variable χ, as follows:

$$\partial_t(\theta+\chi)-\Delta\theta=0, \quad \varepsilon\partial_t\chi-\varepsilon\Delta\chi+\tfrac{1}{\varepsilon}\chi(\chi^2-1)=\theta, \quad \text{in } Q.$$

Here $0<\varepsilon\ll 1$ is a relaxation parameter. The function χ is almost constant in most of the domain and varies abruptly within a narrow transition layer of width $\mathcal{O}(\varepsilon)$.

We also mention domain decomposition methods for phase-change problems (see [81,82]); they rely on interesting estimates of the speed of propagation of the discrete interfaces, which allows one to split the domain in liquid, solid, and transition regions.

Let $\Omega\subset\mathbf{R}^2$ be polygonal and convex. Let $\theta_0\in W_0^{1,\infty}(\Omega)\cap W^{2,\infty}(\Omega\backslash I_0)$, and let I_0 be a Lipschitz curve. Let $\tau=\frac{T}{N}$ denote the time step. Let $\{\mathcal{S}^n\}_{n=1}^N$ a set of graded partitions of Ω into triangles, that are shape regular and weakly acute. Given a triangle $S\in\mathcal{S}^n$, h_S stands for its size and satisfies $\lambda\tau\le h_S\le\Lambda\tau^{1/2}$, where $0<\lambda,\Lambda$ are fixed constants. We also denote by $\mathcal{S}^0$ a quasi-uniform partition of Ω into triangles of size $O(\tau^{1/2})$. Let $\mathcal{E}^n=\{e\}$ be the set of interelement boundaries of $\mathcal{S}^n$, and let h_e denote the length of e. Let $\{\mathbf{x}_j^n\}_{j=1}^{J^n}$ indicate the set of internal nodes and $\{\phi_j^n\}_{j=1}^{J^n}$ the basis functions. Let $\mathcal{R}^n=\cup_{S\in\mathcal{S}^n:h_S=\mathcal{O}(\tau)}S$ indicate the refined region. Let $V^n\subset H_0^1(\Omega)$ indicate the space of continuous piecewise linear functions over $\mathcal{S}^n$. Let $\Pi^n: C^0(\bar\Omega)\to V^n$ be the usual Lagrange interpolation operator. The adaptive sheme reads as follows:

Let $U^0\in V^0$ be defined by $U^0(\mathbf{x}_j^0)=\limsup_{\mathbf{x}\to\mathbf{x}_j^0}u_0(\mathbf{x})$, and set $\Theta^0=\Pi^0\theta_0=\Pi^0\beta(U^0)$. For any $1\le n\le N$, given a mesh $\mathcal{S}^{n-1}$ and discrete functions $U^{n-1},\Theta^{n-1}\in V^{n-1}$, select $\mathcal{S}^n$ and find $U^n,\Theta^n\in V^n$ such that

$$\Theta^n=\Pi^n\beta(U^n), \tag{4.1}$$

$$\langle U^n,\phi\rangle^n+\tau\langle\nabla\Theta^n,\nabla\phi\rangle=\langle U^{n-1},\phi\rangle^n \quad \forall\,\phi\in V^n. \tag{4.2}$$

Here $\langle\cdot,\cdot\rangle^n$ is defined by

$$\langle\zeta,\phi\rangle^n=\textstyle\int_\Omega\Pi^n(\zeta\phi)d\mathbf{x} \quad \forall\,\zeta,\phi\in C^0(\bar\Omega).$$

We define the phase variable X^n by $X^n=U^n-\Theta^n$ and observe that $0\le X^n\le 1$. Hence X^n is an order parameter which determines the solid or crystal phase $\mathcal{C}^n=\{\mathbf{x}\in\Omega: X^n(\mathbf{x})=0\}$, the liquid phase $\mathcal{L}^n=\{\mathbf{x}\in\Omega: X^n(\mathbf{x})=1\}$, and the transition region $\mathcal{T}^n=\text{closure}(\{\mathbf{x}\in\Omega: 0<X^n(\mathbf{x})<1\})$. The discrete interface $F^n=\{\mathbf{x}\in\Omega:\Theta^n(\mathbf{x})=0\}$ satisfies $F^n\subset\mathcal{T}^n$ and $\mathcal{T}^n$ is the union of all triangles of $\mathcal{S}^n$ crossed by F^n. Note that $S\cap\{\mathbf{x}\in\Omega: 0<U^n(\mathbf{x})<1\}\neq\emptyset$ for all $S\subset\mathcal{T}^n$.

It is to be stressed that the nodes of $\mathcal{S}^n$ are used in the right hand side of (4.2) for the numerical integration of piecewise linear functions defined in V^{n-1}. The interpolation error $U^{n-1}-\Pi^nU^{n-1}$ so incurred has to be controlled because it eventually accumulates in time and may destroy convergence as well as stability which, in turn, are not obvious and need investigation. The crucial feature of the adaptive method (4.1), (4.2) is the mesh selection strategy, that is the control of the admissibility of $\mathcal{S}^{n-1}$ and, in the case a new mesh is invoked, the definition of a meshsize function h^n over $\mathcal{S}^{n-1}$, with which $\mathcal{S}^n$ is next generated by an automatic mesh generator (see [121]).

4.1. Interpolation for noncompatible meshes

Our present concern is to recall discrete error estimates in L^p for noncompatible meshes, say $\mathcal{S}$ and $\hat{\mathcal{S}}$, as those produced by a mesh change in the adaptive algorithm. Let $V, \hat{V} \subset H_0^1(\Omega)$ be the usual piecewise linear finite element spaces over $\mathcal{S}, \hat{\mathcal{S}}$, respectively, and let $\Pi, \hat{\Pi}$ be the Lagrange interpolation operators over $V, \hat{V}$. Let $\xi : \mathbf{R} \to \mathbf{R}$ be so that $\xi' \in W^{1,\infty}(\mathbf{R})$; L_ξ and $L_{\xi'}$ indicate the Lipschitz constant of ξ and ξ', respectively. If $\varphi \in W^{2,\infty}(\Omega)$, the following error estimate is well known:

$$\|\xi(\varphi) - \hat{\Pi}\xi(\varphi)\|_{L^\infty(\hat{S})} \leq C h_{\hat{S}}^2 \big(L_\xi |\varphi|_{W^{2,\infty}(\hat{S})} + L_{\xi'} |\varphi|^2_{W^{1,\infty}(\hat{S})} \big) \quad \forall\, \hat{S} \in \hat{\mathcal{S}}.$$

The L^p interpolation theory developed in [103] extends this estimate to piecewise linear functions over $\mathcal{S}$.

For any given set $A \subseteq \bar{\Omega}$, we set $\mathcal{S}_A = \{S \in \mathcal{S} : S \cap A \neq \emptyset\}$, $\mathcal{E}_A = \{e \in \mathcal{E} : e \subset \partial S, S \in \mathcal{S}_A\}$, and $\tilde{A} = \cup_{S \in \mathcal{S}_A} S$. For any $\phi \in V$, set $d_S = |\nabla \phi|_S|$ for all $S \in \mathcal{S}$ and $D_e = \frac{|[\![\nabla \phi]\!]_e|}{h_e}$ for all $e \in \mathcal{E}$, where $[\![\cdot]\!]_e$ indicates the jump operator across e. Note that these quantities are easy to evaluate in practice.

We start by pointing out that there exists a constant $0 < \sigma < 1$, depending only on the regularity of $\mathcal{S}$, such that $\mathrm{dist}(x, S) \geq \sigma h_S$ for all $\mathbf{x} \notin \tilde{S}$. We can then distinguish between two opposite (and mutually exclusive) situations in terms of the relative size of triangles in both meshes $\mathcal{S}$ and $\hat{\mathcal{S}}$. The derefinement case is defined as follows:

$$\text{given } \hat{S} \in \hat{\mathcal{S}}, \quad \sigma h_S < h_{\hat{S}} \quad \text{for all } S \in \mathcal{S}_{\hat{S}}. \tag{4.3}$$

By contrast, the refinement situation reads as follows:

$$\text{given } \hat{S} \in \hat{\mathcal{S}}, \quad \text{there exists} \quad S \in \mathcal{S}_{\hat{S}} \quad \text{such that} \quad \sigma h_S \geq h_{\hat{S}}, \tag{4.4}$$

which, in turn, yields $\hat{S} \subset \tilde{S}$. Let $\hat{\mathcal{S}}_d$ and $\hat{\mathcal{S}}_r$ indicate the set of all $\hat{S}$'s satisfying (4.3) and (4.4), respectively. We then have the following two crucial interpolation estimates without assumptions on the relative size or location of triangles of $\mathcal{S}$ and $\hat{\mathcal{S}}$. Their rather technical proofs can be found in [103]. For all $\hat{S} \in \hat{\mathcal{S}}_d$ we have

$$\begin{aligned} \|\Pi\xi(\phi) - \hat{\Pi}\xi(\phi)\|_{L^\infty(\hat{S})} &\leq C h_{\hat{S}}^2 \big(L_\xi \max\nolimits_{e \in \mathcal{E}_{\hat{S}}} D_e + L_{\xi'} \max\nolimits_{S \in \mathcal{S}_{\hat{S}}} d_S^2 \big), \\ \|\nabla(\phi - \hat{\Pi}\phi)\|_{L^2(\hat{S})} &\leq C h_{\hat{S}} \big(\textstyle\sum_{e \in \mathcal{E}_{\hat{S}}} h_e^2 D_e^2 \big)^{1/2}. \end{aligned} \tag{4.5}$$

The need of control on interpolation errors even for the refinement situation (4.4) and $\xi = I$ arises from the noncompatibility of $\mathcal{S}$ and $\hat{\mathcal{S}}$. For all $\hat{S} \in \hat{\mathcal{S}}_r$ we get

$$\begin{aligned} \|\Pi\xi(\phi) - \hat{\Pi}\xi(\phi)\|_{L^\infty(\hat{S})} &\leq C h_{\hat{S}} L_\xi \max\nolimits_{e \in \mathcal{E}_{\hat{S}}} h_e D_e + C L_{\xi'} \max\nolimits_{S \in \mathcal{S}_{\hat{S}}} h_{\hat{S}}^2 d_S^2), \\ \|\nabla(\phi - \hat{\Pi}\phi)\|_{L^2(\hat{S})} &\leq C h_{\hat{S}} \big(\textstyle\sum_{e \in \mathcal{E}_{\hat{S}}} h_e^2 D_e^2 \big)^{1/2}. \end{aligned} \tag{4.6}$$

The above two estimates quantify the interpolation errors produced by mesh changes. The proper control of such errors, which may eventually accumulate in time and thus compromise accuracy, is reflected in the local refinement strategy of the next section.

4.2. Adaptive strategy and implementation

Based on the current mesh $\mathcal{S}^{n-1}$, certain local mesh parameters are used to construct the next mesh $\mathcal{S}^n$. They are obtained from the discrete temperature Θ^{n-1} and contain information about the discrete regularity of Θ^{n-1} as well as the local velocity of the transition region $\mathcal{T}^{n-1}$ (see Sect. 4.2.2). The key idea hinges upon equidistributing pointwise interpolation errors in space for Θ^{n-1}, in order that the maximum norm error produced by a mesh change be $\mathcal{O}(\tau)$. Three tests are performed to guarantee mesh quality (see Sect. 4.2.3). A typical mesh is designed to be admissible for at least $\mathcal{O}(N^{1/2})$ time steps which, in turn, reflects both a theoretical constraint and a computational goal.

4.2.1. Heuristic motivation

The relation between the local parameters in Sect. 4.2.2 below and the time step τ stems from matching the truncation error in time with the interpolation error in space, both measured in the maximum norm.

We start by examining the desired meshsize near the transition region $\mathcal{T}^{n-1}$. For the travelling waves of Sect. 2.1.1.1 suppose there is an underlying uniform mesh of size h, on which we want to interpolate $\theta_\tau = \beta(u_\tau) \in C^{1,1}(\mathbb{R})$. In Sect. 4.4 we will see that $\beta(U^n)$ is in fact a crucial variable. Since $\theta_\tau''(x) = \frac{1}{\tau} + \mathcal{O}(1)$ in $(0, V\tau)$, we have

$$\|\beta(u_\tau) - \Pi_h\beta(u_\tau)\|_{L^\infty(0,V\tau)} \leq \tfrac{h^2}{2\tau} + \mathcal{O}(h^2).$$

If we now try to balance this spatial error with the corresponding error in time, namely $\mathcal{O}(\tau) = \|\theta - \theta_\tau\|_{L^\infty(0,V\tau)} \geq V^2\frac{\tau}{2} + \mathcal{O}(\tau^2)$, we infer that $h = \mathcal{O}(|V|\tau)$. Such a condition is reflected in (4.9) below.

About the velocity estimation, in Sect. 2.1.1.1 we observed that the first derivative of θ_τ does not exhibit a jump discontinuity at $x = 0$ but a variation of magnitude V within the interval $(0, V\tau)$. Thus, integration of $u_\tau(x) - u_\tau(x - V\tau) = \tau\theta_\tau''(x)$ on $[x_1, x_2]$ yields $V = [\![\theta_\tau']\!]_{x_1}^{x_2} + \mathcal{O}(\tau)$ for $x_1 < 0 < x_2$, $-x_1, x_2 \approx V\tau$, which can be viewed as an appropriate semidiscretization of the usual Stefan condition. Consequently, an overrefinement near the interface is dangerous.

Let us now examine the interpolation error $U^{n-1} - \Pi^n U^{n-1}$ produced by a mesh change $\mathcal{S}^{n-1} \to \mathcal{S}^n$. Since X^{n-1} varies very rapidly within $\mathcal{T}^{n-1}$, letting $S \subset \mathcal{T}^{n-1}$ be free to change would cause an inadmissible error $\|U^{n-1} - \Pi^n U^{n-1}\|_{L^\infty(S)} = \mathcal{O}(1)$ and a subsequent optimal lower bound $\|U^{n-1} - \Pi^n U^{n-1}\|_{L^1(\Omega)} \geq C\tau$, which breaks down the strong stability of the method. We thus enforce the constraint that all $S \subset \mathcal{T}^{n-1}$ will belong to $\mathcal{S}^n$, as expressed in (4.12).

Away from the transition region $\mathcal{T}^{n-1}$ the problem is strictly parabolic, hence the discretization parameters should verify the usual parabolic contraint $h_S = \mathcal{O}(\tau^{1/2})$ for all $S \in \mathcal{S}^{n-1} : S \not\subset \mathcal{T}^{n-1}$. In light of (4.1), we can replace U^{n-1} by either Θ^{n-1} or $\Theta^{n-1} + 1$ in $\Omega\backslash\mathcal{T}^{n-1}$ and then argue as follows for all $\hat{S} \in \mathcal{S}^n : \hat{S} \not\subset \mathcal{T}^{n-1}$. We indicate with d_S and D_e the corresponding discrete derivatives of Θ^{n-1}. We first deal with the derefinement case which corresponds to the set $\mathcal{S}_d^n$ in Sect. 4.1. The estimate (4.5) yields

$$\|\Pi^{n-1}\xi(U^{n-1}) - \Pi^n\xi(U^{n-1})\|_{L^\infty(\hat{S})} \leq C \max_{e\in\mathcal{E}_{\hat{S}}^{n-1}} h_{\hat{S}}^2 D_e + \max_{S\in\mathcal{S}_{\hat{S}}^{n-1}} h_{\hat{S}}^2 d_S^2 \leq C\tau,$$

provided $h_{\hat{S}} \le \hat{h}_e = \mu_1 \frac{\tau^{1/2}}{D_e^{1/2}}$ and $h_{\hat{S}} \le \hat{h}_S = \mu_2 \frac{\tau^{1/2}}{d_S}$ which, in turn, are built into (4.8) and (4.13) below. In addition, we have

$$\sum_{\hat{S}\in\mathcal{S}_d^n} \|\nabla(U^{n-1} - \Pi^n U^{n-1})\|^2_{L^2(\hat{S})} \le C \sum_{\hat{S}\in\mathcal{S}_d^n} \sum_{e\in\mathcal{E}_{\hat{S}}^{n-1}} h_{\hat{S}}^2 D_e h_e^2 D_e$$
$$\le C\tau \sum_{e\in\mathcal{E}^{n-1}} h_e^2 D_e \le C\tau,$$

provided $\sum_{e\in\mathcal{E}^{n-1}} h_e^2 D_e \le C$. This bound is a discrete analogue of $\Delta\theta \in L^\infty(0,T;M(\Omega))$; it is consistent with numerical evidence and will be assumed in the sequel (see (4.14)). Note that we could replace doubled summations by a single one because, as a consequence of (4.3), $\text{card}\{\hat{S} \in \mathcal{S}_d^n : e \cap \hat{S} \ne \emptyset\} = \mathcal{O}(1)$ for all $e \in \mathcal{E}^{n-1}$. On the other hand, for the refinement case, $\hat{S} \in \mathcal{S}_r^n$, we assume that $h_e \le \mu_1^* \hat{h}_e$ for all $e \in \mathcal{E}_{\hat{S}}^{n-1}$ and $h_S \le \mu_2^* \hat{h}_S$ for all $S \in \mathcal{S}_{\hat{S}}^{n-1}$, which avoids strong refinement operations; this condition is enforced in (4.10). Therefore, for all $\hat{S} \in \mathcal{S}_r^n$, using (4.6) we arrive at

$$\|\Pi^{n-1}\xi(U^{n-1}) - \Pi^n\xi(U^{n-1})\|_{L^\infty(\hat{S})} \le C \max_{e\in\mathcal{E}_{\hat{S}}^{n-1}} h_{\hat{S}} h_e D_e + C \max_{S\in\mathcal{S}_{\hat{S}}^{n-1}} h_{\hat{S}}^2 d_S^2 \le C\tau,$$

and

$$\sum_{\hat{S}\in\mathcal{S}_r^n} \|\nabla(U^{n-1} - \Pi^n U^{n-1})\|^2_{L^2(\hat{S})} \le C \sum_{\hat{S}\in\mathcal{S}_r^n} \sum_{e\in\mathcal{E}_{\hat{S}}^{n-1}} (h_{\hat{S}} h_e D_e)^2$$
$$\le C\tau \sum_{\hat{S}\in\mathcal{S}_r^n} \sum_{e\in\mathcal{E}_{\hat{S}}^{n-1}} h_{\hat{S}} h_e D_e \le C\tau \sum_{e\in\mathcal{E}^{n-1}} h_e D_e \sum_{\hat{S}\in\mathcal{S}_r^n : e\cap\hat{S}\ne\emptyset} h_{\hat{S}}$$
$$\le C\tau \sum_{e\in\mathcal{E}^{n-1}} h_e^2 D_e \le C\tau.$$

Hence we can conclude that

$$\|\Pi^{n-1}\xi(U^{n-1}) - \Pi^n\xi(U^{n-1})\|_{L^\infty(\Omega)} \le C\tau, \quad \|\nabla(U^{n-1} - \Pi^n U^{n-1})\|_{L^2(\Omega)} \le C\tau^{1/2}. \tag{4.7}$$

It is worth noting that the above estimates are sharp according to the discrete regularity being dealt with. Since meshes $\mathcal{S}^{n-1}$ and $\mathcal{S}^n$ are noncompatible, we cannot expect a pointwise error estimate for gradients to hold. In fact, consider the refinement case, $\hat{S} \in \mathcal{S}_r^n$, for which $\|\nabla(U^{n-1} - \Pi^n U^{n-1})\|_{L^\infty(\hat{S})} \le C \max_{e\in\mathcal{E}_{\hat{S}}^{n-1}} h_e D_e$. This yields $\|\nabla(U^{n-1} - \Pi^n U^{n-1})\|_{L^\infty(\hat{S})} = \mathcal{O}(1)$ provided $D_e = \mathcal{O}(h_e^{-1})$, as expected to happen near the transition region $\mathcal{T}^{n-1}$.

Since this error may accumulate in time, we will impose a restriction on the maximum number of mesh changes, namely $\mathcal{O}(N^{1/2})$, for the final error to be $\mathcal{O}(\tau^{1/2})$.

4.2.2. Local mesh parameters

See [103,104,106]. We introduce two local parameters in the parabolic zone $\Omega\backslash\mathcal{T}^{n-1}$:

$$\hat{h}_e = \mu_1 \frac{\tau^{1/2}}{D_e^{1/2}} \quad \forall\, e \in \mathcal{E}^{n-1} : e \not\subset \mathcal{T}^{n-1},$$
$$\hat{h}_S = \mu_2 \frac{\tau^{1/2}}{d_S} \quad \forall\, S \in \mathcal{S}^{n-1} : S \not\subset \mathcal{T}^{n-1}. \tag{4.8}$$

Set

$$\mathcal{B}^{n-1} = \cup_{S\in\mathcal{S}^{n-1} : \min_{e\subset\partial S}(\hat{h}_S, \hat{h}_e) < \lambda\tau} S,$$

and note that $\mathcal{B}^{n-1}$ represents the set where the discrete derivatives d_S, D_e are too large. Note that second derivatives D_e and first derivatives d_S may blow up without violating $\hat{h}_e \ge \lambda\tau$ and $\hat{h}_S \ge \lambda\tau$ as far as $D_e \le \frac{\mu_1^2}{\lambda^2\tau}$ and $d_S \le \frac{\mu_2}{\lambda\tau^{1/2}}$, respectively.

In order to determine a local parameter near the transition region $\mathcal{T}^{n-1}$, we first determine a proper velocity indicator (see Sect. 4.2.1.). For each $S \subset \mathcal{T}^{n-1}$ let $S_\pm$ indicate an element displayed on the direction of $\pm\nabla\Theta^{n-1}|_S$ which satisfy $\mathrm{dist}(S_\pm, S) \geq h_S$, and let $\mathbf{b}_\pm$ be the barycenter of $S_\pm$. We define the velocity $\mathbf{V}_S$ as

$$\mathbf{V}_S = \frac{\nabla\Theta^{n-1}|_{S_-} - \nabla\Theta^{n-1}|_{S_+}}{U^{n-1}(\mathbf{b}_+) - U^{n-1}(\mathbf{b}_-)}.$$

We next consider a cone $\mathcal{C}_S$ of axis $\mathbf{V}_S$, vertex at S, opening $\frac{\pi}{2}$ and height $\mu_3^n|\mathbf{V}_S|\tau^{1/2}$. The union of all these cones is the region most likely to contain the evolution of $\mathcal{T}^{n-1}$ for at least $\mathcal{O}(N^{1/2})$ time steps. This is so because $\mathcal{S}^n$ must meet the quality tests of Sect. 4.2.3 during that number of time steps. The local parameter associated with the transition region is thus defined by

$$\hat{h}_S = \tau \min\left(V^*, \max(\lambda, |\mathbf{V}_S|)\right) \quad \forall\, S \subset \mathcal{T}^{n-1}. \tag{4.9}$$

The above constants $\mu_1, \mu_2, \mu_3^n, V^*$ as well as λ, Λ are arbitrary at this stage. Empirical rules of selection apply (see [104]).

4.2.3. Mesh selection algorithm

Assuming that we have a mesh $\mathcal{S}^{n-1}$, we would like to discuss the various tests to be performed on the computed solution $\{U^{n-1}, \Theta^{n-1}\}$ to either accept or discard $\mathcal{S}^{n-1}$.

The first test consists of checking whether the discrete transition region $\mathcal{T}^{n-1}$ is within the refined region $\mathcal{R}^{n-1}$ or not. In the event $\mathcal{T}^{n-1}$ escapes from $\mathcal{R}^{n-1}$, we say that the test has failed. Since the local width of the refined region is proportional to both $\tau^{1/2}$ and the local interface velocity, we thus expect a mesh to be admissible for $\mathcal{O}(N^{1/2})$ time steps, as desired. Rejection of $\mathcal{S}^{n-1}$ is mostly dictated by failure of this test. Also, the restriction $(\mathcal{C}^{n-2} \cap \mathcal{L}^{n-1}) \cup (\mathcal{C}^{n-1} \cap \mathcal{L}^{n-2}) \subset \mathcal{R}^{n-1}$ is imposed.

The second test ascertains that interpolation errors are still equidistributed correctly:

$$\begin{aligned} h_e &\leq \mu_1^* \hat{h}_e \quad \forall\, e \in \mathcal{E}^{n-1} : e \not\subset \mathcal{T}^{n-1} \cup \mathcal{B}^{n-1}, \\ h_S &\leq \mu_2^* \hat{h}_S \quad \forall\, S \in \mathcal{S}^{n-1} : S \not\subset \mathcal{T}^{n-1} \cup \mathcal{B}^{n-1}. \end{aligned} \tag{4.10}$$

This rules out the possibility of an excessive refinement induced by large discrete derivatives. However, the new local meshsize might be much smaller than the current one, if influenced by the refined region $\mathcal{R}^n$.

The third test finally verifies whether or not the truncation error in time matches the meshsize near the transition region $\mathcal{T}^{n-1}$. Since the velocity of $\mathcal{T}^{n-1}$ may vary substantially during an $\mathcal{O}(N^{1/2})$ period of time, thus making (4.9) inadequate as a truncation error indicator, we enforce the constraints:

$$\mu_3^- \hat{h}_S \leq h_S \leq \mu_3^+ \hat{h}_S \quad \forall\, S \subset \mathcal{T}^{n-1}. \tag{4.11}$$

In (4.10) and (4.11), $\mu_1^*, \mu_2^* > 1, \mu_3^- < 1 < \mu_3^+$ are given constants.

If any one of the above tests fails, then the current mesh $\mathcal{S}^{n-1}$ is rejected along with the solution $\{\Theta^{n-1}, U^{n-1}\}$, which is overwritten with the previously computed solution. A new graded mesh $\mathcal{S}^n$ with the following properties is then generated. To preserve the constraint $h_{\hat{S}} \geq \lambda\tau$, we must keep $\mathcal{B}^{n-1}$ fixed because discrete derivatives are too large. Likewise, in order to avoid inadmissible interpolation errors for U^{n-1} we must not modify $\mathcal{T}^{n-1}$. Hence,

$$S \in \mathcal{S}^n \quad \forall\, S \subset \mathcal{T}^{n-1} \cup \mathcal{B}^{n-1} \tag{4.12}$$

is the first restriction on $\mathcal{S}^n$. The second one reads

$$\lambda\tau \le h_{\hat{S}} \le \min_{\substack{S'\subset\mathcal{T}^{n-1}:\mathcal{C}_{S'}\cap\hat{S}\neq\emptyset \\ e\in\mathcal{E}^{n-1}:e\cap\hat{S}\neq\emptyset,\, e\not\subset\mathcal{B}^{n-1}\cup\mathcal{T}^{n-1} \\ S\in\mathcal{S}^{n-1}:S\cap\hat{S}\neq\emptyset,\, S\not\subset\mathcal{B}^{n-1}\cup\mathcal{T}^{n-1}}} \left(\Lambda\tau^{1/2}, \hat{h}_{S'}, \hat{h}_e, \hat{h}_S\right) \quad \forall\, \hat{S}\in\mathcal{S}^n. \tag{4.13}$$

This accounts for the equidistribution of pointwise interpolation errors for U^{n-1} as well as the definition of refined region $\mathcal{R}^n$, which satisfies $\mathcal{T}^{n-1}\subset\mathcal{R}^n$. The effective implementation of (4.13) is discussed in [104]. The three parameters defined in (4.8) and (4.9) are suitably postproccessed for $\mathcal{S}^n$ to satisfy (4.13). We resort to the automatic mesh generator of [121], whose computational complexity $O(J^n \log J^n)$ is quasi-optimal, to minimize the computer time spent in mesh generation. For $\mathcal{S}^n$ to be weakly acute, it is necessary to postprocess $\mathcal{S}^n$ by simply switching the common egde of adjacent triangles whenever the sum of the opposite angles exceeds π. This process is known to converge after finite iterations and leads to a weakly acute triangulation.

Keeping all triangles of $\mathcal{T}^{n-1}\cup\mathcal{B}^{n-1}$ fixed is nearly the best we can do in a general setting. For computational purposes, however, it is always preferable to remove the contraint (4.12). This is feasible for sharp discrete interfaces.

4.3. Stability for the adaptive scheme

As a consequence of the mesh selection algorithm and interpolation theory for noncompatible meshes, the adaptive discrete problem possesses the natural *a priori* estimates, which resemble those in Sect. 3.2.1.

Numerical experiments indicate that a strip $\mathcal{O}(\tau)$-wide behind the discrete interface F^n occurs where $|[\![\nabla\Theta^n]\!]_e| = C\frac{h_e}{\tau}$; thus they suggest the validity of the following L^1 *a priori* estimate, which mimics $\Delta\theta\in L^\infty(0,T;M(\Omega))$:

$$\textstyle\sum_{e\in\mathcal{E}^n} h_e|[\![\nabla\Theta^n]\!]_e| \le C. \tag{4.14}$$

This is a structural assumption on the discrete solutions to be made in the sequel and imposes a severe regularity restriction on the discrete interface. In fact, since the local meshsize near F^n should be $h_e = C\tau$, (4.14) leads to

$$\text{length}(F^n) \le \textstyle\sum_{e\in\mathcal{E}^n: e\cap F^n\neq\emptyset} h_e \le C\sum_{e\in\mathcal{E}^n} h_e|[\![\nabla\Theta^n]\!]_e| \le C.$$

Such a condition is quite reasonable for practical purposes but is not known to hold though, in a general setting. We stress that without some kind of additional regularity it is probably hopeless to improve upon the fixed mesh method of Sect. 3.2. In this light, (4.14) is always assumed at the mesh changes, though it constitutes a limitation of the adaptive method. It is however partially justified by (4.18) below which, being implicitly guaranteed by the scheme, combines with (4.2) to yield

$$\textstyle\sum_{j=1}^{J^n} \left|\sum_{e\cap\mathbf{x}_j^n\neq\emptyset} h_e[\![\nabla\Theta^n]\!]_e\cdot\nu_e\right| \le \frac{2}{\tau}\int_\Omega \Pi^n|U^n - U^{n-1}| \le C,$$

for all time steps n between consecutive mesh changes. We then see that only a cancellation in the above summation could lead to a bound weaker than (4.14). This seems to be unlikely for locally smooth interfaces as well as for cusps because of their local character. This somehow justifies the fact that (4.14) was never violated in our numerical experiments. Designing an algorithm for which (4.14) is implicitly guaranteed constitutes a challenging open problem though. Regarding first derivatives, instead, the following L^2-type *a priori* estimate is implicitly guaranteed (see (4.16) below):

$$\sum_{S\in\mathcal{S}^n} h_S^2 |\nabla\Theta^n|_S|^2 \leq C.$$

THEOREM 4.1. *Let $\{\mathcal{S}^n\}_{n=0}^N$ and $\{U^n,\Theta^n\}_{n=0}^N$ be the meshes and solutions to the adaptive scheme (4.1), (4.2). Then the following a priori estimates hold:*

$$\max_{1\leq n\leq N}\|U^n\|_{L^\infty(\Omega)} + \max_{1\leq n\leq N}\|\Theta^n\|_{L^\infty(\Omega)} \leq C, \tag{4.15}$$

$$\sum_{n=1}^N \|U^n - \Pi^n U^{n-1}\|_{L^2(\Omega)}^2 + \sum_{n=1}^N \tau\|\nabla\Theta^n\|_{L^2(\Omega)}^2 \leq C, \tag{4.16}$$

$$\begin{aligned}\tfrac{1}{\tau}\sum_{n=1}^N \|\Theta^n - \Pi^n\Theta^{n-1}\|_{L^2(\Omega)}^2 + \tfrac{1}{\tau}\sum_{n=1}^N \|U^n - \Pi^n U^{n-1}\|_{L^2(\Omega\setminus\mathcal{R}^n)}^2 \\ + \max_{1\leq n\leq N}\|\nabla\Theta^n\|_{L^2(\Omega)} \leq C,\end{aligned} \tag{4.17}$$

$$\max_{1\leq n\leq N}\|U^n - \Pi^n U^{n-1}\|_{L^1(\Omega)} \leq C\tau. \tag{4.18}$$

The maximum norm estimate (4.15) follows from the discrete maximum principle (3.12) because all meshes are weakly acute. The first term in (4.16) is a discrete-time $H^{1/2}$-estimate that accounts for the global behavior of U^n which, in the limit, is discontinuous. The second term in (4.17) states, instead, a discrete-time H^1 regularity of U^n in $(\mathcal{C}^n\cap\mathcal{C}^{n-1})\cup(\mathcal{L}^n\cap\mathcal{L}^{n-1}) \supset \Omega\setminus\mathcal{R}^n$, where $U^n-\Pi^n U^{n-1}$ and $\Theta^n-\Pi^n\Theta^{n-1}$ are equivalent variables. Estimate (4.17) needs the structural assumption (4.14). The *a priori* estimate (4.18) is a trivial consequence of (4.14) and clearly resembles $\partial_t u \in L^\infty(0,T;M(\Omega))$. We stress the importance of the stability estimate (4.16) in intermediate spaces, as well as the *a priori* bounds (4.15) and (4.18) in nonenergy spaces, as they play an essential role in the error analysis. We just sketch the proof of (4.16) to show the importance of the interpolation error estimates between consecutive mesh changes.

Proof of (4.16). Take $\phi = U^n \in V^n$ in (4.2) and next add the resulting expressions over n from 1 to $1 \leq m \leq N$. we have

$$\sum_{n=1}^m \langle U^n - U^{n-1}, U^n\rangle^n + \sum_{n=1}^m \tau\langle\nabla\Theta^n, \nabla U^n\rangle = 0.$$

Using the elementary identity (3.23), the first term can be further split as follows (note that we can assume $\Pi^0 = \Pi^1$):

$$\begin{aligned}2\sum_{n=1}^m \langle U^n - U^{n-1}, U^n\rangle^n &= \sum_{n=1}^m \int_\Omega \big(\Pi^n(U^n)^2 - \Pi^n(U^{n-1})^2 + \Pi^n(U^n - U^{n-1})^2\big)\\ &= \sum_{n=1}^m \int_\Omega \big(\Pi^n(U^n)^2 - \Pi^{n-1}(U^{n-1})^2\big) + \sum_{n=1}^m \int_\Omega \Pi^n(U^n - U^{n-1})^2\\ &\quad + \sum_{n=2}^m \int_\Omega \big(\Pi^{n-1}(U^{n-1})^2 - \Pi^n(U^{n-1})^2\big) = \mathrm{I}_1 + \mathrm{I}_2 + \mathrm{I}_3.\end{aligned}$$

In view of (3.2), the first two terms can be bounded from below by

$$\mathrm{I}_1 + \mathrm{I}_2 \geq \|U^m\|_{L^2(\Omega)}^2 + \sum_{n=1}^m \|U^n - \Pi^n U^{n-1}\|_{L^2(\Omega)}^2 - C\|U^0\|_{L^2(\Omega)}^2.$$

The remaining term, which occurs only when the mesh is changed, is bounded by $\mathrm{I}_3 \geq -C\sum_{n=2}^m \tau \geq -C$, as clearly follows by (4.7) with $\xi(s) = s^2$. Indeed, since U^n satisfies (4.15), ξ can be suitably modified outside the range of U^n in such a way that $\xi' \in W^{1,\infty}(\mathbf{R})$. To complete the proof, note that (3.13) in conjunction with (4.1) yields

$$\sum_{n=1}^m \tau\langle\nabla\Theta^n, \nabla U^n\rangle \geq \sum_{n=1}^m \tau\|\nabla\Theta^n\|_{L^2(\Omega)}^2. \qquad \square$$

4.4. Error analysis for the adaptive scheme

In this section we illustrate how the above mesh selection algorithm and subsequent stability results lead to an essentially $\mathcal{O}(\tau^{1/2})$-rate of convergence provided a mild restriction on the number of mesh changes is enforced. They are in fact limited to $\mathcal{O}(N^{1/2})$ which, in turn, accounts for the cumulative effect of interpolation errors, that is $\sum_{n=1}^{N} \|U^{n-1} - \Pi^n U^{n-1}\|_{L^2(\Omega)} \leq C\tau^{1/2}$.

We recall the interpolation estimate for the vertex quadrature rule, for any $S \in \mathcal{S}^n$:

$$(4.19) \qquad \int_S |\zeta\phi - \Pi^n|_S(\zeta\phi)|dx \leq Ch_S \|\zeta\|_{L^2(S)} \|\nabla\phi\|_{L^2(S)} \quad \forall\, \zeta, \phi \in V^n.$$

The discrete Green's operator $G^n : H^{-1}(\Omega) \to V^n$ is defined, for any $\psi \in H^{-1}(\Omega)$, by

$$(4.20) \qquad \langle \nabla G^n \psi, \nabla\phi \rangle = \langle \psi, \phi \rangle \quad \forall\, \phi \in V^n,$$

and satisfies, in addition to (3.11), the following error estimate [103]:

$$(4.21) \qquad \|(G - G^n)\psi\|_{L^\infty(\Omega)} \leq Ch_n^2 |\log h_n|^7 \|\psi\|_{L^\infty(\Omega)} \quad \forall\, \psi \in L^\infty(\Omega),$$

where $h_n = \max_{S \in \mathcal{S}^n} h_S$. Let us define the error functions $\tilde{u}$ and $\tilde{\theta}$ as in Sect. 3.2.3.

THEOREM 4.2. *Let (4.14) hold and the total number of mesh changes be limited to $\mathcal{O}(N^{1/2})$. Then*

$$(4.22) \qquad \|\tilde{u}\|_{L^\infty(0,T;H^{-1}(\Omega))} + \|\tilde{\theta}\|_{L^2(Q)} \leq C\tau^{1/2} |\log \tau|^{7/2}.$$

The major novelty in Theorem 4.2 is to be interpreted in terms of properly distributed mesh nodes: for sharp discrete interfaces, only $\sum_{n=1}^{N} J^n = \mathcal{O}(N^{5/2})$ total degrees of freedom are necessary for an $\mathcal{O}(\tau^{1/2})$ global accuracy, which compare quite favorably with quasi-uniform meshes that require $NJ = \mathcal{O}(N^3)$ degrees of freedom because $h_S = \mathcal{O}(\tau)$ (see Theorem 3.2).

Proof. We proceed as in Theorem 3.2. Let us introduce the error equation:

$$(4.23) \qquad \begin{aligned} &\langle \tilde{u}^n - \tilde{u}^{n-1}, \varphi \rangle + \int_{J^n} \langle \nabla\tilde{\theta}(t), \nabla\varphi \rangle dt = \tau \langle \nabla\Theta^n, \nabla(\phi - \varphi) \rangle + \langle U^n - \Pi^n U^{n-1}, \phi - \varphi \rangle \\ &\qquad + \langle U^n - \Pi^n U^{n-1}, \phi \rangle^n - \langle U^n - \Pi^n U^{n-1}, \phi \rangle + \langle U^{n-1} - \Pi^n U^{n-1}, \varphi \rangle \end{aligned}$$

for all $\varphi \in H_0^1(\Omega)$, $\phi \in V^n$. Take first $\varphi = G\tilde{u}^n \in H_0^1(\Omega)$ and $\phi = G^n\tilde{u}^n \in V^n$ in (4.23), next add on n from 1 to $1 \leq m \leq N$ and use (1.22) and (4.20) to arrive at

$$\begin{aligned} \sum_{n=1}^m \langle \tilde{u}^n - \tilde{u}^{n-1}, G\tilde{u}^n \rangle + \sum_{n=1}^m \int_{J^n} \langle \tilde{\theta}(t), \tilde{u}^n \rangle dt &= \sum_{n=1}^m \langle U^n - \Pi^n U^{n-1}, (G^n - G)\tilde{u}^n \rangle \\ &+ \sum_{n=1}^m \left(\langle U^n - \Pi^n U^{n-1}, G^n\tilde{u}^n \rangle^n - \langle U^n - \Pi^n U^{n-1}, G^n\tilde{u}^n \rangle \right) \\ &+ \sum_{n=1}^m \langle U^{n-1} - \Pi^n U^{n-1}, G\tilde{u}^n \rangle, \end{aligned}$$

that is I + II = III + IV + V. The rest of the proof consists simply of evaluating these five terms separately. In view of the restriction $h_n \leq \Lambda\tau^{1/2}$, the analysis of I goes exactly as in Theorem 3.2. The same happens with II with the only exception of II_2. Since β is piecewise linear, we only have to consider $S \subset \mathcal{T}^n$, for otherwise the contribution vanishes. But then $h_S = \mathcal{O}(\tau)$ and so

$$|\mathrm{II}_2| \leq C\tau \sum_{n=1}^m \int_{J^n} \|\nabla\Theta^n\|_{L^2(\Omega)} \|\tilde{u}(t)\|_{L^2(\Omega)} dt \leq C\tau.$$

Term III can be handled by means of (1.11), (4.15), (4.18), and (4.21), as follows:

$$|\mathrm{III}| \leq \sum_{n=1}^m \|U^n - \Pi^n U^{n-1}\|_{L^1(\Omega)} \|(G - G^n)\tilde{u}^n\|_{L^\infty(\Omega)} \leq C\tau |\log\tau|^7.$$

In analyzing term IV, we decompose the underlying integrals over all triangles $S \subset \mathcal{R}^n$, where $h_S = \mathcal{O}(\tau)$, and $S \not\subset \mathcal{R}^n$, for which $h_S \leq \Lambda \tau^{1/2}$. Inequality (4.19) then yields

$$\begin{aligned}|\mathrm{IV}| &\leq \sum_{n=1}^m \sum_{S\in\mathcal{S}^n} \int_S \left|(U^n - \Pi^n U^{n-1})G^n\tilde{u}^n - \Pi^n|_S(U^n - \Pi^n U^{n-1})G^n\tilde{u}^n\right| d\mathbf{x} \\ &\leq C\sum_{n=1}^m \left(\tau\|U^n - \Pi^n U^{n-1}\|_{L^2(\Omega)} + \tau^{1/2}\|U^n - \Pi^n U^{n-1}\|_{L^2(\Omega\setminus\mathcal{R}^n)}\right)\|\nabla G^n\tilde{u}^n\|_{L^2(\Omega)}.\end{aligned}$$

Hence, applying property $\|\nabla G^n\psi\|_{L^2(\Omega)} \leq C\|\psi\|_{H^{-1}(\Omega)}$ for all $\psi \in H^{-1}(\Omega)$, together with (4.16), (4.17), and the elementary inequality (1.26), we get

$$\begin{aligned}|\mathrm{IV}| &\leq \tfrac{1}{8}\max_{1\leq n\leq m}\|\tilde{u}^n\|^2_{H^{-1}(\Omega)} + C\Big(\tau\sum_{n=1}^m \|U^n - \Pi^n U^{n-1}\|^2_{L^2(\Omega)} \\ &\qquad + \sum_{n=1}^m \|U^n - \Pi^n U^{n-1}\|^2_{L^2(\Omega\setminus\mathcal{R}^n)}\Big) \\ &\leq \tfrac{1}{8}\max_{1\leq n\leq m}\|\tilde{u}^n\|^2_{H^{-1}(\Omega)} + C\tau.\end{aligned}$$

In dealing with term V we resort to the crucial estimate $\|U^{n-1} - \Pi^n U^{n-1}\|_{L^\infty(\Omega)} \leq C\tau$ and the constraint $\mathcal{O}(N^{1/2})$ on the admissible number of mesh changes. In fact, we have

$$\begin{aligned}|\mathrm{V}| &\leq C\tau^{-1/2}\max_{1\leq n\leq m}\|\tilde{u}^n\|_{H^{-1}(\Omega)}\|U^{n-1} - \Pi^n U^{n-1}\|_{L^\infty(\Omega)} \\ &\leq \tfrac{1}{8}\max_{1\leq n\leq m}\|\tilde{u}^n\|^2_{H^{-1}(\Omega)} + C\tau.\end{aligned}$$

Therefore, we have obtained the expression

$$\max_{1\leq n\leq N}\|\tilde{u}^n\|^2_{H^{-1}(\Omega)} + \|\tilde{\theta}\|^2_{L^2(Q)} \leq C\tau|\log\tau|^7.$$

The assertion (4.22) then follows from property $u \in C^{0,1/2}(0,T;H^{-1}(\Omega))$. □

REMARK 4.1. The information between consecutive meshes, necessary to advance the algorithm in time, is transfered via a simple interpolation process. Apart from its nontrivial implementation, that requires quadtree data structures to reach a nearly optimal computational complexity $\mathcal{O}(J^{n-1}\log J^n)$, it leads to a nonuniform distribution of the spatial degrees of freedom $J^n = \mathcal{O}(N^{3/2})$. In fact, most of the nodes are concentrated near discrete interfaces for the refined region to be a strip $\mathcal{O}(\tau^{1/2})$-wide. This, in turn, comes from restricting the number of mesh changes to $\mathcal{O}(N^{1/2})$ as a consequence of accuracy considerations. The local meshsize within the refined region is $\mathcal{O}(\tau)$ and discrete interfaces cannot escape from it without causing mesh modification. As numerical evidence indicates, most of the computer time is spent in solving the ensuing strongly nonlinear algebraic systems. Such a task is extremely sensitive to J^n because its total computational complexity is roughly $\mathcal{O}(\sum_{n=1}^N (J^n)^{4/3})$ [104]. On the other hand, each mesh generation takes about an optimal number of operations, namely $\mathcal{O}(J^n\log J^n)$ [121]. It would thus be preferable to have a narrower refined region, say $\mathcal{O}(\tau)$-wide, at the expense of changing the mesh more frequently, say every $\mathcal{O}(1)$ time steps. Hence, the resulting meshes would have $J^n = \mathcal{O}(N)$, that is a quasi-uniform distribution of spatial degrees of freedom, which appears to be optimal as corresponds to the expected value for the (linear) heat equation (see [55]).

The above constraint on the number of mesh changes can be removed, thus allowing $\mathcal{O}(N)$ remeshings without compromising accuracy, if the interpolation process between consecutive meshes is substituted by a more delicate elementwise L^2-projection. Let W^n be the space of discontinuous piecewise linear functions over $\mathcal{S}^n$. Let $P^n : L^2(\Omega) \to W^n$ be the (local) L^2-projection operator defined, for all $\zeta \in L^2(\Omega)$, by

$$\langle P^n\zeta, \phi\rangle = \langle\zeta, \phi\rangle = \sum_{S\in\mathcal{S}^n}\langle\zeta, \phi\rangle_S \quad \forall\, \phi \in W^n.$$

Since no continuity requirements are imposed on W^n, $P^n\zeta$ can be computed elementwise by inverting a 3×3 linear system. The construction of the right-hand side is, however, a difficult task and amounts, for every triangle of the new mesh, to compute exactly the integral of a piecewise, on the old mesh, quadratic function. The use of quadtree data structures is again essential to reach a nearly optimal computational complexity $\mathcal{O}(J^n \log J^n)$, but the process is intrinsically more expensive than a simple interpolation.

Given the new mesh $\mathcal{S}^n$ and a discrete energy density $U^{n-1} \in V^n$, we define

$$\breve{U}^{n-1} \in V^n: \quad \breve{U}^{n-1}(\mathbf{x}_j^n) = \sum_{S \cap \mathbf{x}_j^n \neq \emptyset} \frac{\text{meas}(S)}{\text{meas}(\text{supp}\phi_j^n)} (P^n U^{n-1})|_S(\mathbf{x}_j^n) \quad \forall\, 1 \leq j \leq J^n.$$

Equation (4.2) is then replaced by

$$\langle U^n, \phi \rangle^n + \tau \langle \nabla \Theta^n, \nabla \phi \rangle = \langle \breve{U}^{n-1}, \phi \rangle^n \quad \forall\, \phi \in V^n.$$

Since $\langle \breve{U}^{n-1}, \phi \rangle^n = \langle P^n U^{n-1}, \phi \rangle^n$ for all $\phi \in V^n$, we can make use of the superconvergence error estimate (see [105]):

$$\|U^{n-1} - P^n U^{n-1}\|_{H^{-1}(\Omega)} \leq C\tau^{3/2}.$$

Therefore, the treatment of term V in Theorem 4.2 can be modified as follows:

$$\begin{aligned} |\mathrm{V}| &\leq C\tau^{-1} \max_{1 \leq n \leq m} \|\tilde{u}^n\|_{H^{-1}(\Omega)} \|U^{n-1} - \Pi^n U^{n-1}\|_{H^{-1}(\Omega)} \\ &\leq \tfrac{1}{8} \max_{1 \leq n \leq m} \|\tilde{u}^n\|^2_{H^{-1}(\Omega)} + C\tau. \end{aligned}$$

No restrictions on the number of mesh changes is required.

4.5. Numerical experiments

To illustrate the superior performance of the adaptive method with respect to the fixed mesh method, we present a couple of severe tests. The first experiment correspond to the evolution of an oscillating smooth interface whereas the second example illustrates the formation of a cusp (see [104]).

EXAMPLE 1. *Oscillating interface.* It is a classical two-phase Stefan problem with an interface that moves up and down. The exact temperature is given by the following expression:

$$\theta(\mathbf{x}, t) = \begin{cases} 0.75(r^2 - 1) & \text{if } r < 1 \\ (1.5 - \alpha'(t)\frac{x_2 - \alpha(t)}{r})(r - 1) & \text{if } r \geq 1, \end{cases}$$

where $r = \left(x_1^2 + (x_2 - \alpha(t))^2\right)^{1/2}$, $\alpha(t) = 0.5 + \sin(1.25t)$, and $\Omega = (0,5)^2$, $T = \pi/1.25$. The normal velocity $V(\mathbf{x}, t)$ to the interface $I(t) = \{\mathbf{x} \in \Omega : r = 1\}$ exhibits a significant variation along the front, which is partially solidifying and partially melting; in turn, this example is an extremely difficult test for our numerical method. Comparison results between the adaptive method and the fixed mesh method are presented in Table 4.1; they show a superior performance of the adaptive method in that it requires less computational labor for a desired global accuracy. The L^2-error for temperature behaves linearly in τ, thus much better than predicted. We also have a (linear) pointwise error that is far from being theoretically explainable. The same happens with the L^2-error for energy density and the interface error $\tilde{I} = \max_{1 \leq n \leq N} \text{dist}\big(I(t^n) - F^n\big)$. The improvement gained in L^∞ is clearly more pronounced than that in L^2. The free boundary is located within one single element, thus providing the best possible approximation. This is clearly depicted in Figures 4.4,4.5. Since approximability and nondegeneracy are

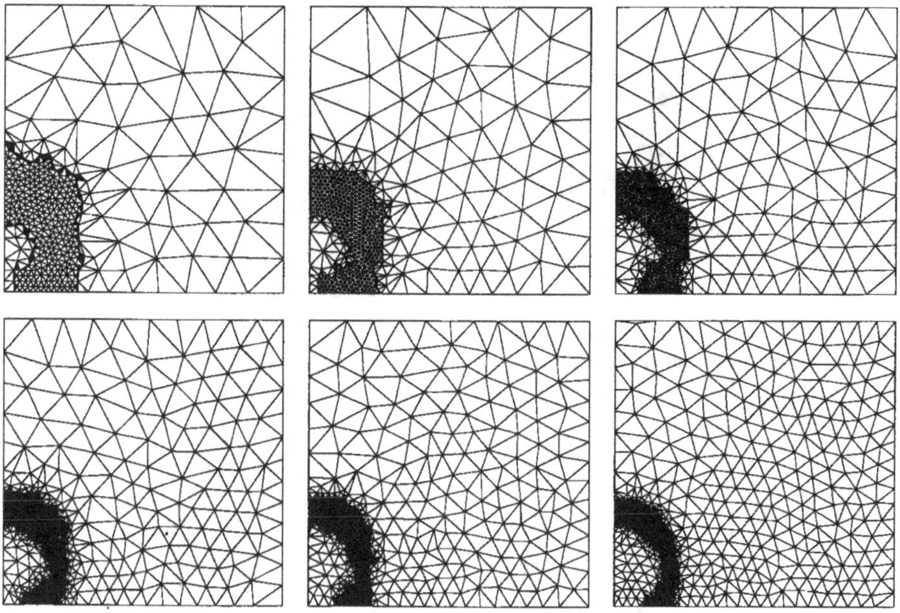

FIGURE 4.1. Example 1. Adaptive method: first mesh for $N = 40, 60, 80, 120, 160, 240$.

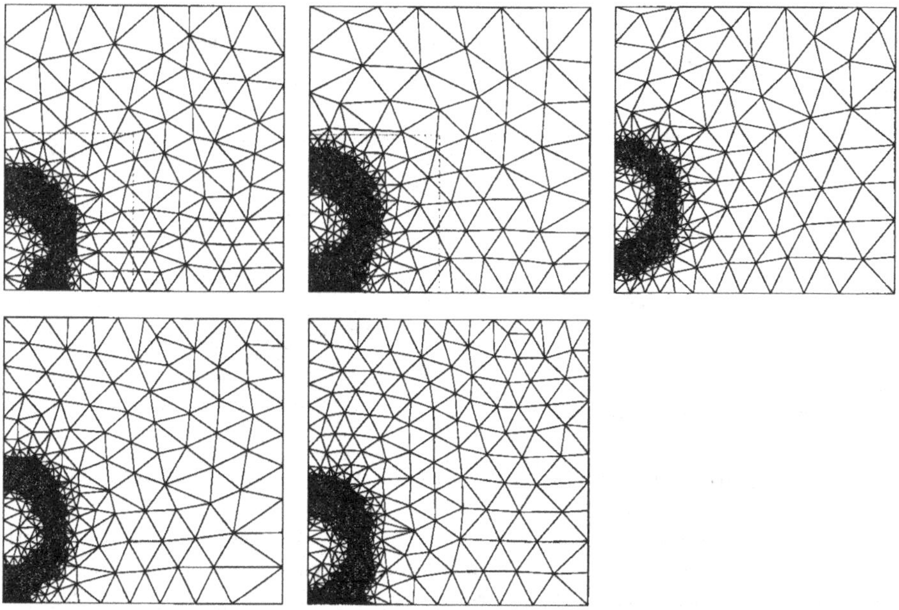

FIGURE 4.2. Example 1. Adaptive method: consecutive meshes for $N = 80$.

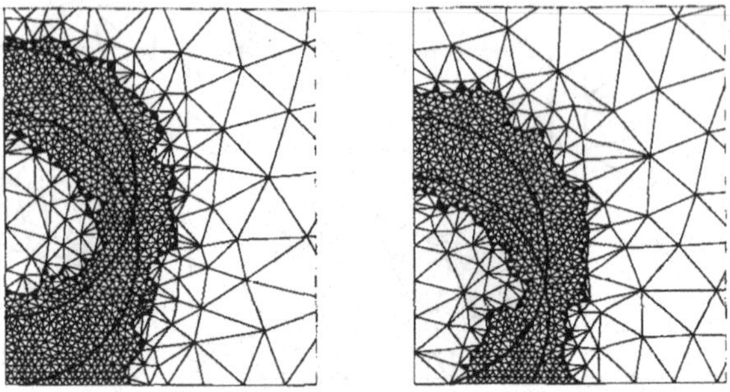

FIGURE 4.3. Example 1. Adaptive method: refined region with the corresponding first and last interface (zoom of Figures 4.2a and 4.2b).

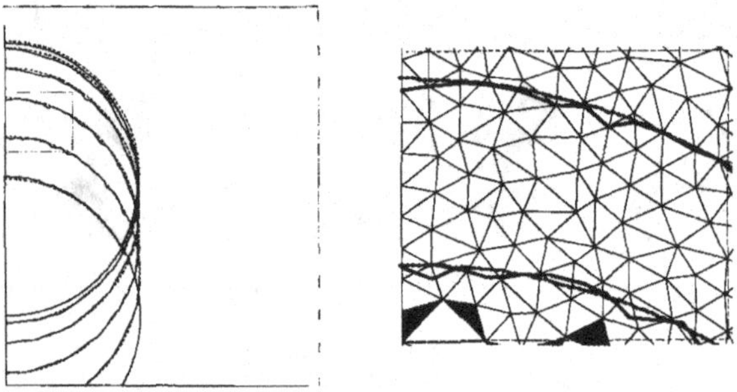

FIGURE 4.4. Example 1. Adaptive method: exact (dashed lines) and computed interfaces at $n = 8l$ $(0 \leq l \leq 5)$ for $N = 80$; zoom of mesh II.

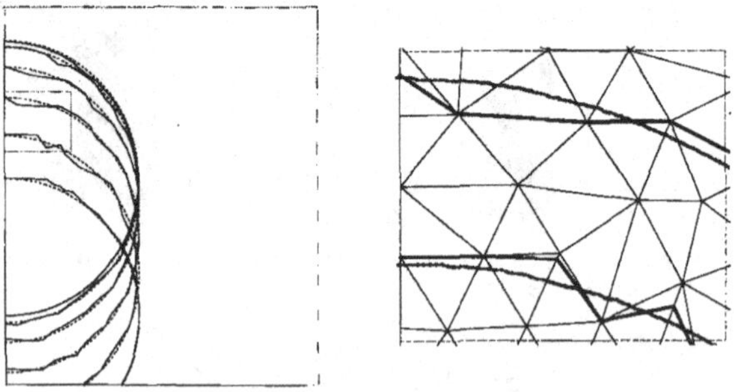

FIGURE 4.5. Example 1. Fixed mesh method: exact (dashed lines) and computed interfaces at $n = 10l$ $(0 \leq l \leq 5)$ for $N = 100$; zoom of the mesh.

tied together, it is worth noting that the nondegeneracy property is not uniform in the present case. Figure 4.1 shows the first mesh for different values of the time step N, whereas Figure 4.2 illustrates the sequence of all meshes for $N = 80$. The black triangles in Figure 4.3 constitute the boundary of the refined region.

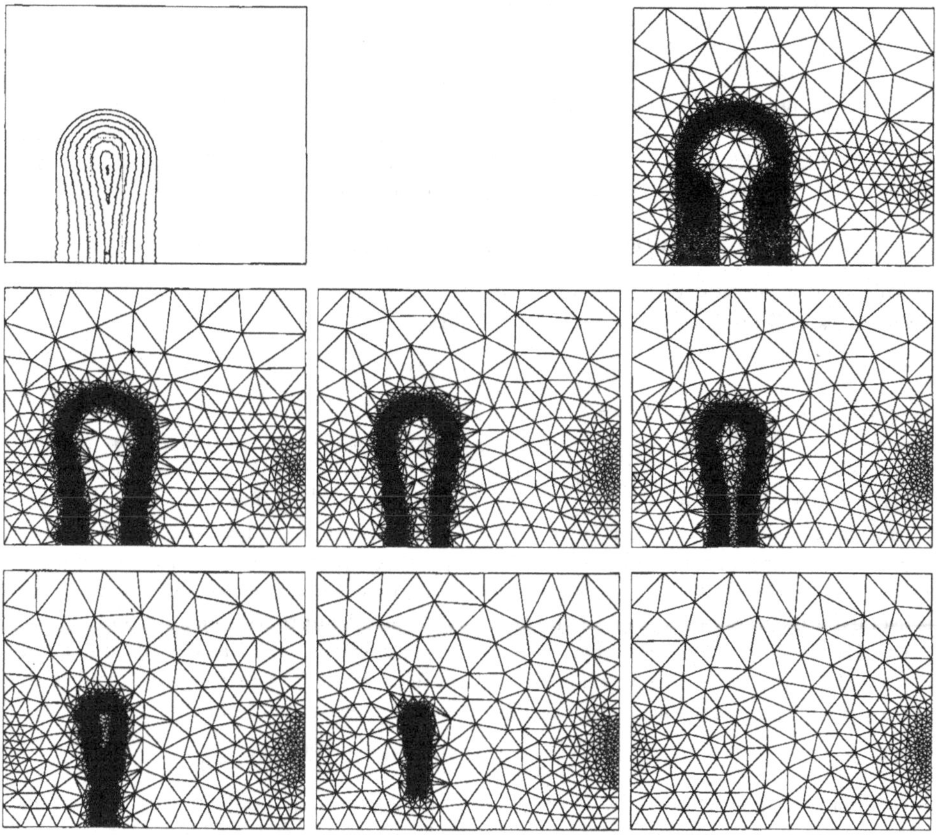

FIGURE 4.6. Example2. Adaptive method: computed interfaces at $n = 0, 10, 20, 30, 40, 50, 60, 67$ and consecutive meshes for $N = 80$.

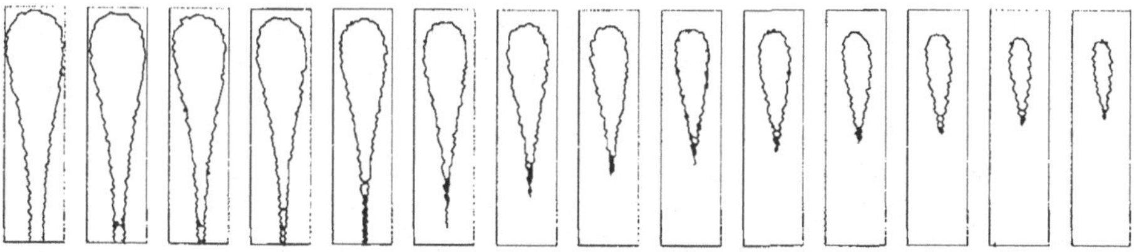

FIGURE 4.7. Example 2. Adaptive method: cusp formation (zoom); steps $49 \leq n \leq 62$ for $N = 80$.

EXAMPLE 2. *Cusp formation.* This a two-phase problem with unknown exact solution and $\Omega = (-2, 4) \times (0, 5)$, $T = 1$. A cusp is expected to develop at $(0, 0)$ as numerical simulation corroborates. The cusp formation is depicted in Figure 4.7, whereas Figure 4.6 contains all consecutive meshes until the solid phase disappears as well as a sequence of interfaces.

adaptive method				fixed mesh method			
$\sum_{n=1}^{N} J^n$	$\|\tilde{\theta}\|_{L^\infty(Q)}$	$\tilde{I}$	CPU	$N \times J$	$\|\tilde{\theta}\|_{L^\infty(Q)}$	$\tilde{I}$	CPU
40×339	10.7	5.83	50	100×1812	12.4	6.88	292
60×592	7.42	5.01	126	150×4107	7.78	5.65	919
80×818	5.17	3.51	235	200×7361	6.31	3.53	2264

TABLE 4.1. Example 1: comparison of pointwise accuracy.

5. Motion by mean curvature

In this section we study the geometric motion of a surface $\Sigma(t) \subset \mathbf{R}^n$ for $t > 0$, that propagates in its normal direction with velocity

$$V(\mathbf{x},t) = \kappa(\mathbf{x},t) + g(\mathbf{x},t) \quad \forall\, \mathbf{x} \in \Sigma(t);$$

hereafter $\kappa(\mathbf{x},t)$ indicates the sum of the principal curvatures of $\Sigma(t)$ at $\mathbf{x}$ and g is a given forcing term. In addition to its intrinsic geometric interest, this problem has relevant applications in the theory of phase transitions in material sciences, flame propagation, crystal growth, etc.; we refer to [24,33] for references and applications. In contrast to its simple classical formulation, such an evolution exhibits a complex structure because topological changes, such as breaking and merging, extinction, and even nonuniqueness may occur [5,17,131]. Geometric front motions can be defined in a generalized (viscosity) sense, even past singularities, via the so-called level set approach [36,57,62,119].

Let $\Sigma_0 \subset \mathbf{R}^n$ be a smooth closed oriented manifold of codimension 1, and let d_0 denote the signed distance function to Σ_0 that is positive outside Σ_0. Let ω denote the (continuous) viscosity solution of the nonlinear degenerate parabolic equation

$$\partial_t\omega - |\nabla\omega|\mathrm{div}\Big(\frac{\nabla\omega}{|\nabla\omega|}\Big) - |\nabla\omega| g = 0 \quad \text{in } \mathbf{R}^n \times (0,\infty), \tag{5.1}$$

satisfying $\omega(\mathbf{x},0) = d_0(\mathbf{x})$ [36,41,57,62]. Such an expression actually says that level sets of $\omega(\cdot,t)$ evolve formally in their inner normal direction with velocity $V = \frac{\partial_t\omega}{|\nabla\omega|} = \kappa + g$. Let $\Sigma(t) = \{\mathbf{x} \in \mathbf{R}^n : \omega(\mathbf{x},t) = 0\}$. Since $\Sigma(t)$ is defined uniquely provided $\{\mathbf{x} \in \mathbf{R}^n : \omega(\mathbf{x},0) = 0\} = \Sigma_0$, it is called generalized evolution by mean curvature (plus forcing); $\Sigma(t)$ coincide with the classical evolution for as long as the latter is smooth. For $g \equiv 0$, it is a consequence of the maximum principle that $\Sigma(t)$ is contained in the convex hull of Σ_0, $\mathrm{conv}(\Sigma_0)$, and that $\Sigma(t)$ eventually disappears in finite time [57,Thm.7.1b;30,66]. Thus, let $\Omega \supset \mathrm{conv}(\Sigma_0)$ be a smooth domain containing the entire evolution for $t \in [0,T]$, and set $Q = \Omega \times (0,T)$. Let $I(t)$ be the inside and $O(t)$ the outside of $\Sigma(t)$ defined in terms of ω by $I(t) = \{\mathbf{x} \in \Omega : \omega(\mathbf{x},t) < 0\}$ and $O(t) = \{\mathbf{x} \in \Omega : \omega(\mathbf{x},t) > 0\}$. Thus $\Omega = \Sigma(t) \cup I(t) \cup O(t)$. Let $d(\cdot,t)$ denote the signed distance function to the front $\Sigma(t)$, $|d(\mathbf{x},t)| = \mathrm{dist}(\mathbf{x},\Sigma(t))$, that is positive in $O(t)$; thus $d(\mathbf{x},0) = d_0(\mathbf{x})$. Such a function d satisfies the following property in the viscosity sense [14,p.446]:

$$\partial_t d(\mathbf{x},t) - \Delta d(\mathbf{x},t) - g\big(\mathbf{x} - d(\mathbf{x},t)\nabla d(\mathbf{x},t),t\big) \geq 0 \quad \text{in } \{d > 0\}, \tag{5.2}$$

and ≤ 0 in $\{d < 0\}$, at least provided $\Sigma(t)$ has empty interior (no fattening). Inequality (5.2) means that whenever $\varphi \in C^\infty(Q)$ *is so that* $d - \varphi$ *has a minimum at* $(\mathbf{x}_0,t_0) \in Q$ *where* $d(\mathbf{x}_0,t_0) = \varphi(\mathbf{x}_0,t_0) > 0$, *then*

$$\partial_t\varphi(\mathbf{x}_0,t_0) - \Delta\varphi(\mathbf{x}_0,t_0) - g\big(\mathbf{x}_0 - \varphi(\mathbf{x}_0,t_0)\nabla\varphi(\mathbf{x}_0,t_0),t_0\big) \geq 0.$$

In addition the following regularity properties of d hold [56,p.1101]:

d is lower semicontinuous in $\{d > 0\}$, upper semicontinuous in $\{d < 0\}$, continuous from below in time, and Lipschitz continuous in space.

For smooth evolutions, the distance d is smooth in a neighborhoud $\mathcal{T} = \{|d| \leq D\}$ of $\Sigma = \{(\mathbf{x}, t) \in Q : \mathbf{x} \in \Sigma(t)\}$ and, for any $(\mathbf{x}, t) \in \mathcal{T}$, $\mathbf{s}(\mathbf{x}, t) = \mathbf{x} - d(\mathbf{x}, t)\nabla d(\mathbf{x}, t) \in \Sigma(t)$ is the unique point of $\Sigma(t)$ at distance $|d(\mathbf{x}, t)|$. Thus $\nabla d(\mathbf{x}, t)$ is the outward unit vector normal to $\Sigma(t)$ at $\mathbf{s}(\mathbf{x}, t)$. In addition d satisfies $\partial_t d(\mathbf{x}, t) = \kappa(\mathbf{x}, t) + g(\mathbf{x}, t)$ for all $\mathbf{x} \in \Sigma(t)$, where the sum of the principal curvatures $\kappa(\mathbf{x}, t)$ is positive if $I(t)$ is convex at $\mathbf{x} \in \Sigma(t)$. Hence, if $\kappa^s(\mathbf{x}, t)$ denotes the sum of the squares of the principal curvatures of $\Sigma(t)$ at $\mathbf{x} \in \Sigma(t)$, property (5.2) becomes (see [63,64,122]):

$$(5.3) \quad \partial_t d(\mathbf{x}, t) - \Delta d(\mathbf{x}, t) - g(\mathbf{s}(\mathbf{x}, t), t) = d(\mathbf{x}, t)\kappa^s(\mathbf{s}(\mathbf{x}, t), t) + \mathcal{O}(d^2(\mathbf{x}, t)) \quad \forall\, (\mathbf{x}, t) \in \mathcal{T}.$$

5.1. Approximation via singularly perturbed double obstacles

Based on the Landau-Ginzburg theory of phase transitions [1], $\Sigma(t)$ can be approximated as a limit as $\varepsilon \downarrow 0$ of the zero level set $\Sigma_\varepsilon(t) = \{\mathbf{x} \in \Omega : \chi_\varepsilon(\mathbf{x}, t) = 0\}$ of the solution χ_ε to the singularly perturbed reaction-diffusion equation:

$$(5.4) \qquad \varepsilon\partial_t\chi_\varepsilon - \varepsilon\Delta\chi_\varepsilon + \tfrac{1}{2\varepsilon}\Psi'(\chi_\varepsilon) = \tfrac{c_0}{2}g \quad \text{in } Q.$$

Here $\Psi(s) = (s^2 - 1)^2$ is a double equal well potential and $c_0 = \int_{-1}^{1} \sqrt{\Psi(s)}ds = \frac{4}{3}$ is a proper scaling factor. Equation (5.4) can also be viewed as the gradient flow for the functional $\int_\Omega (\varepsilon|\nabla\varphi|^2 + \frac{1}{\varepsilon}\Psi(\varphi))$, which is known to Γ-converge to the area functional [94]. Convergence of $\Sigma_\varepsilon(t)$ to $\Sigma(t)$ as $\varepsilon \downarrow 0$ has been proved in [34] for smooth evolutions, and in [14,56] even past singularities, in case of no fattening. See also [47] for a number of conjectures on the geometric motion of fronts. Asymptotic analyses were carried out prior to those convergence results, but apply only to smooth evolutions [30,48,78].

After a short time initial layer, of order $\mathcal{O}(\varepsilon^2|\log\varepsilon|)$, the solution $\chi_\varepsilon(\cdot, t)$ exhibits a transition layer of width $\mathcal{O}(\varepsilon|\log\varepsilon|)$ across $\Sigma_\varepsilon(t)$, but never attains the equilibrium values $+1$ or -1. This is numerically inconvenient because (5.4) must be solved in the entire domain, even though the action takes place in a relatively small region. Computational evidence shows that the location of the approximate interface may be very sensitive to the boundary condition at $\partial\Omega$ which, in turn, leads to instabilities. Since the key condition on Ψ, to achieve the geometric law $V = \kappa + g$ in the limit, is that the two wells possess equal depth, motivated by numerical considerations, we consider the following double obstacle potential:

$$\Psi(s) = \begin{cases} 1 - s^2 & \text{if } s \in [-1, 1] \\ +\infty & \text{if } s \notin [-1, 1]. \end{cases}$$

Equation (5.4) becomes the following double obstacle problem:

$$(5.5) \qquad \varepsilon\partial_t\chi_\varepsilon - \varepsilon\Delta\chi_\varepsilon - \tfrac{1}{\varepsilon}\chi_\varepsilon + \operatorname{sgn}^{-1}(\chi_\varepsilon) \ni \tfrac{c_0}{2}g \quad \text{in } Q,$$

subject to the boundary conditions $\chi_\varepsilon = 1$ on $\partial\Omega \times (0, T)$ and the initial condition $\chi_\varepsilon(\mathbf{x}, 0) = \chi_{0\varepsilon}(\mathbf{x})$ for $\mathbf{x} \in \Omega$, where $\{\mathbf{x} \in \Omega : \chi_{0\varepsilon}(\mathbf{x}) = 0\} = \Sigma_0$. Here $c_0 = \int_{-1}^{+1} \sqrt{\Psi(s)}ds = \frac{\pi}{2}$. In contrast to the usual reaction-diffusion approach with a quartic-like nonlinearity, the solution χ_ε to (5.5) attains the values $+1$ or -1, irrespective of g, outside the $\mathcal{O}(\varepsilon)$-wide transition region where $|\chi_\varepsilon| < 1$. Since the resulting problem has to be solved within such a thin noncoincidence set, where all the action takes place, we realize that this approach retains the geometric (or local) structure of the original problem while taking advantage of the variational structure of (5.5) and being insensitive to singularity formation. This property is essential for numerical purposes, and it bears some intrinsic interest as well.

In Sect. 5.1.3 we prove that the zero level sets $\Sigma_\varepsilon(t)$ of χ_ε converge to $\Sigma(t)$ past singularities provided no fattening occurs. Under a nondegeneracy assumption on ω, $|\nabla\omega(\mathbf{x},t)| > 0$ for $\mathbf{x} \in \Sigma(t)$, then a linear rate of convergence $\mathcal{O}(\varepsilon)$ for interfaces is valid [116] (see also [35]). The nondegeneracy property is always valid before the onset of singularities and also past singularities for a number of geometric flows (e.g., for rotationally symmetric surfaces — see [5,131]). If the evolution is smooth, then an optimal rate of convergence $\mathcal{O}(\varepsilon^2)$ for interfaces can be proved (see [109,110]). The interface error estimates are a consequence of the maximum principle and the explicit construction of sub and supersolutions which, in turn, are inspired by, and indeed rely on, the formal asymptotics developed in [122]. The main differences respect to the convergence proofs with a quartic-like potential, given in [14,56], are the presence of obstacles, that entail lack of regularity of χ_ε even for smooth initial data, and the use of explicit traveling waves.

The local character of the double obstacle formulation, together with its convergence properties even beyond singularities, leads to a robust but effective numerical tool: the dynamic mesh algorithm of [102,112]. This is a finite element solver that solely triangulates the transition layer and then updates it, after having solved the discrete problem, to advance the algorithm in time. Such a simple but crucial idea results in savings of computing time and memory allocation, and shows the potential to handle stiff systems, as the phase field model or phase-change with Gibbs-Thomson curvature effects. In addition, singularity resolution can be enhanced on introducing a space-time dependent relaxation parameter $\varepsilon(\mathbf{x},t) = \varepsilon\alpha(\mathbf{x},t)$:

$$\varepsilon\partial_t(\alpha\chi_\varepsilon) - \varepsilon\nabla\cdot(\alpha\nabla\chi_\varepsilon) - \tfrac{1}{\varepsilon\alpha}\chi_\varepsilon + \operatorname{sgn}^{-1}(\chi_\varepsilon) \ni \tfrac{c_0}{2}g \quad \text{in } Q. \tag{5.6}$$

The control of layer thickness and related local accuracy in terms of $\varepsilon\alpha(\mathbf{x},t)$ has been confirmed both numerically [102,112,122] and theoretically by a rigorous convergence result and error estimates before singularities [111].

5.1.1. Formal asymptotics and travelling waves

We propose here a brief discussion of formal asymptotics for the simplest situation, the radially symmetric case. Even though this presentation is only formal, it motivates several crucial aspects of the supersolution construction of Sect. 5.1.2. See [109,110,122].

We first collect some preliminaries. The unique absolute minimizer γ of the functional $\mathcal{F}(\varphi) = \int_{\mathbf{R}} \left(|\varphi'(x)|^2 + \Psi(\varphi(x))\right)dx$, such that $\gamma(0) = 0$, $|\gamma(x)| \leq 1$ in $\mathbf{R}$, and $\lim_{x\to\pm\infty}\gamma(x) = \pm 1$, is given by

$$\gamma(x) = \begin{cases} -1 & \text{if } x < -\frac{\pi}{2} \\ \sin x & \text{if } x \in [-\frac{\pi}{2}, \frac{\pi}{2}] \\ +1 & \text{if } x > \frac{\pi}{2} \end{cases}$$

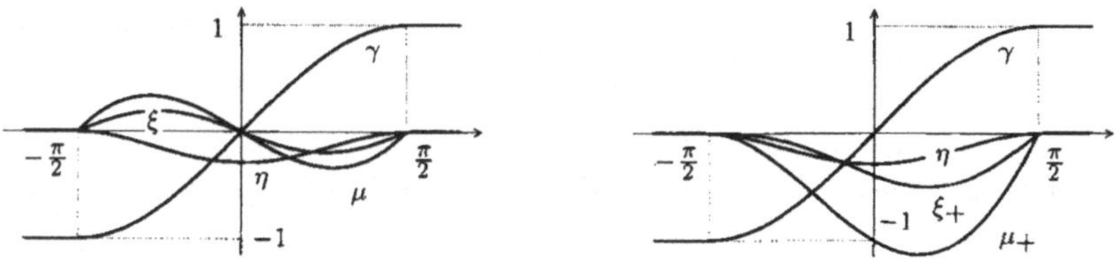

FIGURE 5.1. (a) Functions γ, η, ξ and μ; (b) Functions γ, η, ξ_+ and μ_+.

(see Figure 5.1). Hence, $\gamma \in C^{1,1}(\mathbf{R})$ and solves the (elliptic) double obstacle problem $\gamma''(x) - \frac{1}{2}\Psi'(\gamma(x)) \ni 0$ in $\mathbf{R}$. In particular,

$$\gamma''(x) + \gamma(x) = 0 \quad \forall\, x \in (-\tfrac{\pi}{2}, \tfrac{\pi}{2}). \tag{5.7}$$

Note that $c_0 = \int_{\mathbf{R}}(\gamma'(x))^2 dx$. Let us consider the linear operator $\mathcal{L}\varphi(x) = \varphi''(x) + \varphi(x)$ for $x \in (-\frac{\pi}{2}, \frac{\pi}{2})$, subject to vanishing Dirichlet boundary conditions at $x = \mp\frac{\pi}{2}$. For the problem $\mathcal{L}\varphi = q$ to be solvable, the compatibility condition $\int_{-\pi/2}^{\pi/2} q(x)\gamma'(x)dx = 0$ must be enforced (Fredholm alternative), because $\text{span}\{\gamma'\} = \text{kern}(\mathcal{L})$. Let us define the even function $\eta \in C^{1,1}(\mathbf{R})$:

$$\eta(x) = \begin{cases} \frac{1}{2}\left(x\gamma(x) - c_0 + \gamma'(x)\right) & \text{if } |x| \le \frac{\pi}{2} \\ 0 & \text{if } |x| > \frac{\pi}{2} \end{cases}$$

(see Figure 5.1), which satisfies $\eta(\mp\frac{\pi}{2}) = \eta'(\mp\frac{\pi}{2}) = 0$ and solves the problem

$$\mathcal{L}\eta(x) = \gamma'(x) - \tfrac{c_0}{2} \quad \forall\, x \in (-\tfrac{\pi}{2}, \tfrac{\pi}{2}). \tag{5.8}$$

Finally, let us introduce the odd functions $\xi, \mu \in C^{0,1}(\mathbf{R})$ defined by

$$\xi(x) = \begin{cases} \frac{1}{16}\left(4x\gamma'(x) + (4x^2 - \pi^2)\gamma(x)\right) & \text{if } |x| \le \frac{\pi}{2} \\ 0 & \text{if } |x| > \frac{\pi}{2}, \end{cases} \qquad \mu(x) = \begin{cases} \left(x - \frac{\pi}{2}\gamma(x)\right) & \text{if } |x| \le \frac{\pi}{2} \\ 0 & \text{if } |x| > \frac{\pi}{2} \end{cases}$$

(see Figure 5.1a). They satisfy $\xi(\mp\frac{\pi}{2}) = \mu(\mp\frac{\pi}{2}) = 0$, and

$$\mathcal{L}\xi(x) = x\gamma'(x), \quad \mathcal{L}\mu(x) = x, \quad \forall\, x \in (-\tfrac{\pi}{2}, \tfrac{\pi}{2}), \tag{5.9}$$

but ξ' and μ' jump at $\mp\frac{\pi}{2}$, precisely, $[\![\xi']\!](\mp\frac{\pi}{2}) = \pm\frac{\pi}{8}$, $[\![\mu']\!](\mp\frac{\pi}{2}) = \pm 1$.

Consider in $\mathbf{R}^2$ a circle of radius $\rho(t)$ which evolves with velocity

$$V = -\rho'(t) = \tfrac{1}{\rho(t)} + g\left(\rho(t), t\right) = \kappa + g, \quad \rho(0) = 1. \tag{5.10}$$

If the initial datum $\chi_{0\varepsilon}$ of the double obstacle problem is radially symmetric, so is the solution $\chi_\varepsilon(\cdot, t)$. Using polar coordinates, then (5.5) reads

$$\varepsilon\partial_t\chi_\varepsilon - \tfrac{\varepsilon}{r}\partial_r(r\partial_r\chi_\varepsilon) - \tfrac{1}{\varepsilon}\chi_\varepsilon = \tfrac{c_0}{2}g \quad \text{in } \{|\chi_\varepsilon(r,t)| < 1\}. \tag{5.11}$$

We denote by $r = \rho_\varepsilon(t)$ the zero level set of χ_ε, that is $\chi_\varepsilon(\rho_\varepsilon(t), t) = 0$, and assume $\rho_\varepsilon(0) = 1$. Let us introduce the stretched variable $y = \frac{r - \rho_\varepsilon(t)}{\varepsilon}$ and denote

$$X_\varepsilon(y, t) = \chi_\varepsilon\left(\rho_\varepsilon(t) + \varepsilon y, t\right),$$

whence $\chi_\varepsilon(r, t) = X_\varepsilon\left(\frac{r - \rho_\varepsilon(t)}{\varepsilon}, t\right)$. With the notation $X_\varepsilon^{(k)} = \partial_y^k X_\varepsilon$, we realize that

$$\varepsilon\partial_t\chi_\varepsilon = \varepsilon\partial_t X_\varepsilon - \rho_\varepsilon' X_\varepsilon' \quad \text{and} \quad \tfrac{\varepsilon}{r}\partial_r(r\partial_r\chi_\varepsilon) = \tfrac{1}{r}X_\varepsilon' + \tfrac{1}{\varepsilon}X_\varepsilon''. \tag{5.12}$$

Suppose that both X_ε and ρ_ε can be expressed in terms of ε (inner expansion) as

$$X_\varepsilon(y,t) = \sum_{i=0}^{\infty} \varepsilon^i X_i(y,t), \quad \rho_\varepsilon(t) = \sum_{i=0}^{\infty} \varepsilon^i \rho_i(t).$$

Determining boundary conditions for the inner expansion is a delicate issue that typically involves matching with the outer expansion. The presence of obstacles, however, creates an interesting difference with respect to usual matched asymptotics: the boundary of the transition region should be allowed to move independently of the zero level set, and so adjust for the solution to be globally $C^{1,1}$. Such a formally correct asymptotics, involving a further expansion for the transition region width, is an intriguing open problem that is worth exploring. We simply overcome this difficulty upon imposing continuity of the truncated inner expansion $X_\varepsilon^l = \sum_{i=0}^{l} \varepsilon^i X_i$ with the far field $\chi_\varepsilon = \pm 1$ at $y = \pm\frac{\pi}{2}$, which in turn yields homogeneous Dirichlet conditions for X_i at $y = \pm\frac{\pi}{2}$ for $i \geq 1$ and $X_0(\mp\frac{\pi}{2}, t) = \mp 1$. This arbitrary choice leads to a uniform transition layer $\{|y| \leq \frac{\pi}{2}\}$, and provides the essential information near the interface at the expense of giving up $C^{1,1}$ regularity of X_ε^l. The resulting formal series may not converge, not even in the sense of [48], namely for $\varepsilon \downarrow 0$ but fixed l. Adding further corrections partially compensates for the lack of regularity, as explained at the end of this section, and results in the rigorous construction of barriers of Sect. 5.1.2.

It readily follows that

$$\frac{1}{r} = \left(\varepsilon y + \sum_{i=0}^{\infty} \varepsilon^i \rho_i(t)\right)^{-1} = \frac{1}{\rho_0} - \varepsilon \frac{y + \rho_1}{\rho_0^2} + \mathcal{O}(\varepsilon^2), \tag{5.13}$$

and, with the notation $g_0^{(k)}(t) = (\partial_r^k g)(\rho_0(t), t)$, that

$$g(r,t) = g\left(\varepsilon y + \sum_{i=0}^{\infty} \varepsilon^i \rho_i(t), t\right) = g_0 + \varepsilon(y + \rho_1) g_0' + \mathcal{O}(\varepsilon^2). \tag{5.14}$$

We now intend to derive explicit expressions of, and compatibility conditions for, the first three terms of these formal asymptotic expansions. Using (5.12), together with (5.13) and (5.14), we can substitute X_ε into (5.11) and collect all resulting terms containing like powers of ε. In the sequel we skip details and simply examine the resulting summands in increasing order and equate them to zero, thus starting with the $\frac{1}{\varepsilon}$-term:

$$X_0''(y,t) + X_0(y,t) = 0.$$

Imposing $X_0(0,t) = 0$, in view of (5.7) we clearly obtain

$$X_0(y,t) = \gamma(y) \quad \forall\, y \in [-\tfrac{\pi}{2}, \tfrac{\pi}{2}].$$

The ε^0-term yields

$$X_1''(y,t) + X_1(y,t) = -\left(\rho_0'(t) + \tfrac{1}{\rho_0(t)}\right) X_0'(y) - \tfrac{c_0}{2} g_0(t) = q_1(y,t).$$

The Fredholm alternative enforces the compatibility constraint

$$0 = \int_{-\pi/2}^{\pi/2} q_1(y,t) X_0'(y)\, dy = -c_0\left(\rho_0'(t) + \tfrac{1}{\rho_0(t)} + g_0(t)\right),$$

which, on imposing $\rho_0(0) = 1$, shows that ρ_0 must satisfy (5.10). Thus $\rho_0(t) = \rho(t)$ and $\mathcal{L}X_1(y,t) = g_0(t)\left(X_0'(y) - \frac{c_0}{2}\right)$; in view of (5.8), we then realize that

$$X_1(y,t) = g_0(t)\eta(y) \quad \forall\, y \in [-\tfrac{\pi}{2}, \tfrac{\pi}{2}].$$

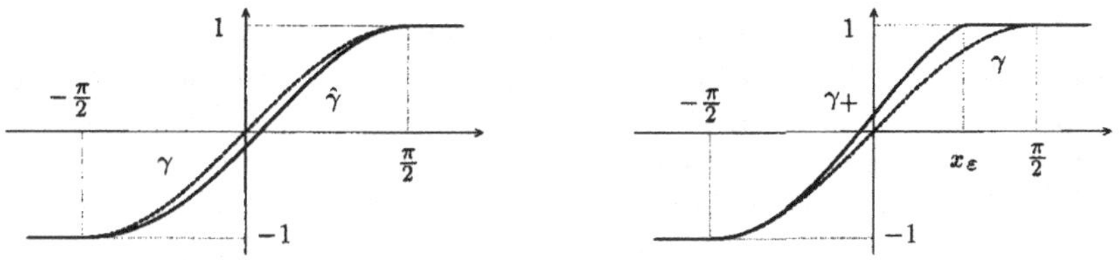

FIGURE 5.2. Function γ (dashed line) and supersolutions (a) $\hat{\gamma}$ (b) γ_+.

Since $\partial_t X_0 = 0$ and $\rho_0' + \frac{1}{\rho_0} = -g_0$, the ε-term leads to

$$X_2''(y,t)+X_2(y,t) = -\big(\rho_1'(t)-\tfrac{y+\rho_1(t)}{\rho_0^2(t)}\big)X_0'(y)+g_0(t)X_1'(y,t)-\tfrac{c_0}{2}g_0'(t)\big(y+\rho_1\big) = q_2(y,t),$$

and corresponding compatibility constraint

$$0 = \int_{-\pi/2}^{\pi/2} q_2(y,t)X_0'(y)dy = -c_0\big(\rho_1'(t) - \tfrac{\rho_1(t)}{\rho_0^2(t)} + \rho_1(t)g_0'(t)\big).$$

Imposing $\rho_1(0) = 0$, we readily get $\rho_1(t) = 0$ and, since $X_1'(y,t) = \frac{1}{2}g_0(t)yX_0'(y)$, X_2 solves $\mathcal{L}X_2(y,t) = \big(\frac{1}{\rho_0^2(t)} + \frac{g_0^2(t)}{2}\big)yX_0'(y) - \frac{c_0}{2}g_0'(t)y$. The y and t dependences can be separated using (5.9), because

$$X_2(y,t) = \big(\tfrac{1}{\rho_0^2(t)} + \tfrac{g_0^2(t)}{2}\big)\xi(y) - \tfrac{c_0}{2}g_0'(t)\mu(y) \quad \forall\, y \in [-\tfrac{\pi}{2}, \tfrac{\pi}{2}].$$

We have now the first clear indication that the rate of convergence for the interfaces should be $\mathcal{O}(\varepsilon^2)$, because $\rho_\varepsilon(t) = \rho(t) + \mathcal{O}(\varepsilon^2)$.

The candidate to viscosity super and subsolutions suggested by the asymptotic is:

$$\hat{\gamma}(x;\mathbf{x},t) = \gamma(x) + \varepsilon g(\mathbf{x},t)\eta(x) \quad \forall\, x \in \mathbf{R},\ (\mathbf{x},t) \in Q. \tag{5.15}$$

Note that $\hat{\gamma}(\cdot;\mathbf{x},t) \in C^{1,1}(\mathbf{R})$ is strictly increasing in $(-\frac{\pi}{2}, \frac{\pi}{2})$ and $\hat{\gamma}'(x) > \frac{1}{2}\gamma'(x)$ for ε sufficiently small, because $\eta'(x) = \frac{1}{2}x\gamma'(x)$. Hereafter we use the notation $\hat{\gamma}^{(k)} = \partial_x^k\hat{\gamma}$.

In order to construct more stringent barriers, we need to incorporate more shape corrections to the standing wave γ. Since ξ' and μ' exhibit jump discontinuities at both $\mp\frac{\pi}{2}$, they cannot enter in the definition of supersolution and further corrections compensate for their lack of regularity. Hence, we introduce two auxiliary functions $\xi_+, \mu_+ \in C^{0,1}(\mathbf{R})$, which can be interpreted as suitable shifts of the original ξ and μ. They are defined by

$$\xi_+(x) = \xi(x) - \tfrac{\pi}{8}\gamma'(x), \quad \mu_+(x) = \mu(x) - \gamma'(x), \quad \forall\, x \in \mathbf{R},$$

and still solve (5.9). Furthermore, ξ_+' and μ_+' are discontinuous at $\frac{\pi}{2}$ because $[\![\xi_+']\!](\frac{\pi}{2}) = -\frac{\pi}{4}$, $[\![\mu_+']\!](\frac{\pi}{2}) = -2$, but $\xi_+'(-\frac{\pi}{2}) = \mu_+'(-\frac{\pi}{2}) = 0$ (see Figure 5.1b). We can now define the candidate to variational supersolution, that will be examined in the next section:

$$\gamma_+(x;\mathbf{x},t) = \min\Big(1, \gamma(x)+\varepsilon(\bar{g}+\varepsilon^2 a)\eta(x)+\varepsilon^2\big(\bar{\kappa}^s+\tfrac{1}{2}\bar{g}^2\big)\xi_+(x)-\tfrac{c_0}{2}\varepsilon^2\bar{g}'\mu_+(x)\Big), \tag{5.16}$$

for all $x \in \mathbf{R}$, $(\mathbf{x},t) \in Q$, where $a > 0$ is to be selected; a may depend on t. Here we set $\bar{\kappa}^s(\mathbf{x},t) = \kappa^s\big(\mathbf{s}(\mathbf{x},t),t\big)$ and $\bar{g}(\mathbf{x},t) = g\big(\mathbf{s}(\mathbf{x},t),t\big)$, $\bar{g}'(\mathbf{x},t) = (\nabla g)\big(\mathbf{s}(\mathbf{x},t),t\big) \cdot \nabla d(\mathbf{x},t)$. Note that $\gamma_+(\cdot;\mathbf{x},t) \in C^{0,1}(\mathbf{R})$, because $[\![\gamma_+']\!](\frac{\pi}{2}) \neq 0$; γ_+ is strictly increasing and satisfies $|\gamma_+| < 1$ in $(-\frac{\pi}{2}, x_\varepsilon)$ for a proper $0 < x_\varepsilon \leq \frac{\pi}{2}$ (see Figure 5.2).

5.1.2. Supersolutions and subsolutions

We intend to use either (5.2) or (5.3) to motivate the construction of supersolutions to the double obstacle problem (5.5). This problem has both a variational [61] and viscosity interpretation [41,146]. We say that a function χ_+ *is a viscosity supersolution to the double obstacle problem (5.5) if and only if* $\chi_+ \geq -1$ *and, if* $\chi_+ - \varphi$ *attains a minimum at* $(\mathbf{x}_0, t_0) \in Q$ *for* $\varphi \in C^\infty(Q)$ *and* $\chi_+(\mathbf{x}_0, t_0) = \varphi(\mathbf{x}_0, t_0) < 1$, *then*

$$\mathcal{J}\varphi = \varepsilon\partial_t\varphi - \varepsilon\Delta\varphi - \tfrac{1}{\varepsilon}\varphi - \tfrac{c_0}{2}g \geq 0 \quad at\ (\mathbf{x}_0, t_0).$$

Similarly we can define a viscosity subsolution. A function χ_ε is called *viscosity solution* of (5.5) if it is both a super and a subsolution. The following lemma is a crucial tool in comparing viscosity super and subsolutions to (5.5), which in particular implies uniqueness for (5.5) (the proof is rather technical and can be found in [116]).

LEMMA 5.1. *Let* χ_+ *be a lower semicontinuous viscosity supersolution and* χ_- *be an upper semicontinuous viscosity subsolution. If* $\chi_+ \geq \chi_-$ *at* $t = 0$ *and on* $\partial\Omega \times (0,T)$, *then* $\chi_+ \geq \chi_-$ *for all* $(\mathbf{x}, t) \in Q$.

Let $K = \{\varphi \in H^1(\Omega) : |\varphi| \leq 1 \text{ in } \Omega, \varphi = 1 \text{ on } \partial\Omega\}$. The variational formulation to the double obstacle problem (5.5) reads:

$$\begin{aligned} &\chi_\varepsilon \in L^2(0,T;K) \cap H^1(0,T;H^{-1}(\Omega)), \\ &\chi_\varepsilon(\cdot,0) = \chi_{0\varepsilon}(\cdot) \quad \text{a.e. in } \Omega, \\ &\varepsilon\langle\partial_t\chi_\varepsilon, \varphi-\chi_\varepsilon\rangle + \varepsilon\langle\nabla\chi_\varepsilon, \nabla(\varphi-\chi_\varepsilon)\rangle - \tfrac{1}{\varepsilon}\langle\chi_\varepsilon, \varphi-\chi_\varepsilon\rangle - \tfrac{c_0}{2}\langle g, \varphi-\chi_\varepsilon\rangle \geq 0 \end{aligned} \tag{5.17}$$

for all $\varphi \in K$ and a.e. $t \in (0,T)$. We say that a function $\chi_+ \in L^2(0,T;H^1(\Omega)) \cap H^1(0,T;H^{-1}(\Omega))$ *is a variational supersolution to the double obstacle problem (5.17) if* $\chi_+ \geq -1$ *and, for a.e.* $t \in (0,T)$,

$$\varepsilon\langle\partial_t\chi_+, \varphi\rangle + \varepsilon\langle\nabla\chi_+, \nabla\varphi\rangle - \langle f, \varphi\rangle \geq 0 \quad \forall\, 0 \leq \varphi \in H_0^1(\Omega), \tag{5.18}$$

with $f \geq \frac{1}{\varepsilon}\chi_+ + \frac{c_0}{2}g$ *a.e. in* $\{\chi_+ < +1\}$. We recall an elementary comparison lemma for variational supersolutions (see [35,109,111]).

LEMMA 5.2. *Let* χ_+ *be a variational supersolution. If* $\chi_+(\cdot,0) \geq \chi_{0\varepsilon}(\cdot)$ *a.e. in* Ω *and* $\chi_+ \geq 1$ *on* $\partial\Omega \times (0,T)$, *then* $\chi_+ \geq \chi_\varepsilon$ *for all* $(\mathbf{x},t) \in Q$.

To see this, take $\varphi = \min(\chi_+, \chi_\varepsilon)$ in (5.17) and $\varphi = \min(\chi_+ - \chi_\varepsilon, 0)$ in (5.18), subtract the resulting expressions and integrate them on $(0, \bar{t})$, and finally use Gronwall's lemma.

REMARK 5.1. We only deal with supersolutions χ_+, because the construction of a subsolution χ_- can be carried out as follows. Let $\hat{\chi}_+$ indicate the supersolution of the auxiliary problem corresponding to $\hat{g} = -g$ and $\hat{d} = -d$. We clearly obtain that $\chi_- = -\hat{\chi}_+$ is a subsolution for the original problem.

i) Viscosity supersolution. Let $\hat{\gamma}$ as in (5.15); set $c = \|\nabla g\|_{L^\infty(Q)} + 1$, $\sigma(t) = be^{2C(T-t)}$.

LEMMA 5.3. *For* $b > 0$ *sufficiently large and shift* $\sigma(t) + \frac{\pi}{2}$, *the following function is a viscosity supersolution to (5.5):*

$$\chi_+(\mathbf{x}, t) = \hat{\gamma}\big(\tfrac{d(\mathbf{x},t)}{\varepsilon} - \tfrac{\pi}{2} - \sigma(t); \mathbf{x}, t\big) \quad \forall\, (\mathbf{x}, t) \in Q. \tag{5.19}$$

Proof. We give here a formal argument and refer to [116] for the rigorous proof. Since $\chi_+ \geq -1$, we only have to show that $\mathcal{J}\chi_+ = \varepsilon\partial_t\chi_+ - \varepsilon\Delta\chi_+ - \frac{1}{\varepsilon}\chi_+ - \frac{c_0}{2}g \geq 0$. Set

$$y(\mathbf{x},t) = \tfrac{d(\mathbf{x},t)}{\varepsilon} - \tfrac{\pi}{2} - \sigma(t) \quad \forall\, (\mathbf{x},t) \in Q.$$

Since

$$\begin{aligned} \varepsilon\partial_t\chi_+ &= \hat{\gamma}'\partial_t d - \varepsilon\sigma'(t)\hat{\gamma}' + \varepsilon\partial_t\hat{\gamma}, \\ \varepsilon\Delta\chi_+ &= \tfrac{1}{\varepsilon}\hat{\gamma}'' + \hat{\gamma}'\Delta d + 2\nabla\hat{\gamma}'\cdot\nabla d + \varepsilon\Delta\hat{\gamma}, \end{aligned}$$

in view of (5.15) we have

$$\begin{aligned} \mathcal{J}\chi_+ = &-\tfrac{1}{\varepsilon}\big(\hat{\gamma}'' + \hat{\gamma} - \varepsilon g(\hat{\gamma}' - \tfrac{c_0}{2})\big) + \hat{\gamma}'\big(\partial_t d - \Delta d - g\big) \\ &- \varepsilon\sigma'(t)\hat{\gamma}' - 2\varepsilon\eta'\nabla g\cdot\nabla d + \varepsilon^2\eta(\partial_t g - \Delta g). \end{aligned}$$

We only have to examine the case $-1 \leq \chi_+(\mathbf{x},t) < 1$, that is $y < \frac{\pi}{2}$. If $y < -\frac{\pi}{2}$, then $\hat{\gamma} = -1$ and $\eta = \eta' = \hat{\gamma}' = \hat{\gamma}'' = 0$, whence $\mathcal{J}\chi_+ = \frac{1}{\varepsilon} - \frac{c_0}{2}g > 0$ for ε sufficiently small. If $-\frac{\pi}{2} < y < \frac{\pi}{2}$, instead, observe that (5.15), in conjunction with (5.7) and (5.8), yields

$$-\tfrac{1}{\varepsilon}\big(\hat{\gamma}'' + \hat{\gamma} - \varepsilon g(\hat{\gamma}' - \tfrac{c_0}{2})\big) = \varepsilon g^2\eta'.$$

Since

(5.20) $$|g(\mathbf{x},t) - g(\mathbf{x} - d(\mathbf{x},t)\nabla d(\mathbf{x},t),t)| \leq \|\nabla g\|_{L^\infty(Q)} d(\mathbf{x},t) \leq \varepsilon c\big(\sigma(t) + \pi\big),$$

and $d > 0$ for $|y| < \frac{\pi}{2}$, by virtue of (5.2) we have

$$\hat{\gamma}'\big(\partial_t d - \Delta d - g\big) \geq -\varepsilon c\hat{\gamma}'\big(\sigma(t) + \pi\big).$$

Hence, noting that $|\eta|, |\eta'| < \gamma'$ in $(-\frac{\pi}{2}, \frac{\pi}{2})$ and recalling that $\hat{\gamma}' > \frac{1}{2}\gamma'$, owing to the definition of $\sigma(t)$ we get $-\sigma'(t) = 2c\sigma(t)$ and

(5.21) $$\tfrac{1}{\varepsilon}\mathcal{J}\chi_+ \geq \hat{\gamma}'\big(c\sigma(t) + \mathcal{O}(1)\big).$$

Hence $\mathcal{J}\chi_+ \geq 0$ for b sufficiently large but independent of ε and ε sufficiently small. Since $\hat{\gamma}(\cdot;\mathbf{x},t) \in C^{1,1}(\mathbf{R})$ by construction, we conclude that $\hat{\gamma}''(y(\mathbf{x},t);\mathbf{x},t)$ is a bounded function and so a distribution without mass concentrated on $\{y = -\frac{\pi}{2}\}$. The proof is thus complete. □

REMARK 5.2. We see that the effect of shift σ is the term $c\sigma(t)$ that controls $\mathcal{O}(1)$ in (5.21). A similar effect is obtained in [14] upon considering the distance to a perturbed motion $V = \kappa + g + a\varepsilon$, instead of $V = \kappa + g$, but with a time-independent shift. This is only feasible in proving convergence without error estimates; see Remark 5.7 below.

REMARK 5.3. It is apparent from (5.20) that the exponential form of σ cannot be avoided unless g is independent of $\mathbf{x}$. In such a case, we could consider the linear shift $\sigma(t) = b(T-t)$, with $b > 0$ to be chosen sufficiently large for χ_+ to be again a supersolution. When $g \equiv 0$, the shift becomes constant, namely $\sigma(t) = 0$. The above argument in Lemma 5.3 still applies and is much simpler than that in [14,56].

ii) Variational supersolution. Let γ_+ as in (5.16) and set $c = \|\kappa^s\|_{L^\infty(\Sigma)} + \|\nabla g\|_{L^\infty(Q)} > 0$.

LEMMA 5.4. *For $a, b > 0$ sufficiently large and shift $S(t) = be^{3Ct}$, the following function is a variational supersolution to (5.17):*

(5.22) $$\chi_+(\mathbf{x},t) = \gamma_+\big(\tfrac{d(\mathbf{x},t)}{\varepsilon} + \varepsilon S(t); \mathbf{x}, t\big) \quad \forall\, (\mathbf{x},t) \in Q.$$

Proof. Since $|\chi_+| \leq 1$, we only have to demonstrate (5.18). Set

$$y(\mathbf{x},t) = \tfrac{d(\mathbf{x},t)}{\varepsilon} + \varepsilon S(t) \quad \forall\, (\mathbf{x},t) \in Q, \tag{5.23}$$

and define the layer $\mathcal{T}_\varepsilon = \{(\mathbf{x},t) \in Q : \mathbf{x} \in \mathcal{T}_\varepsilon(t)\}$, where $\mathcal{T}_\varepsilon(t) = \{\mathbf{x} \in \Omega : |\chi_+(\mathbf{x},t)| < 1\}$; note that $|y| < \frac{\pi}{2}$ in $\mathcal{T}_\varepsilon$ and, for ε sufficiently small, $\mathcal{T}_\varepsilon \subset \mathcal{T}$. In view of (5.3) and (5.23), we have, for all $(\mathbf{x},t) \in \mathcal{T}_\varepsilon$, $|\nabla y| = \frac{1}{\varepsilon}$ and

$$\varepsilon(\partial_t y - \Delta y) = \bar{g} + \varepsilon y \bar{\kappa}^s + \varepsilon^2 (S'(t) - S(t)\bar{\kappa}^s) + \mathcal{O}(d^2), \tag{5.24}$$

$$g = \bar{g} + \varepsilon y \bar{g}' - \varepsilon^2 S(t) \bar{g}' + \mathcal{O}(d^2). \tag{5.25}$$

Note the occurrence of y instead of d on the right hand sides. Now set

$$f(\mathbf{x},t) = \begin{cases} \frac{1}{\varepsilon}\chi_+(\mathbf{x},t) + \frac{c_0}{2} g(\mathbf{x},t) & \text{if } \chi_+(\mathbf{x},t) < +1 \\ 0 & \text{if } \chi_+(\mathbf{x},t) = +1. \end{cases}$$

Integration by parts in $\mathcal{T}_\varepsilon(t)$ for all $t \in (0,T)$ leads to

$$\varepsilon\langle \partial_t \chi_+, \varphi\rangle + \varepsilon\langle \nabla\chi_+, \nabla\varphi\rangle - \langle f, \varphi\rangle = \int_{\mathcal{T}_\varepsilon(t)} \mathcal{J}\chi_+ \varphi + \varepsilon \int_{\partial\mathcal{T}_\varepsilon(t)} \nabla\chi_+ \cdot \nu\varphi - \int_{\Omega\setminus\mathcal{T}_\varepsilon(t)} f\varphi,$$

for all $0 \leq \varphi \in H_0^1(\Omega)$. Since $\nabla\chi_+ \cdot \nu \geq 0$ on $\partial\mathcal{T}_\varepsilon(t)$ and $f \leq 0$ in $\Omega\setminus\mathcal{T}_\varepsilon(t)$ for ε sufficiently small, it only remains to establish that $\mathcal{J}\chi_+ \geq 0$ in $\mathcal{T}_\varepsilon(t)$. Using (5.22) we can write

$$\varepsilon\partial_t\chi_+ = \varepsilon\gamma_+'\partial_t y + \varepsilon\partial_t\gamma_+,$$
$$\varepsilon\Delta\chi_+ = \tfrac{1}{\varepsilon}\gamma_+'' + \varepsilon\gamma_+'\Delta y + 2\nabla\gamma_+' \cdot \nabla d + \varepsilon\Delta\gamma_+;$$

hence, in view of (5.16) we have

$$\mathcal{J}\chi_+ = -\tfrac{1}{\varepsilon}(\gamma_+'' + \gamma_+) + \varepsilon\gamma_+'(\partial_t y - \Delta y) - \tfrac{c_0}{2} g - 2\varepsilon\eta'\nabla\bar{g}\cdot\nabla d + \mathcal{O}(\varepsilon^2).$$

Since $\nabla\bar{g}(\mathbf{x},t)$ is tangent to $\Sigma(t)$ at $\mathbf{s}(\mathbf{x},t)$ whereas $\nabla d(\mathbf{x},t)$ is orthogonal, we obtain $\nabla\bar{g}\cdot\nabla d = 0$. Using (5.7), (5.8), and (5.9), the first term in the expression above reads

$$-\tfrac{1}{\varepsilon}(\gamma_+'' + \gamma_+) = -(\bar{g} + \varepsilon^2 a)(\gamma' - \tfrac{c_0}{2}) - \varepsilon(\bar{\kappa}^s + \tfrac{1}{2}\bar{g}^2) y\gamma' + \varepsilon\tfrac{c_0}{2}\bar{g}' y.$$

On the other hand, we make use of (5.24) and the fact that $S'(t) = 3cS(t)$ to get

$$\varepsilon\gamma_+'(\partial_t y - \Delta y) = \bar{g}\gamma' + \varepsilon\bar{g}^2\eta' + \varepsilon\bar{\kappa}^s y\gamma' + \varepsilon^2 S(t)(3c - \bar{\kappa}^s)\gamma' + \mathcal{O}(\varepsilon^2) + (a+b)\mathcal{O}(\varepsilon^3),$$

where we use that $\mathcal{O}(d^2) = \mathcal{O}(\varepsilon^2) + b\mathcal{O}(\varepsilon^3)$. Noting that $\eta'(y) = \frac{1}{2} y\gamma'(y)$, owing to (5.25), after reordering and cancellation we arrive at

$$\tfrac{1}{\varepsilon^2}\mathcal{J}\chi_+ = a(\tfrac{c_0}{2} - \gamma') + S(t)(3c - \bar{\kappa}^s)\gamma' + S(t)\tfrac{c_0}{2}\bar{g}' + \mathcal{O}(1) + (a+b)\mathcal{O}(\varepsilon).$$

Select $a = 2cS(t)$ (note that the dependence on t is allowed); since $c \geq \bar{\kappa}^s + |\bar{g}'|$, we get

$$\tfrac{1}{\varepsilon^2}\mathcal{J}\chi_+ \geq a(\tfrac{c_0}{2} - \gamma') + a\gamma' - \tfrac{c_0}{4} a + \mathcal{O}(1) + a\mathcal{O}(\varepsilon) = \tfrac{c_0}{4} a + \mathcal{O}(1) + a\mathcal{O}(\varepsilon) \geq 0,$$

for b sufficiently large but independent of ε and ε sufficiently small. □

5.1.3. Convergence and error estimates of the interfaces

In this section we prove convergence of the zero level sets $\Sigma_\varepsilon(t) = \{\mathbf{x} \in \Omega : \chi_\varepsilon(\mathbf{x},t) = 0\}$ to the generalized geometric motion $\Sigma(t)$, provided no fattening occurs. We also prove interface error estimates of order either one or two provided the flow is nondegenerate or smooth, respectively. Key ingredients are the explicit form of supersolutions and the comparison lemmata of Sect. 5.1.2.

5.1.3.1. Convergence and error estimates past singularities. See [116]. Let $l_\varepsilon = (\sigma(0)+\pi)\varepsilon$ and $\Sigma_\pm(0)$ be the smooth surface $\Sigma_\pm(0) = \{\mathbf{x} \in \Omega : d_0(\mathbf{x}) = \mp l_\varepsilon\}$. We designate by $\Sigma_\pm(t)$ the generalized evolving fronts that starting from $\Sigma_\pm(0)$ are governed by $V = \kappa + g$, namely

$$\Sigma_\pm(t) = \{\mathbf{x} \in \Omega : \omega(\mathbf{x},t) = \mp l_\varepsilon\},$$

where ω is the unique Lipschitz continuous viscosity solution of (5.1). Let $d_\pm(\cdot,t)$ be the corresponding signed distance functions to $\Sigma_\pm(t)$, and $\chi_\pm$ be the super and subsolutions just constructed in Sect. 5.1.2 in terms of $d_\pm$ (see (5.19)). We would like to prove that

$$\chi_-(\mathbf{x},t) \le \chi_\varepsilon(\mathbf{x},t) \le \chi_+(\mathbf{x},t) \quad \forall\, (\mathbf{x},t) \in Q, \tag{5.26}$$

where χ_ε is the solution to (5.5) with initial datum $\chi_{0\varepsilon}(\mathbf{x}) \in \operatorname{sgn}(d_0(\mathbf{x}))$ for $\mathbf{x} \in \Omega$. In light of Lemma 5.1 and the fact that χ_+ is lower semicontinuous and χ_- is upper semicontinuous, because of (5.2), we solely have to show that (5.26) is valid on the parabolic boundary of Q. This is certainly the case on $\partial\Omega \times (0,T)$, where $\chi_- = \chi_+ = \chi_\varepsilon = 1$. In addition, since $d_+(\mathbf{x},0) = d_0(\mathbf{x}) + l_\varepsilon$, we have for $t=0$ and $\mathbf{x} \in \Omega$:

$$\chi_+(\mathbf{x},0) = \hat\gamma\left(\tfrac{d_+(\mathbf{x},0)}{\varepsilon} - \tfrac{\pi}{2} - \sigma(0)\right) = \hat\gamma\left(\tfrac{d_0(\mathbf{x})}{\varepsilon} + \tfrac{\pi}{2}\right) \ge \operatorname{sgn}(d_0(\mathbf{x})) = \chi_{0\varepsilon}(\mathbf{x}) \ge \chi_-(\mathbf{x},0).$$

The desired inequality (5.26) then follows immediately from Lemma 5.1.

THEOREM 5.1. *For $\mathbf{x} \in I(t)$ (respectively $\mathbf{x} \in O(t)$), there exists $\varepsilon_0(\mathbf{x},t) > 0$ such that*

$$\chi_\varepsilon(\mathbf{x},t) = -1 \quad (\textit{respectively } \chi_\varepsilon(\mathbf{x},t) = +1) \quad \forall\, \varepsilon \le \varepsilon_0(\mathbf{x},t).$$

Proof. Let $\mathbf{x} \in I(t) = \{\omega(\cdot,t) < 0\}$. For ε sufficiently small, $\omega(\mathbf{x},t) < -l_\varepsilon$, whence $d_+(\mathbf{x},t) < 0$. This implies $\frac{d_+(\mathbf{x},t)}{\varepsilon} - \frac{\pi}{2} - \sigma(t) < -\frac{\pi}{2}$ and therefore $-1 \le \chi_\varepsilon(\mathbf{x},t) \le \chi_+(\mathbf{x},t) = -1$, because of (5.26) and (5.19). Similar reasoning for χ_- completes the proof. □

REMARK 5.4. If $\Sigma(t)$ has empty interior, then Theorem 5.1 establishes convergence of $\Sigma_\varepsilon(t)$ to $\Sigma(t)$. As far as we know, whether such a condition is always valid for the geometric evolution of smooth initial surfaces remains a conjecture [47,57].

In order to derive interface error estimates we are forced to assume more regularity of $\Sigma(t)$. We say that $\mathbf{x} \in \Sigma^*(t)$, the regular part of $\Sigma(t)$, if $\omega(\cdot,t)$ is of class C^1 in a neighborhood of $\mathbf{x} \in \Sigma(t)$ and satisfies the nondegeneracy condition $|\nabla\omega(\mathbf{x},t)| > 0$. Note that $\Sigma(t) = \Sigma^*(t)$ for as long as the motion is classical, that is before the onset of singularities. However, $\Sigma(t) = \Sigma^*(t)$ is known to hold between consecutive singularities for a number of flows, e.g., for surfaces of rotation [5,131].

THEOREM 5.2. *For $\mathbf{x} \in \Sigma^*(t)$, there exists $\varepsilon_0(\mathbf{x},t) > 0$ such that*

$$\operatorname{dist}\big(\mathbf{x}, \Sigma_\varepsilon(t)\big) \le \frac{2(\sigma(0)+\pi)}{|\nabla\omega(\mathbf{x},t)|}\varepsilon \quad \forall\, \varepsilon \le \varepsilon_0(\mathbf{x},t). \tag{5.27}$$

Proof. Since (5.19) yields $\chi_\pm(\cdot,t) = \mp 1$ on $\Sigma_\pm(t)$ and (5.26) thus implies $\chi_\varepsilon(\cdot,t) = \mp 1$ on $\Sigma_\pm(t)$, it suffices to estimate the distance between $\Sigma_\pm(t)$ and $\mathbf{x} \in \Sigma^*(t)$. Using Taylor's formula about $\mathbf{x}$, we obtain for $\mathbf{x} \mp \delta\frac{\nabla\omega(\mathbf{x},t)}{|\nabla\omega(\mathbf{x},t)|} \in \Sigma_\pm(t)$, and so $\delta > 0$,

$$\mp l_\varepsilon = \omega\big(\mathbf{x} \mp \delta\tfrac{\nabla\omega(\mathbf{x},t)}{|\nabla\omega(\mathbf{x},t)|}, t\big) = \omega(\mathbf{x},t) \mp |\nabla\omega(\mathbf{x},t)|\delta + o(\delta) < (>) \mp \tfrac{1}{2}|\nabla\omega(\mathbf{x},t)|\delta,$$

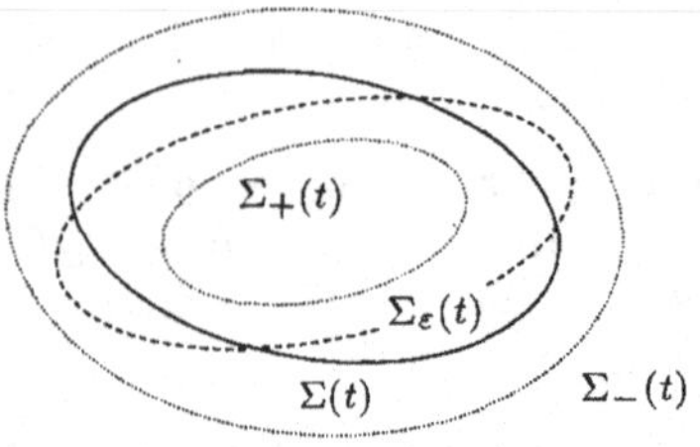

FIGURE 5.3. Interfaces $\Sigma(t) = \{\mathbf{x} \in \Omega : \omega(\mathbf{x},t) = 0\}$, $\Sigma_\varepsilon(t) = \{\mathbf{x} \in \Omega : \chi_\varepsilon(\mathbf{x},t) = 0\}$ (dashed line), and $\Sigma_\pm(t) = \{\mathbf{x} \in \Omega : \omega(\mathbf{x},t) = \mp l_\varepsilon\}$ (dotted lines).

for ε sufficiently small depending on $(\mathbf{x},t)$. Hence $\delta \le \frac{2l_\varepsilon}{|\nabla\omega(\mathbf{x},t)|}$, as asserted in (5.27). □

REMARK 5.5. We can summarize Theorem 5.2 by simply saying that $\Sigma_\varepsilon(t)$ and, obviously, $\Sigma(t)$ lie between the surfaces $\omega(\cdot,t) = \pm l_\varepsilon$ (see Figure 5.3). It is then the profile of ω near a singularity, or equivalently how far the level sets $\Sigma_\pm(t)$ may become apart in the vicinity of a singular point, that determines the rate of convergence.

REMARK 5.6. A linear rate of convergence for the mean curvature flow ($g \equiv 0$) before the onset of singularities, say in $[0,t^*)$, is derived in [35]:

$$\mathrm{dist}_H\big(\Sigma(t), \Sigma_\varepsilon(t)\big) \le C_{t^\#}\varepsilon \quad \forall\, t \le t^\# < t^*;$$

here dist_H stands for the Hausdorff distance. In the smooth regime, however, optimal error estimates of order $\mathcal{O}(\varepsilon^2)$ can be proved [109,110]. The virtue of (5.27) is thus its validity even beyond singularities if the motion is locally smooth.

REMARK 5.7. We stress that perturbing the original motion as in [14], that is replacing g by $g \pm a\varepsilon$, would not immediately lead to (5.27). This is because $d_\pm$ would depend on the viscosity solutions $\omega_\pm$ of the resulting perturbed problems rather than on ω. Even though convergence of $\omega_\pm$ to ω is known as $\varepsilon \downarrow 0$, we would also need uniform nondegeneracy of $\omega_\pm$ to hold, which does not seem to be available.

REMARK 5.8. The exponential blow-up of the constant in (5.27), namely $\sigma(0) = be^{2CT}$, can only be avoided if g is independent of $\mathbf{x}$ (see Remark 5.3). In fact; (5.27) cannot be improved without additional assumptions, as the following radially symmetric flow in two dimensions reveals. Let $g(r,t) = r$ and consider the initial condition $r_\varepsilon(0) = 1+\varepsilon$. The corresponding evolution is given by $r_\varepsilon(t) = \big(1 + (r_\varepsilon^2(0) - 1)\mathrm{e}^{2t}\big)^{1/2}$, which yields $r_\varepsilon(t) - r_0(t) = \mathcal{O}\big(\varepsilon \mathrm{e}^{2t}\big)$.

5.1.3.2. Optimal error estimates for smooth evolutions. See [109,110,111]. We use the more stringent barriers defined in (5.22) for proving optimal error estimates for interfaces before the onset of singularities, say in $[0,t^*)$. Let χ_ε be the solution to (5.5) with initial datum $\chi_{0\varepsilon}(\mathbf{x}) = \gamma\big(\frac{d_0(\mathbf{x})}{\varepsilon}\big)$. First observe that

$$\chi_-(\mathbf{x},t) \le \chi_\varepsilon(\mathbf{x},t) \le \chi_+(\mathbf{x},t) \quad \forall\, (\mathbf{x},t) \in Q. \tag{5.28}$$

In light of Lemma 5.2, since $\chi_- = \chi_+ = \chi_\varepsilon = 1$ on $\partial\Omega \times (0,T)$, we solely have to show that (5.28) is valid for $t = 0$ and $\mathbf{x} \in \Omega$. This is a consequence of the elementary inequality [109]:

$$\gamma(x) - \gamma(x-h) > \tfrac{1}{\pi}h\Big(h + \min\big(\tfrac{\pi}{2} + (x-h), \tfrac{\pi}{2} - x\big)\Big) \qquad \forall\, -\tfrac{\pi}{2} < x-h < x < \tfrac{\pi}{2},\ 0 < h \ll 1,$$

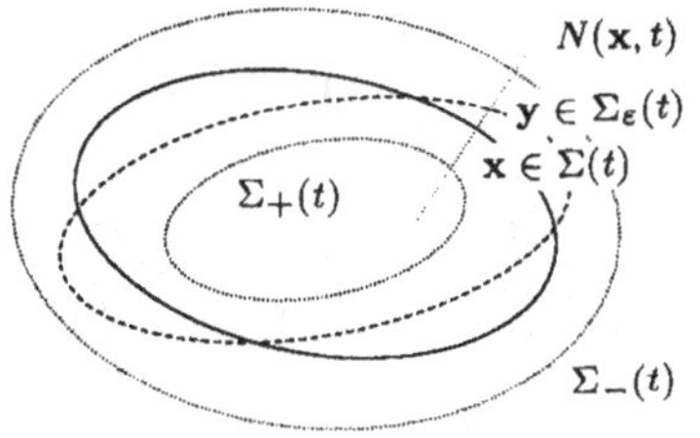

FIGURE 5.4. Interfaces $\Sigma(t)=\{\mathbf{x}\in\Omega : d(\mathbf{x},t)=0\}$, $\Sigma_\varepsilon(t)=\{\mathbf{x}\in\Omega : \chi_\varepsilon(\mathbf{x},t)=0\}$ (dashed line), and $\Sigma_\pm(t)=\{\mathbf{x}\in\Omega : \chi_\pm(\mathbf{x},t)=0\}$ (dotted lines).

which yields

$$\chi_+(\mathbf{x},0)=\gamma_+\left(\tfrac{d_0(\mathbf{x})}{\varepsilon}+\varepsilon S(0)\right)\geq\gamma\left(\tfrac{d_0(\mathbf{x})}{\varepsilon}\right)=\chi_{0\varepsilon}(\mathbf{x})\geq\chi_-(\mathbf{x},0).$$

THEOREM 5.3. *There exist* $0<\varepsilon_0$ *such that for all* $\varepsilon\leq\varepsilon_0$ *and* $0<t\leq t^\#<t^*$ *the following optimal interface error estimate holds:*

$$\operatorname{dist}_H\big(\Sigma(t),\Sigma_\varepsilon(t)\big)\leq C_{t^\#}\varepsilon^2. \tag{5.29}$$

Proof. The proof relies on the following properties of the function γ_+ defined in (5.16), $\gamma_+'(x)>0$ for all $x\in(-\frac{\pi}{2},x_\varepsilon)$ and $\gamma_+'(0)\geq\frac{3}{4}$ for ε sufficiently small, which lead to

$$\big(\nabla\chi_\pm\cdot\nabla d\big)(\mathbf{z},t)=\tfrac{1}{\varepsilon}\gamma_\pm'\left(\tfrac{d(\mathbf{z},t)}{\varepsilon}\pm\varepsilon S(t)\right)+\mathcal{O}(\varepsilon)>0\quad\forall\,\mathbf{z}\in\mathcal{T}_\varepsilon(t), \tag{5.30}$$

$$\big(\nabla\chi_\pm\cdot\nabla d\big)(\mathbf{x},t)\geq\tfrac{1}{2\varepsilon}\quad\forall\,\mathbf{x}\in\Sigma(t). \tag{5.31}$$

For any $\mathbf{x}\in\Sigma(t)$, let $N(\mathbf{x},t)$ denote a straight segment of size $2D$ perpendicular to $\Sigma(t)$ at $\mathbf{x}$, namely $N(\mathbf{x},t)=\{\mathbf{z}=\mathbf{x}+x\nabla d(\mathbf{x},t) : |x|\leq D\}$ (see Figure 5.4). For a given $\mathbf{y}\in\Sigma_\varepsilon(t)$, we claim that $(\mathbf{y},t)\in\mathcal{T}=\{|d(\mathbf{z},t)|\leq D\}$, whence $\mathbf{x}=\mathbf{s}(\mathbf{y},t)\in\Sigma(t)$ is uniquely defined and $\mathbf{y}\in N(\mathbf{x},t)$. In fact, otherwise $(\mathbf{y},t)\notin\mathcal{T}$ and, in view of the definition of $\chi_\pm$, $-1=\chi_-(\mathbf{y},t)\leq\chi_\varepsilon(\mathbf{y},t)=0\leq\chi_+(\mathbf{y},t)=1$, which yields the contradiction $\mathbf{y}\in I(t)\cap O(t)$. Consider now the problem of finding $\mathbf{y}_\pm\in N(\mathbf{x},t)$ such that $\chi_\pm(\mathbf{y}_\pm,t)=0$. We see that $\chi_\pm(\mathbf{x},t)=\gamma\big(\varepsilon S(t)\big)+\mathcal{O}(\varepsilon)=\mathcal{O}(\varepsilon)$ for $\mathbf{x}\in\Sigma(t)$. Upon invoking (5.31) we realize that such $\mathbf{y}_\pm$ do in fact exist and also that they satisfy

$$|\mathbf{x}-\mathbf{y}_\pm|=d(\mathbf{y}_\pm,t)\leq CS(t)\varepsilon^2. \tag{5.32}$$

Moreover both $\mathbf{y}_\pm$ are unique in that (5.30) implies monotonicity of $\chi_\pm$ on $N(\mathbf{x},t)$. With the aid of (5.28), we conclude that all the zeros $\mathbf{y}$ of $\chi_\varepsilon(\cdot,t)$ on $N(\mathbf{x},t)$ must lie between $\mathbf{y}^-$ and $\mathbf{y}^+$, which, in turn, applies to the original $\mathbf{y}\in\Sigma_\varepsilon(t)$, because $\mathbf{y}\in N(\mathbf{x},t)$. Therefore, from (5.32) we readily get

$$\operatorname{dist}\big(\mathbf{x},\Sigma_\varepsilon(t)\big),\operatorname{dist}\big(\mathbf{y},\Sigma(t)\big)\leq|\mathbf{x}-\mathbf{y}|\leq\max_\mp|\mathbf{x}-\mathbf{y}^\mp|\leq CS(t)\varepsilon^2. \tag{5.33}$$

In order to show the optimality of (5.29), we consider a circle evolving by mean curvature, as in Sect. 5.1.1, and prove that $|\rho(t)-\rho_\varepsilon(t)|\geq\frac{1}{2}\varepsilon^2|\rho_2(t)|$ using more stringent barriers [109]. The proof of Theorem 5.3 is thus complete. □

REMARK 5.9. We stress the local character of (5.33) for small times. In turn, the use of a space-time dependent relaxation parameter $\varepsilon\alpha(\mathbf{x},t)$ results in an enhancement of the accuracy where $\alpha(\mathbf{x},t)$ is small [111]:

$$\mathrm{dist}\big(\mathbf{x},\Sigma_\varepsilon(t)\big)\le C_{t\#}\varepsilon^2\alpha^2(\mathbf{x},t)\quad\forall\,\mathbf{x}\in\Sigma(t),\quad \mathrm{dist}\big(\mathbf{y},\Sigma(t)\big)\le C_{t\#}\varepsilon^2\alpha^2(\mathbf{y},t)\quad\forall\,\mathbf{y}\in\Sigma_\varepsilon(t).$$

Remark 5.8 applies also for the exponential blow-up of the constant in (5.33). A quasi-optimal interface error estimate for the quartic potential is proved in [16].

5.2. Fully discrete double obstacle schemes

The key computational feature of χ_ε is that $\chi_\varepsilon(\cdot,t)$ attains the values $+1$ or -1 outside $\mathcal{T}_\varepsilon(t)$ which, using a space-time dependent relaxation parameter, is a layer of local thickness approximately $\frac{\pi}{2}\varepsilon\alpha(\mathbf{x},t)$. A minimum requirement for approximating (5.6) by finite elements is the partitioning of the active set $\mathcal{T}_\varepsilon(t)$. The use of graded meshes guarantees the numerical resolution of $\mathcal{T}_\varepsilon(t)$ and the optimal distribution of spatial degrees of freedom. Let $h(\mathbf{x},t)$ indicates the associated mesh density function, which is supposed to satisfy $h(\mathbf{x},t) = h_0\alpha(\mathbf{x},t)$, with $0 < h_0 < 1$ fixed. For the same local accuracy, the reaction diffusion approach with a constant density function $\min_{(\mathbf{x},t)\in Q}\alpha(\mathbf{x},t)$ would require a higher number of triangles.

For axially symmetric problems, the above ideas are implemented in the two dimensional dynamic mesh algorithm of [102,112]. The actual mesh is a triangulation of just $\mathcal{T}_\varepsilon(t)$, which is significantly smaller than a full mesh, and is updated at every time step to follow the motion of $\mathcal{T}_\varepsilon(t)$, and thus that of $\Sigma_\varepsilon(t)$. Time stepping is explicit, but stability constraints force small time steps only when singularities develop, whereas relatively large time steps are allowed at the beginning or past singularities, when the evolution is smooth and hence $\alpha(\mathbf{x},t)$ is large. Explicit time discretization guarantees that at most one layer of boundary elements of the transition region has to be either added or deleted, thereby making dynamic mesh updating feasible and overruling the use of implicit methods. This method thus retains the local character of the geometric problem at hand while taking advantage of the variational structure of (5.17), and exhibits potential to be exploited in connection with phase transitions. The dynamic mesh procedure thus compares favorably with other explicit marching schemes defined over uniform decompositions of the entire Ω [119,130]. They all can handle singularity formation and topological changes, which, in turn, restrict the applicability of front-tracking techniques to the smooth regime; a typical one is based on discretizing the Laplace-Beltrami operator on the interface [52].

For simplicity we assume $g\equiv 0$ and $\alpha\equiv 1$. We stick to the notation of Sect. 3.1 for finite element spaces. The discrete convex set of piecewise linear functions is $K_h=\{\phi\in V_h^1 : |\phi|\le 1 \text{ in } \Omega, \phi=1 \text{ on } \partial\Omega\}$. We consider the following finite element explicit time stepping of (5.17), which is essential for the dynamic mesh algorithm of Sect. 5.2.1:

Set $X^0 = \Pi_h\gamma(\frac{d_0(\mathbf{x},0)}{\varepsilon})$ and, for all $i\ge 1$, select the time step τ^i and find $X^i\in K_h$ such that, for all $\phi\in K_h$,

$$\varepsilon\langle X^i - X^{i-1},\phi - X^i\rangle_h + \varepsilon\tau^i\langle\nabla X^{i-1},\nabla(\phi - X^i)\rangle - \frac{\tau^i}{\varepsilon}\langle X^{i-1},\phi - X^i\rangle_h \ge 0. \tag{5.34}$$

The time step τ^i is selected adaptively to meet the stability constraint

$$0<\sigma\le\tau^i\|\mathbf{M}^{-1}\mathbf{K}\|_\infty < 1, \tag{5.35}$$

for a fixed σ. This requirement is only necessary, and so verified, in the numerical enlarged transition layer T_e^{i-1} (see Sect. 5.2.1), which is a small region relative to Ω. If τ^i does not satisfy (5.35), then an automatic reduction or increase by a factor σ is performed. This, together with the stability relation $\varepsilon \geq C^* h_S$ for all $S \in \mathcal{S}_h$ [19], in turn, eliminates any possible numerical oscillation.

We stress that the choice of an explicit scheme may not be restrictive in this setting. In fact an implicit scheme would require $\tau^i < \varepsilon^2$ in T_e^{i-1} just for solvability of (5.34) which, together with the relation $h_S = o(\varepsilon)$ necessary for transition layer resolution [18], would anyhow imply $\tau^i \leq Ch_S^2$ for a basically linear relation between h_S and ε.

Let $P : \mathbf{R}^J \to \mathbf{R}^J$ be the componentwise projection on $[-1,1]$ defined, for $\mathbf{q} = (q_j)_{j=1}^J$, by $(P\mathbf{q})_j = \max\big(-1, \min(1, q_j)\big)$. If we identify any function of V_h^1 with the $\mathbf{R}^J$ vector of its nodal values, (5.34) can be written equivalently in matrix form as:

$$X^i = P\big[\mathbf{M}^{-1}\big((1+\tfrac{\tau^i}{\varepsilon^2})\mathbf{M} - \tau^i\mathbf{K}\big)X^{i-1}\big]. \tag{5.36}$$

The actual computation of X^i is thus trivial in that we first have to perform the matrix operations involved on the right hand side of (5.36) (recall that $\mathbf{M}$ is diagonal), and then project each component onto $[-1,1]$.

REMARK 5.10. The following semi-explicit scheme can also be used in conjunction with the dynamic mesh algorithm of Sect. 5.2.1:

$$X^i = P\big[\big((1-\tfrac{\tau^i}{\varepsilon^2})\mathbf{M}\big)^{-1}\big(\mathbf{M} - \tau^i\mathbf{K}\big)X^{i-1}\big]. \tag{5.37}$$

This scheme retains the advantages of (5.36) but exhibits better monotonicity properties.

REMARK 5.11. Convergence and interface error estimates for a fully discrete implicit scheme without numerical quadratures have been studied in [117].

5.2.1. Dynamic mesh algorithm

The dynamic mesh algorithm triangulates solely T_e^{i-1}, a suitably enlarged discrete transition layer, and then updates the resulting mesh and finite element matrices to follow the layer motion. This results in savings of both computing time and memory allocation, and gives more flexibility in selecting the local meshsize (see [102,112]).

We define the discrete transition region (noncoincidence set) and coincidence set by

$$\mathcal{T}^i = \{S \in \mathcal{S}_h : |X^i(\mathbf{b}_S)| < 1\}, \quad T^i = \cup_{S \in \mathcal{T}^i} S,$$
$$\mathcal{C}^i = \{S \in \mathcal{S}_h : |X^i| = 1 \text{ in } S\}, \quad C^i = \big(\cup_{S \in \mathcal{C}^i} S\big)\backslash T^i,$$

whereas the associated dynamic mesh and enlarged transition layer are given by

$$\mathcal{T}_e^i = \{S \in \mathcal{S}_h : S \cap T^i \neq \emptyset\}, \quad T_e^i = \cup_{S \in \mathcal{T}_e^i} S.$$

The following crucial property holds for both schemes (5.36) and (5.37): *if $\mathbf{x}_j$ is a vertex lying in C^{i-1}, then $X^i(\mathbf{x}_j) = X^{i-1}(\mathbf{x}_j) = \pm 1$*, that is $T^i \subset T_e^{i-1}$. In other words, the discrete transition layer T^{i-1} cannot move faster than one triangle per time step and, consequently, X^i does not have to be computed at vertices lying outside T^{i-1}. We then compute the mass and stiffness matrices $\mathbf{M}^i$ and $\mathbf{K}^i$ corresponding to the enlarged transition layer T_e^{i-1}, update τ^i on imposing (5.35), and finally solve (5.36) within T_e^{i-1} only. We stress that $\mathcal{T}_e^{i-1}$ will be the only part of $\mathcal{S}_h$ present at the time step $i-1$, and is minimal in terms of number of triangles. It has to be stressed that $\mathcal{S}_h$ here is fictitious and just used for definition of $\mathcal{T}_e^{i-1}$. Once X^i and $T^i \subset T_e^{i-1}$ have been determined,

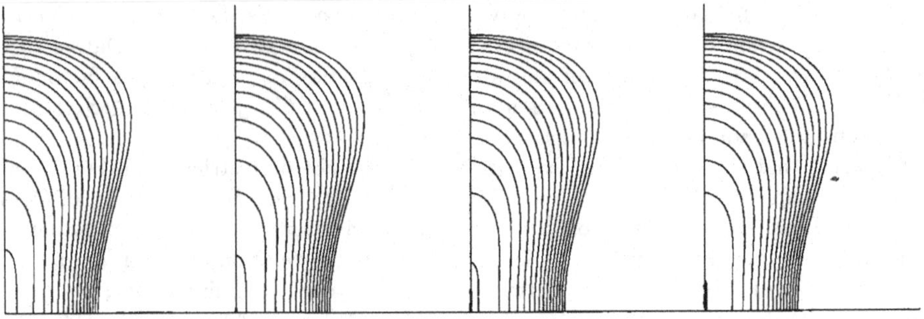

FIGURE 5.5. Front evolution for $c = 0.653$, $c^* = 0.654$, $c = 0.655$, and $c = 0.656$; $t = 0.005i$, $0 \leq i \leq 15$, and t_c^* (bold line).

$\mathcal{T}_e^{i-1}$ is updated as follows. First, all triangles $S \in \mathcal{C}^{i-1}$ satisfying $S \cap T^i = \emptyset$ are removed because their are no longer needed. New triangles are next created at all nodes on the boundary of T^i that are not interior to $\mathcal{T}_e^{i-1}$. The mesh so obtained is $\mathcal{T}_e^i$. A first mesh over $\mathcal{T}_\varepsilon(0)$ is constructed using a mesh front algorithm in the spirit of [121]; the mesh is graded according to $h_0\alpha(\mathbf{x})$, where h_0 is a scaling factor. The dynamic mesh $\mathcal{T}_e^i$, consistent with $h_0\alpha(\mathbf{x})$, is generated using the mesh front algorithm, which allows one to either delete or incorporate efficiently an element on the mesh front.

5.3. Numerical simulation of evolving fronts

We present a numerical experiment for the motion by mean curvature ($g \equiv 0$) that illustrates the practical use of a variable density function $\alpha(\mathbf{x})$ in dealing with large curvatures. We consider the evolution of the axially symmetric dumbbell studied in [5], with initial front

$$\Sigma_0^c = \{x_1^2 + x_2^2 = (1 - x_3^2)(1 - c + cx_3^2)^2\} \subset \mathbb{R}^3.$$

The surface Σ_0^0 is a sphere whereas Σ_0^1 consists of two tear shaped parts with a singularity at the origin. There exists a critical $0 < c^* < 1$ for which $\Sigma^{c^*}(t)$ remains regular and nonconvex until it shrinks to the origin [5]. For $c < c^*$, $\Sigma^c(t)$ becomes convex and shrinks to the origin at $t_c^\dagger$, whereas for $c > c^*$, $\Sigma^c(t)$ focuses at the origin and breaks at t_c^* and eventually disappears at $t_c^\dagger > t_c^*$. For the critical value c^* we thus have $t_{c^*}^\dagger = t_{c^*}^*$. Purpose of the simulation is to compute c^* along with $\Sigma^{c^*}(t)$.

In view of the axial symmetry, we introduce the cylindrical coordinates $r = (x_1^2+x_2^2)^{1/2}$ and $z = x_3$, and define the domain $\Omega = \{(r, z) \in (0, 4)^2\}$ along with a time-independent density function $\alpha(r, z)$ that concentrate the elements near the origin. The numerical results, obtained with $\varepsilon = 0.205$, suggest that $c^* = 0.654$. The sequences of interfaces for $c = 0.653$, $c^* = 0.654$, $c = 0.655$, and $c = 0.656$ are depicted in Figure 5.5a-d, where we observe the nonconvexity of $\Sigma^{c^*}(t)$ for the entire evolution. The boundary elements of the actual graded meshes for $c = c^*$ are drawn in Figure 5.6a-h, together with the corresponding interfaces, for a number of time steps. We point out that the thickness of the transition region varies along the front and is dictated by the local size of $\alpha(r, z)$. This remarkable effect is of great importance in resolving singularities.

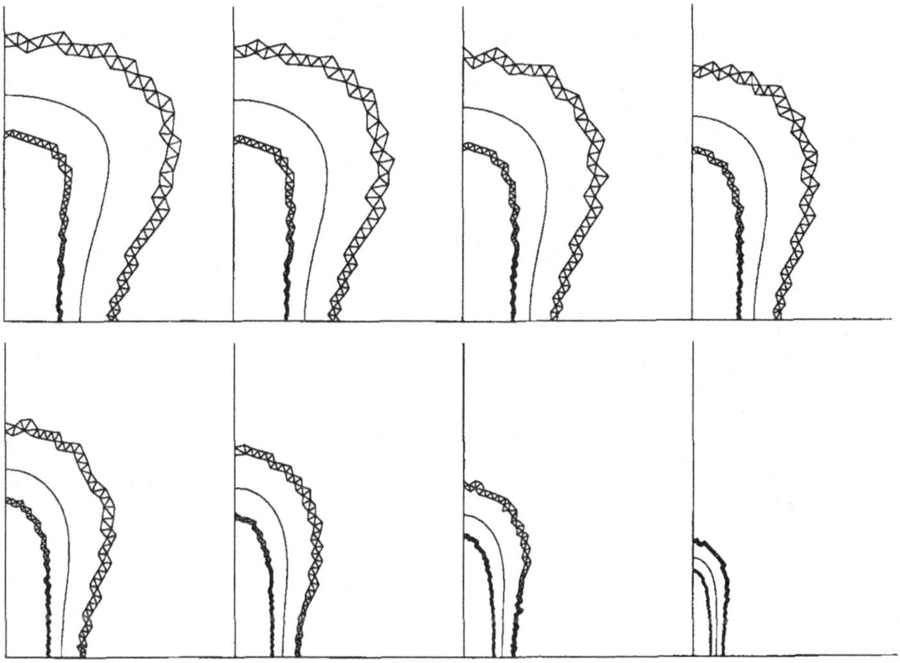

FIGURE 5.6. Graded meshes over the transition region at $t = 0.01i$, $0 \leq i \leq 7$, for $c^* = 0.654$.

References

[1] S.M. ALLEN AND J.W. CAHN, *A macroscopic theory for antiphase boundary motion and its application to antiphase domain coarsing*, Acta Metall. Mater., **27** (1979), pp. 1085–1095.

[2] F. ALMGREN AND L.H. WANG, *Mathematical existence of crystal growth with Gibbs-Thomson curvature effects*, in Proc. Evolving Phase Boundaries (Carnegie Mellon Univ., Pittsburgh, 1991).

[3] R. ALMGREN, *Variational algorithms and pattern formation in dendritic solidification*, J. Comput. Phys., **106** (1993), pp. 337–354.

[4] H.W. ALT AND S. LUCKHAUS, *Quasilinear elliptic-parabolic differential equations*, Math. Z., **183** (1983), pp. 311–341.

[5] S. ALTSCHULER, S. ANGENENT, AND Y. GIGA, *Mean curvature flow through singularities for surfaces of rotation*, Hokkaido Univ. Preprint, **130** (1991).

[6] G. AMIEZ AND P.A. GRÉMAUD, *On a numerical approach of Stefan-like problems*, Numer. Math., **59** (1991), pp. 71–89.

[7] D. ANDREUCCI, *Existence and uniqueness of solution to a concentrated capacity problem with change of phase*, European J. Appl. Math., **1** (1990), pp. 339–351.

[8] D. ANDREUCCI AND C. VERDI, *Existence, uniqueness, and error estimates for a model of polymer crystallization*, Adv. Math. Sci. Appl. (to appear).

[9] D.G. ARONSON, *The porous medium equation*, in Nonlinear Diffusion Problems, LNM 1024 (A. Fasano and M. Primicerio, eds.), Springer-Verlag, Berlin, 1985, pp. 1–46.

[10] O. AXELSSON AND V.A. BARKER, *Finite Element Solution of Boundary Value Problems: Theory and Applications*, Academic Press, Orlando, 1984.

[11] I. BABUŠKA, E.R. DE A. OLIVEIRA, J. GAGO, AND O.C. ZIENKIEWICZ, *Accuracy Estimates and Adaptive Refinements in Finite Element Computations*, Wiley, Chichester, 1986.

[12] C. BAIOCCHI, *Discretization of evolution inequalities*, in Partial Differential Equations and Calculus of Variations, I (F. Colombini, A. Marino, L. Modica, S. Spagnolo, eds.), Birkhäuser, Boston, 1989, pp. 59–92.

[13] C. Baiocchi and G.A. Pozzi, *Error estimates and free boundary convergence for a finite difference discretization of a parabolic variational inequality*, RAIRO Modél. Math. Anal. Numér., **11** (1977), pp. 315–340.

[14] G. Barles, H.M. Soner, and P.E. Souganidis, *Front propagation and phase field theory*, SIAM J. Control Optim., **31** (1993), pp. 439–469.

[15] V.P. Beghishev, S.A. Bolgov, I.A. Keapin, and A.Ya. Malkin, *General treatment of polymer crystallization kinetics. Part 1: a new macrokinetic equation and its experimental verification*, Polym. Engrg. Sci., **24** (1984), pp. 1396–1401.

[16] G. Bellettini and M. Paolini, *Quasi-optimal error estimates for the mean curvature flow with a forcing term*, Differential Integral Equations (to appear).

[17] G. Bellettini and M. Paolini, *Two examples of fattening for the curvature flow with a driving force*, Atti Accad. Naz. Lincei Cl. Sci. Fis. Mat. Natur. Rend. (9) (to appear).

[18] G. Bellettini, M. Paolini, and C. Verdi, *Γ-convergence of discrete approximations to interfaces with prescribed mean curvature*, Atti Accad. Naz. Lincei Cl. Sci. Fis. Mat. Natur. Rend. (9), **1** (1990), pp. 317–328.

[19] G. Bellettini, M. Paolini, and C. Verdi, *Numerical minimization of geometrical type problems related to calculus of variations*, Calcolo, **27** (1991), pp. 251–278.

[20] A.E. Berger, H. Brézis, and J.C.W. Rogers, *A numerical method for solving the problem $u_t - \Delta f(u) = 0$*, RAIRO Modél. Math. Anal. Numér., **13** (1979), pp. 297–312.

[21] A.E. Berger and J.C.W. Rogers, *Some properties of the nonlinear semigroup for the problem $u_t - \Delta f(u) = 0$*, Nonlinear Anal., **8** (1984), pp. 909-939.

[22] J. Berger, A. Köppl, and W. Schneider, *Non-isothermal crystallization, crystallization of polymers, system of rate equations*, Intern. Polymer Processing, **2** (1988), pp. 151–154.

[23] A. Berman and R.J. Plemmons, *Nonnegative Matrices in the Mathematical Sciences*, Academic Press, New York, 1979.

[24] K.A. Brakke, *The Motion of a Surface by its Mean Curvature*, Princeton University Press, Princeton, 1978.

[25] H. Brézis, *Monotonicity methods in Hilbert spaces and some applications to nonlinear partial differential equations*, in Contributions to Nonlinear Functional Analysis (E. Zarantonello ed.), Academic Press, New York, 1971, pp. 101–156.

[26] H. Brézis, *Opérateurs Maximaux Monotones et Semi-groupes de Contractions dans les Espaces de Hilbert*, North-Holland, Amsterdam, 1973.

[27] H. Brézis and A. Pazy, *Convergence and approximation of semigroups of nonlinear operators in Banach spaces*, J. Funct. Anal., **9** (1972), pp. 63–74.

[28] F. Brezzi and L.A. Caffarelli, *Convergence of the discrete free boundaries for finite element approximations*, RAIRO Modél. Math. Anal. Numér., **17** (1983), pp. 385–395.

[29] M. Brokate, *Hysteresis operators*, in [144], to appear.

[30] L. Bronsard and R.V. Kohn, *Motion by mean curvature as the singular limit of Ginzburg-Landau dynamics*, J. Differential Equations, **90** (1991), pp. 211–237.

[31] L.A. Caffarelli, *A remark on the Hausdorff measure of a free boundary and the convergence of coincidence sets*, Boll. Un. Mat. Ital. A (6), **18** (1981), pp. 109–113.

[32] G. Caginalp, *The dynamics of a conserved phase field system: Stefan-like, Hele-Shaw, and Cahn-Hilliard models as asymptotic limits*, IMA J. Appl. Math., **44** (1990), pp. 77–94.

[33] J.W. Cahn, C.A. Handwerker, and J.E. Taylor, *Geometric models of crystal growth*, Acta Metall. Mater., **40** (1992), pp. 1443–1474.

[34] X. Chen, *Generation and propagation of interfaces in reaction-diffusion equations*, J. Differential Equations, **96** (1992), pp. 116–141.

[35] X. Chen and C.M. Elliott, *Asymptotics for a parabolic double obstacle problem*, Proc. Roy. Soc. London Ser. A (to appear).

[36] Y.G. Chen, Y. Giga, and S. Goto, *Uniqueness and existence of viscosity solutions of generalized mean curvature flow equation*, J. Differential Geom., **33** (1991), pp. 749–786.

[37] H.L. Chin, *A finite element front-tracking enthalpy method for Stefan problems*, IMA J. Numer. Anal., **3** (1983), pp. 87–107.

[38] P.G. Ciarlet, *The Finite Element Method for Elliptic Problems*, North-Holland, Amsterdam, 1978.

[39] Ph.G. Ciarlet and P.A. Raviart, *Maximum principle and uniform convergence for the finite element method*, Comput. Methods Appl. Mech. Engrg., **2** (1973), pp. 17–31.

[40] J.F. CIAVALDINI, *Analyse numérique d'un problème de Stefan à deux phases par une méthode d'éléments finis*, SIAM J. Numer. Anal., **12** (1975), pp. 464–487.
[41] M.G. CRANDALL, H. ISHII, AND P.L. LIONS, *User's guide to viscosity solutions of second order partial differential equations*, Bull. Amer. Math. Soc. (N.S.), **27** (1992), pp. 1–67.
[42] M.G. CRANDALL AND T.M. LIGGETT, *Generation of semi-groups of nonlinear transformations on general Banach spaces*, Amer. J. Math., **93** (1971), pp. 265–298.
[43] J. CRANK, *Free and Moving Boundary Problems*, Clarendon Press, Oxford, 1984.
[44] A.B. CROWLEY, *On the weak solution of moving boundary problems*, IMA J. Appl. Math., **24** (1979), pp. 43–57.
[45] A. DAMLAMIAN, *Some results on the multiphase Stefan problem*, Comm. Partial Differential Equations, **2** (1977), pp. 1017–1044.
[46] I.I. DANILYUK, *On the Stefan problem*, Russian Math. Surveys, **40** (1985), pp. 157–223.
[47] E. DE GIORGI, *New ideas in calculus of variations and geometric measure theory*, in Motion by Mean Curvature and Related Topics (G. Buttazzo and A. Visintin, eds.), Gruyter, Berlin, 1994, pp. 63–69.
[48] P. DE MOTTONI AND M. SCHATZMAN, *Geometrical evolution of developped interfaces*, Trans. Amer. Math. Soc. (to appear).
[49] E. DI BENEDETTO AND D. HOFF, *An interface tracking algorithm for the porous medium equation*, Trans. Amer. Math. Soc., **284** (1984), pp. 463–500.
[50] J. DOUGLAS JR. AND T. DUPONT, *Alternating-direction Galerkin methods on rectangles*, in Numerical Solutions of Partial Differential Equations, II (B. Hubbard ed.), Academic Press, New York, 1971, pp. 133–214.
[51] J. DOUGLAS JR., T. DUPONT, AND R. EWING, *Incomplete iteration for time-stepping a Galerkin method for a quasilinear parabolic problem*, SIAM J. Numer. Anal., **16** (1979), pp. 503–522.
[52] G. DZIUK, *An algorithm for evolutionary surfaces*, Numer. Math., **58** (1991), pp. 603–611.
[53] C.M. ELLIOTT, *Error analysis of the enthalpy method for the Stefan problem*, IMA J. Numer. Anal., **7** (1987), pp. 61–71.
[54] J.F. EPPERSON, *Regularization for a class of nonlinear evolution equations*, SIAM J. Math. Anal., **17** (1986), pp. 84–90.
[55] K. ERIKSSON AND C. JOHNSON, *Adaptive finite element methods for parabolic problems I: a linear model problem*, SIAM J. Numer. Anal., **28** (1991), pp. 43–77.
[56] L.C. EVANS, H.M. SONER, AND P.E. SOUGANIDIS, *Phase transitions and generalized motion by mean curvature*, Comm. Pure Appl. Math., **45** (1992), pp. 1097–1123.
[57] L.C. EVANS AND J. SPRUCK, *Motion of level sets by mean curvature. I*, J. Differential Geom., **33** (1991), pp. 635–681.
[58] A. FASANO AND M. PRIMICERIO, *Convergence of Hüber method's for heat conduction problems with change of phase*, Z. Angew. Math. Mech., **53** (1973), pp. 341–348.
[59] A. FASANO, M. PRIMICERIO, AND L. RUBINSTEIN, *A model problem for heat conduction with a free boundary in a concentrated capacity*, IMA J. Appl. Math., **26** (1980), pp. 327–347.
[60] A. FRIEDMAN, *The Stefan problem in several space variables*, Trans. Amer. Math. Soc., **133** (1968), pp. 51–87.
[61] A. FRIEDMAN, *Variational Principles and Free Boundary Problems*, Wiley, New York, 1982.
[62] Y. GIGA, S. GOTO, H. ISHII, AND M.H. SATO, *Comparison principle and convexity preserving properties for singular degenerate parabolic equations on unbounded domains*, Indiana Univ. Math. J., **40** (1991), pp. 443–470.
[63] D. GILBARG AND N.S. TRUDINGER, *Elliptic Partial Differential Equations of Second Order*, Springer-Verlag, Berlin, 1983.
[64] E. GIUSTI, *Minimal Surfaces and Functions of Bounded Variation*, Birkhäuser, Boston, 1984.
[65] I.G. GÖTZ AND B.B. ZALTZMANN, *Nonincrease of mushy region in a nonhomogeneous Stefan problem*, Quart. Appl. Math., **49** (1991), pp. 741–746.
[66] M. GRAYSON, *The heat equation shrinks embedded plane curves to round points*, J. Differential Geom., **26** (1987), pp. 285–314.
[67] P. HANSBO AND C. JOHNSON, *Adaptive finite element methods in computational mechanics*, Comput. Methods Appl. Mech. Engrg., **101** (1992), pp. 143–181.
[68] M. HILPERT, *On uniqueness for evolution problems with hysteresis*, in Mathematical Models for Phase Change Problems, ISNM 88 (J.F. Rodrigues ed.), Birkhäuser, Basel, 1989, pp. 377–388.

[69] K. Hoffman, J. Sprekels, and A. Visintin, *Identification of hysteresis loops*, J. Comput. Phys., **78** (1988), pp. 215–230.

[70] S. Huang, *Regularity of the enthalpy for two-phase Stefan problem in several space variables*, IMA Preprint, Minneapolis (1992).

[71] W. Jäger and J. Kačur, *Solution of porous medium type systems by linear approximation schemes*, Numer. Math., **60** (1991), pp. 407–427.

[72] J.W. Jerome, *Approximation of Nonlinear Evolution Systems*, Academic Press, New York, 1983.

[73] J.W. Jerome and M.E. Rose, *Error estimates for the multidimensional two-phase Stefan problem*, Math. Comp., **39** (1982), pp. 377–414.

[74] X. Jiang and R.H. Nochetto, *A finite element method for a phase relaxation model. Part I: quasi-uniform mesh*, SIAM J. Numer. Anal. (to appear).

[75] X. Jiang and R.H. Nochetto, *Optimal error estimates for a semidiscrete phase relaxation model*, SIAM J. Numer. Anal. (to appear).

[76] X. Jiang, R.H. Nochetto, and C. Verdi, *A $P^1 - P^0$ finite element method for a model of polymer crystallization*, Preprint (1994).

[77] S.L. Kamenomostskaja, *On the Stefan problem*, Math. Sbornik, **53** (1961), pp. 489–514.

[78] J.B. Keller, J. Rubinstein, and P. Sternberg, *Fast reaction, slow diffusion and curve shortening*, SIAM J. Appl. Math., **49** (1989), pp. 116–133.

[79] N. Kenmochi, *Systems of nonlinear PDEs arising from dynamical phase transition*, in [144], to appear.

[80] M.A. Krasnosel'skii and A.V. Pokrowskii, *Systems with Hysteresis*, Springer-Verlag, Berlin, 1989.

[81] Yu.A. Kuznetsov and A.V. Lapin, *Domain decomposition method to realize an implicit difference scheme for the one-phase Stefan problem*, Sov. J. Numer. Anal. Math. Modelling, **3** (1988), pp. 487–504.

[82] A.V. Lapin, *Domain decomposition method for grid approximation of two-phase Stefan problem*, Sov. J. Numer. Anal. Math. Modelling, **6** (1991), pp. 25–42.

[83] S. Luckhaus, *Solutions for the two-phase Stefan problem with the Gibbs-Thomson law for the melting temperature*, European J. Appl. Math., **1** (1990), pp. 101–111.

[84] E. Magenes, *Problemi di Stefan bifase in più variabili spaziali*, Matematiche (Catania), **36** (1981), pp. 65–108.

[85] E. Magenes, *Numerical approximation of nonlinear evolution problems*, in Frontiers in Pure and Applied Mathematics (R. Dautray ed.), North-Holland, Amsterdam, 1991, pp. 193–207.

[86] E. Magenes, *The Stefan problem in a concentrated capacity*, in Problemi attuali dell'Analisi e della Fisica Matematica (dedicated to G. Fichera), 1992, to appear.

[87] E. Magenes, R.H. Nochetto, and C. Verdi, *Energy error estimates for a linear scheme to approximate nonlinear parabolic problems*, RAIRO Modél. Math. Anal. Numér., **21** (1987), pp. 655–678.

[88] E. Magenes and C. Verdi, *Time discretization schemes for the Stefan problem in a concentrated capacity*, Meccanica, **28** (1993), pp. 121–128.

[89] E. Magenes, C. Verdi, and A. Visintin, *Theoretical and numerical results on the two-phase Stefan problem*, SIAMJNA, **26** (1989), pp. 1425–1438.

[90] T.A. Manteuffel, *An incomplete factorization technique for positive definite linear systems*, Math. Comp., **34** (1980), pp. 473–497.

[91] S. Mazzullo, M. Paolini, and C. Verdi, *Polymer crystallization and processing: free boundary problems and their numerical approximation*, Math. Engrg. Indust., **2** (1989), pp. 219–232.

[92] S. Mazzullo, M. Paolini, and C. Verdi, *Numerical simulation of thermal bone necrosis during cementation of femoral prostheses*, J. Math. Biol., **29** (1991), pp. 475–494.

[93] A.M. Meirmanov, *The Stefan Problem*, Gruyter, Berlin, 1992.

[94] L. Modica and S. Mortola, *Un esempio di Γ-convergenza*, Boll. Un. Mat. Ital. B (5), **14** (1977), pp. 285–299.

[95] J. Nitsche, *L^∞-convergence of finite element approximation*, in Mathematical Aspects of Finite Element Method, LNM 606 (I. Galligani and E. Magenes, eds.), Springer-Verlag, Berlin, 1977, pp. 261–274.

[96] R.H. Nochetto, *Error estimates for two-phase Stefan problems in several space variables, I: linear boundary conditions*, Calcolo, **22** (1985), pp. 457–499.

[97] R.H. Nochetto, *A note on the approximation of free boundaries by finite element methods*, RAIRO Modél. Math. Anal. Numér., **20** (1986), pp. 355–368.

[98] R.H. Nochetto, *Error estimates for multidimensional singular parabolic problems*, Japan J. Indust. Appl. Math., **4** (1987), pp. 111–138.

[99] R.H. Nochetto, *A class of non-degenerate two-phase Stefan problems in several space variables*, Comm. Partial Differential Equations, **12** (1987), pp. 21–45.

[100] R.H. Nochetto, *A stable extrapolation method for multidimensional degenerate parabolic problems*, Math. Comp., **53** (1989), pp. 455–470.

[101] R.H. Nochetto, *Finite element methods for parabolic free boundary problems*, in Advances in Numerical Analysis, I: Nonlinear Partial Differential Equations and Dynamical Systems (W. Light ed.), Oxford Academic Press, Oxford, 1991, pp. 34–95.

[102] R.H. Nochetto, M. Paolini, S. Rovida, and C. Verdi, *Variational approximation of the geometric motion of fronts*, in Motion by Mean Curvature and Related Topics (G. Buttazzo and A. Visintin, eds.), Gruyter, Berlin, 1994, pp. 124–149.

[103] R.H. Nochetto, M. Paolini, and C. Verdi, *An adaptive finite elements method for two-phase Stefan problems in two space dimensions. Part I: stability and error estimates. Supplement*, Math. Comp., **57** (1991), pp. 73–108, S1–S11.

[104] R.H. Nochetto, M. Paolini, and C. Verdi, *An adaptive finite elements method for two-phase Stefan problems in two space dimensions. Part II: implementation and numerical experiments*, SIAM J. Sci. Statist. Comput., **12** (1991), pp. 1207–1244.

[105] R.H. Nochetto, M. Paolini, and C. Verdi, *Quasi-optimal mesh adaptation for two-phase Stefan problems in 2D*, in Computational Mathematics and Applications, Publ. 730, IAN-CNR, Pavia, 1989, pp. 313–326.

[106] R.H. Nochetto, M. Paolini, and C. Verdi, *Towards a unified approach for the adaptive solution of evolution phase changes*, in Variational and Free Boundary Problems, IMA VMA 53 (A. Friedman and J. Spruck, eds.), Springer-Verlag, New York, 1993, pp. 171–193.

[107] R.H. Nochetto, M. Paolini, and C. Verdi, *A fully discrete adaptive nonlinear Chernoff formula*, SIAM J. Numer. Anal., **30** (1993), pp. 991–1014.

[108] R.H. Nochetto, M. Paolini, and C. Verdi, *Continuous and semidiscrete travelling waves for a phase relaxation model*, European J. Appl. Math. (to appear).

[109] R.H. Nochetto, M. Paolini, and C. Verdi, *Optimal interface error estimates for the mean curvature flow*, Ann. Scuola Norm. Sup. Pisa Cl. Sci. (4) (to appear).

[110] R.H. Nochetto, M. Paolini, and C. Verdi, *Sharp error analysis for curvature dependent evolving fronts*, Math. Models Methods Appl. Sci., **3** (1993), pp. 711-723.

[111] R.H. Nochetto, M. Paolini, and C. Verdi, *Double obstacle formulation with variable relaxation parameter for smooth geometric front evolutions: asymptotic interface error estimates*, Asymptotic Anal. (to appear).

[112] R.H. Nochetto, M. Paolini, and C. Verdi, *A dynamic mesh method for curvature dependent evolving interfaces*, Preprint (1994).

[113] R.H. Nochetto and C. Verdi, *Approximation of degenerate parabolic problems using numerical integration*, SIAM J. Numer. Anal., **25** (1988), pp. 784–814.

[114] R.H. Nochetto and C. Verdi, *An efficient linear scheme to approximate parabolic free boundary problems: error estimates and implementation*, Math. Comp., **51** (1988), pp. 27–53.

[115] R.H. Nochetto and C. Verdi, *The combined use of a nonlinear Chernoff formula with a regularization procedure for two-phase Stefan problems*, Numer. Funct. Anal. Optim., **9** (1988), pp. 1177–1192.

[116] R.H. Nochetto and C. Verdi, *Convergence of double obstacle problems to the generalized geometric motion of fronts*, SIAM J. Math. Anal. (to appear).

[117] R.H. Nochetto and C. Verdi, *Convergence past singularities for a fully discrete approximation of curvature driven interfaces*, Preprint (1994).

[118] J.M. Ortega and W.C. Rheinboldt, *Iterative Solution of Nonlinear Equations in Several Variables*, Academic Press, New York, 1970.

[119] S. Osher and J.A. Sethian, *Fronts propagating with curvature dependent speed: algorithms based on Hamilton-Jacobi formulations,*, J. Comput. Phys., **79** (1988), pp. 12–49.

[120] M. Paolini, G. Sacchi, and C. Verdi, *Finite element approximation of singular parabolic problems*, Internat. J. Numer. Methods Engrg., **26** (1988), pp. 1989–2007.

[121] M. Paolini and C. Verdi, *An automatic triangular mesh generator for planar domains*, Riv. Inform., **20** (1990), pp. 251–267.

[122] M. Paolini and C. Verdi, *Asymptotic and numerical analyses of the mean curvature flow with a space-dependent relaxation parameter*, Asymptotic Anal., **5** (1992), pp. 553–574.

[123] P. Pietra and C. Verdi, *Convergence of the approximated free boundary for the multidimensional one-phase Stefan problem*, Comput. Mech., **1** (1986), pp. 115–125.

[124] P.A. Raviart, *The use of numerical integration in finite element methods for solving parabolic equations*, in Topics in Numerical Analysis (J.J.H. Miller ed.), Academic Press, London, 1973, pp. 233–264.

[125] J.F. Rodrigues, *Variational methods in the Stefan problem*, in [144], to appear.

[126] J. Rulla, *Error analyses for implicit approximations to solutions to Cauchy problems*, SIAM J. Numer. Anal. (to appear).

[127] J. Rulla and N.J. Walkington, *Optimal rates of convergence for degenerate parabolic problems in two dimensions*, SIAM J. Numer. Anal. (to appear).

[128] G. Savaré, *Approximation and regularity of evolution variational inequalities*, Rend. Accad. Naz. Sci XL 111, **17** (1993), pp. 83–111.

[129] G. Savaré, *Weak solutions and maximal time regularity for abstract evolution inequalities*, Preprint (1994).

[130] J.A. Sethian, *Recent numerical algorithms for hypersurfaces moving with curvature-dependent speed : Hamilton-Jacobi equations, conservation laws*, J. Differential Geom., **31** (1990), pp. 131–162.

[131] H.M. Soner and P.E. Souganidis, *Singularities and uniqueness of cylindrically symmetric surfaces moving by mean curvature*, Comm. Partial Differential Equations, **18** (1992), pp. 859–894.

[132] V. Thomée, *Galerkin Finite Element Methods for Parabolic Problems*, Springer-Verlag, Berlin, 1984.

[133] M.C. Tobin, *Theory of phase transition kinetics with growth site impingement, I*, J. Polym. Sci. Polym. Phys. Ed., **12** (1974), pp. 394–406.

[134] M.C. Tobin, *Theory of phase transition kinetics with growth site impingement, II*, J. Polym. Sci. Polym. Phys. Ed., **14** (1976), pp. 2253–2257.

[135] C. Verdi, *On the numerical approach to a two-phase Stefan problem with nonlinear flux*, Calcolo, **22** (1985), pp. 351–381.

[136] C. Verdi, *Optimal error estimates for an approximation of degenerate parabolic problems*, Numer. Funct. Anal. Optim., **9** (1987), pp. 657–670.

[137] C. Verdi, *BV regularity of the enthalpy for semidiscrete two-phase Stefan problems*, Istit. Lombardo Accad. Sci. Lett. Rend. A, **126** (1992), pp. 29–42.

[138] C. Verdi and A. Visintin, *Numerical approximation of hysteresis problems*, IMA J. Numer. Anal., **5** (1985), pp. 447–463.

[139] C. Verdi and A. Visintin, *Numerical analysis of the multidimensional Stefan problem with supercooling and superheating*, Boll. Un. Mat. Ital. B (7), **1** (1987), pp. 795–814.

[140] C. Verdi and A. Visintin, *Error estimates for a semiexplicit numerical scheme for Stefan-type problems*, Numer. Math., **52** (1988), pp. 165–185.

[141] C. Verdi and A. Visintin, *Numerical approximation of the Preisach model for hysteresis*, RAIRO Modél. Math. Anal. Numér., **23** (1989), pp. 335–356.

[142] A. Visintin, *Stefan problem with phase relaxation*, IMA J. Appl. Math., **34** (1985), pp. 225–245.

[143] A. Visintin, *Differential Models of Hysteresis*, Springer-Verlag, Berlin, to appear.

[144] A. Visintin, *Modelling and Analysis of Phase Transition and Hysteresis Phenomena*, Springer-Verlag, Berlin, to appear.

[145] D.E. Womble, *A front-tracking method for multiphase free boundary problems*, SIAM J. Numer. Anal., **26** (1989), pp. 380–396.

[146] N. Yamada, *Viscosity solutions for a system of elliptic inequalities with bilateral obstacles*, Funkcial. Ekvac., **30** (1987), pp. 417–425.

[147] M. Zlamal, *A finite element solution of the nonlinear heat equation*, RAIRO Modél. Math. Anal. Numér., **14** (1980), pp. 203–216.

C.I.M.E. Session of Phase Transition and Hysteresis

List of Participants

M. BERNARDI, Via Solari 11, 27100 Pavia

B. BIGI, INDAM, Città Universitaria, 00185 Roma

C. BONDIOLI, Via Solari 11, 27100 Pavia

E. COMPARINI, Dip. Mat. Univ., Viale Morgagni 67/A, 50134 Firenze

E. D'AMBROGIO, Dip. di Scienze Matem., P.le Europa 1, 34127 Trieste

A. DAMLAMIAN, Centre de Math., URS-CNRS No.169, Ecole Polyt.,91128 Palaiseau, France

Z. DING, Dip. Mat. Univ., Viale Morgagni 67/A, 50134 Firenze

R. GIANNI, Dip. Mat. Univ., Viale Morgagni 67/A, 50134 Firenze

C.R. GRISANTI, S.N.S., Piazza dei Cavalieri 7, 56126 Pisa

P. MANNUCCI, Dip. Mat. Univ., Viale Morgagni 67/A, 50134 Firenze

T. NADZIEJA, Math. Inst. Univ. Wroclaw, Pl. Grunwaldzki 2/4, 50-384 Wroclaw, Poland

M. PRIMICERIO, Dip. Mat. Univ., Viale Morgagni 67/A, 50134 Firenze

R. RICCI, Dip. Mat. Univ., Via C. Saldini 50, 20122 Milano

M. ROMEO, D.I.B.E., Univ. di Genova, Via Opera Pia 11a, 16145 Genova

G. SAVARE', IAN-CNR, Palazzo dell'Università, Corso Carlo Alberto 5, 27100 Pavia

M. SCHWARZ, Dept. of Civil Eng., Univ. of Essen, 45 117 Essen, FRG

G. TRONEL, Univ. P. et M. Curie, Lab. d'Anal. Num., Tour 55-65,
5ème étage, 4 place Jussieu, 75252 Paris Cedex 05, France

J.M. URBANO, Depto de Mat., Univ. de Coimbra, 3000 Coimbra, Portugal

Z. VOREL, Dept. of Math., Univ. of Southern California, University Park,
Los Angeles, CA 90089-1113, USA

P. ZECCA, Dip. di Sist. e Inf., Univ., Via S. Marta 3, 50139 Firenze

LIST OF C.I.M.E. SEMINARS

		Publisher
1954 -	1. Analisi funzionale	C.I.M.E.
	2. Quadratura delle superficie e questioni connesse	"
	3. Equazioni differenziali non lineari	"
1955 -	4. Teorema di Riemann-Roch e questioni connesse	"
	5. Teoria dei numeri	"
	6. Topologia	"
	7. Teorie non linearizzate in elasticità, idrodinamica,aerodinamica	"
	8. Geometria proiettivo-differenziale	"
1956 -	9. Equazioni alle derivate parziali a caratteristiche reali	"
	10. Propagazione delle onde elettromagnetiche	"
	11. Teoria della funzioni di più variabili complesse e delle funziòni automorfe	"
1957 -	12. Geometria aritmetica e algebrica (2 vol.)	"
	13. Integrali singolari e questioni connesse	"
	14. Teoria della turbolenza (2 vol.)	"
1958 -	15. Vedute e problemi attuali in relatività generale	"
	16. Problemi di geometria differenziale in grande	"
	17. Il principio di minimo e le sue applicazioni alle equazioni funzionali	"
1959 -	18. Induzione e statistica	"
	19. Teoria algebrica dei meccanismi automatici (2 vol.)	"
	20. Gruppi, anelli di Lie e teoria della coomologia	"
1960 -	21. Sistemi dinamici e teoremi ergodici	"
	22. Forme differenziali e loro integrali	"
1961 -	23. Geometria del calcolo delle variazioni (2 vol.)	"
	24. Teoria delle distribuzioni	"
	25. Onde superficiali	"
1962 -	26. Topologia differenziale	"
	27. Autovalori e autosoluzioni	"
	28. Magnetofluidodinamica	"

1963 - 29. Equazioni differenziali astratte "
30. Funzioni e varietà complesse "
31. Proprietà di media e teoremi di confronto in Fisica Matematica "

1964 - 32. Relatività generale "
33. Dinamica dei gas rarefatti "
34. Alcune questioni di analisi numerica "
35. Equazioni differenziali non lineari "

1965 - 36. Non-linear continuum theories "
37. Some aspects of ring theory "
38. Mathematical optimization in economics "

1966 - 39. Calculus of variations Ed. Cremonese, Firenze
40. Economia matematica "
41. Classi caratteristiche e questioni connesse "
42. Some aspects of diffusion theory "

1967 - 43. Modern questions of celestial mechanics "
44. Numerical analysis of partial differential equations "
45. Geometry of homogeneous bounded domains "

1968 - 46. Controllability and observability "
47. Pseudo-differential operators "
48. Aspects of mathematical logic "

1969 - 49. Potential theory "
50. Non-linear continuum theories in mechanics and physics and their applications "
51. Questions of algebraic varieties "

1970 - 52. Relativistic fluid dynamics "
53. Theory of group representations and Fourier analysis "
54. Functional equations and inequalities "
55. Problems in non-linear analysis "

1971 - 56. Stereodynamics "
57. Constructive aspects of functional analysis (2 vol.) "
58. Categories and commutative algebra "

1983 -	90. Complete Intersections	(LNM 1092)	Springer-Verlag
	91. Bifurcation Theory and Applications	(LNM 1057)	"
	92. Numerical Methods in Fluid Dynamics	(LNM 1127)	"
1984 -	93. Harmonic Mappings and Minimal Immersions	(LNM 1161)	"
	94. Schrödinger Operators	(LNM 1159)	"
	95. Buildings and the Geometry of Diagrams	(LNM 1181)	"
1985 -	96. Probability and Analysis	(LNM 1206)	"
	97. Some Problems in Nonlinear Diffusion	(LNM 1224)	"
	98. Theory of Moduli	(LNM 1337)	"
1986 -	99. Inverse Problems	(LNM 1225)	"
	100. Mathematical Economics	(LNM 1330)	"
	101. Combinatorial Optimization	(LNM 1403)	"
1987 -	102. Relativistic Fluid Dynamics	(LNM 1385)	"
	103. Topics in Calculus of Variations	(LNM 1365)	"
1988 -	104. Logic and Computer Science	(LNM 1429)	"
	105. Global Geometry and Mathematical Physics	(LNM 1451)	"
1989 -	106. Methods of nonconvex analysis	(LNM 1446)	"
	107. Microlocal Analysis and Applications	(LNM 1495)	"
1990 -	108. Geoemtric Topology: Recent Developments	(LNM 1504)	"
	109. H_∞ Control Theory	(LNM 1496)	"
	110. Mathematical Modelling of Industrical Processes	(LNM 1521)	"
1991 -	111. Topological Methods for Ordinary Differential Equations	(LNM 1537)	"
	112. Arithmetic Algebraic Geometry	(LNM 1553)	"
	113. Transition to Chaos in Classical and Quantum Mechanics	to appear	"
1992 -	114. Dirichlet Forms	(LNM 1563)	"
	115. D-Modules, Representation Theory, and Quantum Groups	(LNM 1565)	"
	116. Nonequilibrium Problems in Many-Particle Systems	(LNM 1551)	"

1972 - 59. Non-linear mechanics "
60. Finite geometric structures and their applications "
61. Geometric measure theory and minimal surfaces "

1973 - 62. Complex analysis "
63. New variational techniques in mathematical physics "
64. Spectral analysis "

1974 - 65. Stability problems "
66. Singularities of analytic spaces "
67. Eigenvalues of non linear problems "

1975 - 68. Theoretical computer sciences "
69. Model theory and applications "
70. Differential operators and manifolds "

1976 - 71. Statistical Mechanics Ed Liguori, Napoli
72. Hyperbolicity "
73. Differential topology "

1977 - 74. Materials with memory "
75. Pseudodifferential operators with applications "
76. Algebraic surfaces "

1978 - 77. Stochastic differential equations "
78. Dynamical systems Ed Liguori, Napoli and Birhäuser Verlag

1979 - 79. Recursion theory and computational complexity "
80. Mathematics of biology "

1980 - 81. Wave propagation "
82. Harmonic analysis and group representations "
83. Matroid theory and its applications "

1981 - 84. Kinetic Theories and the Boltzmann Equation (LNM 1048) Springer-Verlag
85. Algebraic Threefolds (LNM 947) "
86. Nonlinear Filtering and Stochastic Control (LNM 972) "

1982 - 87. Invariant Theory (LNM 996) "
88. Thermodynamics and Constitutive Equations (LN Physics 228) "
89. Fluid Dynamics (LNM 1047) "

1993 -	117.	Integrable Systems and Quantum Groups	to appear	Springer-Verlag
	118.	Algebraic Cycles and Hodge Theories	to appear	"
	119.	Phase Transitions and Hysteresis	(LNM 1584)	"
1994 -	120.	Recent Mathematical Methods in Nonlinear Wave Propagation	to appear	"
	121.	Dynamical Systems	to appear	"
	122.	Transcendental Methods in Algebraic Geometry	to appear	"

Lecture Notes in Mathematics

For information about Vols. 1–1411
please contact your bookseller or Springer-Verlag

Vol. 1412: V.V. Kalashnikov, V.M. Zolotarev (Eds.), Stability Problems for Stochastic Models. Proceedings, 1987. X, 380 pages. 1989.

Vol. 1413: S. Wright, Uniqueness of the Injective III_1 Factor. III, 108 pages. 1989.

Vol. 1414: E. Ramirez de Arellano (Ed.), Algebraic Geometry and Complex Analysis. Proceedings, 1987. VI, 180 pages. 1989.

Vol. 1415: M. Langevin, M. Waldschmidt (Eds.), Cinquante Ans de Polynômes. Fifty Years of Polynomials. Proceedings, 1988. IX, 235 pages. 1990.

Vol. 1416: C. Albert (Ed.), Géométrie Symplectique et Mécanique. Proceedings, 1988. V, 289 pages. 1990.

Vol. 1417: A.J. Sommese, A. Biancofiore, E.L. Livorni (Eds.), Algebraic Geometry. Proceedings, 1988. V, 320 pages. 1990.

Vol. 1418: M. Mimura (Ed.), Homotopy Theory and Related Topics. Proceedings, 1988. V, 241 pages. 1990.

Vol. 1419: P.S. Bullen, P.Y. Lee, J.L. Mawhin, P. Muldowney, W.F. Pfeffer (Eds.), New Integrals. Proceedings, 1988. V, 202 pages. 1990.

Vol. 1420: M. Galbiati, A. Tognoli (Eds.), Real Analytic Geometry. Proceedings, 1988. IV, 366 pages. 1990.

Vol. 1421: H.A. Biagioni, A Nonlinear Theory of Generalized Functions, XII, 214 pages. 1990.

Vol. 1422: V. Villani (Ed.), Complex Geometry and Analysis. Proceedings, 1988. V, 109 pages. 1990.

Vol. 1423: S.O. Kochman, Stable Homotopy Groups of Spheres: A Computer-Assisted Approach. VIII, 330 pages. 1990.

Vol. 1424: F.E. Burstall, J.H. Rawnsley, Twistor Theory for Riemannian Symmetric Spaces. III, 112 pages. 1990.

Vol. 1425: R.A. Piccinini (Ed.), Groups of Self-Equivalences and Related Topics. Proceedings, 1988. V, 214 pages. 1990.

Vol. 1426: J. Azéma, P.A. Meyer, M. Yor (Eds.), Séminaire de Probabilités XXIV, 1988/89. V, 490 pages. 1990.

Vol. 1427: A. Ancona, D. Geman, N. Ikeda, École d'Eté de Probabilités de Saint Flour XVIII, 1988. Ed.: P.L. Hennequin. VII, 330 pages. 1990.

Vol. 1428: K. Erdmann, Blocks of Tame Representation Type and Related Algebras. XV. 312 pages. 1990.

Vol. 1429: S. Homer, A. Nerode, R.A. Platek, G.E. Sacks, A. Scedrov, Logic and Computer Science. Seminar, 1988. Editor: P. Odifreddi. V, 162 pages. 1990.

Vol. 1430: W. Bruns, A. Simis (Eds.), Commutative Algebra. Proceedings. 1988. V, 160 pages. 1990.

Vol. 1431: J.G. Heywood, K. Masuda, R. Rautmann, V.A. Solonnikov (Eds.), The Navier-Stokes Equations – Theory and Numerical Methods. Proceedings, 1988. VII, 238 pages. 1990.

Vol. 1432: K. Ambos-Spies, G.H. Müller, G.E. Sacks (Eds.), Recursion Theory Week. Proceedings, 1989. VI, 393 pages. 1990.

Vol. 1433: S. Lang, W. Cherry, Topics in Nevanlinna Theory. II, 174 pages. 1990.

Vol. 1434: K. Nagasaka, E. Fouvry (Eds.), Analytic Number Theory. Proceedings, 1988. VI, 218 pages. 1990.

Vol. 1435: St. Ruscheweyh, E.B. Saff, L.C. Salinas, R.S. Varga (Eds.), Computational Methods and Function Theory. Proceedings, 1989. VI, 211 pages. 1990.

Vol. 1436: S. Xambó-Descamps (Ed.), Enumerative Geometry. Proceedings, 1987. V, 303 pages. 1990.

Vol. 1437: H. Inassaridze (Ed.), K-theory and Homological Algebra. Seminar, 1987–88. V, 313 pages. 1990.

Vol. 1438: P.G. Lemarié (Ed.) Les Ondelettes en 1989. Seminar. IV, 212 pages. 1990.

Vol. 1439: E. Bujalance, J.J. Etayo, J.M. Gamboa, G. Gromadzki. Automorphism Groups of Compact Bordered Klein Surfaces: A Combinatorial Approach. XIII, 201 pages. 1990.

Vol. 1440: P. Latiolais (Ed.), Topology and Combinatorial Groups Theory. Seminar, 1985–1988. VI, 207 pages. 1990.

Vol. 1441: M. Coornaert, T. Delzant, A. Papadopoulos. Géométrie et théorie des groupes. X, 165 pages. 1990.

Vol. 1442: L. Accardi, M. von Waldenfels (Eds.), Quantum Probability and Applications V. Proceedings, 1988. VI, 413 pages. 1990.

Vol. 1443: K.H. Dovermann, R. Schultz, Equivariant Surgery Theories and Their Periodicity Properties. VI, 227 pages. 1990.

Vol. 1444: H. Korezlioglu, A.S. Ustunel (Eds.), Stochastic Analysis and Related Topics VI. Proceedings, 1988. V, 268 pages. 1990.

Vol. 1445: F. Schulz, Regularity Theory for Quasilinear Elliptic Systems and – Monge Ampère Equations in Two Dimensions. XV, 123 pages. 1990.

Vol. 1446: Methods of Nonconvex Analysis. Seminar, 1989. Editor: A. Cellina. V, 206 pages. 1990.

Vol. 1447: J.-G. Labesse, J. Schwermer (Eds), Cohomology of Arithmetic Groups and Automorphic Forms. Proceedings, 1989. V, 358 pages. 1990.

Vol. 1448: S.K. Jain, S.R. López-Permouth (Eds.), Non-Commutative Ring Theory. Proceedings, 1989. V, 166 pages. 1990.

Vol. 1449: W. Odyniec, G. Lewicki, Minimal Projections in Banach Spaces. VIII, 168 pages. 1990.

Vol. 1450: H. Fujita, T. Ikebe, S.T. Kuroda (Eds.), Functional-Analytic Methods for Partial Differential Equations. Proceedings, 1989. VII, 252 pages. 1990.

Vol. 1451: L. Alvarez-Gaumé, E. Arbarello, C. De Concini, N.J. Hitchin, Global Geometry and Mathematical Physics.

Montecatini Terme 1988. Seminar. Editors: M. Francaviglia, F. Gherardelli. IX, 197 pages. 1990.

Vol. 1452: E. Hlawka, R.F. Tichy (Eds.), Number-Theoretic Analysis. Seminar, 1988–89. V, 220 pages. 1990.

Vol. 1453: Yu.G. Borisovich, Yu.E. Gliklikh (Eds.), Global Analysis – Studies and Applications IV. V, 320 pages. 1990.

Vol. 1454: F. Baldassari, S. Bosch, B. Dwork (Eds.), p-adic Analysis. Proceedings, 1989. V, 382 pages. 1990.

Vol. 1455: J.-P. Françoise, R. Roussarie (Eds.), Bifurcations of Planar Vector Fields. Proceedings, 1989. VI, 396 pages. 1990.

Vol. 1456: L.G. Kovács (Ed.), Groups – Canberra 1989. Proceedings. XII, 198 pages. 1990.

Vol. 1457: O. Axelsson, L.Yu. Kolotilina (Eds.), Preconditioned Conjugate Gradient Methods. Proceedings, 1989. V, 196 pages. 1990.

Vol. 1458: R. Schaaf, Global Solution Branches of Two Point Boundary Value Problems. XIX, 141 pages. 1990.

Vol. 1459: D. Tiba, Optimal Control of Nonsmooth Distributed Parameter Systems. VII, 159 pages. 1990.

Vol. 1460: G. Toscani, V. Boffi, S. Rionero (Eds.), Mathematical Aspects of Fluid Plasma Dynamics. Proceedings, 1988. V, 221 pages. 1991.

Vol. 1461: R. Gorenflo, S. Vessella, Abel Integral Equations. VII, 215 pages. 1991.

Vol. 1462: D. Mond, J. Montaldi (Eds.), Singularity Theory and its Applications. Warwick 1989, Part I. VIII, 405 pages. 1991.

Vol. 1463: R. Roberts, I. Stewart (Eds.), Singularity Theory and its Applications. Warwick 1989, Part II. VIII, 322 pages. 1991.

Vol. 1464: D. L. Burkholder, E. Pardoux, A. Sznitman, Ecole d'Eté de Probabilités de Saint- Flour XIX-1989. Editor: P. L. Hennequin. VI, 256 pages. 1991.

Vol. 1465: G. David, Wavelets and Singular Integrals on Curves and Surfaces. X, 107 pages. 1991.

Vol. 1466: W. Banaszczyk, Additive Subgroups of Topological Vector Spaces. VII, 178 pages. 1991.

Vol. 1467: W. M. Schmidt, Diophantine Approximations and Diophantine Equations. VIII, 217 pages. 1991.

Vol. 1468: J. Noguchi, T. Ohsawa (Eds.), Prospects in Complex Geometry. Proceedings, 1989. VII, 421 pages. 1991.

Vol. 1469: J. Lindenstrauss, V. D. Milman (Eds.), Geometric Aspects of Functional Analysis. Seminar 1989-90. XI, 191 pages. 1991.

Vol. 1470: E. Odell, H. Rosenthal (Eds.), Functional Analysis. Proceedings, 1987-89. VII, 199 pages. 1991.

Vol. 1471: A. A. Panchishkin, Non-Archimedean L-Functions of Siegel and Hilbert Modular Forms. VII, 157 pages. 1991.

Vol. 1472: T. T. Nielsen, Bose Algebras: The Complex and Real Wave Representations. V, 132 pages. 1991.

Vol. 1473: Y. Hino, S. Murakami, T. Naito, Functional Differential Equations with Infinite Delay. X, 317 pages. 1991.

Vol. 1474: S. Jackowski, B. Oliver, K. Pawałowski (Eds.), Algebraic Topology, Poznań 1989. Proceedings. VIII, 397 pages. 1991.

Vol. 1475: S. Busenberg, M. Martelli (Eds.), Delay Differential Equations and Dynamical Systems. Proceedings, 1990. VIII, 249 pages. 1991.

Vol. 1476: M. Bekkali, Topics in Set Theory. VII, 120 pages. 1991.

Vol. 1477: R. Jajte, Strong Limit Theorems in Noncommutative L_2-Spaces. X, 113 pages. 1991.

Vol. 1478: M.-P. Malliavin (Ed.), Topics in Invariant Theory. Seminar 1989-1990. VI, 272 pages. 1991.

Vol. 1479: S. Bloch, I. Dolgachev, W. Fulton (Eds.), Algebraic Geometry. Proceedings, 1989. VII, 300 pages. 1991.

Vol. 1480: F. Dumortier, R. Roussarie, J. Sotomayor, H. Żoładek, Bifurcations of Planar Vector Fields: Nilpotent Singularities and Abelian Integrals. VIII, 226 pages. 1991.

Vol. 1481: D. Ferus, U. Pinkall, U. Simon, B. Wegner (Eds.), Global Differential Geometry and Global Analysis. Proceedings, 1991. VIII, 283 pages. 1991.

Vol. 1482: J. Chabrowski, The Dirichlet Problem with L^2-Boundary Data for Elliptic Linear Equations. VI, 173 pages. 1991.

Vol. 1483: E. Reithmeier, Periodic Solutions of Nonlinear Dynamical Systems. VI, 171 pages. 1991.

Vol. 1484: H. Delfs, Homology of Locally Semialgebraic Spaces. IX, 136 pages. 1991.

Vol. 1485: J. Azéma, P. A. Meyer, M. Yor (Eds.), Séminaire de Probabilités XXV. VIII, 440 pages. 1991.

Vol. 1486: L. Arnold, H. Crauel, J.-P. Eckmann (Eds.), Lyapunov Exponents. Proceedings, 1990. VIII, 365 pages. 1991.

Vol. 1487: E. Freitag, Singular Modular Forms and Theta Relations. VI, 172 pages. 1991.

Vol. 1488: A. Carboni, M. C. Pedicchio, G. Rosolini (Eds.), Category Theory. Proceedings, 1990. VII, 494 pages. 1991.

Vol. 1489: A. Mielke, Hamiltonian and Lagrangian Flows on Center Manifolds. X, 140 pages. 1991.

Vol. 1490: K. Metsch, Linear Spaces with Few Lines. XIII, 196 pages. 1991.

Vol. 1491: E. Lluis-Puebla, J.-L. Loday, H. Gillet, C. Soulé, V. Snaith, Higher Algebraic K-Theory: an overview. IX, 164 pages. 1992.

Vol. 1492: K. R. Wicks, Fractals and Hyperspaces. VIII, 168 pages. 1991.

Vol. 1493: E. Benoît (Ed.), Dynamic Bifurcations. Proceedings, Luminy 1990. VII, 219 pages. 1991.

Vol. 1494: M.-T. Cheng, X.-W. Zhou, D.-G. Deng (Eds.), Harmonic Analysis. Proceedings, 1988. IX, 226 pages. 1991.

Vol. 1495: J. M. Bony, G. Grubb, L. Hörmander, H. Komatsu, J. Sjöstrand, Microlocal Analysis and Applications. Montecatini Terme, 1989. Editors: L. Cattabriga, L. Rodino. VII, 349 pages. 1991.

Vol. 1496: C. Foias, B. Francis, J. W. Helton, H. Kwakernaak, J. B. Pearson, H_∞-Control Theory. Como, 1990. Editors: E. Mosca, L. Pandolfi. VII, 336 pages. 1991.

Vol. 1497: G. T. Herman, A. K. Louis, F. Natterer (Eds.), Mathematical Methods in Tomography. Proceedings 1990. X, 268 pages. 1991.

Vol. 1498: R. Lang, Spectral Theory of Random Schrödinger Operators. X, 125 pages. 1991.

Vol. 1499: K. Taira, Boundary Value Problems and Markov Processes. IX, 132 pages. 1991.

Vol. 1500: J.-P. Serre, Lie Algebras and Lie Groups. VII, 168 pages. 1992.

Vol. 1501: A. De Masi, E. Presutti, Mathematical Methods for Hydrodynamic Limits. IX, 196 pages. 1991.

Vol. 1502: C. Simpson, Asymptotic Behavior of Monodromy. V, 139 pages. 1991.

Vol. 1503: S. Shokranian, The Selberg-Arthur Trace Formula (Lectures by J. Arthur). VII, 97 pages. 1991.

Vol. 1504: J. Cheeger, M. Gromov, C. Okonek, P. Pansu, Geometric Topology: Recent Developments. Editors: P. de Bartolomeis, F. Tricerri. VII, 197 pages. 1991.

Vol. 1505: K. Kajitani, T. Nishitani, The Hyperbolic Cauchy Problem. VII, 168 pages. 1991.

Vol. 1506: A. Buium, Differential Algebraic Groups of Finite Dimension. XV, 145 pages. 1992.

Vol. 1507: K. Hulek, T. Peternell, M. Schneider, F.-O. Schreyer (Eds.), Complex Algebraic Varieties. Proceedings, 1990. VII, 179 pages. 1992.

Vol. 1508: M. Vuorinen (Ed.), Quasiconformal Space Mappings. A Collection of Surveys 1960-1990. IX, 148 pages. 1992.

Vol. 1509: J. Aguadé, M. Castellet, F. R. Cohen (Eds.), Algebraic Topology - Homotopy and Group Cohomology. Proceedings, 1990. X, 330 pages. 1992.

Vol. 1510: P. P. Kulish (Ed.), Quantum Groups. Proceedings, 1990. XII, 398 pages. 1992.

Vol. 1511: B. S. Yadav, D. Singh (Eds.), Functional Analysis and Operator Theory. Proceedings, 1990. VIII, 223 pages. 1992.

Vol. 1512: L. M. Adleman, M.-D. A. Huang, Primality Testing and Abelian Varieties Over Finite Fields. VII, 142 pages. 1992.

Vol. 1513: L. S. Block, W. A. Coppel, Dynamics in One Dimension. VIII, 249 pages. 1992.

Vol. 1514: U. Krengel, K. Richter, V. Warstat (Eds.), Ergodic Theory and Related Topics III, Proceedings, 1990. VIII, 236 pages. 1992.

Vol. 1515: E. Ballico, F. Catanese, C. Ciliberto (Eds.), Classification of Irregular Varieties. Proceedings, 1990. VII, 149 pages. 1992.

Vol. 1516: R. A. Lorentz, Multivariate Birkhoff Interpolation. IX, 192 pages. 1992.

Vol. 1517: K. Keimel, W. Roth, Ordered Cones and Approximation. VI, 134 pages. 1992.

Vol. 1518: H. Stichtenoth, M. A. Tsfasman (Eds.), Coding Theory and Algebraic Geometry. Proceedings, 1991. VIII, 223 pages. 1992.

Vol. 1519: M. W. Short, The Primitive Soluble Permutation Groups of Degree less than 256. IX, 145 pages. 1992.

Vol. 1520: Yu. G. Borisovich, Yu. E. Gliklikh (Eds.), Global Analysis – Studies and Applications V. VII, 284 pages. 1992.

Vol. 1521: S. Busenberg, B. Forte, H. K. Kuiken, Mathematical Modelling of Industrial Process. Bari, 1990. Editors: V. Capasso, A. Fasano. VII, 162 pages. 1992.

Vol. 1522: J.-M. Delort, F. B. I. Transformation. VII, 101 pages. 1992.

Vol. 1523: W. Xue, Rings with Morita Duality. X, 168 pages. 1992.

Vol. 1524: M. Coste, L. Mahé, M.-F. Roy (Eds.), Real Algebraic Geometry. Proceedings, 1991. VIII, 418 pages. 1992.

Vol. 1525: C. Casacuberta, M. Castellet (Eds.), Mathematical Research Today and Tomorrow. VII, 112 pages. 1992.

Vol. 1526: J. Azéma, P. A. Meyer, M. Yor (Eds.), Séminaire de Probabilités XXVI. X, 633 pages. 1992.

Vol. 1527: M. I. Freidlin, J.-F. Le Gall, Ecole d'Eté de Probabilités de Saint-Flour XX – 1990. Editor: P. L. Hennequin. VIII, 244 pages. 1992.

Vol. 1528: G. Isac, Complementarity Problems. VI, 297 pages. 1992.

Vol. 1529: J. van Neerven, The Adjoint of a Semigroup of Linear Operators. X, 195 pages. 1992.

Vol. 1530: J. G. Heywood, K. Masuda, R. Rautmann, S. A. Solonnikov (Eds.), The Navier-Stokes Equations II – Theory and Numerical Methods. IX, 322 pages. 1992.

Vol. 1531: M. Stoer, Design of Survivable Networks. IV, 206 pages. 1992.

Vol. 1532: J. F. Colombeau, Multiplication of Distributions. X, 184 pages. 1992.

Vol. 1533: P. Jipsen, H. Rose, Varieties of Lattices. X, 162 pages. 1992.

Vol. 1534: C. Greither, Cyclic Galois Extensions of Commutative Rings. X, 145 pages. 1992.

Vol. 1535: A. B. Evans, Orthomorphism Graphs of Groups. VIII, 114 pages. 1992.

Vol. 1536: M. K. Kwong, A. Zettl, Norm Inequalities for Derivatives and Differences. VII, 150 pages. 1992.

Vol. 1537: P. Fitzpatrick, M. Martelli, J. Mawhin, R. Nussbaum, Topological Methods for Ordinary Differential Equations. Montecatini Terme, 1991. Editors: M. Furi, P. Zecca. VII, 218 pages. 1993.

Vol. 1538: P.-A. Meyer, Quantum Probability for Probabilists. X, 287 pages. 1993.

Vol. 1539: M. Coornaert, A. Papadopoulos, Symbolic Dynamics and Hyperbolic Groups. VIII, 138 pages. 1993.

Vol. 1540: H. Komatsu (Ed.), Functional Analysis and Related Topics, 1991. Proceedings. XXI, 413 pages. 1993.

Vol. 1541: D. A. Dawson, B. Maisonneuve, J. Spencer, Ecole d' Eté de Probabilités de Saint-Flour XXI - 1991. Editor: P. L. Hennequin. VIII, 356 pages. 1993.

Vol. 1542: J.Fröhlich, Th.Kerler, Quantum Groups, Quantum Categories and Quantum Field Theory. VII, 431 pages. 1993.

Vol. 1543: A. L. Dontchev, T. Zolezzi, Well-Posed Optimization Problems. XII, 421 pages. 1993.

Vol. 1544: M.Schürmann, White Noise on Bialgebras. VII, 146 pages. 1993.

Vol. 1545: J. Morgan, K. O'Grady, Differential Topology of Complex Surfaces. VIII, 224 pages. 1993.

Vol. 1546: V. V. Kalashnikov, V. M. Zolotarev (Eds.), Stability Problems for Stochastic Models. Proceedings, 1991. VIII, 229 pages. 1993.

Vol. 1547: P. Harmand, D. Werner, W. Werner, M-ideals in Banach Spaces and Banach Algebras. VIII, 387 pages. 1993.

Vol. 1548: T. Urabe, Dynkin Graphs and Quadrilateral Singularities. VI, 233 pages. 1993.

Vol. 1549: G. Vainikko, Multidimensional Weakly Singular Integral Equations. XI, 159 pages. 1993.

Vol. 1550: A. A. Gonchar, E. B. Saff (Eds.), Methods of Approximation Theory in Complex Analysis and Mathematical Physics IV, 222 pages, 1993.

Vol. 1551: L. Arkeryd, P. L. Lions, P.A. Markowich, S.R. S. Varadhan. Nonequilibrium Problems in Many-Particle Systems. Montecatini, 1992. Editors: C. Cercignani, M. Pulvirenti. VII, 158 pages 1993.

Vol. 1552: J. Hilgert, K.-H. Neeb, Lie Semigroups and their Applications. XII, 315 pages. 1993.

Vol. 1553: J.-L- Colliot-Thélène, J. Kato, P. Vojta. Arithmetic Algebraic Geometry. Trento, 1991. Editor: E. Ballico. VII, 223 pages. 1993.

Vol. 1554: A. K. Lenstra, H. W. Lenstra, Jr. (Eds.), The Development of the Number Field Sieve. VIII, 131 pages. 1993.

Vol. 1555: O. Liess, Conical Refraction and Higher Microlocalization. X, 389 pages. 1993.

Vol. 1556: S. B. Kuksin, Nearly Integrable Infinite-Dimensional Hamiltonian Systems. XXVII, 101 pages. 1993.

Vol. 1557: J. Azéma, P. A. Meyer, M. Yor (Eds.), Séminaire de Probabilités XXVII. VI, 327 pages. 1993.

Vol. 1558: T. J. Bridges, J. E. Furter, Singularity Theory and Equivariant Symplectic Maps. VI, 226 pages. 1993.

Vol. 1559: V. G. Sprindžuk, Classical Diophantine Equations. XII, 228 pages. 1993.

Vol. 1560: T. Bartsch, Topological Methods for Variational Problems with Symmetries. X, 152 pages. 1993.

Vol. 1561: I. S. Molchanov, Limit Theorems for Unions of Random Closed Sets. X, 157 pages. 1993.

Vol. 1562: G. Harder, Eisensteinkohomologie und die Konstruktion gemischter Motive. XX, 184 pages. 1993.

Vol. 1563: E. Fabes, M. Fukushima, L. Gross, C. Kenig, M. Röckner, D. W. Stroock, Dirichlet Forms. Varenna, 1992. Editors: G. Dell'Antonio, U. Mosco. VII, 245 pages. 1993.

Vol. 1564: J. Jorgenson, S. Lang, Basic Analysis of Regularized Series and Products. IX, 122 pages. 1993.

Vol. 1565: L. Boutet de Monvel, C. De Concini, C. Procesi, P. Schapira, M. Vergne. D-modules, Representation Theory, and Quantum Groups. Venezia, 1992. Editors: G. Zampieri, A. D'Agnolo. VII, 217 pages. 1993.

Vol. 1566: B. Edixhoven, J.-H. Evertse (Eds.), Diophantine Approximation and Abelian Varieties. XIII, 127 pages. 1993.

Vol. 1567: R. L. Dobrushin, S. Kusuoka, Statistical Mechanics and Fractals. VII, 98 pages. 1993.

Vol. 1568: F. Weisz, Martingale Hardy Spaces and their Application in Fourier Analysis. VIII, 217 pages. 1994.

Vol. 1569: V. Totik, Weighted Approximation with Varying Weight. VI, 117 pages. 1994.

Vol. 1570: R. deLaubenfels, Existence Families, Functional Calculi and Evolution Equations. XV, 234 pages. 1994.

Vol. 1571: S. Yu. Pilyugin, The Space of Dynamical Systems with the C^0-Topology. X, 188 pages. 1994.

Vol. 1572: L. Göttsche, Hilbert Schemes of Zero-Dimensional Subschemes of Smooth Varieties. IX, 196 pages. 1994.

Vol. 1573: V. P. Havin, N. K. Nikolski (Eds.), Linear and Complex Analysis – Problem Book 3 – Part I. XXII, 489 pages. 1994.

Vol. 1574: V. P. Havin, N. K. Nikolski (Eds.), Linear and Complex Analysis – Problem Book 3 – Part II. XXII, 507 pages. 1994.

Vol. 1575: M. Mitrea, Clifford Wavelets, Singular Integrals, and Hardy Spaces. XI, 116 pages. 1994.

Vol. 1576: K. Kitahara, Spaces of Approximating Functions with Haar-Like Conditions. X, 110 pages. 1994.

Vol. 1577: N. Obata, White Noise Calculus and Fock Space. X, 183 pages. 1994.

Vol. 1578: J. Bernstein, V. Lunts, Equivariant Sheaves and Functors. V, 139 pages. 1994.

Vol. 1579: N. Kazamaki, Continuous Exponential Martingales and *BMO*. VII, 91 pages. 1994.

Vol. 1580: M. Milman, Extrapolation and Optimal Decompositions with Applications to Analysis. XI, 161 pages. 1994.

Vol. 1581: D. Bakry, R. D. Gill, S. A. Molchanov, Lectures on Probability Theory. Editor: P. Bernard. VIII, 420 pages. 1994.

Vol. 1582: W. Balser, From Divergent Power Series to Analytic Functions. X, 108 pages. 1994.

Vol. 1583: J. Azéma, P. A. Meyer, M. Yor (Eds.), Séminaire de Probabilités XXVIII. VI, 334 pages. 1994.

Vol. 1584: M. Brokate, N. Kenmochi, I. Müller, J. F. Rodriguez, C. Verdi. Editor: A. Visintin. Phase Transitions and Hysteresis. Montecatini Terme, 1993. VII. 291 pages. 1994.

Vol. 1585: G. Frey, On Artin's Conjecture for Odd 2-dimensional Representations. VII, 148 pages. 1994.

Vol. 1586: R. Nillsen, Difference Spaces and Invariant Linear Forms. ••, ••• pages. 1994.

Vol. 1587: N. Xi, Representations of Affine Hecke Algebras. VIII, 137 pages. 1994.

Vol. 1589: J. Bellissard, M. Degli Esposti, G. Forni, S. Graffi, S. Isola, J. N. Mather, Transition to Chaos in Classical and Quantum Mechanics. Montecatini, 1991. Editor: S. Graffi. VII, 192 pages. 1994.